Complex Analytic Sets

Mathematics and Its Applications (*Soviet Series*)

Volume 46

Complex Analytic Sets

by

E. M. Chirka
Steklov Mathematical Institute, Moscow, U.S.S.R.

Translated from the Russian by R. A. M. Hoksbergen

KLUWER ACADEMIC PUBLISHERS
DORDRECHT / BOSTON / LONDON

Library of Congress Cataloging in Publication Data

Chirka, E. M. (Evgeniĭ Mikhaĭlovich)
[Kompleksnye analiticheskie mnozhestva. English]
Complex analytic sets / by E.M. Chirka ; translated from the Russian by R.A.M. Hoksbergen.
p. cm. -- (Mathematics and its applications (Soviet series))
Translation of: Kompleksnye analiticheskie mnozhestva.
Bibliography: p.
Includes index.
ISBN 0-7923-0234-6 (U.S.)
1. Analytic sets. 2. Manifolds (Mathematics) I. Title.
II. Series: Mathematics and its applications (Kluwer Academic Publishers). Soviet series.
QA331.C49813 1989
515--dc20 89-11161

ISBN 0-7923-0234-6

Published by Kluwer Academic Publishers,
P.O. Box 17, 3300 AA Dordrecht, The Netherlands.

Kluwer Academic Publishers incorporates
the publishing programmes of
D. Reidel, Martinus Nijhoff, Dr W. Junk and MTP Press.

Sold and distributed in the U.S.A. and Canada
by Kluwer Academic Publishers,
101 Philip Drive, Norwell, MA 02061, U.S.A.

In all other countries, sold and distributed
by Kluwer Academic Publishers Group,
P.O. Box 322, 3300 AH Dordrecht, The Netherlands.

printed on acid free paper

Original title: **КОМПЛЕКСНЫЕ АНАЛИТИЧЕСКИЕ МНОЖЕСТВА**
Original publisher: Nauka, Moscow, © 1985

Printed in The Netherlands

SERIES EDITOR'S PREFACE

'Et moi, ..., si j'avait su comment en revenir, je n'y serais point allé.'

Jules Verne

The series is divergent; therefore we may be able to do something with it.

O. Heaviside

One service mathematics has rendered the human race. It has put common sense back where it belongs, on the topmost shelf next to the dusty canister labelled 'discarded non-sense'.

Eric T. Bell

Mathematics is a tool for thought. A highly necessary tool in a world where both feedback and non-linearities abound. Similarly, all kinds of parts of mathematics serve as tools for other parts and for other sciences.

Applying a simple rewriting rule to the quote on the right above one finds such statements as: 'One service topology has rendered mathematical physics ...'; 'One service logic has rendered computer science ...'; 'One service category theory has rendered mathematics ...'. All arguably true. And all statements obtainable this way form part of the raison d'être of this series.

This series, *Mathematics and Its Applications*, started in 1977. Now that over one hundred volumes have appeared it seems opportune to reexamine its scope. At the time I wrote

> "Growing specialization and diversification have brought a host of monographs and textbooks on increasingly specialized topics. However, the 'tree' of knowledge of mathematics and related fields does not grow only by putting forth new branches. It also happens, quite often in fact, that branches which were thought to be completely disparate are suddenly seen to be related. Further, the kind and level of sophistication of mathematics applied in various sciences has changed drastically in recent years: measure theory is used (non-trivially) in regional and theoretical economics; algebraic geometry interacts with physics; the Minkowsky lemma, coding theory and the structure of water meet one another in packing and covering theory; quantum fields, crystal defects and mathematical programming profit from homotopy theory; Lie algebras are relevant to filtering; and prediction and electrical engineering can use Stein spaces. And in addition to this there are such new emerging subdisciplines as 'experimental mathematics', 'CFD', 'completely integrable systems', 'chaos, synergetics and large-scale order', which are almost impossible to fit into the existing classification schemes. They draw upon widely different sections of mathematics."

By and large, all this still applies today. It is still true that at first sight mathematics seems rather fragmented and that to find, see, and exploit the deeper underlying interrelations more effort is needed and so are books that can help mathematicians and scientists do so. Accordingly MIA will continue to try to make such books available.

If anything, the description I gave in 1977 is now an understatement. To the examples of interaction areas one should add string theory where Riemann surfaces, algebraic geometry, modular functions, knots, quantum field theory, Kac-Moody algebras, monstrous moonshine (and more) all come together. And to the examples of things which can be usefully applied let me add the topic 'finite geometry'; a combination of words which sounds like it might not even exist, let alone be applicable. And yet it is being applied: to statistics via designs, to radar/sonar detection arrays (via finite projective planes), and to bus connections of VLSI chips (via difference sets). There seems to be no part of (so-called pure) mathematics that is not in immediate danger of being applied. And, accordingly, the applied mathematician needs to be aware of much more. Besides analysis and numerics, the traditional workhorses, he may need all kinds of combinatorics, algebra, probability, and so on.

In addition, the applied scientist needs to cope increasingly with the nonlinear world and the extra mathematical sophistication that this requires. For that is where the rewards are. Linear models are honest and a bit sad and depressing: proportional efforts and results. It is in the non-linear world that infinitesimal inputs may result in macroscopic outputs (or vice versa). To appreciate what I am hinting at: if electronics were linear we would have no fun with transistors and computers; we would have no TV; in fact you would not be reading these lines.

There is also no safety in ignoring such outlandish things as nonstandard analysis, superspace and anticommuting integration, p-adic and ultrametric space. All three have applications in both electrical engineering and physics. Once, complex numbers were equally outlandish, but they frequently proved the shortest path between 'real' results. Similarly, the first two topics named have already provided a number of 'wormhole' paths. There is no telling where all this is leading - fortunately.

Thus the original scope of the series, which for various (sound) reasons now comprises five subseries: white (Japan), yellow (China), red (USSR), blue (Eastern Europe), and green (everything else), still applies. It has been enlarged a bit to include books treating of the tools from one subdiscipline which are used in others. Thus the series still aims at books dealing with:

- a central concept which plays an important role in several different mathematical and/or scientific specialization areas;
- new applications of the results and ideas from one area of scientific endeavour into another;
- influences which the results, problems and concepts of one field of enquiry have, and have had, on the development of another.

This volume in the series is about analytic sets in the sense of the theory of functions of several complex variables (or analytic geometry as that phrase is used in the modern sense of the word in analogy with algebraic geometry). That is, in this book an analytic set is a set in a complex analytic manifold that is locally given by the zeros of a finite number of holomorphic functions (and it is not a complete separable metric continuous image of the set of irrational numbers or one of the various generalizations of that idea).

The concept has been around for a while and several books, mostly called "several complex variables" or "analytic varieties" devote space to the topic, especially in the form of the local theory and, much related to that, the theory of singularities. As a separate subject within several complex variables, important for a wide range of problems involving the analysis and construction of holomorphic mappings, and stressing - though the local theory remains important - more global and geometric aspects (such as analytic covers, metrical properties) and boundary properties, the field is a rather young one.

It is a pleasure to welcome this systematic and thorough treatment of the topic by a very well known author in the field in this series.

The shortest path between two truths in the real domain passes through the complex domain.

J. Hadamard

La physique ne nous donne pas seulement l'occasion de résoudre des problèmes ... elle nous fait pressentir la solution.

H. Poincaré

Never lend books, for no one ever returns them; the only books I have in my library are books that other folk have lent me.

Anatole France

The function of an expert is not to be more right than other people, but to be wrong for more sophisticated reasons.

David Butler

Bussum, June 1989

Michiel Hazewinkel

Table of Contents

Preface

The theory of complex analytic sets is part of the modern geometrical theory of functions of several complex variables. A wide circle of problems in multidimensional complex analysis, related to holomorphic functions and maps, can be reformulated in terms of analytic sets. In these reformulations additional phenomena may emerge, while for the proofs new methods are necessary. (As an example we can mention the boundary properties of conformal maps of domains in the plane, which may be studied by means of the boundary properties of the graphs of such maps, cf. article [184] by the author.)

The theory of complex analytic sets is a relatively young branch of complex analysis. Basically, it was developed to fulfill the need of the theory of functions of several complex variables, but for a long time its development was, so to speak, within the framework of algebraic geometry - by analogy with algebraic sets. And although at present the basic methods of the theory of analytic sets are related with analysis and geometry, the foundations of the theory are expounded in the purely algebraic language of ideals in commutative algebras.

In the present book I have tried to eliminate this noncorrespondence and to give a geometric exposition of the foundations of the theory of complex analytic sets, using only classical complex analysis and a minimum of algebra (well-known properties of polynomials of one variable). Moreover, it must of course be taken into consideration that algebraic geometry is one of the most important domains of application of the theory of analytic sets, and hence a lot of attention is given in the present book to algebraic sets.

The known monographs on complex analytic sets (Abhyankar [1], Hervé [203], Whitney [150], Gunning and Rossi [35], etc.) are, basically, concerned with the so-called local theory. This direction is related to the theory of singularities of differentiable maps, and is gradually absorbed by it. In the present book, local problems also take an important place, but the main goal was the exposition of global properties of analytic sets as the geometric analog of holomorphic functions and holomorphic maps.

The book consists of four chapters and an appendix. Chapter 1 contains the base of the theory. As already noted, the present exposition is distinguished by the extensive use of geometric and analytic methods. Theorems on removing singularities of holomorphic functions of several variables and simple geometric theorems on proper projections of sets in $\mathbb{C}^n$ lie at their basis. The notion of analytic cover, which was developed in detail in the book by Gunning and Rossi [35], plays an important part in these problems. Canonical defining functions of finite tuples of points in $\mathbb{C}^n$, analogs of the standard polynomials $\prod(z-a_j)$ on the plane, also proved to be a convenient tool. These functions were systematically used in Whitney's book [150].

Chapter 2 is concerned with infinitesimal properties of analytic sets and intersection theory. The paragraphs on tangent cones are written under extensive influence of Whitney's work [148], [149], [150], the exposition of the intersection theory of holomorphic chains is close to Draper's paper [53], but gives proofs of more geometrical nature.

In Chapter 3 we study metrical properties and other problems related to the integration of differential forms around analytic sets. These problems are closely related to multidimensional Nevanlinna theory, which was intensively developed in the last years (cf. [42], [186]).

Chapter 4 is, perhaps, closest of all to function theory. In it we study parallels with theorems on analytic continuation and on boundary properties of holomorphic functions of several variables.

In the Appendix we have collected some necessary information from multidimensional complex analysis; the results, which are included in the well-known handbooks and monographs [32], [35], [175], [185], are moreover expounded very quickly. Since Hausdorff measures play an important part in the exposition given, at the end of the Appendix proofs of the necessary properties of these measures are given.

Some remarks on what I would have liked to include also in the book. In the local problems - Milnor's theory of isolated singularities [87] and its generalizations. In Chapter 3 a paragraph on volumes of tubes over analytic sets and their relations with Chern classes and curvatures, in the spirit of Griffiths' article [41], would have been appropriate. In Chapter 4 I would have liked to include additional results on boundary properties of analytic sets, in particular those from

[184]. Moreover, it was the intention to give an exposition of the properties of holomorphic, weakly holomorphic, and other functions on analytic sets. Unfortunately, shortage of space and time did not allow me to include this material.

In order to give an exposition of the foundations of the theory by a new method a number of technical results had to be proved, using a minimum of prerequisites. All these results are generally known and have been published somewhere or belong to mathematical folklore; it is difficult in these cases to indicate sources. This also pertains to a lot of material of the subsequent chapters: absence of authorship indication simply means that I could not succeed in exactly establishing the author of a given result, and by no means indicates that this result belongs to the author of the present book. For reference purposes an extensive bibliography is given at the end of the book; it includes classical works and works of recent years.

References within the book are divided into paragraphs and sections. Numbering of paragraphs within the main text (Chapters 1-4) is compound, paragraphs are divided into sections, a reference like 'p.11.3' means section 3 of paragraph 11. The Appendix has its own numbering of paragraphs, e.g. 'A5.4' means section 4 of paragraph 5 of the Appendix. A reference like 'Theorem 13.5' (Lemma, Proposition) means the theorem in p.13.5 (if there is only one); within sections there is no such numbering and, e.g., 'Theorem 2' means Theorem 2 of the present section.

The sign ■ indicates the begin and end of a proof.

Finally, for their critical remarks and constant stimulating support I consider it my pleasant duty to thank my colleagues and friends, especially S.I. Pinchuk, A. Sadullaev, and B.V. Shabat.

E.M. Chirka

List of notations

$\mathbb{C}$; the complex plane, the field of complex numbers;

$\mathbb{R}$; the real line, the field of real numbers;

$\mathbb{R}_+ = \{x \in \mathbb{R} : x \geqslant 0\}$;

$\mathbb{Z}$; the integers, $\mathbb{Z}_+ = \mathbb{Z} \cap \mathbb{R}_+$;

$\mathbb{N}$; the natural numbers;

$\mathbb{C}^n = \mathbb{C} \times \cdots \times \mathbb{C}$, $\mathbb{R}^n = \mathbb{R} \times \cdots \times \mathbb{R}$ (n times);

$(\mathbb{C}^n)^*$; the space of $\mathbb{C}$-linear functions on $\mathbb{C}^n$ (the dual of $\mathbb{C}^n$);

$(\mathbb{C}^n)^*_{\mathbb{R}}$; the space of $\mathbb{R}$-linear complex valued functions on $\mathbb{C}^n$;

$\mathbb{C}^n_* = \mathbb{C}^n \setminus \{0\}$;

$\mathbb{P}_n$; complex n-dimensional projective space (see A3.2);

$G(k,m)$; the Grassmannian of complex k-dimensional subspaces in $\mathbb{C}^n$ (see A3.4);

$(z,w) = z_1\bar{w}_1 + \cdots + z_n\bar{w}_n$; the Hermitian scalar product in $\mathbb{C}^n$;

$|z| = (z,z)^{1/2} \geqslant 0$;

$\langle z,w \rangle = z_1 w_1 + \cdots + z_n w_n$;

$\mathbb{C}^n \approx \mathbb{R}^{2n}$; $(z_k = x_k + iy_k)$, $(z_1, \ldots, z_n) \leftrightarrow (x_1, y_1, \ldots, x_n, y_n)$, where the isomorphism $\mathbb{R}^n \subset \mathbb{C}^n$ is the natural inclusion $x = (x_1, \ldots, x_n) \in \mathbb{C}^n$;

$B(a,r)$; the ball in $\mathbb{C}^n$ or $\mathbb{R}^n$ with center a and radius r;

$\mathbb{U}(n)$; the group of unitary transformations of $\mathbb{C}^n$ (see A2.1);

$[z] \doteq [z_0, \ldots, z_n]$; the point of $\mathbb{P}_n$ with homogeneous coordinates $(z_0, \ldots, z_n)$;

$\Pi: \mathbb{C}^{n+1}_* \to \mathbb{P}_n$; the canonical projection $z \to [z]$;

$L^\perp$; the subspace of $\mathbb{C}^n$ orthogonal to $L \subset \mathbb{C}^n$, or the complex plane in $\mathbb{P}_n$ orthogonal to the plane $L \subset \mathbb{P}_n$ (see A3.3);

Notations with multi-indices $I = (i_1, \ldots, i_p)$, $i_k \in \mathbb{Z}_+$.

$\# I$; the number of elements (the length) of I;

$|I| = i_1 + \cdots + i_p \qquad (p = \# I)$;

$z_I = (z_{i_1}, \ldots, z_{i_p})$;

$\mathbb{C}_I = \mathbb{C}_{i_1 \ldots i_p} = \mathbb{C}_{z_I}$; the coordinate plane of the variable z_I;

π_I; the orthogonal projection of $\mathbb{C}^n$ onto $\mathbb{C}_I$;

$z^I = z_1^{i_1} \cdots z_n^{i_n}$ ($\sharp I = n$);

$\dfrac{\partial^{|I|}}{\partial z^I} = \dfrac{\partial^{|I|}}{\partial z_1^{i_1} \cdots \partial z_n^{i_n}}$;

I is ordered if $i_1 < \cdots < i_p$;

$\sum\limits_{\sharp I = p}'$; summation over the ordered multi-indices of length p;

$\overline{E}$; the closure of a set E;

$\partial E = \overline{E} \setminus E$;

$\sharp E$; the number of elements of a set E;

$:=$ (or $=:$); equality by definition;

δ_{jk}; the Kronecker symbol ($= 1$ if $j = k$ and $= 0$ if $j \neq k$);

$E + c = \{x + c \colon x \in E\}$ in $\mathbb{C}^n$ or in $\mathbb{R}^n$;

$\wedge$; exterior product (of polyvectors, forms, linear spaces) (see A4.1);

dist; the distance;

det; the determinant of a matrix;

All manifolds in this book are assumed to be countably compact, i.e. to be representable as a countable union of compact subsets. The phrase "manifold M in Ω" means manifold located in Ω, i.e. a submanifold of a neighborhood of it in Ω.

bM; the boundary of a manifold M (with the induced orientation);

$T_a M$; the tangent space to a manifold M at a point a; for manifolds in $\mathbb{R}^N$ it is usually regarded as a subspace of $\mathbb{R}^N \approx T_a \mathbb{R}^N$;

$T_a^{\mathbb{C}} M = (T_a M) \cap i(T_a M) \subset T_a \Omega$; the complex tangent plane to M, i.e. the complex component of $T_a M$ in $T_a \Omega$ (if Ω is complex), see A2.4;

$\operatorname{rank}_z f$; the rank of a differentiable map f at a point z (the complex rank if f is holomorphic);

Z_f; the zero set of a function f;

$L^1_{\text{loc}}(U)$; the set of functions integrable on compact subsets of U (of locally integrable functions in U);

$C^k(U)$; the set of k times continuously differentiable functions on an open subset U of a sufficiently smooth manifold;

$\mathcal{D}(U)$; the set of differential forms of class C^∞ with compact support in U;

$\phi^{p,q}$; the component of a form ϕ of bidegree (p,q) on a complex manifold (the (p,q)-component of ϕ), see A4.2;

vol_k; the volume of subsets of a k-dimensional Riemannian manifold;

$\mathcal{H}_a$; the Hausdorff measure of order (or dimension) α (in other words, the α-measure), see A6.1;

Complex Analytic Sets

Chapter 1

FUNDAMENTALS OF THE THEORY OF ANALYTIC SETS

1. Zeros of holomorphic functions

1.1. Weierstrass' preparation theorem. *Let a function f be holomorphic in a domain $V = V' \times \{|z_n| < R\}$, where V' is a neighborhood of the coordinate origin $0'$ in $\mathbf{C}^{n-1}$, and let also $f(0', z_n) \not\equiv 0$ in the disc $|z_n| < R$. Let $r < R$ be such that $f(0', z_n)$ does not have zeros on the circle $|z_n| = r$, and let k be the number of its zeros in the disk U_n: $|z_n| < r$, counted with multiplicities. Then f can, in a certain neighborhood $U = U' \times U_n \subset V$ of the coordinate origin in $\mathbf{C}^n$, be represented in the form*

$$f(z) = (z_n^k + c_1(z')z_n^{k-1} + \cdots + c_k(z'))\phi(z), \qquad (\star)$$

where the functions $c_j(z')$ are holomorphic in U', while ϕ is holomorphic and zero free in U.

Thus, the zero set Z_f of f locally coincides with the zero set of a polynomial in one variable, with coefficients that holomorphically depend on the remaining variables, and with leading coefficient 1 (such polynomials are called *Weierstrass polynomials*). The factorization $(\star)$ is a generalization of the well-known property of holomorphic functions of one variable: if $f \not\equiv 0$ in a neighborhood of 0 in $\mathbf{C}$ and if k is the order of the zero of f at 0, then $f(z) = z^k g(z)$, where g is holomorphic and not equal to zero at 0. In view of the extraordinary importance of Weierstrass' theorem for the whole theory of analytic sets, we give a full proof of this Theorem, although it can be found in many easily accessible books (cf. [185], [35], [43]).

■ Since $f(0',z_n) \neq 0$ on the circle $|z_n| = r$, there is, by the continuity of f, a polydisk $U' \ni 0'$ in $\mathbf{C}^{n-1}$ such that $\overline{U'} \times \overline{U}_n \subset V$ and $f(z',z_n) \neq 0$ for $z' \in U'$ and $|z_n| = r$. The Theorem on the logarithmic residue implies that, for fixed z', the number of zeros of the function $f(z',z_n)$ in the disk U_n is

$$n(z') := \frac{1}{2\pi i} \int_{|z_n|=r} \left[\frac{\partial f(z',z_n)}{\partial z_n} / f(z',z_n) \right] dz_n.$$

This integral obviously depends continuously on z'. Since it takes only integer values, $n(z') \equiv k$ in U', i.e. for each fixed $z' \in U'$ the function $f(z',z_n)$ has exactly k zeros in U_n, counted with multiplicities. Suppose that $\alpha_1(z'), \ldots, \alpha_k(z')$ are these zeros, each written as many times as its multiplicity (the order is arbitrary). We put

$$P(z',z_n) := \prod_1^k (z_n - \alpha_j(z')) =: z_n^k + c_1(z')z_n^{k-1} + \cdots + c_k(z')$$

and prove that the coefficients c_j holomorphically depend on z'. For this we need two algebraic theorems: Viète's formulas (for coefficients as symmetric functions of the roots) and the fundamental theorem on symmetric polynomials (cf., e.g., [67]), according to which any symmetric polynomial in $\alpha = (\alpha_1, \ldots, \alpha_k)$ can be written as a polynomial in the elementary symmetric functions $\sigma_m(\alpha) = \sum_1^k \alpha_j^m$. These theorems imply that $c_j(z') = g_j(\sigma_1(\alpha(z')), \ldots, \sigma_k(\alpha(z')))$, where g_j are polynomials in the variables $\sigma_1, \ldots, \sigma_k$ with coefficients depending only on k and j. Hence, in order to prove holomorphy of the functions c_j it is sufficient to prove holomorphy of the functions $s_m(z') := \sum_1^k \alpha_j(z')^m$. Here we use the generalized formula for the logarithmic residue from the theory of functions of one complex variable (see, e.g., [185]): if a function h is holomorphic in a disk $|\zeta| < r$, and a function g is holomorphic in it and has no zeros on the circle $|\zeta| = r$, then

$$\frac{1}{2\pi i} \int_{|\zeta|=r} h(\zeta) \frac{g'(\zeta)}{g(\zeta)} d\zeta = \sum_1^k h(\alpha_j),$$

where $\alpha_1, \ldots, \alpha_k$ are the zeros of g in the disk $|\zeta| < r$ (each written as many times as its multiplicity). This theorem can be easily derived from Cauchy's residue theorem. Applying it with $\zeta = z_n$, $h(\zeta) = z_n^m$, and $g(\zeta) = f(z', z_n)$ for fixed $z' \in U'$, we obtain

$$s_m(z') = \sum_1^k \alpha_j(z')^m = \frac{1}{2\pi i} \int_{\partial U_n} z_n^m \left[\frac{\partial f(z', z_n)}{\partial z_n} / f(z', z_n) \right] dz_n$$

and hence $s_m \in \mathcal{O}(U')$ by Lemma 1, A1.1.[1]) Thus, all c_j are holomorphic in U' and hence $P(z', z_n)$ is a holomorphic function in the domain $U' \times \mathbb{C}_n$.

Finally we study the function $\phi = f/P$ in the polydisk U. Since for fixed $z' \in U'$, f and ϕP have the same zeros in z_n (counted with multiplicities), for fixed z' all zeros of P are removable singularities for $\phi(z', z_n)$. We extend ϕ to these points by continuity, and then the function $\phi(z', z_n)$ will be, for fixed z', holomorphic with respect to z_n in the closed disk $\overline{U}_n$. Since ϕ is holomorphic (with respect to z) in the neighborhood $U' \times \partial U_n$, Lemma 2, A1.1, implies that it is holomorphic in U. By construction, the function ϕ has no zeros in U. ■

R e m a r k. The condition that $f(0', z_n) \not\equiv 0$ in a neighborhood of 0 is not an essential restriction. For a holomorphic function f in the ball $B\colon |z| < r$ it can be simply realized by a linear change of coordinates (arbitrarily close to $z \mapsto z$). In any case, if $f|_B \not\equiv 0$, then there is a complex line $L \ni 0$ (arbitrarily close to the $\mathbb{C}_n$-axis) such that $f|_{L \cap B} \not\equiv 0$; such a line can be taken as the $\mathbb{C}_n$-axis in new coordinates.

1.2. Dependence of roots on parameters. The roots of a polynomial, especially multiple roots, depend in a fairly bad way on the coefficients. However, if under a holomorphic transformation of the coefficients the number of geometrically distinct roots does not change, then these roots also holomorphically depend on the parameters. First we single out an important simple case.

P r o p o s i t i o n 1. *Let a function f be holomorphic in a polydisk $U = U' \times U_n$ and suppose that for each fixed $z' \in U'$ it has a unique zero*

[1]) For the notation $\mathcal{O}(U')$ see A1.1. (Translator's note.)

$z_n = \alpha(z')$ *in the disk* U_n. *Then the function* $\alpha(z')$ *is holomorphic in* U'.

■ Hurwitz' Nullstellen theorem (or Rouché's theorem) for the family of functions $f(z',z_n)$ depending on the parameter $z' \in U'$ implies that $\alpha(z')$ is continuous; in particular, the set Z_f, being the graph of $\alpha(z')$, does not have limit points on $U' \times \partial U_n$. By the proof of Weierstrass' preparation theorem, the function f can be represented in U in the form

$$f(z) = (z_n - \alpha(z'))^k \phi(z) = (z_n^k + c_1(z')z_n^{k-1} + \cdots)\phi(z),$$

where k is the order of the zero of $f(0',z_n)$ at the point $z_n = \alpha(0')$ and ϕ is holomorphic in U. Since $c_1(z') = k\alpha(z')$, this theorem implies $\alpha(z') \in \mathcal{O}(U')$. ■

R e m a r k. We have proved also that under the conditions of the Proposition $f(z) = (z_n - \alpha(z'))^k \phi(z)$, where ϕ is holomorphic and zero free in U.

P r o p o s i t i o n 2. *Let* f *be holomorphic in a domain* $D = D' \times D_n$ *and suppose that for each fixed* $z' \in D'$ *it has in* $D_n \subset \mathbf{C}_n$ *exactly* m *geometrically distinct zeros. Then, if the domain* D' *is simply connected, these zeros depend holomorphically on* z'; *more precisely, there exist functions* $\alpha_1(z'), \ldots, \alpha_m(z')$ *holomorphic in* D, *natural numbers* $k_1, \ldots, k_m$, *and a function* $\phi \in \mathcal{O}(D)$ *that does not vanish anywhere in* D, *such that*

$$f(z) = (z_n - \alpha_1(z'))^{k_1} \cdots (z_n - \alpha_m(z'))^{k_m} \phi(z), \quad z \in D.$$

■ As in Proposition 1, Z_f has no limit points on $D' \times \partial D_n$. Let $\zeta' \in D'$, let $\alpha_1(\zeta'), \ldots, \alpha_m(\zeta')$ be the zeros of $f(\zeta',z_n)$ in D_n, and let U_{nj}: $|z_n - \alpha_j(\zeta')| < r_{nj}$ be pairwise nonintersecting disks in D_n. Since Z_f is closed in D there exists a polydisk $U' \ni \zeta'$ such that $U' \times \partial U_{nj} \subset D \setminus Z_f$ for all $j = 1, \ldots, m$. The Theorem on the logarithmic residue implies (as in p.1.1) that for each $z' \in U'$ the function $f(z',z_n)$ has at most one zero in U_{nj}. Since the number of distinct zeros of $f(z',z_n)$ in D_n is m by assumption, there is, for fixed $z' \in U'$, in each U_{nj} exactly one zero $\alpha_j(z')$ of $f(z',z_n)$. By Proposition 1 (in the polydisk $U' \times U_{nj}$), $\alpha_j(z')$ is a holomorphic function in U'.

This argument can be repeated in a neighborhood of any point $\zeta' \in D'$. By construction, the analytic elements obtained coincide, after a suitable renumbering, on the intersections of the corresponding polydisks in D', and can therefore be analytically continued along any path in D'. Since D' is simply connected, the monodromy Theorem (which can be proved in $\mathbb{C}^n$ as is done in $\mathbb{C}$) implies that these elements make up the m holomorphic functions $\alpha_j(z')$.

In each polydisk $U^j = U' \times U_{nj}$ described, we represent f as $f(z) = (z_n - \alpha_j(z'))^{k_j}\phi_j(z)$, where ϕ_j is holomorphic and zero free in U^j, and we put $\phi(z) = f(z) / \prod_1^m (z_n - \alpha_j(z'))^{k_j}$. Then ϕ is defined and holomorphic outside the zeros of f. In $U^j \setminus Z_f$ it is equal to $\phi_j / \prod_{i \neq j}(z_n - \alpha_i(z'))^{k_i}$, and therefore it can be holomorphically defined in all of U^j. Since the U^j cover $Z_f \cap \{z' \in U'\}$, ϕ is holomorphic and zero free in $U' \times D_n$. By covering D' with such polydisks U' we obtain that ϕ is holomorphic and zero free in the whole domain D. ■

1.3. Discriminant set.

Theorem. *Let a function f be holomorphic in a bounded domain $D = D' \times D_n$ in $\mathbb{C}^n$ and suppose that its zero set Z_f does not have limit points on $D' \times \partial D_n$. Let $m < \infty$ be the maximal number of geometrically distinct zeros of the function $f(z', z_n)$ in D_n for fixed $z' \in D'$. Then the set of points $z' \in D'$ above which this maximum is attained is open and everywhere dense in D'. Moreover, there is a function $\Delta(z') \not\equiv 0$ holomorphic in D', such that the set of exceptional points $\zeta' \in D'$ (for which $\# Z_f \cap \{z' = \zeta'\} < m$) coincides with the zero set Z_Δ.*

Here, the condition $m < \infty$ can be omitted, since it clearly follows from the uniqueness Theorem and Rouché's theorem. The set Z_Δ is called the *discriminant set* of the function f (more precisely, of the ramified cover $Z_f \to D'$).

■ Let us denote by G' the set of those $\zeta' \in D'$ for which the number of points of Z_f in the fiber $z' = \zeta'$ is maximal ($=m$). Since $m < \infty$, G' is not empty. We show that G' is open.

Let $\zeta' \in G'$ and let $\{(\zeta', \zeta_{nj})\}_{j=1}^m = Z_f \cap \{z' = \zeta'\}$ be the fiber of Z_f above ζ'. We cover the ζ_{nj} by pairwise nonintersecting disks $\overline{U}_{nj} \subset D_n$ and denote by U_n the union of the interiors of there disks. Since f is zero free on the compact set $\{\zeta'\} \times \partial U_n$ these exists, by continuity, a polydisk $U' \ni \zeta'$ in D' such that the set $\overline{U'} \times \partial U_n$ also does not intersect Z_f. The Theorem on the logarithmic residue implies that the number of zeros of $f(z', z_n)$ in each disk U_{nj} does not depend on $z' \in U'$ (as in Weierstrass' preparation theorem), and since $f(\zeta', \zeta_{nj}) = 0$, for each fixed $z' \in U'$ there is at least one zero of f in each such disk. This and the maximality of m obviously imply that above each point $z' \in U'$ the set Z_f has exactly

m geometrically distinct points; in other words, $U' \subset G'$.

By Proposition 2, p.1.2, above each polydisk $U' \subset G'$ the set Z_f coincides with the zero set of the Weierstrass polynomial $\prod_1^m (z_n - \alpha_j(z')) = z_n^m + c_1(z')z_n^{m-1} + \cdots + c_m(z')$, which has simple roots for each fixed $z' \in U'$. In intersections of such sets $U' \subset D'$ the corresponding Weierstrass polynomials coincide (the zeros with respect to z_n are the same and simple), therefore the whole set $Z_f \cap (G' \times D_n)$ is given by one Weierstrass polynomial $z_n^m + c_1(z')z_n^{m-1} + \cdots + c_m(z')$, $c_j \in \mathcal{O}(G')$, with simple zeros with respect to z_n. We form the discriminant of this polynomial, i.e. the function $\Delta(z') = \prod_{i \neq j} (\alpha_i(z') - \alpha_j(z'))^2$, where $\alpha_j(z')$, $j = 1, \ldots, m$, are the roots of the polynomial. This expression does not depend on the order of the roots for fixed z', since it is symmetric with respect to α_j. As was proved above, in a neighborhood of each point $\zeta' \in G'$ these roots can be chosen to holomorphically depend on z', therefore $\Delta(z')$ is holomorphic in G'. Since Z_f does not have limit points on $D' \times \partial D_n$ and since $\sharp Z_f \cap \{z' = \zeta'\} < m$ for points $\zeta' \in \partial G' \cap D'$ (by the definition of G), when moving to the boundary of G' some roots of our Weierstrass polynomial will stick together, and hence all limit values of $\Delta(z')$ on $\partial G' \cap D'$ vanish (the $\alpha_j(z')$ are uniformly bounded inside D', since D_n is bounded). Thus, by defining $\Delta(z')$ to be zero in $D' \setminus G'$ we obtain a function continuous in D' and holomorphic outside its zeros. By Radó's theorem (A1.5) this extended function $\Delta(z')$ is holomorphic in D'. By construction, $\Delta(z') \not\equiv 0$ and $D' \setminus G' = Z_\Delta$. ■

C o r o l l a r y 1. *Let a function f be holomorphic in a bounded domain $D = D' \times D_n$ in $\mathbb{C}^n$ and suppose that its zero set Z_f does not have limit points on $D' \times \partial D_n$. Then there is a Weierstrass polynomial F such that $Z_F = Z_f$, and such that for any fixed z' not belonging to the discriminant set of f the polynomial $F(z', z_n)$ has only simple roots (with respect to z_n).*

■ In the proof of the Theorem we already constructed such a polynomial F for $z' \in D' \setminus Z_\Delta$. Since D_n is bounded, the coefficients $c_j(z')$ are uniformly bounded on $D' \setminus Z_\Delta$ (by Viète's formulas). Since Z_Δ is a polar subset in D', the functions c_j can be analytically continued to D'. Since Z_f does not have limit points on $D' \times \partial D_n$ and since Z_Δ is nowhere dense in D' (by the uniqueness Theorem for $\Delta(z')$), Rouché's theorem implies that the intersections of Z_f and Z_F with $(D' \setminus Z_\Delta) \times D_n$ are everywhere dense in Z_f and Z_F, respectively. Above

$D' \setminus Z_\Delta$ the zeros of f and F coincide, hence they coincide also above Z_Δ. ■

C o r o l l a r y 2. *The discriminant set* Z_f *is the zero set of the discriminant of the polynomial* F *constructed above, i.e. it is the projection of the set* $\{F = \partial F / \partial z_n = 0\}$ *in* D'.

We also recall that the discriminant of a polynomial of one variable can be represented in the form of a special determinant in the coefficients of the polynomial and its derivative (see e.g., [67]). If the coefficients, as in our case, depend holomorphically on parameters, the discriminant is a holomorphic function of these parameters; but this was in another way already proved in the Theorem.

1.4. Factorization into irreducible factors. Let a function f be holomorphic in a bounded domain $D = D' \times D_n$ in $\mathbb{C}^n$ and suppose that the set Z_f does not have limit points on $D' \times \partial D_n$. Let π: $(z',z_n) \mapsto z'$, let $\sigma \subset D'$ be the discriminant set for f, and let S be a connected component of $Z_f \setminus \pi^{-1}(\sigma)$. Rouché's theorem and the definition of σ readily imply that $\pi(S) = D' \setminus \sigma$ and that the number of points in the fibers $\pi^{-1}(z') \cap S$, $z' \notin \sigma$, is constant. If $z' \in D' \setminus \sigma$, then above a certain neighborhood $U' \subset D'$ of z' the set Z_f is the union of m complex manifolds which are graphs of functions. Therefore, if $\alpha_1(z'), \ldots, \alpha_s(z')$ are all the points of the fibers $\pi^{-1}(z') \cap S$, the polynomial $\prod_1^s (z_n - \alpha_j(z'))$ in z_n has coefficients that holomorphically depend on z' in $D' \setminus \sigma$. Since these coefficients are uniformly bounded (because of the boundedness of D_n) and σ is polar subset in D', the polynomial F_S constructed can be analytically continued to $D' \times \mathbb{C}_n$, and its zero set coincides with $\bar{S} \cap (D' \times D_n)$. By construction, for each fixed $z' \in D' \setminus \sigma$ the zeros of F_S are simple. Hence there exists a maximal natural number j_S such that for each fixed $z' \in D' \setminus \sigma$ the function $f(z', z_n)$ can be holomorphically divided by $(F_S)^{j_S}$. Lemma 2, A1.1, implies that the function $f \cdot (F_S)^{-j_S}$ is holomorphic in $D \setminus \pi^{-1}(\sigma)$, and Lemma A1.3 gives that this function is holomorphic in $D' \times D_n$. The connectivity of S and the maximality of j_S readily imply that $f \cdot (F_S)^{-j_S}$ is zero free on S. Since the number of connected components of $Z_f \setminus \pi^{-1}(\sigma)$ is at most m, by repeating this process we obtain, in the end, a factorization $f = F_1^{j_1} \cdots F_k^{j_k} \cdot g$, where F_i are Weierstrass polynomials and g is a function holomorphic in D and zero free in $D \setminus \pi^{-1}(\sigma)$. Since $D \setminus \pi^{-1}(\sigma)$ is everywhere dense in D and $f(z', \cdot) \not\equiv 0$ in all fibers $\pi^{-1}(z') \cap D$, Rouché's theorem implies that g is zero free in D itself.

A function f, holomorphic in a domain $D \subset \mathbf{C}^n$, is called *irreducible* (in D) if it cannot be represented in the form $f = f_1 \cdot f_2$, where f_1, f_2 are holomorphic functions in D with nonempty zero sets (in D). An example of such a function is the polynomial F_S constructed above and corresponding to a connected component of the set $Z_f \setminus \pi^{-1}(\sigma)$: if $F_S = f_1 \cdot f_2$, then the connectivity of S and the simplicity of the roots of F_S on S imply that one of f_j is zero free on S, hence on all of D. An irreducible function dividing a given holomorphic function f is called an *irreducible factor* of f. We have proved above that under definite geometric conditions such factors exist. From the local consideration it is obvious that the above constructed factorization into irreducible factors, $f = F_1^{j_1} \cdots F_k^{j_k} \cdot g$, is unique up to renumbering and factors that do not vanish anywhere in D.

A function f, holomorphic in a neighborhood of a point a, is called irreducible at this point if there is a fundamental system of neighborhoods $U_j \ni a$ such that $f \mid_{U_j}$ is irreducible in U_j for each j. Suppose we have chosen coordinates such that $a = 0$, f is holomorphic in a polydisk $D' \times D_n$ and $f(0', z_n) \not\equiv 0$. By shrinking D' and D_n we obtain the situation of the Weierstrass preparation theorem, and, moreover, we may assume that the fiber $\pi^{-1}(0') \cap Z_f$ consists of the point $z = 0$ only. Let σ be the discriminant set for $f|_{D' \times D_n}$. Then 0 is a limit point for any connected component of the set $Z_f \setminus \pi^{-1}(\sigma)$, since any such component covers all of $D' \setminus \sigma$. This readily implies that for any arbitrary polydisk $U' \subset D'$ with center at 0 the number of components of $Z_f \cap \pi^{-1}(U' \setminus \sigma)$ is one and the same; let us denote this number by k. If $F_1, \ldots, F_k$ are the Weierstrass polynomials corresponding to these components in $D' \times D_n$, this means that all F_j are irreducible at 0 and, hence, the factorization $f = F_1^{j_1} \cdots F_k^{j_k} \cdot g$ is a factorization into factors irreducible at 0 (and, simultaneously, in D).

1.5. Multiplicity of zeros. Divisor of a holomorphic function. Let a function f be holomorphic and $\not\equiv 0$ in a neighborhood of a in $\mathbf{C}^n$ and let $f(z) = \sum_0^\infty (f)_j(z-a)$ be its expansion into homogeneous polynomials in $z - a$ (see A1.1). By the uniqueness Theorem there exists a minimal k such that $(f)_k \not\equiv 0$. This number is called the *multiplicity* (*order*) of the zero[1)] of the function f at the point a, and is denoted by $\operatorname{ord}_a f$. In other words, $\operatorname{ord}_a f = k$ if all

[1)]Also: multiplicity (order) of vanishing. (Translator's note.)

partial derivatives of f at a up to order $k-1$ inclusive vanish and if some k-th derivative does not vanish. The polynomial $(f)_k(z-a)$ is called the *initial homogeneous polynomial* of f at a. If $f(a) \neq 0$, then $\operatorname{ord}_a f := 0$, and in case $f \equiv 0$ in a neighborhood of a we may put $\operatorname{ord}_a f = +\infty$. Clearly, $\operatorname{ord}_a (f \cdot g) = \operatorname{ord}_a f + \operatorname{ord}_a g$ and $\operatorname{ord}_a (f+g) \geqslant \min(\operatorname{ord}_a f, \operatorname{ord}_a g)$. For any function f, holomorphic in a domain G, the function $\operatorname{ord}_z f$, $z \in G$, is obviously upper semicontinuous, i.e. sets of smaller values $\operatorname{ord}_z f < t$ are open in G (in view of the continuity of derivatives). The pair $(Z_f, \operatorname{ord}_z f)$ is called the *divisor* of the holomorphic function f, and is denoted by D_f or (f). Since under multiplication of functions the orders are added, the divisors of holomorphic functions in a domain G form a commutative and associative semigroup under addition

$$(f)+(g) := (Z_f \cup Z_g, \operatorname{ord}_z f + \operatorname{ord}_z g) = (f \cdot g).$$

In order to introduce this operation we formally do not need the set Z_f, therefore the divisor of f is sometimes said to be the function $\operatorname{ord}_z f$ itself (cf., e.g., [155]).

The order of a function f at a point a can be defined in terms of the orders of its restrictions to complex lines. By A1.1, for any vector $v \in \mathbb{C}^n_\star$ the numbers $(f)_j(v)$ are the Taylor coefficients of the function $f_v(\lambda) = f(a+\lambda v)$ of the single variable λ. Therefore the following Proposition holds.

P r o p o s i t i o n 1. *The order of the zero of a function f at a point a is equal to the minimum of the orders of its restrictions to complex lines passing through a; more precisely, $\operatorname{ord}_a f \leqslant \operatorname{ord}_0 f_v$ for each $v \in \mathbb{C}^n_\star$. Moreover, $\operatorname{ord}_0 f_v = \operatorname{ord}_a f =: k$ if and only if $(f)_k(v) \neq 0$.*

We call vectors $v \in \mathbb{C}^n_\star$ along which this minimum is attained (i.e. $\operatorname{ord}_0 f_v = \operatorname{ord}_a f$) *growth vectors* of f at a. Since $(f)_k \not\equiv 0$ by the definition of order, the uniqueness Theorem implies that the set of growth vectors of f at a is open and everywhere dense in $\mathbb{C}^n$. Its complement is the cone of zeros of the homogeneous polynomial $(f)_k(z)$, tangent to Z_f at a (see p.8.4).

P r o p o s i t i o n 2. *Let a function f be holomorphic and $\not\equiv 0$ in a neighborhood of a point a in $\mathbb{C}^n$. Then the multiplicity of its zero $\operatorname{ord}_a f$ is the maximal number $\eta \in \mathbb{R}$ such that the function $|f(z)| / |z-a|^\eta$ is bounded in a neighborhood of a.*

■ Let $a = 0$, $f = \Sigma (f)_j$ the expansion in homogeneous polynomials in z, and

$k = \mathrm{ord}_0 f$. If f is holomorphic in the ball $|z| \leqslant r$, Cauchy's inequality (A1.1) implies $|(f)_j(z)| \leqslant M(|z|/r)^j$, $M < \infty$, and hence $|f(z)|/|z|^k \leqslant M\sum_k^\infty |z|^{j-k}/r^j < \infty$ for $|z| < r$. If now $\eta > k$ and v is a growth vector of f, then $|f(\lambda v)| \geqslant |\lambda|^k \ (|(f)_k(v)| - 2M|\lambda|)$ for $|\lambda| < r/2$, hence $|f(\lambda v)|/|\lambda|^\eta \to \infty$ as $\lambda \to 0$. ■

We note that in the situation of Theorem 1.3 (when f is holomorphic in $D' \times D_n$ and Z_f does not have limit points on $D' \times \partial D_n$), the vector $(0, \ldots, 0, 1)$ is a growth vector of f at all points $z \in Z_f \setminus \pi^{-1}(\sigma)$, where $\sigma \in D'$ is the discriminant set for f. Hence, from the preceding Section we easily obtain the following Theorem on division of holomorphic functions.

Proposition 3. *Let functions f and g be holomorphic in a domain D, $f \not\equiv 0$, and $(g) \leqslant (f)$, i.e. $\mathrm{ord}_z g \leqslant \mathrm{ord}_z f$ for all $z \in D$ (in particular, $Z_g \subset Z_f$). Then there are a natural number m and a function h holomorphic in D such that $f = g^m \cdot h$; moreover, $\mathrm{ord}_a h < \mathrm{ord}_a g$ at at least one point $a \in Z_g$.*

■ If Z_g is empty, the proof is easy since $1/g$ is holomorphic in D. Suppose therefore that Z_g is nonempty and let m be the infimum of the integer parts of the numbers $\mathrm{ord}_z f / \mathrm{ord}_z g$, $z \in Z_g$. On any family of natural numbers an infimum is attained, hence there exists a point $a \in Z_g$ such that $\mathrm{ord}_a f - m\, \mathrm{ord}_a g < \mathrm{ord}_a g$. The remaining part easily follows from local factorization into irreducible factors. ■

Remark. The condition of Proposition 3 can, of course, be substantially weakened. E.g., in the situation of Theorem 1.3 it is sufficient to require that $Z_g \subset Z_f$ and that there is at least one z' outside the discriminant set of g for which the order of the zero of f with respect to z_n is smaller than the order of the zero of g with respect to z_n (this also easily follows from the factorization into irreducible factors in D).

2. Definition and simplest properties of analytic sets. Sets of codimension 1

2.1. Complex analytic sets are complex manifolds with possible singularities of special type. We define them in almost the same way as embedded complex manifolds (A2.2), only the number of defining functions is not restricted and the condition on the rank is removed.

D e f i n i t i o n 1. Let Ω be a complex manifold (e.g. a domain in $\mathbb{C}^n$ or in $\mathbb{P}_n$). A set $A \subset \Omega$ is called a (complex) *analytic subset* of Ω if for each point $a \in \Omega$ there are a neighborhood $U \ni a$ and functions $f_1, \ldots, f_N$ holomorphic in this neighborhood such that

$$A \cap U = \{z \in U: f_1(z) = \cdots = f_N(z) = 0\}$$

(the functions f_j are locally defining functions of the set A). In other words, locally A is the set of common zeros of a finite tuple of holomorphic functions.

The locality of the definition is important not only for the generality. With this definition all complex submanifolds are automatically analytic subsets (furthermore, on many complex manifolds there are no global holomorphic functions at all, except for constants).

From the definition it follows that an analytic subset of a manifold Ω must be closed in Ω. In many questions only being locally closed is important, therefore it is expedient to also consider certain more general objects.

D e f i n i t i o n 2. A set A on a complex manifold Ω is called a (local) *analytic set* if it is, in a neighborhood of each of its points, the set of common zeros of a certain finite family of holomorphic functions.

E.g., any domain $D \subset \mathbb{C}^n$ is an analytic set in $\mathbb{C}^n$, but it is an analytic subset in $\mathbb{C}^n$ only if $D = \mathbb{C}^n$. The difference between these notions is nevertheless of little importance, since it clearly follows from the definitions that

Every (local) analytic set on a complex manifold is an analytic subset of a certain neighborhood of it.

Thus, since being closed is inessential, we prefer to speak of analytic sets. Yet one more convention: if A, A' are analytic subsets in Ω and $A' \subset A$, then A' will be called an analytic subset of A.

From the definition it is not clear that an analytic set is a "complex manifold with singularities". The precise meaning of these words will be clarified only at the end of this Chapter, and we begin the study of the structure of analytic sets with the most trivial properties. First some simple consequences of the definitions.

1. *The intersection of a finite number of analytic sets is also an analytic set* (obvious).

2. *The union of a finite number of analytic subset is also an analytic subset.*

■ If in a domain $U \subset \Omega$ the sets A_j are defined by the systems of holomorphic functions $\{f_{jk}\}_{k=1}^{N_j}$, respectively, then $(\cup_1^m A_j) \cap U$ is the set of common zeros of all functions of the form $\prod_{j=1}^m f_{jk_j}$, where $1 \leqslant k_j \leqslant N_j$. ■

Here, being closed is essential: e.g., the union of the two analytic sets $\{|z| < 1\}$ and $\{z = 1\}$ in $\mathbf{C}$ already is not an analytic set, since it is not closed in any neighborhood of the point $z = 1$. We also note that the union or intersection of two complex submanifolds need not be a submanifold (it may have singularities). Thus, the class of all analytic subsets of a given complex manifold is conveniently distinguished from the class of all its complex submanifolds by the fact that it is closed under (finite) intersection and union.

3. *Under a holomorphic map $\phi: X \to Y$ of complex manifolds the pre-image of any analytic set $A \subset Y$ is an analytic set in X.*

■ Let $\phi(b) = a \in A$, and let $f_1, \ldots, f_N$ be holomorphic functions in a neighborhood $U \ni a$ defining $A \cap U$. Then $V = \phi^{-1}(U)$ is a neighborhood of b and in it the set $\phi^{-1}(A)$ coincides with the set of common zeros of the holomorphic functions $f_1 \circ \phi, \ldots, f_N \circ \phi$. ■

The image of an analytic set (and also that of an analytic subset) under a holomorphic map need not at all be an analytic set. E.g., under the map $\zeta \mapsto (\zeta^2 - \zeta, \zeta^3 - \zeta)$ of the unit disk $\{|\zeta| < 1\} \subset \mathbf{C}$ into $\mathbf{C}^2$, the image is not analytic in any neighborhood of the point (0,0), which is contained in it.

4. *The direct product $A_1 \times A_2$ of analytic subsets $A_1 \subset \Omega_1$, $A_2 \subset \Omega_2$ is an analytic subset in $\Omega_1 \times \Omega_2$.*

■ If π_j is the projection of $\Omega_1 \times \Omega_2$ onto Ω_j, $j = 1,2$, then $A_1 \times \Omega_2 = \pi_1^{-1}(A_1)$ and $\Omega_1 \times A_2 = \pi_2^{-1}(A_2)$ are analytic subsets in $\Omega_1 \times \Omega_2$ (by 3); the set $A_1 \times A_2$ is their intersection. ■

2.2. Simplest topological properties. The uniqueness Theorem for holomorphic functions implies the following simple, but important property, which will be called the uniqueness Theorem for analytic sets in the sequel.

Proposition 1. *If the complex manifold Ω is connected and the analytic set $A \subset \Omega$ contains a nonempty open subset in Ω, then $A = \Omega$.*

■ Let A^0 be the set of interior points of A (with respect to Ω); by assumption A^0 is nonempty, and by definition it is open. Let a be a limit point for A^0. Then $a \in A$ (since A is closed in Ω) and hence in some connected coordinate neighborhood $U \ni a$ the set $A \cap U$ is the set of common zeros of holomorphic functions $f_1, \ldots, f_N$. Since all f_j vanish on the nonempty open set $A^0 \cap U$, $f_j \equiv 0$ in U, and hence $a \in U \subset A^0$. Thus, A^0 is also closed in Ω, and, since Ω is connected, $A^0 = \Omega$. ■

Corollary. *If A is a proper analytic subset of a connected complex manifold Ω (i.e. $A \neq \Omega$), then A is nowhere dense in Ω.*

Proposition 2. *Any proper analytic subset A on a Riemann surface*[*)] *Ω is locally finite, i.e. $\sharp A \cap K < \infty$ for every compact set $K \subset \Omega$.*

■ If $a \in A$ and $f_1, \ldots, f_N$ define A in a connected coordinate neighborhood $U \ni a$, then at least one $f_j \not\equiv 0$ in U (otherwise $U \subset A$ and $A = \Omega$ by Proposition 1). The uniqueness Theorem for functions of one complex variable implies that f_j is zero free in some punctured neighborhood $V \setminus \{a\}$ of a. This means that $V \cap A = \{a\}$, hence it has been proved that A is a discrete set. Since it is also a closed set, $A \cap K$ is a discrete compact set, i.e. a finite set. ■

The following property of analytic sets is in essence a complex property (for $\mathbb{R}$-analytic sets it is clearly false).

Proposition 3. *Let Ω be a connected complex manifold and A a proper analytic subset in Ω. Then the set $\Omega \setminus A$ is arcwise connected.*

It is just this property that we have in mind when we say that "an analytic set does not separate a domain".

[*)] A Riemann surface is a connected one-dimensional complex manifold (cf. [155]).

■ Let (U,z) be a coordinate neighborhood on Ω such that $z(U)$ is a convex domain in $\mathbb{C}^n$. For simplicity we assume $U \subset \mathbb{C}^n$. Let $a,b \in U \setminus A$ and let $L: z = a + \lambda(b-a)$, $\lambda \in \mathbb{C}$, be the complex line through a and b. By Proposition 2, $A \cap L \cap U$ is a discrete set without limit points in $L \cap U$, therefore there is in $(L \cap U) \setminus A$ a path (a polygon line) joining a and b.

Now we consider the general case. Since Ω is connected, it is arcwise connected; hence for any two points $a,b \in \Omega \setminus A$ there is a smooth path $\gamma: [0,1] \to \Omega$ such that $\gamma(0) = a$ and $\gamma(1) = b$. Let $U_0, \ldots, U_N$ be a covering of $\gamma([0,1])$ by coordinate neighborhoods that are convex with respect to the corresponding coordinates and are such that $U_0 \ni a$, $U_N \ni b$, and $U_{j-1} \cap U_j$ nonempty for $j = 1, \ldots, N$ (such a chain of neighborhoods obviously exists). We arbitrarily choose points a_j in $(U_{j-1} \cap U_j) \setminus A$ and put $a_0 = a$, $a_N = b$. As has already been proved, there is in each domain $U_j \setminus A$ a path joining a_j and a_{j+1}; the composite (union) of these paths is a path joining a and b in $\Omega \setminus A$. ■

2.3. Regular and singular points. In order to compare analytic sets with complex manifolds, the points of an analytic set are conveniently divided into two classes.

D e f i n i t i o n. A point a of an analytic set A on a complex manifold Ω is called *regular* if there is a neighborhood $U \ni a$ in Ω such that $A \cap U$ is a complex submanifold of Ω. The (complex) dimension of this submanifold is said to be the dimension of the set A at its regular point a, and is denoted by $\dim_a A$. Isolated points of A are regular by definition (they are zero-dimensional manifolds), the dimension of A at them is taken to be 0. The set of all regular points of an analytic set A is denoted by $\operatorname{reg} A$. Every point in the complement $A \setminus \operatorname{reg} A =: \operatorname{sng} A$ is called a *singular* point of A.

The set $\operatorname{reg} A$ is open in A, by definition, hence $\operatorname{sng} A$, the set of singular points, is closed in A (if A is an analytic subset, $\operatorname{sng} A$ is also closed in Ω). The connected components of $\operatorname{reg} A$ are complex manifolds, embedded in Ω (in general, of different dimensions - from 0 to $\dim \Omega$).

P r o p o s i t i o n 1. *Let A be an analytic subset of a (not necessarily connected) one-dimensional complex manifold Ω. Then all points of A are regular (i.e.* $\operatorname{reg} A = A$*), and A can be represented in the form $A_1 \cup A_0$, where A_1 is the union of certain connected components of Ω and A_0 is a closed discrete subset in $\Omega \setminus A_1$.*

The *proof* follows in an obvious manner from Propositions 1, 2, p.2.2.

We will consider typical examples of singularities (singular points) of analytic sets in $\mathbb{C}^2$. First of all, such can be the points of intersection of several complex manifolds, e.g., $A: z_1 z_2 = 0$ or $A: z_1(z_2 - z_1^2) = 0$ at the coordinate origin. Another type of singularity is formed by analytic sets $A: z_2^l = z_1^k$, where k, l are natural numbers, $k, l > 1$. In this case, $\operatorname{reg} A = A \setminus \{0\}$, and 0 is a singular point (see below). If k and l are coprime, $A \setminus \{0\}$ is a connected manifold. The simplest example is the semicubic parabola $z_2^2 = z_1^3$ (see A2.3). If k and l are not coprime, A splits into several analytic subsets in $\mathbb{C}^2$ with connected regular sets; if $k > l$ is not divisible by l, each of these analytic subsets in $\mathbb{C}^2$ also has 0 as singularity.

In $\mathbb{C}^n$ for $n > 2$ singular points of analytic sets need not be isolated, e.g., the equations in z_1, z_2 given above define in $\mathbb{C}^n$ analytic sets with singularities on $(n-2)$-dimensional planes. However, there are isolated singularities in higher dimensions too. E.g., the cone $z_1^2 + \cdots + z_n^2 = 0$ in $\mathbb{C}^n$ has as unique singular point the coordinate origin; all other points are regular, since $d(\sum z_j^2) \neq 0$ at them.

To prove each time that a given point of an analytic set is singular, is rather tedious, so we give the following simple

P r o p o s i t i o n 2. *If 0 is a regular point of an analytic set $A \subset \mathbb{C}^n$, then there are a coordinate neighborhood $U \ni 0$ in $\mathbb{C}^n$ and a coordinate plane $\mathbb{C}_I$, $I = (i_1, \ldots, i_p)$, such that $A \cap U$ can be bijectively projected onto $\mathbb{C}_I \cap U$ (under $z \mapsto z_I$).*

■ By the definition of regularity, there is a neighborhood $U_0 \ni 0$ such that $A \cap U_0$ is a p-dimensional complex submanifold in U_0, where $p = \dim_0 A$. The tangent space $T_0 A$ to it is a p-dimensional complex subspace in $\mathbb{C}^n$. Thus there exists an $I = (i_1, \ldots, i_p)$ such that $T_0 A$ can be bijectively projected onto $\mathbb{C}_I$. Since A is tangent to $T_0 A$ at 0, the projection $z_I|_A$ is also single-valued in some neighborhood of 0. ■

In the sequel we will prove that this property is also sufficient for regularity (see p.3.3), while we remark here that the given Proposition gives a simple condition for points to be singular. E.g., the semicubic parabola $z_2^2 = z_1^3$ is, in any neighborhood of 0, projected two-sheetedly onto the $\mathbb{C}_1$-axis and three-sheetedly onto the $\mathbb{C}_2$-axis. Therefore, for it the coordinate origin is not a regular point.

Precisely the same is true for the cone $A: z_1^2 + \cdots + z_n^2 = 0$ in $\mathbf{C}^n$: it is clearly projected infinitely-sheetedly on the p-dimensional coordinate planes with $p < n-1$, and two-sheetedly on an $(n-1)$-dimensional plane $z_j = 0$, in an arbitrarily small neighborhood of 0. Hence 0 is a singular point of the cone A.

The following Theorem is the first (topological) statement about the "smallness" of the set of singular points.

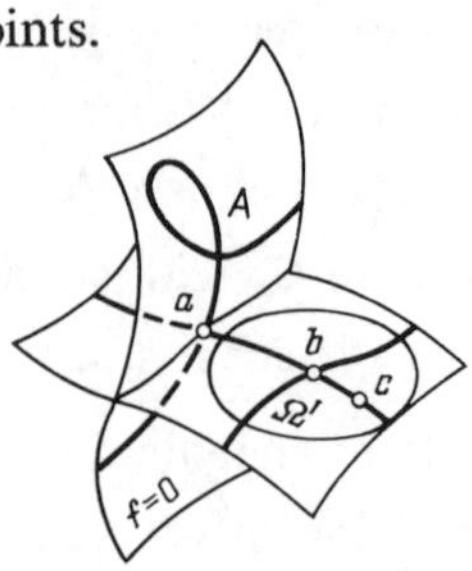

Figure 1.

T h e o r e m. *The set of regular points of an arbitrary analytic set A is everywhere dense in A, and its set of singular points is closed and nowhere dense in A.*

■ The proof proceeds by induction with respect to the dimension of the ambient manifold $\Omega \supset A$. For $n = 1$ all points of A are regular (Proposition 1).

$n-1 \Rightarrow n$. Let $a \in A$, and let U be an arbitrary connected coordinate neighborhood of a in Ω, in which $A \cap U$ is defined by holomorphic functions $f_1, \ldots, f_N$. It suffices to prove that $(\operatorname{reg} A) \cap U$ is nonempty. If all $f_j \equiv 0$ on U, then $U \subset A$ and $a \in \operatorname{reg} A$. We will therefore assume that $f_1 \not\equiv 0$ in U. The function f_1 vanishes on $A \cap U$; it may happen that all first-order, second-order, etc., partial derivatives of f_1 (with respect to the coordinates in U) also vanish identically on $A \cap U$. However, since $f_1 \not\equiv 0$ on U and U is connected, the uniqueness Theorem (A1.1) implies that there is a multi-index $I = (i_1, \ldots, i_n)$ such that the function $f = \partial^{|I|} f_1 / \partial z^I$ vanishes identically on $A \cap U$ but a certain derivative $\partial f / \partial z_j$ of it differs from zero at some point $b \in A \cap U$. The implicit function Theorem then implies that there exists a neighborhood $V \ni b$ in U such that the set $\Omega' = \{z \in V: f(z) = 0\}$ is a complex submanifold of V, and $\dim \Omega' = n-1$ (Fig. 1). Since $f = 0$ on $A \cap V$, we have $A \cap V \subset \Omega'$, i.e. $A \cap V$ is an analytic subset on the $(n-1)$-dimensional complex manifold Ω'. By the induction hypothesis, there are a point $c \in A \cap V$ and an $(n-1)$-dimensional neighborhood $W' \ni c$ in Ω' such that $A \cap W'$ is a complex submanifold of Ω'. If W is an arbitrary neighborhood of c in V such that $W \cap \Omega' = W'$, then $A \cap W$ obviously is a complex submanifold in W, hence c is a regular point of A. ■

2.4. Dimension. The Theorem proved allows us to introduce the following important definition.

Definition 1. Let A be an analytic set on a complex manifold Ω. The *dimension* of A at an arbitrary point $a \in A$ is the number

$$\dim_a A := \overline{\lim_{\substack{z \to a \\ z \in \operatorname{reg} A}}} \dim_z A.$$

The dimension of A is, by definition, the maximum of its dimensions at points:

$$\dim A := \max_{z \in A} \dim_z A = \max_{z \in \operatorname{reg} A} \dim_z A.$$

At points $a \notin A$ it is convenient to put $\dim_A A = -1$. The *codimension* of an analytic set $A \subset \Omega$ is, by definition, equal to $\dim \Omega - \dim A$.

Since in a neighborhood of each of its regular points the dimension of A is constant, at regular points this definition coincides with the one adopted earlier. We obtain from the definition:

The function $\dim_z A$ *is upper semicontinuous on* A, *i.e. all sets* $\{z \in A: \dim_z A \geqslant p\}$ *are closed in* A.

We will put $A_{(p)} := \{z \in A: \dim_z A = p\}$. An analytic set is called homogeneous (in dimension) if its dimensions at all points coincide. A homogeneous analytic set of dimension p is also called a *pure p-dimensional analytic set*; we will use the latter name.

Examples. The analytic set $z_1 z_2 = z_1 z_3 = 0$ in $\mathbf{C}^3$ has dimension 2 at all points of the complex plane $\mathbf{C}_{23}$ and dimension 1 at the points of the punctured one-dimensional plane $\mathbf{C}_1 \setminus \{0\}$. The analytic set $z_1 z_2 = z_1 z_3 = 1$ in $\mathbf{C}^3$ is pure one-dimensional; it coincides with the set $z_1 z_2 = 1$, $z_2 = z_3$, and is bijectively mapped onto the "hyperbola" $z_1 z_2 = 1$ under projection on $\mathbf{C}_{12}$.

In the initial stage of studying the local structure of analytic sets, next to singular points, points at which the dimension of the analytic set is less than maximal cause a lot of trouble. Such "troublesome" points are conveniently collected into one set.

Definition 2. Let A be a p-dimensional analytic set. Its *singular locus* is the set

$$\mathbb{S}(A) := (\operatorname{sng} A) \cup \{z \in A: \dim_z A < \dim A\} = (\operatorname{sng} A) \cup \bigcup_{k<p} A_{(k)}.$$

Since in a small neighborhood of a regular point an analytic set is homogeneous in dimension, all limit points of the set $\{z \in A: \dim_z A < p\}$ in A belong to $\operatorname{sng} A$, hence *the singular locus of any analytic set is closed in this set.* We will use this unsuitable term a short while, until we have proved that any analytic set splits into pure ones, and then we will consider pure-dimensional sets only (for those, $\mathbb{S}(A) = \operatorname{sng} A$).

If $\dim A = \dim \Omega$, all points of A of interest fall within the category "singular", but it is easy to reduce the dimension in this case.

Proposition. *Let A be an analytic subset of a (not necessarily connected) n-dimensional complex manifold Ω, and let* $\dim A = n$. *Then $A_{(n)}$ is the union of certain connected components of Ω and A can be represented as $A_{(n)} \cup A'$, where A' is an analytic subset of the complex manifold $\Omega' = \Omega \setminus A_{(n)}$; moreover,* $\dim A' < n$.

■ Let Ω_1 be the union of those components of Ω in which there are points $a \in \operatorname{reg} A$ such that $\dim_a A = n$. The uniqueness Theorem (p.2.2) implies that each such component belongs to A, hence $A_{(n)} = \Omega_1$. The set $\Omega' = \Omega \setminus \Omega_1$ is open in Ω, $A' := A \cap \Omega$ is an analytic subset of Ω' and $\dim A' < n - 1$, because $A' \cap A_{(n)}$ is empty. ■

Thus, n-dimensional analytic sets on an n-dimensional manifold trivially lead to analytic sets of lower dimension; therefore we will not consider such n-dimensional analytic sets in the sequel. The uniqueness Theorem also implies that

An analytic set A on an n-dimensional complex manifold Ω is nowhere dense in Ω if and only if $\dim A \leqslant n - 1$.

2.5. Regularity in $\mathbb{P}_n$ and $\mathbb{C}_*^{n+1}$. As an illustration of the theme regularity and dimension we consider how these notions are related for analytic sets in $\mathbb{P}_n$ and their pre-images in $\mathbb{C}_*^{n+1}$ under the canonical projection $\Pi: \mathbb{C}_*^{n+1} \to \mathbb{P}_n$.

P r o p o s i t i o n. *Let A be an analytic set in $\mathbb{P}_n$ and let $\tilde{A} = \Pi^{-1}(A)$. Then*

$$\operatorname{reg} A = \Pi(\operatorname{reg} \tilde{A}), \quad \operatorname{sng} A = \Pi(\operatorname{sng} \tilde{A}), \quad \mathbb{S}(A) = \Pi(\mathbb{S}(\tilde{A})),$$

and at each point $\tilde{a} \in \tilde{A}$ the dimension of $\tilde{A}$ is higher by 1 than the dimension of A at the point $a = \Pi(\tilde{a}) = [\tilde{a}]$.

■ Let $a \in \operatorname{reg} A$ and $\dim_a A = p$. Without loss of generality we may assume that a lies in the coordinate neighborhood U_0: $z_0 \neq 0$ with coordinates $\zeta = (z_1 / z_0, \ldots, z_n / z_0)$. Since in a neighborhood of a the set A is a p-dimensional submanifold, there exist functions $f_1, \ldots, f_{n-p}$, holomorphic in a neighborhood $U \ni a$, $U \subset U_0$, defining in it the set A, and such that $\operatorname{rank} (\partial f / \partial \zeta)(a) = n - p$. Let $\tilde{U} = \Pi^{-1}(U)$, then $\tilde{A} \cap \tilde{U}$ is defined by the holomorphic functions $f_j(z_1 / z_0, \ldots, z_n / z_0) =: g_j(z_0, \ldots, z_n)$. Since for $k > 0$, $\partial g_j / \partial z_k = (1 / z_0)(\partial f_j / \partial z_k)$ (at corresponding points) and $z_0 \neq 0$ at points $\tilde{a} = \Pi^{-1}(a)$, we have $\operatorname{rank} (\partial g / \partial z)(\tilde{a}) = n - p$, and hence $\tilde{a} \in \operatorname{reg} \tilde{A}$. Since the number of variables is increased by 1 and the number of defining functions remains the same, $\dim_{\tilde{a}} \tilde{A} = \dim_a A + 1$. Thus, we have proved that $\Pi^{-1}(\operatorname{reg} A) \subset \operatorname{reg} \tilde{A}$ or, in other words, $\Pi(\operatorname{sng} \tilde{A}) \subset \operatorname{sng} A$.

Conversely, let $\tilde{a} \in \operatorname{reg} \tilde{A}$ and $\dim_{\tilde{a}} \tilde{A} = p + 1$, and let us assume that the z_0-coordinate of $\tilde{a}$ is equal to $a_0 \neq 0$. Since the tangent plane to the manifold $\tilde{A}$ at $\tilde{a}$ contains the complex line $\mathbb{C}\tilde{a}$ passing through 0, the hyperplane $z_0 = a_0$ at $\tilde{a}$ intersects $\tilde{A}$ transversally. Hence there is a neighborhood $\tilde{U} \ni \tilde{a}$ such that $A' := \tilde{A} \cap \tilde{U} \cap \{z_0 = a_0\}$ is a p-dimensional complex submanifold in $\tilde{U}$. Since $a_0 \neq 0$ we may assume that $\tilde{U}$ is disjoint from the plane $z_0 = 0$, and we may also assume that $\tilde{U}$ is a cone, i.e. $\tilde{U} = \Pi^{-1}(U)$ for some neighborhood $U \ni a = \Pi(\tilde{a})$ in $\mathbb{P}_n$. If $b = \Pi(\tilde{b}) \in A \cap U$, then $b' = (a_0 / b_0)\tilde{b}$ belongs to A', hence $\Pi(A') = A \cap U$. But A' lies on the plane $z_0 = a_0$, on which Π is biholomorphic (Π maps it into the coordinate neighborhood $U_0 \subset \mathbb{P}_n$). The image of a p-dimensional complex submanifold under a biholomorphic map is also such a submanifold, hence $a = \Pi(\tilde{a}) \in \operatorname{reg} A$.

Thus, we have proved that $\Pi(\operatorname{reg} \tilde{A}) = \operatorname{reg} A$, and $\dim_{\tilde{a}} \tilde{A} = \dim_a A + 1$ if $a \in \operatorname{reg} A$. The remaining part obviously follows from these relations. ■

2.6. Principal analytic sets. A nonempty analytic set on a complex manifold Ω is called *principal* if there is a function $f \in \mathcal{O}(\Omega)$, not identically vanishing on any component of Ω, such that $A = Z_f$: $f = 0$. An analytic set is called *locally principal* if it is principal in a neighborhood of each of its points.

P r o p o s i t i o n. *The dimension of a locally principal analytic set A on an n-dimensional complex manifold at any point $a \in A$ is equal to $n-1$.*

■ Let $A = Z_f$. The uniqueness Theorem implies that A is nowhere dense in Ω, hence $\dim A \leqslant n-1$ (see p.2.4). Fix an arbitrary point $a \in A$ and a coordinate neighborhood (U,z) of it. The uniqueness Theorem for functions implies that there is a multi-index J such that $g|_{A \cap U} = 0$ for $g = \partial^{|J|} f / \partial z^J$ while $dg \neq 0$ at some point $b \in A \cap U$. Thus, in a neighborhood of b the set A is an $(n-1)$-dimensional manifold, hence $\dim_a A = n-1$ by the arbitrariness of U. ■

In p.2.8 we will prove that, conversely, every pure $(n-1)$-dimensional analytic set on an n-dimensional complex manifold is locally principal.

The following estimate of Hausdorff dimensions plays an important role in the study of the structure of sets of singular points.

L e m m a. *Let A be a locally principal analytic set on an n-dimensional complex manifold. Then $\mathcal{H}_\alpha(A) = 0$ for $\alpha > 2n-2$ and (even more important) $\mathcal{H}_\beta(\operatorname{sng} A) = 0$ for $\beta > 2n-4$.*

■ The proof proceeds by induction with respect to n. For $n = 1$ the set A is discrete (p.2.2), hence $\mathcal{H}_\alpha(A) = 0$ for all $\alpha > 0$; for $n = 1$ the set $\operatorname{sng} A$ is empty.

$n-1 \Rightarrow n$. The statement is local, so we may assume that A is defined in a neighborhood $U \ni 0$ in $\mathbf{C}^n$ as the zero set of a holomorphic function f for which $f(0',z_n) \not\equiv 0$ (see p.1.1). According to the Weierstrass preparation theorem and p.1.3 we may assume that f is a Weierstrass polynomial, with coefficients that are holomorphic in a domain $U' \subset \mathbf{C}^{n-1}$, which has for each fixed z' outside the discriminant set $\sigma \subset U'$ only simple zeros with respect to z_n. Let $\pi\colon (z',z_n) \mapsto z' \in \mathbf{C}^{n-1}$ (Figure 2). By p.1.3, σ is a principal analytic subset of U' or is empty. By the induction hypothesis, $\mathcal{H}_\gamma(\operatorname{sng}\sigma) = 0$ for $\gamma > 2n-6$ (if $n = 2$, then $\sigma \subset \mathbf{C}^1$ and thus $\operatorname{sng}\sigma$ is empty). Therefore, $\mathcal{H}_\beta(\pi^{-1}(\operatorname{sng}\sigma)) = 0$ for $\beta > 2n-4$, and $\pi^{-1}(\operatorname{reg}\sigma) = \operatorname{reg}\sigma \times \mathbf{C}_n$ is an $(n-1)$-dimensional submanifold in $\mathbf{C}^n$ (or is empty). Since f is a polynomial in z_n with leading coefficient 1, the set $A : f = 0$ is nowhere dense on the manifold $\pi^{-1}(\operatorname{reg}\sigma)$, hence $A \cap \pi^{-1}(\operatorname{reg}\sigma)$ is empty or is a principal subset of this manifold. By the induction hypothesis,

$\mathcal{H}_\beta(A \cap \pi^{-1}(\text{reg } \sigma)) = 0$ for $\beta > 2n - 4$, hence

$$\mathcal{H}_\beta(A \cap \pi^{-1}(\sigma)) \leqslant \mathcal{H}_\beta(A \cap \pi^{-1}(\text{reg } \sigma)) + \mathcal{H}_\beta(A \cap \pi^{-1}(\text{sng } \sigma)) = 0,$$

if $\beta > 2n - 4$. Since $\partial f / \partial z_n \neq 0$ at the points of $A \setminus \pi^{-1}(\sigma)$, the set $A \setminus \pi^{-1}(\sigma)$ belongs to reg A and is an $(n-1)$-dimensional complex manifold in $\mathbb{C}^n$. Thus sng $A \subset A \cap \pi^{-1}(\sigma)$ and therefore $\mathcal{H}_\beta(\text{sng } A) = 0$ for $\beta > 2n - 4$ and $\mathcal{H}_\alpha(A) \leqslant \mathcal{H}_\alpha(A \setminus \pi^{-1}(\sigma)) + \mathcal{H}_\alpha(A \cap \pi^{-1}(\sigma)) = 0$ for $\alpha > 2n - 2$. ■

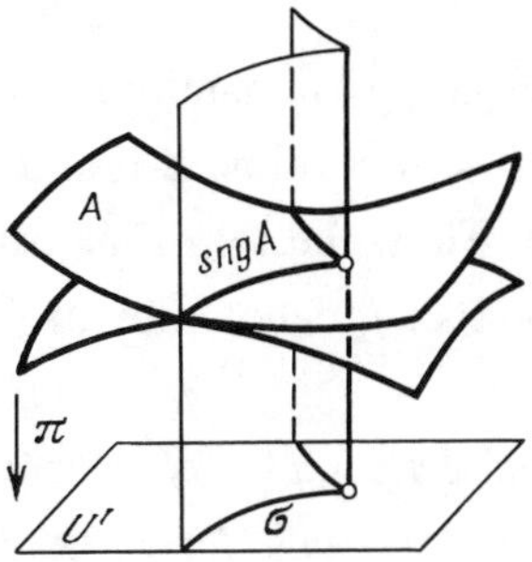

Figure 2.

Taking into account the Proposition, this Lemma can be formulated also as:

C o r o l l a r y 1. *The Hausdorff dimension of a locally principal analytic set on an n-dimensional complex manifold is equal to* $2n - 2$, *and the Hausdorff dimension of the set of its singular points is at most* $2n - 4$.

C o r o l l a r y 2. *Let A be an analytic set on an n-dimensional complex manifold. Then: (1) if* dim $A < n$, *then the Hausdorff dimension of A does not exceed* $2n - 2$; *and (2) if* dim $A < n - 1$, *then the Hausdorff dimension of A is at most* $2n - 4$.

■ The statement is local. (1) Let A be given by holomorphic functions $f_1, \ldots, f_N$ in a connected neighborhood U of a point a of it, and let, say, $f_1 \not\equiv 0$ in U. Then $A \cap U$ belongs to the principal analytic set A_1: $f_1 = 0$ in U, hence $\mathcal{H}_\alpha(A \cap U) \leqslant \mathcal{H}_\alpha(A_1) = 0$ for all $\alpha > 2n - 2$.

(2) As before, $A \cap U$ belongs to the principal analytic set A_1. By the Lemma, $\mathcal{H}_\alpha(A \cap \text{sng } A_1) = 0$ for $\alpha > 2n - 4$. The remaining part of $A \cap U$, the set $A \cap \text{reg } A_1$, is an analytic subset of the $(n-1)$-dimensional complex manifold reg A_1. Therefore, by (1), $\mathcal{H}_\alpha(A \cap \text{reg } A_1) = 0$ for $\alpha > 2n - 4$. ■

C o r o l l a r y 3. *Let A be an analytic set on a n-dimensional complex manifold and let* $\dim A < n$. *Then the Hausdorff dimension of its singular locus* $\mathcal{S}(A)$ *is at most* $2n - 4$.

2.7. Critical points. Let A be an analytic set on a complex manifold X. A map $f: A \to Y$ into a complex manifold Y is called holomorphic (on A) if for each point $a \in A$ there is a neighborhood $U \ni a$ in X and a holomorphic map F: $U \to Y$, such that $F|_{A \cap U} = f|_{A \cap U}$. Obviously, in such a situation $f|_{\operatorname{reg} A}$ is a holomorphic map between complex manifolds. A point $a \in \operatorname{reg} A$ is called a *critical point* of f if $\operatorname{rank}_a f < \min(\dim_a A, \dim Y)$. By definition, all singular points of A are critical for f. Thus, the set of all critical points of a holomorphic map f: $A \to Y$ (the *branch locus* of f, denoted by $\operatorname{br} f$) is closed in A.

L e m m a. *Let A be an analytic subset of a bounded domain $D' \times D_n$ in $\mathbb{C}^n$ without limit points on $D' \times \partial D_n$, and let π: $(z', z_n) \mapsto z'$. Then the Hausdorff dimension of* $\operatorname{br} \pi|_A$ *does not exceed* $2n - 4$.

■ Since $\mathcal{H}_\alpha(\mathcal{S}(A)) = 0$ for $\alpha > 2n - 4$, it suffices to prove the statement in a neighborhood of the "$(n-1)$-dimensional" singular points. Let $a \in (\operatorname{br} \pi|_A) \cap \operatorname{reg} A$ with $\dim_a A = n - 1$. By the implicit function Theorem, there are an integer $j < n$ and a polydisk neighborhood $U \ni a$ such that $A \cap U$ is the graph $z_j = \phi(z_j')$ of a holomorphic function of the variable $z_j' = (z_j, \ldots, z_{j-1}, z_{j+1}, \ldots, z_n)$. By a unitary change of coordinates z' (which has no influence on the conditions of the Lemma) we have $j = 1$. In this case the functions $z_2, \ldots, z_n$ are local coordinates on $A \cap U$. In these coordinates the projection $\pi|_A$ has the form

$$(z_2, \cdots, z_n) \mapsto (\phi(z_2, \cdots, z_n), \ z_2, \cdots, z_n).$$

Up to the sign, its Jacobian is $\partial\phi / \partial z_n$, so that $(\operatorname{br} \pi|_A) \cap U = \{z \in A \cap U: \partial\phi / \partial z_n = 0\}$. If $\partial\phi / \partial z_n \equiv 0$ on $A \cap U$, then ϕ is independent of z_n and, hence, $A \cap U \cap \pi^{-1}(z') = \{z'\} \times U_n$ for each $z' \in \pi(A \cap U)$. On the other hand, $A \cap \pi^{-1}(z')$ is an analytic subset of the one-dimensional complex domain $\{z'\} \times D_n$, and, moreover, it has no limit points on the boundary of this domain. Therefore (Proposition 2, p.2.2), $A \cap \pi^{-1}(z')$ is a finite set. This contradiction shows that $\partial\phi / \partial z_n \not\equiv 0$ on $A \cap U$, hence $(\operatorname{br} \pi|_A) \cap U$ is a proper analytic subset of the connected $(n-1)$-dimensional complex manifold $A \cap U$. By p.2.6, its Hausdorff dimension does not exceed $2n - 4$. ■

2.8. Local representation of sets of codimension 1.

T h e o r e m. *Let A be an $(n-1)$-dimensional analytic subset of a polydisk $U = U' \times U_n$ in $\mathbb{C}^n$ without limit points on $U' \times \partial U_n$, and let π: $(z', z_n) \mapsto z'$. Denote by σ the image of the set $\mathcal{S}(A) \cup (\operatorname{br} \pi|_A)$ under this map π. Then σ is a closed subset of U' of Hausdorff dimension at most $2n - 4$, and*

(1) *π: $A \setminus \pi^{-1}(\sigma) \to U' \setminus \sigma$ is a locally biholomorphic k-sheeted cover $(1 \leqslant k < \infty)$;*

(2) *$\pi^{-1}(\sigma)$ is nowhere dense in $A_{(n-1)} := \{z: \dim_z A = n-1\}$;*

(3) *there is a Weierstrass polynomial $F(z', z_n)$ of degree k in z_n such that $A_{(n-1)} = \{z \in U: F(z', z_n) = 0\}$; in particular, $A_{(n-1)}$ is a principal analytic subset in U.*

■ Since A has no limit points on $U' \times \partial U_n$, σ is closed in U'. Since the Hausdorff dimension of $\mathcal{S}(A) \cup (\operatorname{br} \pi|_A)$ is at most $2n - 4$, $\mathcal{H}_\alpha(\sigma) = 0$ for $\alpha > 2n - 4$.

(1) Since A has Hausdorff dimension $2n - 2$, it contains points not belonging to $\mathcal{S}(A) \cup (\operatorname{br} \pi|_A)$. At each of those, $\operatorname{rank} \pi|_A = n - 1$, hence $\pi|_A$ is locally biholomorphic at them. In particular, $\pi(A)$ has Hausdorff dimension $2n - 2$. Hence there is a point $a' \in U' \setminus \sigma$ such that $\pi^{-1}(a') \cap A$ is nonempty, say equal to $\{a^1, \ldots, a^k\}$; the number k is finite by Proposition 2, p.2.2, since A has no limit points on $U' \times \partial U_n$. At each point a^j the rank of $\pi|_A$ is equal to $n - 1$, hence there is a neighborhood $W_j \ni a^j$ and a neighborhood $V' \ni a'$, such that π: $A \cap W_j \to V'$ is a biholomorphic map. Since A is closed in U and does not go beyond $U' \times \partial U_n$, V' can be chosen small such that $\pi^{-1}(z') \cap U \subset \cup W_j$ for all $z' \in V'$. Hence we have proved that the number of points in a fiber, $\# \pi^{-1}(z') \cap A$, is locally constant in $U' \setminus \sigma$. Since $U' \setminus \sigma$ is connected (Proposition 6, A.6.1), $\# \pi^{-1}(z') \cap A \equiv k$ in $U' \setminus \sigma$ and statement (1) is proved.

(2) The set $A_{(n-1)}$ consists of the $(n-1)$-dimensional complex manifold $A^o = A_{(n-1)} \cap \operatorname{reg} A$, which is open and everywhere dense in $A_{(n-1)}$ (by the definition of dimension at a point) and of the nowhere dense, closed subset $A_{(n-1)} \cap \operatorname{sng} A$. As proved above, the Hausdorff dimension of σ does not exceed $2n - 4$. Outside $\operatorname{br} \pi|_A$ the projection $\pi|_{A_{(n-1)}}$ is locally biholomorphic, hence the Hausdorff dimension of the set $\Sigma = \pi^{-1}(\sigma) \cap (A^o \setminus \operatorname{br} \pi|_A)$ also does

not exceed $2n-4$. Since $\mathfrak{S}(A)$ does not intersect A^o at all, $\pi^{-1}(\sigma) \cap A^o \subset \Sigma \cup (\mathrm{br}\,\pi|_A)$ is nowhere dense on the $(n-1)$-dimensional complex manifold A^o (p.2.1), hence on all of $A_{(n-1)}$.

(3) Let $z' \in U' \setminus \sigma$. Then (see above) there is a neighborhood $V' \ni z'$ such that $\pi^{-1}(V') \cap A$ consists of k manifolds S_j: $z_n = \alpha_j(z')$, where $\alpha_j \in \mathcal{O}(V')$. Put

$$f(z',z_n) = \prod_{j=1}^{k} (z_n - \alpha_j(z')) = z_n^k + c_1(z')z_n^{k-1} + \cdots + c_k(z').$$

By Viète's formulas, the c_j are symmetric functions in $\alpha_1, \ldots, \alpha_k$, and are, moreover, defined and holomorphic in all of $U' \setminus \sigma$. Since all values $\alpha_j(z') \in U_n$, the c_j are uniformly bounded in $U' \setminus \sigma$. The Theorem on removable singularities (A1.4) implies that there exist $C_j \in \mathcal{O}(U')$ such that $C_j|_{U' \setminus \sigma} = c_j$. We denote the zero set of the Weierstrass polynomial $F(z',z_n) = z_n^k + C_1(z')z_n^{k-1} + \cdots + C_k(z')$ in U by Z. By construction, above $U' \setminus \sigma$ the sets Z and $A_{(n-1)}$ coincide. Rouché's theorem for functions of one variable z_n obviously implies that Z coincides with the closure of $Z \cap \pi^{-1}(U \setminus \sigma)$ in U. This also holds, according to (2), for $A_{(n-1)}$, hence $A_{(n-1)} = Z$. ■

C o r o l l a r y 1. *Every pure $(n-1)$-dimensional analytic set on an n-dimensional complex manifold (= a complex hypersurface) is locally principal. More precisely, if $a \in A$ and if local coordinates z have been chosen such that a is an isolated point of the set $A \cap \{z'=a'\}$, then there are a neighborhood $U \ni a$ of the form $U' \times U_n$ and a Weierstrass polynomial $F(z',z_n)$, such that $A \cap U = Z_F$; moreover, for almost all $z' \in U'$ the polynomial F has simple roots with respect to z_n.*

C o r o l l a r y 2. *Let A be an analytic set on an n-dimensional complex manifold and suppose* $\dim A = n-1$. *Then for each point $a \in A$ there is a neighborhood $U \ni a$ such that $A \cap U$ can be represented as $(A_{(n-1)} \cap U) \cup A'$, where A' is an analytic subset of U of dimension $< n-1$.*

■ Let, in a neighborhood of a, A be defined by holomorphic functions $f_1, \ldots, f_N$. Let $g_1, \ldots, g_k$ be the irreducible factors of these functions at a that are common for all f_j. Then $f_j = g_1^{m_{jk}} \cdots g_k^{m_{jk}} \cdot h_j$, where the h_j cannot be divided by any g_j (in any neighborhood of a) and the $h_1, \ldots, h_N$ have no common irreducible factors at a. The set $(\cap Z_{h_j})_{(n-1)}$ does no contain a, since otherwise all h_j could be divided by the Weierstrass polynomial for this hyperplane (by Corollary

1). Therefore there is a neighborhood $U \ni a$ in which all functions constructed above and all factorizations are defined, while the set $A' = (\cap Z_{h_j}) \cap U$ has dimension $< n - 1$. In this neighborhood the set $A = \cap Z_{f_j}$ can be represented as $A = (\cup_{i=1}^{k} Z_{g_i}) \cup (\cap_{j=1}^{N} Z_{h_j}) = (A_{(n-1)} \cap U) \cup A'$. ■

2.9. Minimal defining functions. Let A be a principal analytic set on a complex manifold Ω, i.e. $A = Z_f$ for a certain $f \in \mathcal{O}(\Omega)$ (see p.2.6). The function f is called a *minimal* defining function of the set A if for every open set $U \subset \Omega$ and every function $g \in \mathcal{O}(U)$ such that $g|_{A \cap U} = 0$, there exists an $h \in \mathcal{O}(U)$ such that $g = f \cdot h$ in U, i.e. (locally) f divides every holomorphic function vanishing on A. By p.1.4, the Weierstrass polynomial F constructed in Theorem 2.8, and defining $A_{(n-1)}$, has this property. Hence we have

P r o p o s i t i o n 1. *Every complex hypersurface (= pure $(n-1)$-dimensional analytic set on an n-dimensional complex manifold) locally has a minimal defining function.*

Minimal defining functions are characterized by the following interesting property:

P r o p o s i t i o n 2. *A defining function f of a principal analytic set $A = Z_f$ is minimal if and only if the set $\{z \in A\colon (df)_z = 0\}$ is nowhere dense in A. For any minimal defining function f the set $\{f = df = 0\}$ coincides with* sng A.

■ Let $a \in \operatorname{reg} A$. Then there is a coordinate neighborhood $U \ni a$ in which A can be represented as a connected graph $z_n = \alpha(z')$ (by the implicit function Theorem). By p.1.2, there are a natural number k and a function $\phi \in \mathcal{O}(U)$ such that $f(z) = (z_n - \alpha(z'))^k \phi(z)$ and ϕ is zero free in U, in particular $\phi(a) \neq 0$. If $(df)_a = 0$, then $k > 1$ and, hence, the function $z_n - \alpha(z') \in \mathcal{O}(U)$, vanishes on $A \cap U$, and cannot be divided by f; hence f is not minimal. In the same way it is proved that if f is a defining function for A, then either $\{f = df = 0\}$ is somewhere dense in A, or $(df)_z \neq 0$ at all points $z \in \operatorname{reg} A$. By the implicit function Theorem, $(df)_z = 0$ at all singular points of A, hence for any defining function f either $\{f = df = 0\} = \operatorname{sng} A$, or $\{f = df = 0\}$ is somewhere dense in A.

Suppose that $\{f = df = 0\}$ is nowhere dense in A, and let F be a minimal defining function for A in the domain U. Then $f = F \cdot h$, where $h \in \mathcal{O}(U)$ is zero free outside A. Since $df = h \cdot dF + F \cdot dh$ and $F|_{A \cap U} = 0$, the set Z_h is nowhere dense in A, hence, as was proved above, $h(z) \neq 0$ for any $z \in \operatorname{reg} A \cap U$, i.e. $Z_h \subset \operatorname{sng} A$. Since $\operatorname{sng} A$ has Hausdorff dimension $\leqslant 2n - 4$, this implies that Z_h is empty and, hence, $1/h \in \mathcal{O}(U)$. Thus, f divides F (in U), and f is a minimal defining function. ■

This Proposition can also be formulated as follows: a defining function f of an analytic set $A = Z_f$ is minimal if and only if the order of the zero of f at any regular point equals 1.

C o r o l l a r y 1. *If A is a pure $(n-1)$-dimensional analytic subset of an n-dimensional complex manifold Ω, then* $\operatorname{sng} A$ *is also an analytic subset in Ω, of dimension $\leqslant n - 2$.*

■ If f is a minimal defining function for A in a coordinate neighborhood (U,z), then $(\operatorname{sng} A) \cap U$ is the set of common zeros of the holomorphic functions $f, \partial f/\partial z_1, \ldots, \partial f/\partial z_n$. The dimension of the analytic set $\operatorname{sng} A$ is strictly less than $n - 1$, so that its Hausdorff dimension does not exceed $2n - 4$. ■

C o r o l l a r y 2. *Let A be a pure $(n-1)$-dimensional analytic subset of a domain $D' \times D_n \subset \mathbb{C}^n$ without limit points on $D' \times \partial D_n$, and let $\pi: (z', z_n) \mapsto z'$. Then* $\operatorname{br} \pi|_A$ *is an analytic subset of A of dimension $\leqslant n - 2$.*

■ If f is a minimal defining function for A, then $\operatorname{br} \pi|_A = \{f = \partial f/\partial z_n = 0\}$. ■

3. Proper projections

In this paragraph we introduce the technique of properly projecting analytic sets, which allows us to reduce the codimension and to reduce general analytic sets to sets of codimension 1.

3.1. Proper maps. A continuous map $f: X \to Y$ of topological spaces is called *proper* if the pre-image of every compact set $K \subset Y$ is a compact set in X. The spaces X and Y are assumed to be Hausdorff and locally compact (every point has a fundamental system of neighborhoods with compact closure; for us these will be locally closed subsets of a complex manifold). We give some simple properties of proper maps, easily following from the definition.

1) A proper map is closed, i.e. the image of a closed set is closed.

2) Let $X \to^{f} Y \to^{g} Z$ be continuous maps such that $h = g \circ f$ is proper. Then f and $g|_{f(X)}$ are also proper maps.

3) Let $D \subset X$ and $G \subset Y$ be subsets with $\overline{G}$ compact. The restriction of the projection $(x,y) \mapsto x$ onto a closed subset $A \subset D \times G$ is proper if and only if A does not have limit points on the set $D \times \partial G$.

4) Localization lemma. *Let $f: X \to Y$ be a proper map between locally compact Hausdorff spaces, let L be a compact set in Y, and let K be a subset of $f^{-1}(L)$ that is simultaneously open and closed (e.g. a connected component of $f^{-1}(L)$). Then there are fundamental systems of neighborhoods $U_j \supset K$ and $V_j \supset L$ such that all restrictions f: $U_j \to V_j$ are proper maps.*

A map f is called *finite* if the pre-image of each point consists of a finite number of points. A particular instance of 4) is the following statement.

5) If $f: X \to Y$ is a proper finite map, and $a \in X$, then there are fundamental systems of neighborhoods $U_j \ni a$ and $V_j \ni f(a)$ such that all f: $U_j \to V_j$ are proper maps.

3.2. Exception of variables. Exception of variables is one of the first stages in the solution of complicated systems of equations. We are mainly interested in the geometrical side of the method. A system of holomorphic equations in (z',z'') corresponds to an analytic set in the space of variables (z',z''), while exception of variables z'' means transition to the projection of this set in the space of variables z'. We will prove that under "good" projection analyticity of a set, dimension of a set, and certain other properties will be preserved.

Theorem. *let $G = G' \times G''$, where $G' \subset \mathbb{C}^p$, $G'' \subset \mathbb{C}^m$ are open subsets, $p + m = n$, and let π: $(z',z'') \mapsto z'$. Let A be an analytic subset in G such that π: $A \to G$ is a proper map. Then $A' = \pi(A)$ is an analytic subset in G', and the* *number of pre-images, $\sharp\pi^{-1}(z') \cap A$, is locally finite in G'. If, moreover, $G = \mathbb{C}^n$, $G' = \mathbb{C}^p$, and A is an algebraic subset in $\mathbb{C}^n$, then A' is also an algebraic subset in $\mathbb{C}^p$.*

■ Since A' is closed in G' (p.3.1) it suffices to prove the analyticity of A' in a neighborhood of an arbitrary point $a' \in A'$. We proceed by induction with respect to m.

The basic step $m = 1$. Since $\pi|_A$ is proper, $\pi^{-1}(a') \cap A$ is a compact analytic set in the on-dimensional complex plane, hence (p.2.2) a finite set, say equal to $\{(a', a_{nj})\}$. Let $\overline{U}_j$ be pairwise disjoint closed disks with centers a_{nj} in $G_n = G'' \subset \mathbb{C}_n$. Then, since A is closed in G, there is a polydisk $U' \ni a'$ in G' such that $A \cap (U' \times \partial U_j) = \varnothing$ for all j. The polydisks $U^j = U' \times U_j$ can be chosen small such that $A \cap U^j$ is defined in U^j by a system of holomorphic equations $f_{jk} = 0$, and it can moreover be assumed that $f_{j1}(a', z_n) \not\equiv 0$ in U_j. In view of Weierstrass' preparation theorem we may assume that all f_{j1} are Weierstrass polynomials. By replacing f_{jk} with $f_{jk} + cf_{j1}$ (if $f_{jk}(a', z_n) \equiv 0$) we may assume that all f_{jk} are Weierstrass polynomials also. Hence, in $U' \times \mathbb{C}$ the set $A \cap \pi^{-1}(U')$ is given by the system of equations $\prod_j f_{jk_j} = 0$, which are polynomial in z_n. We renumber these, and then find that $A \cap \pi^{-1}(U')$ is the set of common zeros in $U' \times \mathbb{C}$ of Weierstrass polynomials $f_1, \ldots, f_l$. The degree of f_1(in z_n) will be assumed maximal, and is denoted by d.

We replace the system $f_2, \ldots, f_l$ by the one-parameter family of functions $F_t = \sum_2^l f_k t^k$, $t \in \mathbb{C}$; the zero sets in $U' \times \mathbb{C}_n$ of these families coincide. As is well-known, the polynomials f_1 and F_t in one variable z_n have common zeros if and only if the resultant $R(f_1, F_t)$ vanishes. Therefore $A' \subset \{z' \in U': R(f_1, F_t) \equiv^t 0\}$. We write the resultant (being the corresponding determinant in the coefficients of f_1 and F_t, see, e.g., [67]) in powers of t: $R = \sum_0^{ld} R_k(z') t^k$. The condition $R \equiv^t 0$ is equivalent to $R_k(z') = 0$ for all k, hence $A' \subset \{z' \in U': R_k(z') = 0, k = 0, \ldots, ld\}$.

Conversely, suppose all $R_k(b') = 0$, and let $b_{n1}, \ldots, b_{nd} \in \mathbb{C}$ be the roots of the polynomial $f_1(b', z_n)$. The resultant $R(f_1(b', \cdot), F_t(b', \cdot)) \equiv^t 0$, hence for each $t \in \mathbb{C}$ the polynomial $F_t(b', z_n)$ vanishes at at most one of the points b_{nj}. This can be written as $\bigcup_{j=1}^d \{t: F_t(b', b_{nj}) = 0\} = \mathbb{C}_t$. On the other hand, $F_t(b', b_{nj})$ is a polynomial in t, and if a polynomial does not vanish identically on $\mathbb{C}_t$, then the number of its roots is finite. Hence, for some j_0 the polynomial

$\sum_2^l f_k(b',b_{nj_0})t^k \equiv 0$. This is equivalent to $f_k(b',b_{nj_0}) = 0$, $k = 2,\ldots,l$, and since $f_1(b',b_{nj_0}) = 0$ by the definition of b_{nj}, the point $b = (b',b_{nj_0})$ is a common zero of $f_1,\ldots,f_l$, i.e. b belongs to A. Thus, if all $R_k(b') = 0$, then $b' \in A'$, and we have proved that $A' \cap U'$ coincides with the set of common zeros of the functions R_k, which are holomorphic in U' (and of polynomials if all f_k are polynomials in z).

Since $\pi^{-1}(U') \cap A \subset Z_{f_1}$ and f_1 is of degree d in z_n, $\#\pi^{-1}(z') \cap A \leqslant d$ above U', the number of pre-images of $\pi|_A$ is locally finite for $z' \in U'$.

$m-1 \Rightarrow m$. We represent π in the form $\pi = \pi_2 \circ \pi_1$, where $\mathbb{C}^p \times \mathbb{C}^m \xrightarrow{\pi_1} \mathbb{C}^p \times \mathbb{C}^{m-1} \xrightarrow{\pi_2} \mathbb{C}^p$. Since $\pi|_A$ is proper, $\pi_1|_A$ is proper also (p.3.1). According to the case $m = 1$, $\pi_1(A)$ is an analytic subset in $\pi_1(G)$, and the number of points in the fibers of $\pi_1|_A$ is locally finite above $\pi_1(G)$. For the same reasons, $\pi_2|_{\pi_1(A)}$ is a proper map and, by the induction hypothesis, $\pi_2(\pi_1(A)) = \pi(A)$ is an analytic subset in G'; moreover, $\#\pi^{-1}(z') \cap A$ is locally finite. ■

3.3. Corollaries. Theorem 3.2, together with the Theorem on the existence of proper projections (p.3.4), is one of the main local tools in the theory of analytic sets. We prove several corollaries immediately following from Theorem 3.2.

P r o p o s i t i o n 1. *An analytic set $A \subset \mathbb{C}^n$ is compact if and only if A is a finite set.*

■ For $n = 1$ this is the uniqueness Theorem (p.2.2). The transition from $n-1$ to n is realized using the projection $\pi\colon z \mapsto z' \in \mathbb{C}^{n-1}$. Since A is compact, $\pi(A)$ is a compact analytic set in $\mathbb{C}^{n-1}$, hence a finite set. Since the fibers of $\pi|_A$ are locally finite, $\#A < \infty$. ■

Yet another useful statement - preservation of dimension under proper finite maps.

P r o p o s i t i o n 2. *Let A be an analytic subset of a complex manifold X and let $f\colon A \to Y$ be a proper finite holomorphic map. Then, at every $w \in f(A)$,*

$$\dim_w f(A) = \max\{\dim_z A : f(z)=w\};$$

in particular, $\dim A = \dim f(A)$.

■ According to the localization Lemma we may assume that X, Y are open subsets in $\mathbb{C}^n$, $\mathbb{C}^p$. By passing from the map to its graph we see that it suffices to investigate the case of proper projection $\pi: A \to U' \subset \mathbb{C}^p$ of an analytic set $A \subset U' \times U''$ in $\mathbb{C}^{p+m}$. We proceed by induction with respect to $m = \dim U''$. The basic step $m = 1$. Let $A' = \pi(A)$, $a' \in \operatorname{reg} A'$, and let a neighborhood $V' \ni a'$ be such that $A' \cap V'$ is a connected complex manifold of, say, dimension $q = \dim_{a'} A'$. Then $A_1 = A \cap \pi^{-1}(V')$ is a proper analytic subset of the connected $(q+1)$-dimensional complex manifold $(A' \cap V') \times \mathbb{C}$. By the uniqueness Theorem (p.2.2) $\dim A_1 \leqslant q$; in particular, $\dim_a A \leqslant q$ at all $a \in \pi^{-1}(a')$.

Since the $\mathcal{H}_{2q}$-measure does not increase under projection, the Hausdorff dimension of A_1 is at least $2q$, hence (p.2.6) $\dim A_1 = q$. Since V' can be chosen arbitrarily small, there exists a sequence of points $z^j \in A$ such that $\pi(z^j) \to a'$ and $\dim_{z^j} A = q$. Since $\pi|_A$ is proper, we can extract a subsequence converging to some point $a \in A$ for which $\pi(a) = a'$; the definition of dimension implies that $\dim_a A = q$.

$m-1 \Rightarrow m$. We represent $\pi: U \to U'$ in the form $\pi = \pi_2 \circ \pi_1$, where $\pi_1: z \to (z_1, \ldots, z_{n-1})$, the case already considered, and $\pi_1: (z_1, \ldots, z_{n-1}) \to z'$. By p.3.1, $\pi_1|_A$ and $\pi_2|_A$ are proper. Both preserve dimension: for the first this has already been proved, for the second this is the induction hypothesis. ■

The following statement is the geometrical analog of the Theorem on holomorphy of an inverse map (A2.2).

P r o p o s i t i o n 3. *Let $G' \subset \mathbb{C}^p$, $G'' \subset \mathbb{C}^m$ be open subsets, A an analytic subset in $G = G' \times G''$, and $\pi: (z', z'') \mapsto z'$. Assume that $A' = \pi(A)$ is a complex submanifold in G' and that $\pi: A \to A'$ is one-to-one. Then A is a complex submanifold in G and $\pi: A \to A'$ is a biholomorphic map.*

This Proposition can be regarded as a theorem on removing singularities of a map $f: A' \to G''$ with analytic graph: it is stated that f must be holomorphic.

■ The statement is local. After a suitable biholomorphic change of coordinates in G' (locally), we may assume that $A' = G' \cap \{z_{q+1} = \cdots = z_p = 0\} = G' \cap \mathbb{C}^q$. Put $'z = (z_1, \ldots, z_q)$. For each $j = p+1, \ldots, p+m$ the restriction to A of the projection $\pi_j(z', z'') \mapsto ('z, z_j)$ is a one-to-one and, obviously, proper map from A onto a certain analytic subset A_j in $\pi_j(G) \subset \mathbb{C}^q \times \mathbb{C}_j$. Since A_j can be bijectively projected onto $G' \cap \mathbb{C}^q$, its dimension is q. By Theorem 2.8, A_j is the zero set of a Weierstrass polynomial of degree one in z_j, i.e. A_j: $z_j = \phi_j('z)$ where $\phi_j \in \mathcal{O}(G' \times \mathbb{C}^q)$. Thus, A is contained in the common graph Γ: $z'' = \phi('z)$ of the functions ϕ_j, where $\phi = (\phi_{p+1}, \ldots, \phi_{p+m})$, above $G' \cap \mathbb{C}^q$. Since A and Γ can be bijectively projected onto $G' \cap \mathbb{C}^q$, $A = \Gamma$ is a complex submanifold in G and $\pi|_A$ is a biholomorphism. ■

C o r o l l a r y . *Let A be an analytic set in $\mathbb{C}^n$. The point $0 \in A$ is regular if and only if there are a neighborhood $U \ni 0$ in $\mathbb{C}^n$ and a coordinate plane $\mathbb{C}_I$ such that the projection π_I: $A \cap U \to \mathbb{C}_I \cap U$ is one-to-one.*

3.4. Existence of proper projections. The following simple Lemma is the basic theorem on the existence of locally proper maps of analytic sets.

L e m m a 1. *Let A be an analytic set in $\mathbb{C}^n$, $0 \in A$, and $\dim_0 A \leqslant p$. Then there are a unitary transformation $l \in \mathbb{U}(n)$ and an arbitrarily small neighborhood $U_l = U_l' \times U_l'' \ni 0$, where $U_l' \subset \mathbb{C}^p_{z'}$, $U_l'' \subset \mathbb{C}^{n-p}_{z''}$, such that the orthogonal projection π: $l(A) \cap U_l \to U_l'$ (π: $z \to z'$) is a proper map. The set of such l is open and everywhere dense in $\mathbb{U}(n)$.*

■ It suffices to consider the case $p = n - 1$ (the rest follows by induction). Since A is nowhere dense in the ball $B = B(0,r)$ (chosen small such that $A \cap B$ is closed in B), the set of complex lines $L \ni 0$ such that $L \cap B \subset A$ is closed and nowhere dense in $\mathbb{P}_{n-1}$. If $L \cap B \not\subset A$, then $A \cap L \cap B$ is a proper analytic subset of the disk $L \cap B$, i.e. it is a locally finite set (p.2.2). Let l be the unitary transformation of $\mathbb{C}^n$ mapping L to the $\mathbb{C}_n$-axis; then the set $l(A) \cap \mathbb{C}_n \cap B$ is locally finite, hence there is an $r_n \in (0,r)$ such that the circle $\mathbb{C}_n \cap \{|z_n| = r_n\}$ does not intersect $l(A)$. Since $l(A) \cap B$ is closed in B there exists an $r' > 0$ such that the domain U_l: $|z'| < r'$, $|z_n| < r_n$ belongs to B and $l(A)$ does not intersect the set $|z'| \leqslant r'$, $|z_n| = r_n$. By p.3.1, the restriction to $l(A) \cap U_l$ of the projection $(z', z_n) \mapsto z'$ is a proper map. ■

L e m m a 2. *Let A be an analytic set in $\mathbb{C}^n$, $0 \in A$, and $\dim_0 A \leqslant p$. Let $\Lambda_1, \ldots, \Lambda_k$ be an arbitrary finite tuple of p-dimensional subspaces in $\mathbb{C}^n$ and let π^j: $\mathbb{C}^n \to \Lambda_j$ be the orthogonal projections. Then there are a unitary transformation $l \in \mathbb{U}(n)$ and neighborhoods $U_j \ni 0$, $j = 1, \ldots, k$, such that all restrictions π^j: $l(A) \cap U_j \to \Lambda_j \cap U_j$ are proper maps.*

■ The unitary transformations l that have the property stated in Lemma 1 with π replaced by a π^j form, by Lemma 1, an everywhere dense open subset in $\mathbb{U}(n)$. The intersection of a finite number of such subsets is also open and everywhere dense in $\mathbb{U}(n)$. Any l from this intersection is appropriate. ■

A description of zero-dimensional analytic sets can easily be obtained from the statements proved above.

P r o p o s i t i o n. *An analytic subset A of a complex manifold Ω is zero-dimensional if and only if it is locally finite in Ω, i.e. $\# A \cap K < \infty$ for every compact set $K \subset \Omega$.*

■ The statement is local, hence we may assume that Ω is a ball in $\mathbb{C}^n$. By Lemma 1, after a suitable unitary change of coordinates the projection $\pi: A \cap U \to U' \subset \mathbb{C}_1$ will be a proper map. By p.3.2 and p.3.3, $\pi(A \cap U)$ is a zero-dimensional analytic set in $U' \subset \mathbb{C}_1$, hence (p.2.2) a locally finite set in U'. Since $\pi|_{A \cap U}$ has locally finite fibers (p.3.2), $\# A \cap K < \infty$ for every compact set $K \subset U$. The converse is obviously true. ■

3.5. On the dimension. In p.3.4 it was proved that every p-dimensional analytic set $A \subset \mathbb{C}^n$ locally allows proper projection into a p-dimensional subspace in $\mathbb{C}^n$. Since dimension is preserved under such a projection there is no subspace of smaller dimension having this property. If $\pi: A \cap U \to U' \subset \mathbb{C}^p$ is a proper projection, each point $a \in A \cap U$ is an isolated point in the fiber $\pi^{-1}(a') \cap A$. Conversely, if a is an isolated point in $A \cap L$ where L is a complex plane in $\mathbb{C}^n$, then there is a neighborhood $U \ni a$ such that the projection of $A \cap U$ along L into some open subset $U' \subset L^{\perp}$ is a proper mapping (see p.3.1). This simple remark, together with the existence Theorem in p.3.4, can be formulated in the following way.

P r o p o s i t i o n 1. *Let A be an analytic set in $\mathbb{C}^n$ and let $\dim_a A = p$. Then there is a complex plane L of dimension $q \leqslant n - p$ such that a is an isolated point of the set $A \cap L$; the value $q = n - p$ is the maximum possible value.*

C o r o l l a r y. *Let A be an analytic set in $\mathbb{C}^n$ and let $\operatorname{codim}_a A := n - \dim_a A$. Then* $\operatorname{codim}_a A = \max\{\dim L$: *$L \ni a$ is a complex plane and a is an isolated point of $A \cap L$*$\}$.

If $\dim L > n - q$ and $a \in L$, then a is not isolated in $A \cap L$, hence $\dim_a A \cap L = r > 0$. And more precise? Let $a = 0$ and $L = \mathbf{C}^q \subset \mathbf{C}^n$. By Proposition 1, there is a subspace $L' \subset \mathbf{C}^q$ of dimension $q - r$ such that 0 is an isolated point of $(A \cap \mathbf{C}^q) \cap L' = A \cap L'$. Since $(n - p)$ is the maximum possible dimension of such L', we have $q - r \leqslant n - p$, i.e. $r \geqslant (p + q) - n$, or, in more detail, $\dim_a A \cap L \geqslant \dim_a A + \dim_a L - n$. This property also holds in a more general situation.

Proposition 2. *Let A_1, A_2 be analytic sets on an n-dimensional complex manifold and let $a \in A_1 \cap A_2$. Then*

$$\dim_a A_1 \cap A_2 \geqslant \dim_a A_1 + \dim_a A_2 - n.$$

■ The statement is local, therefore we may assume that Ω is a domain in $\mathbf{C}^n$. Above we have clarified the case when one of the A_j is a complex plane. The general case is reduced to this one as follows. We consider in $\mathbf{C}^n_z \times \mathbf{C}^n_w$ the intersection of the analytic set $A_1 \times A_2$ with the diagonal Δ: $z = w$. It is clear that $(z,w) \in (A_1 \times A_2) \cap \Delta$ if and only if $z = w \in A_1 \cap A_2$. Hence, $A_1 \cap A_2$ is the projection of the analytic set $(A_1 \times A_2) \cap \Delta$ under biholomorphic projection of Δ onto $\mathbf{C}^n_z$, so that the dimensions of these analytic sets at corresponding points a and (a,a) coincide. But $\dim_{(a,a)} A_1 \times A_2 = \dim_a A_1 + \dim_a A_2$ (consider regular points), and Δ is a complex plane of dimension n in $\mathbf{C}^{2n}$. As was already proved,

$$\dim_a (A_1 \cap A_2) \geqslant ((\dim_a A_1 + \dim_a A_2) + n) - 2n. \quad ■$$

Obviously, by induction we obtain the more general inequality

$$\dim_a \bigcap_{j=1}^{k} A_1 \geqslant \left[\sum_{j=1}^{k} \dim_a A_j \right] - (k-1)n,$$

which is more conveniently stated in terms of codimensions.

Proposition 3. *Let $A_j \ni a$ be analytic sets on a complex manifold. Then*

$$\operatorname{codim}_a \bigcap_j A_j \leqslant \sum_j \operatorname{codim}_a A_j.$$

C o r o l l a r y. *Let functions* $f_1, \ldots, f_k$ *be holomorphic in a neighborhood of a point* a *on an* n*-dimensional complex manifold and suppose that all* $f_j(a) = 0$. *Then the set of their common zeros,* $\bigcup_j Z_{f_j}$, *has dimension at* a *at least* $n - k$.

These are essentially "complex" statements, in some way analogous to the principle of the argument, which are a source of various existence theorems (see, e.g., p.10.4); of course, the analogous statements for $\mathbb{R}$-analytic sets are not true.

Finally we consider how dimension changes under limit transitions.

P r o p o s i t i o n 4. *Let* $\{A_j\}$ *be a family of analytic subsets of a domain* D *in* $\mathbb{C}^n$, *and let* A *be its limit set (i.e. the set of all points of the form* $a = \lim a_{j_\nu} \in D$, $a_{j_\nu} \in A_{j_\nu}$, $j_\nu \to \infty$). *Assume that* A *is also an analytic set and that there is a point* $a \in A$ *such that in any neighborhood* $U \ni a$ *the inequality* $\dim A_j \cap U \geqslant p$ *holds for an infinite number of indices* j. *Then* $\dim_a A \geqslant p$.

■ Let $a = 0$ and $\dim_a A = m$. By Proposition 1 there is an $(n-m)$-dimensional plane L such that 0 is isolated in $A \cap L$; in particular, there is a ball $B \ni 0$ such that A is at positive distance r from the compact set $L \cap \partial B$. Since A is the limit set of $\{A_j\}$, for $j > j'$ all A_j also do not intersect the $(r/2)$-neighborhood of $L \cap \partial B$. Since $0 = \lim a_{j_\nu}$, where $\dim_{j_\nu} A_{j_\nu} \geqslant p$, for $j \geqslant j''$ the sets $A_{j_\nu} \cap B$ are nonempty. By construction, for $j \geqslant j'''$ the set $(L + a_{j_\nu}) \cap A_{j_\nu} \cap B$ is compact, hence a_{j_ν} is isolated in it. Since $\dim L = n - m$, by Proposition 1, $p = \dim_{a_{j_\nu}} A_{j_\nu} \leqslant m = \dim_0 A$. ■

3.6. Almost single-sheeted projections.

L e m m a. *Let* $U = U' \times U''$, *where* U', U'' *are domains in* $\mathbb{C}^p$ *and* $\mathbb{C}^m$, *let* $\pi: (z', z'') \mapsto z'$, *and let* A *be an analytic subset in* U *such that* $\pi: A \to U'$ *is a proper map. Then there is a linear function* $\lambda(z'') \in (\mathbb{C}^m)^*$ *such that the holomorphic map* $\pi_\lambda: (z', z'') \mapsto (z', \lambda(z'')) \in \mathbb{C}^{p+1}$ *has the following properties* ($A_\lambda := \pi_\lambda(A)$):

(1) $\pi_\lambda: A \to U_\lambda = U' \times \lambda(U'')$ *is a proper map;*

(2) $A \cap \pi_\lambda^{-1}(\operatorname{reg} A_\lambda) \subset \operatorname{reg} A$ *and* $\pi_\lambda: A \cap \pi_\lambda^{-1}(\operatorname{reg} A_\lambda) \to \operatorname{reg} A_\lambda$ *is a biholomorphic map;*

(3) $\pi_\lambda^{-1}(\operatorname{sng} A_\lambda)$ *contains* $\operatorname{sng} A$ *and is nowhere dense in* A. *The set of such functions* λ *is everywhere dense and of the second category in* $(\mathbb{C}^m)^*$.

■ Let $\{a_1, \ldots, a_2, \ldots\}$ be a countable, everywhere dense subset in A, and let $a_j' = \pi(a_j)$. Then $\{a_1', \ldots, a_2', \ldots\}$ is a countable, everywhere dense subset in A'. Since $\pi|_A$ is proper (Figure 3), for every j the number of pre-images, $\sharp\pi^{-1}(a_j') \cap A$, is finite, hence $E = \cup_j \pi^{-1}(a_j') \cap A$ is a countable subset in A. Let $E'' = \{z'': z \in E\}$. For arbitrary $\zeta'' \neq \eta''$ from E'' the set of functions $\lambda \in (\mathbf{C}^m)^*$ such that $\lambda(\zeta'') = \lambda(\eta'')$ is a complex hyperplane; in particular, it is closed and nowhere dense in $(\mathbf{C}^m)^*$. Since the union of a countable family of nowhere dense closed subsets is a set of the first (Baire) category, its complement is an everywhere dense subset of the second category in $(\mathbf{C}^m)^*$. Take λ from this complement; then $\lambda(\zeta'') \neq \lambda(\eta'')$ for arbitrary $\zeta, \eta \in E$ with $\zeta'' \neq \eta''$. Put $a_{\lambda_j} = \pi_\lambda(a_j)$; then, by construction, $\pi_\lambda^{-1}(a_{\lambda_j}) \cap A = \{a_j\}$, i.e. above each a_{λ_j} in A there is precisely one point a_j (if $a \in A$ and $a_\lambda = a_{\lambda_j}$, then $a \in E$, $a' = a_j'$, $\lambda(a'') = \lambda(a_j'')$, hence $a = a_j$). We will show that π_λ has all the properties listed.

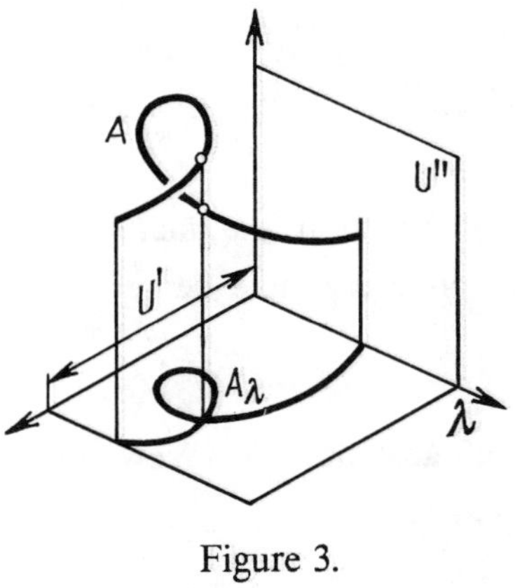

Figure 3.

(1) This is property 2) of proper maps from p.3.1.

(2) By Proposition 3, p.3.3, it suffices to prove bijectivity of $\pi_\lambda|_A$ above $\operatorname{reg} A_\lambda$. Let $b_\lambda \in \operatorname{reg} A_\lambda$ and let $\pi_\lambda^{-1}(b_\lambda) \cap A = \{b^1, \ldots, b^k\}$, where the b^j are distinct. Then we can find pairwise disjoint neighborhoods $V^j \ni b^j$ such that $\pi_\lambda(V^j) = V_\lambda$ and all maps $\pi_\lambda\colon A \cap V^j \to V_\lambda$ are proper. We may assume that $A_\lambda \cap V_\lambda$ is a connected complex submanifold in V_λ (since $b_\lambda \in \operatorname{reg} A_\lambda$). Since $A_\lambda \cap V_\lambda = \bigcup_1^k \pi_\lambda(A \cap V^j)$ is a finite union of analytic subsets, at least one of these is somewhere dense in $A_\lambda \cap V_\lambda$. By the uniqueness Theorem for analytic sets, $A_\lambda \cap V_\lambda = \pi_\lambda(A \cap V^i)$ for some i. Now suppose that $k = \sharp\pi_\lambda^{-1}(b_\lambda) \cap A > 1$. Since $\{a_1, a_2, \ldots\}$ is everywhere dense in A, there is a

sequence $a_{j_\nu} \to b^l$ with $l \neq i$. The points $a_{\lambda_{j_\nu}} \in V_\lambda$ and $a_{\lambda j_\nu} \to b_\lambda$. Since $\pi_\lambda(A \cap V^i) = A_\lambda \cap V_\lambda$ and $\pi_\lambda^{-1}(a_{\lambda j}) \cap A = \{a_j\}$ for each j, all $a_{j_\nu} \in A \cap V^i$. Hence they cannot tend to $b^l \notin \overline{V^i}$. This contradiction shows that $\# \pi_\lambda^{-1}(b_\lambda) \cap A = 1$ for every $b_\lambda \in \operatorname{reg} A_\lambda$.

(3) The inclusion $\pi_\lambda^{-1}(\operatorname{sng} A_\lambda) \supset \operatorname{sng} A$ follows from (2). We will show that all a_j are limit points of $A \cap \pi_\lambda^{-1}(\operatorname{reg} A_\lambda)$, i.e. that $A \cap \pi_\lambda^{-1}(\operatorname{reg} A_\lambda)$ is an everywhere dense subset of A (hence, $\pi_\lambda^{-1}(\operatorname{sng} A_\lambda)$ is a nowhere dense closed subset of A). Since $\operatorname{reg} A_\lambda$ is everywhere dense in A_λ, there is a sequence $a_{\lambda i_k} \to a_{\lambda j}$ such that $a_{\lambda i_k} \in \operatorname{reg} A_\lambda$. Since the map $\pi_\lambda|_A$ is proper, we can extract a subsequence from $\{a_{i_k}\}$ converging to a point $a \in A$ such that $a_\lambda = a_{\lambda j}$. But $A \cap \pi_\lambda^{-1}(a_{\lambda j}) = \{a_j\}$ and thus $a_j = a$ is a limit point of $A \cap \pi_\lambda^{-1}(\operatorname{reg} A_\lambda)$. ■

Finite maps having the properties (1) - (3) will be called *almost single-sheeted.*

C o r o l l a r y. *Under the conditions of the Lemma there is a unitary transformation of* $\mathbb{C}^n$ *after whose performance all projections* $(z',z'') \mapsto (z',z_j)$, $j > p$, *become almost single-sheeted on* A.

The *proof* obviously follows from the fact that the functions $\lambda \in (\mathbb{C}^m)^*$ fulfilling the Lemma form on everywhere dense set of the second category in $(\mathbb{C}^m)^*$.

3.7. Local representation of analytic sets. Now we can prove the analog of Theorem 2.8 for general analytic sets.

T h e o r e m. *Let* A *be an analytic set in* $\mathbb{C}^n$, $\dim_a A = p$, $0 < p < n$, *U a neighborhood of a, and* $\pi: A \cap U \to U' \subset \mathbb{C}^p$ *a proper projection. Then there is an analytic subset* $\sigma \subset U'$ *of dimension* $<p$ *and a natural number k such that*

(1) $\pi: A \cap U \setminus \pi^{-1}(\sigma) \to U' \setminus \sigma$ *is a locally biholomorphic k-sheeted cover, in particular,* $\# \pi^{-1}(z') \cap A \cap U = k$ *for all* $z' \in U' \setminus \sigma$;

(2) $\pi^{-1}(\sigma)$ *is nowhere dense in* $A_{(p)} \cap U$.

■ We will assume that $A \subset U$. Let π_λ be an almost single-sheeted projection on A "covering" π, say $\pi_\lambda: (z',z'') \mapsto (z', z_{p+1})$. Then $A_\lambda = \pi_\lambda(A)$ is an analytic subset of the domain $U_\lambda \subset \mathbb{C}^{p+1}$, moreover, $\dim A_\lambda = \dim A = p$, i.e. the codimension of A is 1. By p.2.8, $A_\lambda = (A_\lambda)_{(p)} \cup A_0$, where A_0 is an analytic subset in U_λ of dimension $<p$ and $(A_\lambda)_{(p)}$ is the zero set of a Weierstrass polynomial $F(z',z_{p+1})$, whose degree k is equal to the maximal number of sheets of $(A_\lambda)_{(p)}$ above the points $z' \in U'$. Put $\pi = \pi' \circ \pi_\lambda$, $\sigma_1 = \pi'(A_0)$, and let σ_2 be the discriminant set of F. Then $\sigma = \sigma_1 \cup \sigma_2$ is an analytic subset in U' of dimension $<p$ and $\pi': A_\lambda \setminus (\pi')^{-1}(\sigma) \to U' \setminus \sigma$ is a locally biholomorphic k-sheeted cover. Since $(\pi')^{-1}(\sigma) \supset \operatorname{sng} A_\lambda$ and $\pi_\lambda|_A$ is biholomorphic above $\operatorname{reg} A_\lambda$, $\pi: A \setminus \pi^{-1}(\sigma) \to U' \setminus \sigma$ is also a locally biholomorphic k-sheeted cover.

By Proposition 2, p.3.3, the analytic set $\Sigma = A \cap \pi^{-1}(\sigma)$ has dimension $<p$, and is thus nowhere dense among the points $a \in \operatorname{reg} A$ at which $\dim_a A = p$. Such points form in $A_{(p)}$ an open, everywhere dense subset, hence Σ is nowhere dense in $A_{(p)}$. ■

Up till now we have not said anything about analyticity of $A_{(p)}$, this will be done in the next paragraph. Now we give two simple corollaries concerning dimensions.

L e m m a. *If $A_1 \subset A$ are analytic subsets of a complex manifold and if A_1 is nowhere dense in $A_{(p)}$, where $p = \dim A$, then $\dim A_1 < p$. If A_1 is also nowhere dense in A, then $\dim_z A_1 < \dim_z A$ at all $z \in A_1$.*

■ The statement is local, therefore we confine ourselves to the standard situation in $\mathbb{C}^n$. Let $\pi: A \cap U \to U' \subset \mathbb{C}^p$ be a proper projection and let $\sigma \subset U'$ be the critical analytic set from the Theorem. If $a' \in U' \setminus \sigma$, then there is a neighborhood $V' \ni a'$ such that $\pi^{-1}(V') \cap A$ consists of k manifolds W_j, each of which belongs to $A_{(p)}$ and can be biholomorphically projected onto V'. Since A_1 is nowhere dense in $A_{(p)}$, we find that $\pi(A_1 \cap U) \cap V' = s \cup ubj\pi(A_1 \cap W_j)$ is also nowhere dense in V'. Since a' is arbitrary this means that the analytic set $\pi(A_1 \cap U)$ is nowhere dense in U', hence its dimension $<p$. By p.3.3, the dimension of $A_1 \cap U$ is also strictly less than p. The second assertion of the Lemma follows from the first and from the definition of dimension at a point. ■

P r o p o s i t i o n. *The Hausdorff dimension of an analytic set A is equal to* $2 \dim A$.

■ The proof proceeds by induction with respect to $p = \dim A$. For $p = 0$ the statement is true, since the Hausdorff dimension of a locally finite set is zero.

$p-1 \Rightarrow p$. The statement is local, therefore it is sufficient to consider the standard situation in $\mathbb{C}^n$. Let $\pi: A \to U' \subset \mathbb{C}^p$ be a proper projection with critical set $\sigma \subset U'$ (as in the Theorem). Then the analytic set $\pi^{-1}(\sigma) \cap A$ is nowhere dense in $A_{(p)}$, hence has dimension $<p$ by the Lemma. By the induction hypothesis, its Hausdorff dimension does not exceed $2p-2$. But $\pi^{-1}(\sigma)$ contains $\mathbb{S}(A)$, and $A \setminus \mathbb{S}(A) = (\operatorname{reg} A) \cap A_{(p)}$ is a p-dimensional complex manifold, whose Hausdorff dimension is equal to $2p$. ■

C o r o l l a r y. *If A is an analytic set of dimension p, then the set $\mathbb{S}(A) = (\operatorname{sng} A) \cup \{z \in A : \dim_z A < p\}$ has Hausdorff dimension $\leqslant 2p-2$.*

■ Locally, $\mathbb{S}(A) \subset \pi^{-1}(\sigma)$, and the analytic sets A and $A_1 = A \cap \pi^{-1}(\sigma)$ satisfy the condition of the Lemma. ■

3.8. Images of analytic sets. In conclusion we will consider some properties of images of analytic sets under arbitrary holomorphic maps. Let A be an analytic set on a complex manifold X and let $f: X \to Y$ be a holomorphic map into another complex manifold (cf. p.2.7). An important geometric characteristic of the map f is the dimension of its fibers, i.e. of pre-images of points. We denote by $\dim_z f$ the codimension of the fiber, i.e. $\dim_z A - \dim_z f^{-1}(f(z))$, and put $\dim f := \max_{z \in A} \dim_z f$. Proposition 4, p.3.5, implies that the function $\dim_z f^{-1}(f(z))$ on A is upper semicontinuous. Hence, if A is pure p-dimensional, then the function $\dim_z f$ is lower semicontinuous, in particular, a set of maximal values of it is open in A. This implies $\dim f \geqslant \dim (f|_{\operatorname{sng} A})$. Using the analyticity of $\mathbb{S}(A)$ (which will be proved in p.5.2), it is not difficult to show that $\dim f \geqslant \dim (f|_{\mathbb{S}(A)})$ in the general case (for its proof it suffices that $\mathbb{S}(A)$ is contained in an analytic set of dimension $<p$, and this clearly follows from Theorem 3.7). Since at regular points (outside an analytic set of smaller dimension) $\dim_z f = \operatorname{rank}_z f$, in the case of a pure p-dimensional set $\dim f$ is the maximal rank of f on $\operatorname{reg} A$. If $\operatorname{reg} A$ is connected, by the uniqueness Theorem this implies the equality $\dim_z f = \dim f$ on $(\operatorname{reg} A) \setminus A'$, where A' is an analytic subset in $\operatorname{reg} A$ of dimension $<p$. The rank Theorem obviously implies that also $\dim f \leqslant \dim Y$.

The following Proposition gives an idea of the image of an analytic set under a, not necessarily proper, holomorphic map. Remmert's theorem on proper holomorphic maps, which is affiliated to it in subject-matter, will be proved later, in p.5.8.

P r o p o s i t i o n. *Let* $f\colon A \to Y$ *be a holomorphic map of a p-dimensional analytic set A into a complex manifold. Then* $f(A)$ *is contained in an (at most) countable union of analytic sets on Y, of dimensions not exceeding* $\dim f$. *Moreover, for any* $q \geqslant p - \dim f$ *the set* $\{w \in Y\colon \dim f^{-1}(w) \geqslant q\}$ *is contained in a countable union of analytic sets of dimensions not exceeding* $p - q$.

■ It suffices to prove the second statement (the first is obtained from it for $q = p - \dim f$).

The statement is local, hence we may assume that A is an analytic subset of a neighborhood of 0 in $\mathbb{C}^n$ and that $\mathbb{S}(A)$ is contained in an analytic subset $A' \subset A$ of dimension $<p$. Then $A \setminus A'$ is a (not necessarily connected) p-dimensional complex manifold. Let S be a connected component of $A \setminus A'$ and let r_S denote the maximal rank of f at the points of S. Then $C_S = \{z \in S\colon \operatorname{rank}_z f < r_S\}$ is an analytic subset in S of dimension $<p$, and $C = \cup C_S$ is an analytic subset in $A \setminus A'$ of dimension also $<p$. The set $E_q \subset Y$ above which $\dim f^{-1}(w) \geqslant q$ can be represented as the union of three sets:

$$E'\colon \dim f^{-1}(w) \cap A' \geqslant q,$$

$$E''\colon \dim f^{-1}(w) \cap C \geqslant q, \quad E^0 = E_q \setminus (E' \cup E'').$$

The codimension of the map (i.e. $\dim A - \dim f$) is the minimal dimension of a fiber, hence if $w \in E'$ or $w \in E''$, then $q \geqslant \dim A' - \dim f|_{A'}$, or, respectively, $q \geqslant \dim C - \dim f|_C$. By induction with respect to the dimension of the analytic set we obtain that E' is contained in a countable union of analytic sets on Y of dimensions $\leqslant \dim A' - q < p - q$, while E'' is contained in a countable union of analytic sets of dimensions $\leqslant \dim C - q < p - q$. It remains to investigate the set E^0. Let $w \in E^0$; then $\dim (f^{-1}(w) \setminus (A' \cup C)) \geqslant q$, hence there is a connected component S in $A \setminus A'$ such that $\dim (f^{-1}(w) \cap (S \setminus C)) \geqslant q$. Since $\operatorname{rank}_z f = f_S$ on $S \setminus C$, the rank Theorem implies that all fibers of $f|_{S \setminus C}$ have dimension $p - r_S$, hence $p - r_S \geqslant q$. According to the same Theorem, the image of $S \setminus C$ is contained in a countable union of complex manifolds in Y of dimension $r_S \leqslant p - q$. It remains to note that the number of connected components of $A \setminus A'$ is at most countable, and that $E^0 \subset \cup_S f(S \setminus C)$. ■

C o r o l l a r y 1. *Let* $f\colon A \to Y$ *be a holomorphic map,* $q \geqslant p - \dim f$, *and* $K \subset A$ *a compact set. Then the set* $\{w \in Y\colon \max_{z \in K} \dim_z f^{-1}(w) \geqslant q\}$ *is a compact set of Hausdorff dimension at most* $2(p-q)$. *In particular, if* $q > p - \dim f$, *this set has in Y Hausdorff codimension* $\geqslant 2$.

■ By Proposition 4, p.3.5, the set indicated above is closed in Y. Since it belongs to the compact set $f(K)$, it is also compact. The statement about the dimension follows from the Proposition, in view of the countable semi-additivity of Hausdorff measures. The last statement is then obtained from the inequality $p - q < \dim f \leqslant \dim Y$. ■

Corollary 2. *Let A be an analytic set in* $\mathbb{C}^n$ *of dimension p and let* $k > n-p$. *Then the set of planes* $L \in G(k,n)$ *such that* $\dim A \cap L > p+k-n$ *is contained in a countable union of analytic sets in* $G(k,n)$ *of codimension* $\geqslant 1$. *If* $0 \in A$, *then* $\{L: \dim_0 A \cap L > p+k-n\}$ *is a compact set in* $G(k,n)$ *of Hausdorff codimension* $\geqslant 2$.

■ Let $I = \{(z,L): z \in L\}$ be the incidence manifold in $\mathbb{C}^n \times G(k,n)$ (see A3.5) and let $\tilde{A} = \{(z,L): z \in A \cap L\}$ be the "lift" of A to this manifold. For a fixed $z \in \mathbb{C}^n_*$ the fiber of I above z is isomorphic to $G(k-1,n-1)$; in particular, it has dimension $(k-1)(n-k)$. Therefore $\tilde{A}$ is an analytic set on I of dimension $p+(k-1)(n-k) = \dim G(k,n) + (p+k-n)$. Let π be the projection of $\tilde{A}$ in $G(k,n)$. Since $\pi^{-1}(L) = \{(z,L): z \in A \cap L\}$ is biholomorphically projected in $\mathbb{C}^n$ onto the set $A \cap L$, the Proposition implies that the set $\{L: \dim A \cap L > p+k-n\}$ is contained in a countable union of analytic sets of dimensions less than $\dim \tilde{A} - (p+k-n) = \dim G(k,n)$. Further, if $0 \in A$, then the compact set $\{0\} \times G(k,n)$ belongs to $\tilde{A}$, and $\dim_{(0,L)} \pi^{-1}(L) = \dim_0 A \cap L$. Therefore, by Corollary 1, the set of planes L for which $\dim_0 A \cap L > p+k-n$ is a compact set of Hausdorff codimension $\geqslant 2$. ■

4. Analytic covers

4.1. Definitions. Analytic covers are a generalization of locally biholomorphic covers. They are also called ramified analytic covers and differ from locally biholomorphic covers by the presence of singularities of special type.

Definition 1. Let A be a locally closed set on a complex manifold X and let $f : A \to Y$ be a continuous, proper, finite map into another complex manifold Y. The triple (A,f,Y) is called an *analytic cover* (above Y) if:

(1) there exist an analytic subset $\sigma \subset Y$ (possibly, empty) of dimension $< \dim Y$ and a natural number k such that $A \setminus f^{-1}(\sigma)$ is a complex manifold in X and $f : A \setminus f^{-1}(\sigma) \to Y \setminus \sigma$ is a locally biholomorphic k-sheeted cover (i.e. a locally biholomorphic map each fiber of which consists of k points);

(2) the set $f^{-1}(\sigma)$ is nowhere dense in A.

An analytic cover is often written as a map f: $A \to Y$ or $A \xrightarrow{f} Y$; if the nature of f (and Y) are clear, the set A itself is also called an analytic cover. Since f is proper and σ is nowhere dense in Y, it easily follows that $f(A) = Y$. The localization Lemma (p.3.1) and property (2) easily imply that the number of points in each fiber does not exceed k: $\sharp f^{-1}(w) \leqslant k$, $w \in Y$. Above the critical set σ of an analytic cover $f : A \to Y$ certain sheets of the cover may "stick together".

We have encountered analytic covers already several times: in Weierstrass' preparation theorem, in Theorem 2.8, in Theorem 3.7. Part of the last theorem may be formulated as follows:

Every pure p-dimensional analytic set can be locally represented as an analytic cover above a domain in $\mathbb{C}^p$.

The basic aim of the present paragraph is to prove the converse statement for somewhat more general objects.

A closed set σ on a complex manifold Y is called *removable* if σ is nowhere dense in Y and if every function h holomorphic and bounded on $Y \setminus \sigma$ can be holomorphically continued onto all of Y. Examples of such sets have been considered in §1. A set σ is called *locally removable* if it is closed and if it is removable in a neighborhood of each of its points.

D e f i n i t i o n 2. A *generalized k-sheeted analytic cover* is a triple (A,f,Y), where A is a locally closed set on some complex manifold, $f : A \to Y$ is a continuous and proper map into another complex manifold Y, and where there exists a locally removable set $\sigma \subset A$ such that $f^{-1}(\sigma)$ is nowhere dense in A, $A \setminus f^{-1}(\sigma)$ is a complex manifold, and $f : A \setminus f^{-1}(\sigma) \to Y \setminus \sigma$ is a locally biholomorphic k-sheeted cover. A (minimal) set σ having the properties listed is called a (the) *critical set* of the generalized analytic cover $f : A \to Y$.

Below we will prove that every generalized analytic cover in an affine space is also an analytic set and an (ordinary) analytic cover with analytic critical set. For the case dim Y = dim $X - 1$ this was done in p.2.8: we have locally constructed a minimal defining function (a Weierstrass polynomial), which, being bounded, could be holomorphically continued at the points of the discriminant set σ. In the general situation we proceed analogously.

4.2. Canonical defining functions. In the case of codimension 1 we have used in an essential way the fact that every set in the plane $\mathbf{C}$ consisting of k points $\alpha_1, \ldots, \alpha_k$ is the zero set of a polynomial, $\prod_1^k (z - \alpha_j)$, whose coefficients could be polynomially expressed in the coordinates of these points. We want to have an analogous representation for finite sets in $\mathbf{C}^m$ with m arbitrary. Such sets can, as will turn out, also be given by standard systems of polynomials; we have borrowed their definition and use from Whitney's book [150].

Let $\alpha^1, \ldots, \alpha^k$ be points in $\mathbf{C}^m$, not necessarily distinct. From the bilinear functions $<w, z - \alpha^j>$, where $<a,b> := a_1 b_1 + \cdots + a_m b_m$, we compose the polynomial

$$P_\alpha(z,w) = <w, z - \alpha^1> \cdots <w, z - \alpha^k>$$

in the variable $(z,w) \in \mathbf{C}^{2m}$. Assume that $P_\alpha(z,w) \equiv^w 0$ for some fixed $z \in \mathbf{C}^m$. Then, at each point $w \in \mathbf{C}^m$ at least one of the factors of P_α vanishes. The uniqueness Theorem implies that there is an index j such that $<w, z - \alpha^j> \equiv^w 0$, hence $z = \alpha^j$. Thus, $P_\alpha(w,z) \equiv^w 0$ if and only if z is one of the points $\alpha^1, \ldots, \alpha^k$. Write P_α in powers of w: $P_\alpha(z,w) = \sum_{|I|=k} \phi_I(z;\alpha) w^I$, where $I = (i_1, \ldots, i_n)$. The condition $P_\alpha(z,w) \equiv^w 0$ is equivalent to the system of equations $\phi_I(z;\alpha) = 0$, $|I| = k$. In this way we have constructed a tuple of $\binom{k+m-1}{m-1}$ polynomials $\phi_I(z;\alpha)$ in the variable $z = (z_1, \ldots, z_m)$ whose common zeros in $\mathbf{C}^m$ coincide with the given system of points $\alpha = \{\alpha^1, \ldots, \alpha^k\}$; these polynomials $\phi_I(z;\alpha)$ will be called *canonical defining functions* for the system α. We will clarify how they depend on the coordinates of the points α^j. From the definition of P_α it is clear that the $\phi_I(z;\alpha)$ are polynomials in z of degree $\leqslant k$, whose coefficients can also be polynomially expressed in terms of the coordinates of the points $\alpha^1, \ldots, \alpha^k$. Renumbering the points α^j does not change P_α, hence the $\phi_I(z;\alpha)$ do not change also. Therefore, $\phi_I(z;\alpha) = \sum_{|J| \leqslant k} \psi_{IJ}(\alpha) z^J$, where

$\psi_{IJ}(\alpha)$ are polynomials in the $k-m$ variables α_i^j, symmetric with respect to superscripts (i.e. not depending on the order of the $\alpha^1,\ldots,\alpha^k$), and with integer coefficients depending only on I,J (this can be regarded as a multidimensional generalization of Viète's formula). The above-said implies, in particular, that $|\psi_{IJ}(\alpha)| \leqslant c(1+|\alpha^1|+\cdots+|\alpha^k|)^k$ for some constant $c = c(k,n)$. From this we obtain the main property of canonical defining functions of interest to us: if the points $\alpha^1,\ldots,\alpha^k$ holomorphically depend on some parameter z', then the coefficients of these polynomials, being standard polynomials in the coordinates of the α^j, also holomorphically depend on z'. The explicit form of the coefficients ψ_{IJ} is of little interest to us, although this information is extremely useful in the study of concrete systems α, which in the sequel will be fibers of analytic covers.

In certain problems it is useful to know how the roots of a system of canonical defining polynomials change under variation of the coefficients of the polynomials. The following simple statement is sufficient for our purposes.

L e m m a 1. *Let $\phi_I(z;\alpha)$, $|I| = k$, be canonical defining polynomials for some system of points $\{\alpha^1,\ldots,\alpha^k\}$ in $\mathbf{C}^m$. Then the set*

$$\{z \in \mathbf{C}^m\colon\ |\phi_I(z;\alpha)| < \epsilon^k,\ \ |I| = k\}$$

is contained in the union of the balls $|z-\alpha^j| < C\epsilon$, $j = 1,\ldots,k$, where C is a constant depending only on m and k.

■ Fix a point z in the given set. Let $P_\alpha(z,w) = \prod_1^k \langle w, z-\alpha^i\rangle = \sum_{|I|=k}\phi_I(z;\alpha)w^I$; then, by the requirement,

$$|P_\alpha(z,w)| < \epsilon^k \sum_{|I|=k} |w^I| \leqslant \epsilon^k \left[\sum_1^m |w_j|\right]^k \leqslant (\epsilon\sqrt{m})^k$$

for all w, $|w| \leqslant 1$. Put $a^j = (z-\alpha^j)/|z-\alpha^j|$. The points w for which $\langle w,a^j\rangle = 0$ form in $\mathbf{C}_w^m$ a plane. The cones $|\langle w,a^j\rangle| \leqslant c$ contract to this plane as $c \to 0$, hence there is a constant $c(k,n)$, depending on k and m only, such that the volume of the intersection of the cone $|\langle w,a^j\rangle| \leqslant c(k,m)$ with the ball $|w| \leqslant 1$ does not exceed $c_m/2k$, where c_m denotes the volume of the ball $|w| \leqslant 1$. The number $c(k,n)$ does not depend on j, since for a fixed c all cones $|\langle w,a^j\rangle| \leqslant c$ can be transformed into each other by unitary transformations. Since we have $\prod_1^k |\langle w,z-\alpha^j\rangle| < (\epsilon\sqrt{m})^k$ for all w, $|w| \leqslant 1$, the cones

$|<w,z-\alpha^j>| < \epsilon\sqrt{(m)}$, or, equivalently, $|<w,a^j>| < \epsilon\sqrt{(m)}/|z-\alpha^j|$, considered all together cover the whole ball $|w| \leqslant 1$. By the choice of $c(k,m)$ there is at least one j such that $\epsilon\sqrt{(m)}/|z-\alpha^j| > c(k,n)$, or, equivalently, $|z-\alpha^j| < \epsilon\sqrt{m}/c(k,n)$. ■

Corollary. *Let $\alpha = \{\alpha^1, \ldots, \alpha^k\}$ and $\beta = \{\beta^1, \ldots, \beta^k\}$ be two systems of points in a domain D in $\mathbb{C}^m$, and let $\phi_I(z;\alpha)$, $\phi_I(z;\beta)$, $|I| = k$, be corresponding canonical defining polynomials. If $|\phi_I(z;\alpha)-\phi_I(z;\beta)| < \epsilon^k$ for all I, $|I| = k$, and $z \in D$, then the systems α and β lie in $C\epsilon$-neighborhoods of each other.*

■ If $z \in \beta$, then $\phi_I(z;\beta) = 0$, hence $|\phi_I(z;\alpha)| < \epsilon^k$ for all I. By the Lemma, z lies in the $C\epsilon$-neighborhood of the set α. The same holds with β replaced by α, vice versa. ■

The limit behavior of canonical defining functions and corresponding systems of points is described by the following Lemma, which easily follows from the definitions.

Lemma 2. *Let $\alpha_j = \{\alpha_j^1, \ldots, \alpha_j^k\}$ be a sequence of systems of k points in a bounded domain D in $\mathbb{C}^m$, converging to $\alpha = \{\alpha^1, \ldots, \alpha^k\}$ in the sense that $\alpha_j^\nu \to \alpha^\nu$ as $j \to \infty$, $\nu = 1, \ldots, k$. Then $\phi_I(z;\alpha_j) \to \phi_I(z;\alpha)$, $|I| = k$, uniformly on compact subsets in $\mathbb{C}^m$. Conversely, if the polynomials $\phi_I(z;\alpha_j)$ converge in this way to a polynomial $\phi_I(z)$, $|I| = k$, then the sequence of systems α_j has, under a suitable ordering of points, a limit $\alpha = \{\alpha^1, \ldots, \alpha^k\}$, and $\phi_I(z) = \phi_I(z;\alpha)$, $|I| = k$.*

■ The first statement obviously follows from the definition of the functions ϕ_I. For the proof of the converse statement we note that, since D is bounded, we can choose a subsequence $\{\alpha_{j_\nu}\}$ from $\{\alpha_j\}$ converging to some system $\alpha = \{\alpha^1, \ldots, \alpha^k\}$. The direct statement implies that $\phi_I(z) = \phi_I(z;\alpha)$, $|I| = k$. If $\beta = \{\beta^1, \ldots, \beta^k\}$ is another limit system for $\{\alpha_j\}$, then, of course, $\phi_I(z;\alpha) = \phi_I(z;\beta)$, $|I| = k$. But then $P_\beta(z,w) \equiv P_\alpha(z,w)$ (cf. the definition of ϕ_I), hence the systems α and β coincide, up to the order of their points. Thus, any convergent subsequence from $\{\alpha_j\}$ has the same limit α, up to the order of the points of this limit. Since all $|\alpha_j^\nu|$ are uniformly bounded, under a suitable ordering of the points of the systems α_j the sequence $\{\alpha_j\}$ also converges to α. ■

4.3. Analytic covers as analytic sets.

T h e o r e m. *Let A be a closed subset of bounded domain $U = U'\times U''$ in $\mathbb{C}^n$, where $U' \subset \mathbb{C}^p$, $U'' \subset \mathbb{C}^m$, and let π: $(z',z'')\mapsto z' \in \mathbb{C}^p$. Suppose that the restriction π: $A \to U'$ is a generalized k-sheeted analytic cover. Then A is a pure p-dimensional analytic subset in U and π: $A \to U'$ is an analytic cover.*

■ Let $\sigma \subset U'$ be the critical set for π: $A \to U'$. By the definition of generalized analytic cover, for each point $a' \in U'\setminus\sigma$ there is a neighborhood $V' \ni a'$ in U' such that $A \cap \pi^{-1}(V')$ is the union of k manifolds, being the graphs of certain vector functions α^j holomorphic in V'. For each fixed $z' \in V'$ we construct canonical defining functions of the system of points $\alpha(z') = \{\alpha^1(z'), \ldots, \alpha^k(z')\}$ in $\mathbb{C}^m$; these functions are polynomials in z''. Their coefficients $\psi_{IJ}(\alpha(z'))$ are holomorphic in z' and do not depend on the order of $\alpha^1(z), \ldots, \alpha^k(z')$. Therefore, being constructed locally in a neighborhood of each point of $U'\setminus\sigma$, they can be "glued" to global holomorphic functions $\tilde{\psi}_{IJ}(z')$ on $U'\setminus\sigma$. These functions are uniformly bounded on $U'\setminus\sigma$, since all $\alpha^j(z')$ belong to the bounded domain U'' in $\mathbb{C}^m$. Since σ is a removable subset in U', all $\tilde{\psi}_{IJ}$ can be continued to corresponding functions $\phi_{IJ}(z')$ holomorphic in U'. Thus we can define in $U'\times\mathbb{C}^m$ the holomorphic functions

$$\Phi_I(z',z'') = \sum_{|J|\leqslant k} \phi_{IJ}(z')\cdot(z'')^J, \quad |I| = k,$$

which are, for each fixed $z' \in U'\setminus\sigma$, canonical defining functions of the system of k points $\alpha^1(z'), \ldots, \alpha^k(z') \in \mathbb{C}^m_{z''}$. Denote by $\tilde{A}$ the set of common zeros of the functions $\Phi_I(z',z'')$ in $U'\times\mathbb{C}^m$; we will prove that $\tilde{A} = A$.

Since these sets coincide above $U'\setminus\sigma$, and $\pi^{-1}(U'\setminus\sigma)$ is everywhere dense in A (by the definition of generalized cover), it suffices to prove that $\pi^{-1}(U'\setminus\sigma)$ is everywhere dense in $\tilde{A}$ also. Let $a' \in \sigma$ and $a_i' \to a'$, $a_i' \in U'\setminus\sigma$. Form in $\mathbb{C}^m$ the systems $\alpha_i = \{\alpha_i^1, \ldots, \alpha_i^k\}$ of z''-coordinate of the points of the corresponding sets $\pi^{-1}(a_i') \cap A$. By passing to subsequences we may assume that the α_i converge to a system $\alpha = \{\alpha^1, \ldots, \alpha^k\}$, i.e. $\alpha_i^j \to \alpha^j$ as $i \to \infty$, $j = 1, \ldots, k$. By Lemma 2, p.4.2, the $\Phi_I(a',z'')$, $|I| = k$, are canonical defining functions for the system α. In other words, the set $\pi^{-1}(a') \cap \tilde{A}$ coincides with the set of (not necessarily distinct) points of the system $\{(a',\alpha^j)\}$; in particular, each of its points is a limit point for $\pi^{-1}(U'\setminus\sigma) \cap A$.

So, we have proved that A is an analytic subset of U. The fact that π: $A \to U'$ is an analytic cover (i.e. an analytic subset in U' can be taken for σ) follows from

Theorem 3.7. ■

The functions $\Phi_I(z',z'')$, $|I| = k$, constructed in the proof of the Theorem will be called *canonical defining functions* of the analytic cover $\pi: A \to U'$. For sets of codimension 1 it is always a single function - the Weierstrass polynomial from Theorem 2.8. As is obvious from the construction, for each fixed $a' \in U'$ the functions $\Phi_I(a',z'')$ are canonical defining functions for a system $\alpha(a') = \{\alpha^1(a'), \ldots, \alpha^k(a')\}$, formed from the z''-coordinates of the points of the fiber $\pi^{-1}(a') \cap A$. The number of occurrences of a given point $a \in \pi^{-1}(a') \cap A$ in the system $\{(a', \alpha^j(a')\}$ is called the *multiplicity* of the analytic cover $\pi: A \to U'$ at a. It is obvious from the proof that this multiplicity is equal to the number of points in the fiber $\pi^{-1}(z') \cap A$, $z' \in U' \setminus \sigma$, converging to a as $z' \to a'$ (for more details see p.10.1).

C o r o l l a r y. *For every p-dimensional analytic set A, the set* $A_{(p)} = \{z \in A: \dim_z A = p\}$ *is also an analytic set.*

■ By p.3.4 and Theorem 3.7, each point $a \in A$ has a neighborhood U of the form $U' \times U''$ (in suitable coordinates), such that the projection $\pi: A_{(p)} \cap U \to U' \subset \mathbb{C}^p$ is a generalized analytic cover. As proved above, $A_{(p)} \cap U$ is an analytic subset in U. ■

4.4. The theorem of Remmert-Stein-Shiffman. Theorem 4.3 can be interpreted as a result on the removal of singularities of an analytic set $A \cap \pi^{-1}(U' \setminus \sigma)$ under the condition of "smallness" of the critical set σ. Its various applications are again and again related to the removal of singularities. One of such applications is the following Theorem of B. Shiffman [190].

T h e o r e m. *Let E be a closed subset of a complex manifold* Ω *and let A be a pure p-dimensional analytic subset in* $\Omega \setminus E$. *If* $\mathcal{H}_{2p-1}(E) = 0$, *then the closure* $\bar{A}$ *of A in* Ω *is a pure p-dimensional analytic subset in* Ω.

■ It suffices to prove that $\bar{A}$ is analytic in a neighborhood of an arbitrary point $a \in \bar{A}$, hence we may assume that Ω is a domain in $\mathbb{C}^n$ and that $a = 0$. Since the Hausdorff dimension of A is $2p$ (p.3.7) and $\bar{A} \subset A \cup E$, $\mathcal{H}_{2p+1}(\bar{A}) = 0$. Let $\{a^1, a^2, \ldots\}$ be a countable, everywhere dense subset in $\operatorname{reg} A$, and let $T_j = T_{a^j} A \ni 0$ be the tangent space to A at a^j. Since $\dim T_j = p$, the set $C = \overline{A} \cup (\bigcup_j T_j)$ has $\mathcal{H}_{2p+1}$-measure zero. By A6.4 there is an $(n-p)$-dimensional plane $L \ni 0$ such that $\mathcal{H}_1(C \cap L) = 0$. Without loss of generality we assume that $L = \mathbf{C}_{p+1\ldots n}$. Since $\mathcal{H}_1(\overline{A} \cap L) = 0$ and $\overline{A}$ is closed in Ω, we find in a standard manner domains $U' \subset \mathbf{C}_{1\ldots p}$, $U'' \subset \mathbf{C}_{(p+1)\ldots n}$ such that $0 \in U = U' \times U'' \subset \Omega$ and such that the projection $\pi: A \cap U \to U'$ is proper. Since $T_j \cap \mathbf{C}_{(p+1)\ldots n} = \{0\}$, the map $\pi|_A$ is locally biholomorphic in a neighborhood of each point a^j, hence br $\pi|_A$ is nowhere dense in A. It follows now from p.3.7 that the Hausdorff dimension of br $\pi|_A$ does not exceed $2p-2$. Therefore, if $\Sigma = (\overline{A} \cap E) \cup (\text{br}\, \pi|_A)$, then Σ is closed in $\overline{A}$ and $\mathcal{H}_{2p-1}(\Sigma) = 0$. Denote $\pi(\Sigma \cap U) = \sigma$. Then σ is also closed in U' and $\mathcal{H}_{2p-1}(\sigma) = 0$; in particular, $U' \setminus \sigma$ is connected and σ is a removable set. The map $\pi|_{\overline{A} \cap U}$ is locally biholomorphic above $U' \setminus \sigma$, hence a finitely-sheeted cover. The set $\pi^{-1}(\sigma) \cap \overline{A} \cap U$ is nowhere dense in $\overline{A} \cap U$, since the set $\{a^1, a^2, \ldots\}$ is dense in $\overline{A}$ and a neighborhood of each $a^j \in U$ is single-sheetedly projected into an open set in U'. Thus we have proved that $\pi: \overline{A} \cap U \to U'$ is a generalized analytic cover. By Theorem 4.3, $\overline{A} \cap U$ is a pure p-dimensional analytic subset in U. ■

An important particular instance of Shiffman's theorem is the following well-known Theorem of Remmert-Stein [102], [35]:

C o r o l l a r y. *Let S be an analytic subset of a complex manifold Ω and let A be a pure p-dimensional analytic subset in $\Omega \setminus S$. If now* $\dim S < p$, *then the closure of A in Ω is a pure p-dimensional analytic subset in Ω.*

It is easy to see that in this Theorem the dimension of S cannot be increased. Consider, e.g., in $\mathbf{C}^2_{z,w} \setminus \{z=0\}$ the graph $A: w = e^{1/z}$ of the holomorphic function $f(z) = e^{1/z}$, which has an essential singularity at 0 in $\mathbf{C}$. Its closure in $\mathbf{C}^2$ coincides with $A \cup \mathbf{C}_w$ (by Sokhotskiĭ's theorem). If $\overline{A}$ were an analytic set, then, first, its dimension would be 1, and, secondly, all points of $\mathbf{C}_w$ would be singular for $\overline{A}$, which contradicts Corollary 3, p.2.6, on the Hausdorff dimension of the set of singular points.

4.5. Analyticity of sng A. Canonical defining functions of an analytic cover can be used to describe sets of singular points, in the same way as minimal defining functions of principal analytic sets. We restrict ourselves here to sets that are homogeneous with respect to dimension.

L e m m a. *Let $U = U' \times U'' \subset \mathbb{C}^n$ be a domain and A a pure p-dimensional analytic subset in U such that the projection $\pi\colon A \to U' \subset \mathbb{C}^p$ is a proper map. Then the set of critical points* br $\pi|_A$ *is also analytic, and* dim (br $\pi|_A) < p$.

■ Let $\Phi_I(z',z'')$, $|I| = k$, be canonical defining functions of the analytic cover $\pi\colon A \to U'$. We first show that at all points of the set br $\pi|_A$ the rank of the matrix $(\partial\Phi_I / \partial z'')$ with $m = n - p$ columns and $\binom{k+m-1}{m-1} \geqslant m$ rows is strictly less than m. Namely, for $a \in \operatorname{sng} A$ this follows from the implicit function Theorem. For $a \in$ br $\pi|_A$ a regular point of A, T_aA contains a nonzero vector $v = (0',v'')$. Since all $\Phi_I|_A = 0$, we have $(\partial\Phi_I / \partial v)(a) = 0$ (derivatives along v), hence the columns of the matrix indicated above are linearly dependent.

Now we prove that the rank of the matrix $(\partial\Phi_I / \partial z'')$ at all other points equals m. Recall that the holomorphic functions Φ_I are defined by the equation $\langle w,z''-\alpha^1(z')\rangle \cdots \langle w,z''-\alpha^k(z')\rangle = \sum_{|I|=k} \Phi_I(z)w^I$, where $w \in \mathbb{C}^m$ and $(z',\alpha^j(z'))$ are the points of the fiber $\pi^{-1}(z') \cap A$, written taking multiplicities into account (cf. p.4.3). If $a \in A \setminus (\text{br } \pi|_A)$, then there is a neighborhood $V = V' \times V'' \ni a$ such that the map $\pi\colon A \cap V \to V'$ is biholomorphic. Therefore, for $z' \in V'$ there is one factor in the product $\prod_1^k := \prod_1^k \langle w,z''-\alpha^j(z')\rangle$, say $\langle w,z''-\alpha^1(z')\rangle$, such that $\alpha^1(a') = a''$, and which is holomorphic in all variables. The product $\prod_1^k$ can be represented in the form $\langle w,z''-\alpha^1(z')\rangle\prod^{k_2}$, where the function $\prod_2^k$ is holomorphic in $V' \times \mathbb{C}^m_{z''} \times \mathbb{C}^m_w$ also. Assume that our statement concerning the rank is not true. Then there is a vector $v = (0',v'') \neq 0$ such that all $(\partial\Phi_I / \partial v)(a) := \sum_{p+1}^n v_j(\partial\Phi_I / \partial z_j)(a) = 0$, hence $\sum_{|I|=k}(\partial\Phi_I / \partial v)(a)w^I \overset{w}{\equiv} 0$. However, at the lefthand side of this identity stands the derivative of $\prod_1^k$ (as a function of z) in the direction v at the point a. Since $a'' = \alpha^1(a')$, this derivative is equal to $\langle w,v''\rangle \cdot \prod_2^k \langle w,a''-\alpha^j(a')\rangle$. But (by construction) $a'' \neq \alpha^j(a')$ for $j > 1$, and $v'' \neq 0$, therefore this polynomial in w cannot vanish identically. The contradiction thus obtained shows that the rank of the matrix $(\partial\Phi_I / \partial z'')(a)$ is equal to m. Thus, we have proved that

$$\text{br } \pi|_A = \left\{ z \in A : \operatorname{rank} \frac{\partial\Phi_I}{\partial z''}(z) < n - p \right\},$$

i.e. br $\pi|_A$ is defined in U by a finite system of holomorphic equations. The dimension of this analytic set is strictly less than p, since this set is nowhere dense in A. ■

T h e o r e m. *Let A be a pure p-dimensional analytic subset of a complex manifold Ω. Then* sng A *is also an analytic subset in Ω, of dimension $<p$.*

■ The statement is local, it is therefore sufficient to consider the situation in a neighborhood of 0 in $\mathbf{C}^n$. By p.3.4 we may assume that neighborhoods $U_I \ni 0$ exist such that the coordinate projections $\pi_I: A \cap U_I \to \mathbf{C}_I \cap U_I$ are proper maps for all I with $\# I = p$. Put $U = \cap_I U_I$. Then $(\operatorname{sng} A) \cap U$ belongs to all sets $\operatorname{br} \pi_I|_A$. On the other hand, if $a \in (\operatorname{reg} A) \cap U$, then the tangent plane T_aA can be bijectively projected onto at least one $\mathbf{C}_I$, hence $a \notin \operatorname{br} \pi_I|_A$ for this I. Hence, $(\operatorname{sng} A) \cap U = \cap_{\# I=p} (\operatorname{br} \pi_I|_A) \cap U$, and, by the Lemma, $(\operatorname{sng} A) \cap U$ is an analytic subset in U. Since $\operatorname{sng} A$ is nowhere dense in A, its dimension is strictly less than p. ■

Finally we give one simple Proposition regarding the dimension of $\operatorname{br} \pi|_A$ at points.

P r o p o s i t i o n. *Let A be a pure p-dimensional analytic subset of a domain $U = U' \times U''$ in $\mathbf{C}^n$ such that the projection $\pi: A \to U \subset \mathbf{C}^p$ is proper. Then* $\dim(\operatorname{br} \pi|_A) \leqslant p-1$. *If* $\dim_a(\operatorname{br} \pi|_A) < p-1$ *and $a \in A$, then there is a neighborhood $V = V' \times V'' \ni a$ such that the set $A \cap V$ is the union of a finite number of complex manifolds M_j, the maps $\pi: M_j \to V'$ are biholomorphic, and* $\operatorname{br} \pi|_{A \cap V} = \cup_{i \neq j}(M_i \cap M_j)$.

■ We choose a neighborhood $V \ni a$ such that $\dim(\operatorname{br} \pi|_A) \cap V < p-1$ (this is possible by the definition of dimension), and such that V' is a ball in U'. Then the critical set σ of the cover $\pi: A \cap V \to V'$ has (complex) codimension $\geqslant 2$. Since V' is a ball, the domain $V' \setminus \sigma$ is simply connected (for $p = 2$ this is trivial, since then σ is locally finite; in the general case the proof of this topological fact is not difficult also). The monodromy Theorem states that every connected cover above a simply connected domain is single-sheeted, hence the connected components of $A \cap V \setminus \pi^{-1}(\sigma)$ have the form $M_j^0: z'' = \phi_j(z')$, where the vector functions ϕ_j are holomorphic in $V' \setminus \sigma$. By continuation of them across the removable set σ (A1.4) we obtain manifolds $M_j: z'' = \phi_j(z')$, $z' \in V$, lying in $A \cap V$. Since $\pi^{-1}(\sigma)$ is nowhere dense in $A \cap V$, we find $\cup M_j = A \cap V$. ■

5. Decomposition into irreducible components and its consequences

5.1. Connected components of reg A. Since reg A is open in A, each connected component S_j of reg A is also open in A. From the definition of regular point (and connected component) it obviously follows that if $S_j \neq S_k$ then $S_j \cap \bar{S}_k$ is empty and $\bar{S}_j \cap \bar{S}_k \subset \operatorname{sng} A$ (closure in A). We investigate this decomposition of reg A in more detail.

T h e o r e m. *Let A be an analytic subset of a complex manifold Ω. Then:*

(1) The decomposition of reg $A = \cup_j S_j$ *into connected components is locally finite, i.e. every compact set $K \subset \Omega$ intersects only finitely many sets S_j;*

(2) if $\{S_j\}_{j \in J}$ is some family of connected components of reg A *and $S = \cup_{j \in J} S_j$, then the closure $\bar{S}$ is an analytic subset in Ω of dimension* $\max_{j \in J} \dim S_j$.

■ The proof proceeds by induction with respect to $p = \dim A$. For $p = 0$ the statement is trivial, since in this case A is a locally finite subset in Ω.

$p-1 \Rightarrow p$. The statement is local, hence it suffices to prove it in a coordinate neighborhood of an arbitrary point $a \in \bar{S}$, and hence we may assume that Ω is a domain in $\mathbb{C}^n$. Represent S in the form $S^p \cup S'$, where S^p is the union of all $S_j \subset S$ of dimension p and $S' = S \setminus S^p$. We will separately prove that $\overline{S^p}$ and $\overline{S'}$ are analytic subsets in Ω of the dimensions required (in the course of reasoning (1) will also be proved).

Let $a \in \overline{S^p}$, and suppose that coordinates have been chosen such that $a = 0$ and such that for some neighborhood $U = U' \times U'' \ni 0$ the projection $\pi: A \cap U \to U' \subset \mathbb{C}^p$ is proper, with critical analytic set $\sigma \subset U'$. Since $S^p \cap U$ is the union of certain connected components of $(\operatorname{reg} A) \cap U$ and since $S(A) \cap U \subset \pi^{-1}(\sigma)$, the set $S_0^p = (S^p \cap U) \setminus \pi^{-1}(\sigma)$ is the union of certain components of the manifold $A \cap U \setminus \pi^{-1}(\sigma)$ (being simultaneously an open and a closed subset). Since the analytic set $\pi^{-1}(\sigma) \cap A \cap U$ has dimension $<p$, the set $\pi^{-1}(\sigma)$ is nowhere dense in $S^p \cap U$, hence the closure of S_0^p in U coincides with $\overline{S^p} \cap U$. Since $\pi: A \cap U \setminus \pi^{-1}(\sigma) \to U' \setminus \sigma$ is a locally biholomorphic k-sheeted cover, $k < \infty$, the number of connected components of S_0^p is finite ($\leqslant k$), and $\pi: S_0^p \to U' \setminus \sigma$ is also a locally biholomorphic cover. Therefore the closure of S_0^p in U, which is equal to $\overline{S^p} \cap U$, is an analytic cover above U' (it belongs to A

hence does not intersect $U' \times \partial U''$). By Theorem 4.3, $\overline{S^p} \cap U$ is a pure p-dimensional analytic subset in U.

Let now $a \in S'$. By p.3.7 there are a neighborhood $V \ni a$ and an analytic subset $A_0 \subset V$ of dimension $<p$ such that $A \cap V = (A_{(p)} \cap V) \cup A_0$. Clearly, $S' \cap V \subset A_0$. Let $S_j \subset S'$ be such that $S_j \cap V$ is nonempty. Then $S_j \cap V \subset \operatorname{reg} A_0$ and, therefore, each connected component $S_{ji} \subset S_j \cap V$ is an open subset of some component S_{ji}^0 of $\operatorname{reg} A_0$. By the induction hypothesis, $\overline{\cup_{j,i} S_{ji}^0} \cap V$ is an analytic subset in V of dimension $\max_{j,i} \dim S_{ji} < p$. Thus it is sufficient to prove that each S_{ji} is dense in S_{ji}^0. Since S_{ji} does not intersect $A_{(p)}$, the set $S_{ji}^0 \cap A_{(p)}$ is a proper analytic subset of the connected complex manifold S_{ji}^0, it is nowhere dense in S_{ji}^0, and has connected complement $S_{ji}^0 \setminus A_{(p)} \supset S_{ji}$. Since S_{ji} are components in $(\operatorname{reg} A) \cap V \setminus A_{(p)}$, we have $S_{ji} = S_{ji}^0 \setminus A_{(p)}$. Hence $\overline{S'} \cap V = \overline{\cup_{j,i} S_{ji}^0} \cap V$ is an analytic subset in V of dimension $\max \dim S_{ji} \cap V$. ■

The following statement can be regarded as a theorem on removing singularities of analytic sets. It was implicit form encountered earlier, and will be used in the sequel on precisely this level.

C o r o l l a r y. *Let A, A' be analytic subsets of a complex manifold Ω. Then $\overline{A \setminus A'}$ is also an analytic subset in Ω.*

■ Let $\operatorname{reg} A = \cup_{j \in J} S_j$ be the decomposition into connected components, and let $I \subset J$ be the set of those j for which $S_j \setminus A'$ is nonempty. Since S_j is a connected manifold, the uniqueness Theorem implies that $S_j \cap A'$ is nowhere dense in S_j for $j \in I$, hence $\overline{S_j \setminus A'} = \overline{S_j}$. By the Theorem, $A^0 = \cup_{j \in I} \overline{S_j}$ is an analytic subset in Ω. Since obviously $\cup_{j \in I} S_j$ is everywhere dense in $A \setminus A'$, we have $\overline{A \setminus A'} = \overline{\cup S_j \setminus A'} = \cup_{j \in I} \overline{S_j} = A^0$ (the second equality follows from the local finiteness of the union). ■

5.2. Decomposition by dimension. Analyticity of $\operatorname{sng} A$ **and** $\mathbb{S}(A)$. Theorem 5.1 implies a decomposition of an arbitrary analytic set as a union of analytic sets that are homogeneous with respect to dimension.

T h e o r e m 1. *Let A be a p-dimensional analytic subset of a complex manifold Ω. Then, for any $k \geqslant 0$, the closure in Ω of the set $A_{(k)} = \{z \in A : \dim_z A = k\}$ is* *either empty or is a pure k-dimensional analytic subset in* Ω*; hence* $A = \cup_0^p \overline{A}_{(k)}$ *is a finite union of analytic subsets in* Ω *that are homogeneous in dimension. Moreover, for any k the sets* $\cup_{j \geq k} A_{(j)}$ *and* $\overline{\cup_{j<k} A_{(j)}}$ *are also analytic; they are equal to* $\cup_k^p \overline{A}_{(j)}$ *and* $\cup_{j<k} \overline{A}_{(j)}$*, respectively.*

■ Since the decomposition reg $A = \cup S_j$ into connected components is locally finite, $\overline{A}_{(k)} = \cup \{\overline{S}_j\colon \dim S_j = k\}$ is a pure k-dimensional analytic subset in Ω by Theorem 5.1. For the same reason, the closure of $\cup_{j<k} A_{(j)}$ coincides with $\cup \{\overline{S}_j\colon \dim S_j < k\}$, and the set $\cup_{j \geq k} A_{(j)}$ is closed because $\dim_z A$ is semicontinuous. ■

This Theorem allows one to restrict in many questions the consideration to analytic sets that are homogeneous with respect to dimension. In the sequel we will do so without special discussion.

T h e o r e m 2. *Let A be an analytic subset of a complex manifold* Ω. *Then* sng A *and* $\mathbb{S}(A)$ *are also analytic subsets in* Ω*; moreover,* $\dim \mathbb{S}(A) < \dim A$ *and* $\dim_z (\operatorname{sng} A) < \dim_z A$ *at all* $z \in \operatorname{sng} A$.

■ For pure-dimensional analytic sets this has been proved already in p.4.5. In the general case, the equation

$$\operatorname{sng} A = (\cup \operatorname{sng} \overline{A}_{(k)}) \cup \bigcup_{j \neq k} (\overline{A}_{(j)} \cap \overline{A}_{(k)})$$

obviously holds. It implies, by Theorem 1, that sng A is an analytic subset in Ω. Since it is nowhere dense in A, the dimension of sng A at any of its points is strictly less than the dimension of A at this point. Further, if $\dim A = p$, then $\mathbb{S}(A) = (\operatorname{sng} A) \cup \cup_0^{p-1} A_{(k)}$, and this equals $(\operatorname{sng} A) \cup \overline{\cup_0^{p-1} A_{(k)}}$, since $\mathbb{S}(A)$ is closed. Thus, $\mathbb{S}(A)$ is an analytic set; its dimension is strictly less than p, since $\mathbb{S}(A)$ is nowhere dense in $A_{(p)}$. ■

5.3. Irreducibility. An analytic subset A of a complex manifold Ω is called *reducible* (in Ω) if $A = A_1 \cup A_2$ where A_1, A_2 are also analytic subsets in Ω, distinct from A. If A cannot be represented in this form, A is called *irreducible* (in Ω). An analytic set $A \subset \Omega$ is called irreducible if it is irreducible in some neighborhood of it (this property does obviously not depend on the choice of a neighborhood in which A is closed). By p.5.2, irreducible analytic sets are homogeneous in dimension.

P r o p o s i t i o n. *An analytic set A is irreducible if and only if the set* reg A *is connected.*

■ We may assume that A is closed in Ω. Let S be a connected component in reg A. Then (p.5.1) $A_1 = \bar{S}$ and $A_2 = \overline{(\text{reg } A) \setminus S}$ are analytic subsets in Ω, and $A = A_1 \cup A_2$ since reg A is dense in A. If A is irreducible, either $A_1 = A$ or $A_2 = A$. Since $A_2 \cap S$ is empty, $A_1 = A$, i.e. S is dense in A, hence in reg A. Since S is closed in reg A this implies that reg $A = S$, i.e. reg A is connected.

Conversely, if reg $A = S$ is connected and A_1, A_2 are analytic subsets in Ω such that $A = A_1 \cup A_2$, then $S_j = A_j \cap S$ are analytic subsets of the connected complex manifold S; moreover, $S = S_1 \cup S_2$, hence one of S_j, say S_1, is somewhere dense in S. The uniqueness Theorem now implies $S = S_1$. But then $A = \bar{S} = \bar{S}_1 \subset A_1$, hence $A = A_1$. ■

This implies the following simple, but important

C o r o l l a r y 1. *Let A, A' be analytic subsets of a complex manifold, where A is irreducible and is not contained in A'. Then the set $A \cap A'$ is nowhere dense in A (and* $\dim A \cap A' < \dim A$*).*

■ If $A \cap A'$ is somewhere dense in A, then $A' \cap$ reg A is an analytic subset of the connected complex manifold reg A, containing an open subset in reg A. By the uniqueness Theorem, $A' \supset$ reg A, hence $A' \supset \overline{\text{reg} A} = A$, contradicting the assumptions made. ■

This Corollary can be reformulated as the following uniqueness Theorem for analytic sets.

C o r o l l a r y 2. *Let A, $\tilde{A}$ be analytic subsets of a complex manifold, where A is irreducible. Now if $A \cap \tilde{A}$ contains a nonempty open subset in A (or, equivalently, if* $\dim A \cap \tilde{A} = \dim A$*), then $A \subset \tilde{A}$. In particular, if* $\dim A = \dim \tilde{A} = \dim A \cap \tilde{A}$ *and the sets A, $\tilde{A}$ are irreducible, then $A = \tilde{A}$.*

In relation with local problems it is convenient to have, besides global irreducibility, the following notions. An analytic set A is called *irreducible at a point* $a \in A$ if there is a fundamental system of neighborhoods $U_j \ni a$ such that all analytic sets $A \cap U_j$ are irreducible in the corresponding U_j (e.g., A is irreducible at each of its regular points). The set A is called *locally irreducible* if it is irreducible at all points of it.

E x a m p l e s. (a) The semicubic parabola A: $z_2^2 = z_1^3$ in $\mathbb{C}^2$ is globally and locally irreducible (in polydisks U_r: $|z_1| < r^2$, $|z_2| < r^3$ the sets $(\operatorname{reg} A) \cap U_r$ are connected).

An example of a locally irreducible, but globally reducible analytic set is given by an arbitrary unconnected complex manifold.

(b) The set $A: z_2^2 = z_1^3 + z_1^2$ in $\mathbb{C}^2$ is the image of $\mathbb{C}_\zeta$ under the map ϕ: $\zeta \mapsto (\zeta^2 - 1,\ \zeta(\zeta^2 - 1))$. The pre-image of the point $z = 0$ consists of the two points $\zeta = \pm 1$, at other points ϕ is single-sheeted (Fig. 4.a). The set $\operatorname{reg} A = A \setminus \{0\}$ is connected, hence A is irreducible. On the other hand, if a neighborhood $V \ni 0$ lies in the polydisk $|z_1| < 1$, $|z_2| < 1$, then $(\operatorname{reg} A) \cap V$ has at least two connected components $S_\pm$ such that $\phi^{-1}(S_\pm)$ lies in distinct branches of the lemniscate $|\zeta^2 - 1| < 1$: $\phi^{-1}(S_+) \ni 1$ and $\phi^{-1}(S_-) \ni -1$. Hence A is not irreducible at $z = 0$.

(c) The singular points of the analytic set $A: z_1^2 = z_3 z_2^2$ in $\mathbb{C}^3$ form the $\mathbb{C}_3$-axis, while $\operatorname{reg} A$ is the graph $z_3 = (z_1 / z_2)^2$ above $\mathbb{C}^2 \setminus \{z_2 = 0\}$ (Fig. 4.b). Hence $\operatorname{reg} A \cap V$ is connected in any polydisk neighborhood $V \ni 0$, and A is irreducible, globally and at 0. However, it is clear that $A \cap V$ is reducible in any domain $V \ni (0, 0, z_3)$ if $z_3 \neq 0$ and if the diameter of V does not exceed $|z_3|$. Thus, A is not locally irreducible. This example also shows that the set of irreducibility points of an analytic set A (i.e. points at which A is irreducible) need not be open or closed in A.

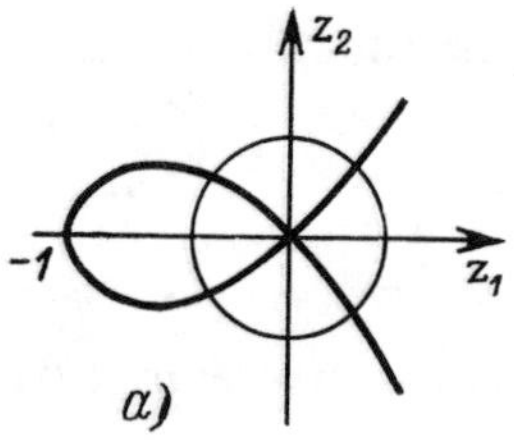

Figure 4a.

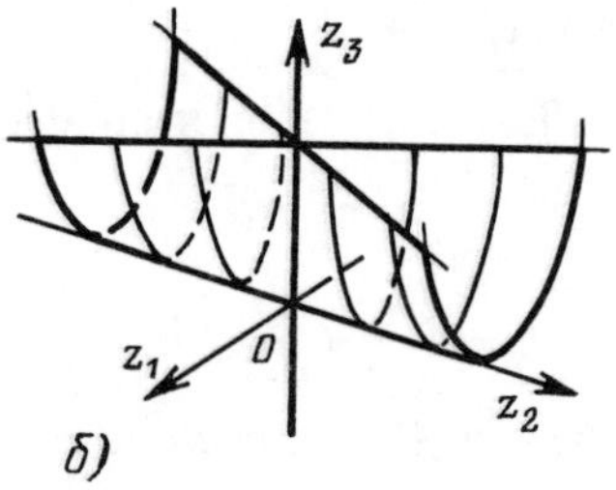

Figure 4b.

5.4. Irreducible components. An irreducible analytic subset A' of an analytic set A is called an *irreducible component* of A if every analytic subset $A'' \subset A$ such that $A'' \neq A'$ and $A' \subset A''$ is reducible.

L e m m a. *Let A be an analytic set and S a connected component of* reg A. *Then the closure $\overline{S}$ of S in A is an irreducible component of A.*

■ We may assume that A is an subset of a complex manifold Ω. By p.5.1, $\overline{S}$ is also an analytic subset in Ω. Let A'' be another analytic subset of A, containing $\overline{S}$. Since S is a dense open subset of reg $\overline{S}$, we find that reg $\overline{S}$ is connected, hence $\overline{S}$ is an irreducible analytic set. Since S is open in A it is also open in A'', hence there is a connected component S'' of reg A' containing S as an open subset. But then $\overline{S} \cap S''$ is an analytic subset of the connected complex manifold S'', containing the nonempty open subset S. The uniqueness Theorem implies that $\overline{S} \supset S''$, and thus $\overline{S} = \overline{S''}$. If A'' is irreducible, then, by p.5.3, $A'' = \overline{S''}$, i.e. $A'' = \overline{S}$. ■

This Lemma and p.5.1 imply the following important Theorem on decomposition into irreducible components.

T h e o r e m. *Let A be an analytic subset of a complex manifold Ω. Then:*

(1) every irreducible component of A has the form $\overline{S}$, where S is a connected component of reg A;

(2) if reg $A = \cup_{j \in J} S_j$ *is the decomposition into connected components (J is finite or countable, $S_j \cap S_k = \varnothing$ for $j \neq k$), then $A = \cup_{j \in J} \overline{S}_j$, and this is the decomposition of A into irreducible components;*

(3) the decomposition of A into irreducible components is locally finite, i.e. for every compact set $K \subset \Omega$ there is only a finite (or empty) set of indices $j \in J$ such that $K \cap \overline{S}_j$ is nonempty.

■ Let A' be an irreducible component of A and put $A_j' = A' \cap \overline{S_j}$. By p.5.1, A_j' is an analytic subset in Ω for every j, $A' = \cup A_j'$, and the union is locally finite. The definition of irreducibility implies that $A' = A_j'$ for some j, i.e. $A' \subset \overline{S}_j$. The definition of irreducible component and the irreducibility of $\overline{S}_j$ imply that $A' = \overline{S}_j$.

Since $\cup S_j$ is a locally finite union, $\cup \overline{S}_j$ is an analytic subset in Ω and $A = \cup \overline{S}_j$, since reg A is dense in A. By the Lemma, all $\overline{S}_j$ are irreducible components of A.

Obviously, statement (3) follows from Proposition 5.1. ■

Local finiteness of this decomposition implies the following relation between irreducibility at a point and "prelimiting" irreducibility in a neighborhood of it.

P r o p o s i t i o n. *Let A be an analytic set. For each point $a \in A$ there is a neighborhood $U \ni a$ such that all irreducible components of the analytic set $A \cap U$ are irreducible at a.*

■ Let A be a pure p-dimensional analytic subset of a domain $D' \times D''$ in $\mathbb{C}^n$ such that the projection $\pi: A \to D' \subset \mathbb{C}^p$ is a k-sheeted analytic cover with critical analytic set $\sigma \subset D'$. Let $a \in A$. By shrinking D' and D'' it can be assumed that the fiber $\pi^{-1}(a') \cap A$ consists of the single point a. Let $B' = B'(a',r') \subset D'$, where $a' = \pi(a)$. Since $B' \setminus \sigma$ is connected, the number of connected components of the set $A \cap \pi^{-1}(B' \setminus \sigma)$ does not exceed k. When diminishing r', this number does not decrease (this can easily be seen from the definition of connected components and from the fact that $\pi|_A$ is a k-sheeted cover above the domain $D' \setminus \sigma$). Hence there exists an $r_0' > 0$ such that for all $r' \leqslant r_0'$ the number of components is constant and equal to, say, $k_0 \leqslant k$. Moreover, if S_j^0, $j = 1, \ldots, k$, are the connected components of $A \cap \pi^{-1}(B(a',r_0') \setminus \sigma)$, then for any $r' \leqslant r_0'$ the connected components of $A \cap \pi^{-1}(B' \setminus \sigma)$ have the form $S_j^0 \cap (B' \times D'')$. Since $\pi^{-1}(a') \cap A = \{a\}$, for every $r'' > 0$ there is an $r' \in (0, r_0')$ such that $A \cap (B' \times B(a'',r'')) = A \cap (B' \times D'')$. Thus, there is a fundamental system of neighborhoods $U_r = B(a',r') \times B(a'',r'') \ni a$ such that all $S_j^0 \cap U_r$ are connected, hence the analytic sets $\overline{S_j^0} \cap U_r$ are irreducible in U_r, respectively. Taking for U any such U_r we obtain that the analytic set $\overline{S_j^0} \cap U$ is irreducible in U at a. By construction, the $\overline{S_j^0} \cap U$, $j = 1, \ldots, k_0$, are the irreducible components of $A \cap U$. ■

In certain problems it turns out to be convenient to maximally localise the notion of analytic set. As for functions, this is achieved by the notion of germ. Two analytic sets A_1, A_2 on a complex manifold are said to be equivalent at a point a if there is a neighborhood $U \ni a$ such that $A_1 \cap U = A_2 \cap U$. The corresponding equivalence classes are called *germs* of analytic sets at a. The equivalence class containing a set A is usually denoted by $(A)_a$. Each element of a class α is called a representative of α. The set-theoretical operations and relations can be naturally defined for germs: $\alpha \subset \beta$ means that $A \subset B$ for certain representations $A \in \alpha$, $B \in \beta$; $\alpha \cap \beta := (A \cap B)_a$; $\alpha \cup \beta := (A \cup B)_a$; etc. By definition, the dimension of a germ $(A)_a$ is equal to $\dim_a A$, and a germ is called homogeneous with respect to dimension (or, of pure dimension) if it has a representative that is such. A germ of analytic sets α is called reducible if $\alpha = \beta \cup \gamma$ for certain germs β, γ distinct from α, and irreducible if such a representation is impossible. Clearly, a germ α is irreducible if and only if every representative of it is irreducible at the point a. The Proposition proved above implies that every germ of an analytic set can be uniquely represented as a finite union of irreducible germs, which are called its components. The basic properties of dimensions of analytic sets and their intersections (p.3.5 and p.5.6) can be naturally reformulated in terms of germs; this we will not do.

5.5. Stratifications. Now we can clarify in what sense an analytic set is a complex manifold with singularities (of which we spoke at the very beginning of this Chapter). Let A be an arbitrary analytic set, then $M_p = A \setminus \mathbb{S}(A)$ is a complex manifold of dimension $p = \dim A$ (generally, not connected). In turn, $\mathbb{S}(A)$ is also an analytic set, hence $M_{p_1} = \mathbb{S}(A) \setminus \mathbb{S}(\mathbb{S}(A))$ is a complex manifold of dimension $p_1 = \dim \mathbb{S}(A) < p$. By continuing this process, which ends after at most $p + 1$ steps, we obtain a finite decomposition $A = M_p \cup M_{p_1} \cup \cdots \cup M_{p_s}$ of the analytic set into pairwise disjoint complex manifolds (this decomposition is called a partition by dimension). Thus, the structure of an analytic subset on a complex manifold Ω can be pictured as follows: a p-dimensional analytic subset is a p-dimensional complex manifold in $\Omega \setminus \mathbb{S}(A)$, where $\mathbb{S}(A)$ is an analytic subset of dimension $<p$; in turn, $\mathbb{S}(A)$ is $\cdots$, etc. as above. This is so to speak the analytic point of view. The synthetic, constructive point of view is given by the reverse order: zero-dimensional analytic subsets are locally finite subsets in Ω, one-dimensional analytic subsets are obtained by "pasting" to these one-dimensional complex manifolds, two-dimensional analytic subsets are one- or zero-dimensional analytic subsets to which two-dimensional complex manifolds are pasted, etc. Of course, this description is very global, and nothing has been said about the manner of "pasting" manifolds to singularities, i.e. about the structure of analytic sets in a neighborhood of singular points. It is precisely with the study of this problem that a further understanding of the local and global structure of analytic sets is related.

The natural partition of an analytic set into complex manifolds sketched above is far from being ideal. In practice we need partitions having a number of additional properties. We will restrict ourselves to the description of partitions satisfying most simple topological requirements.

D e f i n i t i o n. Let A be an analytic subset of a complex manifold Ω. A partition of A into complex manifolds $M_j \subset \Omega$ (called the strata of the partition) is called a *stratification* of A if:
(1) the strata M_j pairwise do not intersect, $A = \bigcup M_j$, and the union is locally finite in Ω;
(2) the closure $\overline{M}_j$ of each stratum and its boundary (border) $\overline{M}_j \setminus M_j$ are analytic subsets in Ω;
(3) if $M_j \cap \overline{M}_k$ is nonempty and $M_j \neq M_k$, then $M_j \subset \overline{M}_k$ and $\dim M_j < \dim M_k$ (the property of contiguity of strata).

The natural partition by dimension of an analytic set into manifolds does not, in general, satisfy these requirements. E.g., for $A = \mathbb{C}_{12} \cup \mathbb{C}_3$ in $\mathbb{C}^3$ the partition $A = (\mathbb{C}_{12} \setminus \{0\}) \cup \mathbb{C}_3$ does not satisfy (3), and a stratification of A is given by, e.g., $A = (\mathbb{C}_{12} \setminus \{0\}) \cup (\mathbb{C}_3 \setminus \{0\}) \cup \{0\}$. A natural question arises: can an arbitrary analytic set be stratified at all?

P r o p o s i t i o n. *Any analytic subset A of a complex manifold Ω has a stratification.*

■ For the strata of maximal dimension $p = \dim A$ we take the connected components of $A \setminus S(A)$, and we proceed by induction.

Suppose that the strata M_i of dimensions $\geqslant d$ have already been constructed, are connected, satisfy the conditions (1)-(3) except for $\cup M_i = A$, and let the complement in A to their union be an analytic subset $A' \subset A$ of dimension $d' < d$ (for $d = p$ this property is: $A' = S(A)$). Let $A' \setminus S(A') = \cup M_j'$ be the decomposition into connected components. Without condition (3) they could be taken as the strata of dimension d', but here it is necessary to correct. Since $\overline{M_j'}$ and the already constructed $\overline{M}_i$ are irreducible in Ω, either $\overline{M_j'} \subset \overline{M}_i$ or $\dim \overline{M_j'} \cap \overline{M}_i < d'$ (p.5.3). Let A'' be the union of all analytic sets $\overline{M_j'} \cap \overline{M}_i$ of dimensions $< d'$ and put $M_j = M_j' \setminus A''$. Since A'' is an analytic subset in Ω of dimension $< d'$ (the given union is locally finite), the M_j are connected complex manifolds of dimension d' and $\overline{M}_j = \overline{M_j'}$ are analytic subsets in Ω. Furthermore,

$$\overline{M}_j \setminus M_j = (\overline{M_j'} \setminus M_j') \cup (\overline{M}_j \cap A'') = (\overline{M_j'} \cap S(A')) \cup (\overline{M}_j \cap A'')$$

are analytic subsets in Ω of dimension $< d'$ hence (2) is satisfied. Since M_j' are the connected components of $A' \setminus \mathbb{S}(A')$, $M_{j_1} \cap \overline{M_{j_2}} = \varnothing$ if $M_{j_1} \neq M_{j_2}$. For if M_i is a stratum of dimension $\geqslant d$, then $\overline{M_j} \cap M_i$ is empty, since $\overline{M_j} \subset A'$. Assume now that $M_j \cap \overline{M_i}$ is nonempty. The definition of A'' then implies that $\dim \overline{M_j} \cap \overline{M_i} \geqslant d'$, hence $M_j \subset \overline{M_i}$ (by Corollary 5.3). Thus (3) is fulfilled too. Finally, the complement to the union of all already constructed strata is an analytic subset of $\mathbb{S}(A') \cup A''$ of dimension $< d'$. Therefore this construction stops after a finite ($\leqslant p+1$) number of steps (the corresponding A' turns out to be empty), implying that $\cup M_i = A$. ■

In an analogous way it can be proved that every partition of A into strata satisfying (1)-(2) has a refinement which is a stratification in our sense (cf. Whitney [149]). The stratification of an analytic set constructed in the proof is called primary. There are stratifications with important additional properties of regularity of contiguity of tangent structures (satisfying the so-called conditions A and B of Whitney and others), see, e.g., [148], [149], [152].

5.6. Intersections of analytic sets. The properties of irreducible sets (p.5.3 and p.5.4) readily imply the following statements.

P r o p o s i t i o n 1 (uniqueness Theorem). *If two irreducible analytic subsets A_1, A_2 of a complex manifold Ω coincide inside an open set U for which $A_1 \cap U$ is nonempty, then $A_1 = A_2$.*

This clearly follows from Corollary 2, p.5.3.

P r o p o s i t i o n 2. *If A, $\tilde{A}$ are two analytic subsets, $A \subset \tilde{A}$, and if A is irreducible, then A is contained in an irreducible component of $\tilde{A}$.*

■ If $\tilde{A} = \cup A_j$ is the decomposition into irreducible components, then $A = \cup(A \cap A_j)$. Since this union is locally finite, the definition of irreducibility implies that $A = A \cap A_j$ for some j, i.e. $A \subset A_j$. ■

P r o p o s i t i o n 3. *If A, $\tilde{A}$ are pure p-dimensional analytic subsets of a complex manifold and $A \subset \tilde{A}$, then A is the union of a family of irreducible components of $\tilde{A}$.*

■ Each irreducible component $A' \subset A$ is pure p-dimensional and is contained in an irreducible component of $A'' \subset \tilde{A}$, which is also pure p-dimensional

(Proposition 2). Therefore A' contains an open subset in A'', hence $A' = A''$ by Proposition 1. ■

T h e o r e m. *Let $\{A_\alpha\}_{\alpha\in I}$ be an arbitrary family of analytic subsets of a complex manifold Ω. Then $A = \bigcap_{\alpha\in I} A_\alpha$ is also an analytic subset in Ω; moreover, for every $K \subset\subset \Omega$ there is a finite subset $J \subset I$ such that $A \cap K = (\bigcap_{\alpha\in J} A_\alpha) \cap K$.*

■ The proof proceeds by induction with respect to $p = \min_\alpha \dim A_\alpha$. If $p = 0$, then some A_α is zero-dimensional and the statement is trivial.

$p-1 \Rightarrow p$. Let $p = \dim A_{\alpha_0}$ and let $\cup S_j$ be the decomposition of $\operatorname{reg} A_{\alpha_0}$ into connected components. Denote by A^0 the union of all $\overline{S_j}$ that are entirely contained in A, and fix an arbitrary open set U with compact closure in Ω. By Theorem 5.1, only finitely many S_j intersect U, say $S_1, \ldots, S_k$. Distinguish those among them, say $S_1, \ldots, S_m$, for which $S_i \cap U \not\subset A$ (if there are no, $A \cap U = A_{\alpha_0} \cap U$). Fix a point $a_i \in (S_i \cap U) \setminus A$. Since $A = \bigcap_\alpha A_\alpha$, there is for every i an index $\alpha_i \in I$ such that $a_i \notin A_{\alpha_i}$. Put $A' = \cup_1^m (\overline{S_i} \cap A_{\alpha_i})$. Then $\dim A' < p$, since $\dim \overline{S_i} \cap A_{\alpha_i} < \dim \overline{S_i}$. The set $A^1 = \bigcap_{\alpha\in I} A_\alpha \cap A'$ is an intersection of a family of analytic subsets in Ω, to be precise, of the family $\{A', A_\alpha \colon \alpha\in I\}$, which includes the analytic set A' of dimension $\leqslant p-1$. By the induction hypothesis, A^1 is an analytic subset in Ω, and $A^1 \cap U = A' \cap A_{\alpha_1} \cap \cdots \cap A_{\alpha_N} \cap U$ for some finite tuple of indices $\{\alpha_1, \ldots, \alpha_N\} \subset I$, $N \geqslant m$. But then $A \cap U = (A^0 \cup A') \cap U$ is also an analytic subset in U. Furthermore, by construction $A_{\alpha_0} \cap U = (A^0 \cup \cup_1^m \overline{S_i}) \cap U$. Since A^0 belongs to all A_α,

$$A_{\alpha_0} \cap A_{\alpha_1} \cap \cdots \cap A_{\alpha_N} = A^0 \cup (A' \cap A_{\alpha_1} \cap \cdots \cap A_{\alpha_N}) = A^0 \cup A^1$$

inside the set U, hence A coincides in this set with the intersection of the sets A_{α_j}, $j = 0, \ldots, N$. ■

Using the results of this Section the Theorem on local representation of analytic sets as covers (p.3.7) can be reformulated as follows.

P r o p o s i t i o n 4. *Let A be a pure p-dimensional analytic set, let local coordinates and a neighborhood U of a point $a \in A$ be chosen such that $U = U' \times U''$, where $U' \subset \mathbb{C}^p$, $U'' \subset \mathbb{C}^m$, and such that the projection $\pi \colon A \cap U \to U'$ is proper.*

Then there are Weierstrass polynomials $F_j(z',z_{p+j})$ above U', $j = 1,\ldots,m$, such that $A \cap U$ is contained in the pure p-dimensional analytic set $\tilde{A} = \{z \in U: F_j(z',z_{p+j}) = 0,\ j = 1,\ldots,m\}$ and is the union of certain irreducible components of it.

■ If A_j is the projection of $A \cap U$ in $U_j' = U' \times z_{p+j}(U)$, then A_j is an analytic subset in U_j' of codimension 1, and the projection $A_j \to U'$ is proper. By p.2.8, A_j is the zero set of a Weierstrass polynomial $F_j(z',z_{p+j})$ above U'. Obviously, the set $\tilde{A}$ of common zeros of these F_j, regarded as functions on $U' \times \mathbf{C}^m$, is contained in $A \cap U$. Since the leading coefficients of the F_j are 1, the projection $\tilde{A} \to U'$ is proper, hence $\dim_z \tilde{A} \leqslant p$ for all $z \in \tilde{A}$ (p.3.3). On the other hand, $\dim_z \tilde{A} \geqslant p$ by p.3.5, hence $\tilde{A}$ is pure p-dimensional. By Proposition 3, A is the union of certain irreducible components of $\tilde{A}$. ■

5.7. The number of defining functions.

P r o p o s i t i o n. *Let $\mathfrak{F} = \{f_\alpha\}_{\alpha \in I}$ be an arbitrary family of functions, holomorphic on an n-dimensional complex manifold Ω. Then the set Z of their common zeros is an analytic subset in Ω; moreover, there are functions $g_0,\ldots,g_n \in \mathcal{O}(\Omega)$ (infinite linear combinations of the f_α) whose set of common zeros coincides with Z.*

■ The first statement (the analyticity of Z) was proved in p.5.6. Let Ω_i be the connected components of Ω not belonging to Z, and let $a_i \in \Omega_i \setminus Z$ be arbitrarily chosen points. For every i there is a function $f_{\alpha_i} \in \mathfrak{F}$ such that $f_{\alpha_i}(a_i) \neq 0$. Represent Ω as a countable union of compact sets, $\cup K_j$, and choose, by induction with respect to j, a number c_j such that $|c_j f_{\alpha_j}(z)| < 2^{-j}$ for all $z \in K_j$ and $|\sum_{k=1}^{j} c_k f_{\alpha_k}(a_i)| > |c_i f_{\alpha_i}(a_i)|/2$ for $i \leqslant j$ (this can obviously be done). The series $\sum_1^\infty c_k f_{\alpha_k}$ converges uniformly on compact sets in Ω to a holomorphic function, which we will denote by g_n. By construction, $g_n(a_i) \neq 0$ for all i, hence the set $Z_{g_n} \cap \Omega_i$ has dimension $< n$ if $\Omega_i \not\subset Z$. In other words, all n-dimensional irreducible components of the set Z_{g_n} belong to Z.

The remaining g_k will be constructed by induction with respect to decreasing k, with the condition that $g_k|_Z = 0$ and that all irreducible components of $Z_{g_n} \cap \cdots \cap Z_{g_k}$, of dimensions $\geqslant k$, belong to Z. Let $g_n,\ldots,g_{k+1}$ be the countable linear combinations of the f_α already constructed, and let A_i be the irreducible components of $Z_{g_n} \cap \cdots \cap Z_{g_{k+1}}$ not belonging to Z. Choose points

$a_i \in A_i \setminus A$ and construct, as above a holomorphic function g_k (as a countable linear combination of certain f_α) distinct from zero at all a_i, but vanishing on Z. By Lemma 5.5, g_k is the required function. By construction, the set of common zeros of $g_n, \ldots, g_0$ coincides with Z. ■

Now arises a natural question: Is it possible to at least locally define a pure p-dimensional analytic set on an n-dimensional complex manifold by $n-p$ holomorphic functions? In general, the answer is negative. An analytic subset A of pure codimension m on a complex manifold Ω is called a *complete intersection* if there are m functions $f_1, \ldots, f_m$, holomorphic on Ω, such that $A = \cap_1^m Z_{f_k}$ is the set of their common zeros. An analytic set A is called a *local complete intersection* if for each point $a \in A$ there is a neighborhood $U \ni a$ in Ω such that $A \cap U$ is a complete intersection in U. By Theorem 2.8, every analytic set of codimension 1 is a local complete intersection. A global complete intersection is sometimes more conveniently understood to be an analytic set of pure codimension m that can be represented as the intersection of m analytic sets of codimension 1 (hypersurfaces). In $\mathbb{C}^n$ and on complex manifolds for which the second Cousin problem is solvable these definitions coincide, but on, say, compact manifolds it is more natural to use the second definition. Even to give an example of an incomplete intersection is not simple. Up till now, e.g., it is unknown whether an arbitrary smooth curve (a pure 1-dimensional analytic subset without singularities) in $\mathbb{P}_3$ is a complete intersection or not. Complete intersections have a number of additional topological, geometric, and analytic properties; see [156], [194].

E x a m p l e. Two vectors $z, w \in \mathbb{C}^3$ are proportional if and only if the pair $(z,w) \in \mathbb{C}^6$ satisfies the equation $z \wedge w = 0$, or, in matrix notation,

$$\operatorname{rank} \begin{bmatrix} z_1 & z_2 & z_3 \\ w_1 & w_2 & w_3 \end{bmatrix} < 2.$$

The corresponding analytic set A in $\mathbb{C}^6$ is defined by the three functions $f_1 = z_1 w_2 - z_2 w_1$, $f_2 = z_1 w_3 - z_3 w_1$, and $f_3 = z_2 w_3 - z_3 w_2$. The first two define the set $A \cup \{z_1 = w_1 = 0\}$, and f_3 "cuts out" the extra zeros. The set A is pure 4-dimensional with as unique singular point the coordinate origin. It is not difficult to prove that A is locally and globally irreducible. But it is not at all easy to prove that A is not a complete intersection in any neighborhood of the coordinate origin in $\mathbb{C}^6$ (the corresponding 3-dimensional algebraic set in $\mathbb{P}_5$, which is irreducible and without singularities, is also not a global complete intersection).

The known proofs of this fact are far from being elementary (e.g., the short proof in [127] is based on delicate properties of cohomology with holomorphic coefficients).

5.8. A theorem on proper maps. In conclusion of this paragraph we prove Remmert's well-known theorem [100], [101] on the analyticity of the image of an analytic set under a proper holomorphic map. In its contents the theorem is close to §3, but for its proof we need the results of the preceding paragraphs. For terminology see p.3.8.

Theorem. *Let A be an analytic set on a complex manifold and let $f\colon A \to Y$ be a holomorphic map into another complex manifold that is, moreover, proper on A. Then $f(A)$ is an analytic subset in Y of dimension* $\dim f$.

■ The statement concerning $f(A)$ is local, hence we may assume that Y is a domain in $\mathbb{C}^m$ and that $\overline{f(A)}$ is a compact set. This (and the properness of f) imply that it suffices to investigate the case when A is irreducible. The Theorem will be proved by induction with respect to $p = \dim A$. For $p = 0$ the statement is trivial.

$p-1 \Rightarrow p$. By the induction hypothesis, $E_1 = f(\operatorname{sng} A)$ is an analytic subset in A of dimension $\dim f|_{\operatorname{sng} A} \leqslant \dim f =: q$ (cf. p.3.8).

Step 1. Assume that $\dim E_1 = q$ and put $A' = f^{-1}(E_1)$. Then A' is an analytic subset in A, consisting of full fibers of f on A, and $\dim f|_{A'} = \dim E_1 = q$. By the definition of $\dim f|_{A'}$ there is a point $a \in A'$ such that the set $f^{-1}(f(a)) \subset A'$ has dimension $\dim_a A' - q \leqslant p - q$. Since $p - q$ is the minimal dimension of the fibers of f on A (by the definition of $\dim f$), this implies that $\dim A' = p$, hence $A' = A$ by the uniqueness Theorem. Thus, in case $\dim E_1 = q$ the image of A coincides with that of $\operatorname{sng} A$ and, by induction, is an analytic subset in Y of dimension $\dim f|_{\operatorname{sng} A} = q = \dim f$.

Step 2. Now we assume that $\dim E_1 < q$. By p.3.8, the set of points $z \in \operatorname{reg} A$ at which $\operatorname{rank}_z f = q$ is open in $\operatorname{reg} A$, and its complement is an analytic subset $A'' \subset \operatorname{reg} A$ of dimension $< p$ ($\operatorname{reg} A$ is connected since A is irreducible). At the points of A'' the rank of f is strictly less than q, therefore, by Proposition 3.8, the set $E_2 = f(A'')$ in Y has Hausdorff dimension $\leqslant 2(q-1)$. The set $A'' \cup \operatorname{sng} A$ is closed in A, hence $E = E_1 \cup E_2 = f(A'' \cup \operatorname{sng} A)$ is a closed subset of Y of

Hausdorff dimension $\leqslant 2(q-1)$. Put $\tilde{E} = f^{-1}(E)$. Then $\tilde{E} = A' \cup A'' \cup [f^{-1}(E_2) \cap (\operatorname{reg} A \setminus A'')]$. The fibers of f on $\operatorname{reg} A \setminus A''$ have dimension $p - q$, hence the rank Theorem implies that the last term in the decomposition of $\tilde{E}$ has Hausdorff dimension $\leqslant 2(q-1) + 2(p-q) = (2p-1)$. Since the same is true for A' and A'', $\tilde{E}$ is a closed, nowhere dense subset of A, hence E belongs to the closure in Y of $f(A \setminus \tilde{E})$. The set $A \setminus \tilde{E}$ consists of the regular points of A at which the rank of f is equal to q. Since $\tilde{E}$ is closed and consists of the full fibers of f on A, the map $f\colon A \setminus \tilde{E} \to Y \setminus E$ is proper, but in this case it is already a map between complex manifolds, moreover of constant rank.

Step 3. Fix an arbitrary point $w \in f(A \setminus \tilde{E})$. Since $f|_{A \setminus \tilde{E}}$ is proper, $f^{-1}(w)$ is compact in $A \setminus \tilde{E}$. The rank Theorem implies that for each point $z \in f^{-1}(w)$ there is a neighborhood U_z in $A \setminus \tilde{E}$ such that $f(U_z)$ is a complex manifold in Y of dimension $\operatorname{rank}_z f = q$. Cover $f^{-1}(w)$ by a finite number of such U^j; then $U = \cup U^j$ is a neighborhood of $f^{-1}(w)$ in $A \setminus \tilde{E}$, hence $(A \setminus \tilde{E}) \setminus U$ is closed. Since f is proper, the image of this set is closed also; since this image does not contain the point w, its complement is a neighborhood of w. Thus, in a neighborhood of w the set $f(A \setminus \tilde{E})$, and hence $f(A)$, coincides with the finite union of the q-dimensional complex manifolds $f(U^j)$; in particular, it is a q-dimensional analytic set. Thus we have proved that $f(A \setminus \tilde{E})$ is a pure q-dimensional analytic subset in $Y \setminus E$ (it is closed in $Y \setminus E$, since $f|_{A \setminus \tilde{E}}$ is proper). Shiffman's Theorem (p.4.4) implies that the closure of $f(A \setminus \tilde{E})$ in Y is a pure q-dimensional analytic subset in Y. Since all points of E are limit points for $f(A \setminus \tilde{E})$, this closure coincides with $f(A)$. ■

Remmert's theorem is true for proper maps and more general objects than analytic sets, e.g. for holomorphic, and even for c-holomorphic, maps between reduced complex spaces (analytic spaces in the terminology of Gunning and Rossi [35]). The proof for this general situation does not differ from the one give above (cf., e.g., [35]). Analyticity of sets is also preserved by semiproper holomorphic maps (a map $f\colon A \to Y$ is called semiproper if for every compact set $Q \subset Y$ there is a compact set $K \subset A$ such that $f(A) \cap Q = f(K)$); more details can be found in [70], [71], [150].

6. One-dimensional analytic sets

6.1. Local parametrization. One-dimensional analytic sets have an essentially simpler structure than analytic sets of higher dimensions. We begin with the local structure.

Let A be a pure one-dimensional analytic set in $\mathbf{C}^n$, 0 a singular point of it, and $U = U_1 \times U'' \ni 0$ a polydisk neighborhood such that $\pi: A \cap U \to U_1 \subset \mathbf{C}_1$ is a proper projection. Then this is an analytic cover with zero-dimensional critical analytic set $\sigma \subset U_1$. By shrinking U_1 and U'' we may assume that $\sigma = \{0\}$ and that the fiber $\pi^{-1}(0) \cap A \cap U$ consists also of the single point 0. Decompose $A \cap U$ into irreducible components. Some of these will not have a singularity at 0, others will, like $A \cap U$, have 0 as unique singular point. "Intersection" between components is not of interest to us now, and we will here investigate the irreducible situation. So, let there be given an irreducible analytic subset $S \subset U$ such that the projection $\pi: S \setminus \{0\} \to U_1 \setminus \{0\}$ is a locally biholomorphic k-sheeted cover above a punctured disk $0 < |z_1| < r_1$, and $\pi^{-1}(0) \cap S = \{0\}$. Fix a point $a \in S$ such that $a_1 = |a_1| > 0$, and a path $\gamma: [0,r_1) \to S$ passing through a and such that $\pi \circ \gamma(r) = r$. Sine $\pi: S \setminus \{0\} \to U_1 \setminus \{0\}$ is a cover, such a path obviously does exist. Since S is irreducible, $S \setminus \{0\}$ is connected. The union over all $r \in (0,r_1)$ of the connected components of the set $\pi^{-1}(\{|z_1| = r\}) \cap S$ that intersect the arc $\gamma((0,r_1))$ is simultaneously an open and a closed subset of $S \setminus \{0\}$, hence coincides with it. For each $r \in (0,r_1)$ the set $\Gamma_r = \pi^{-1}(\{|z_1| = r\}) \cap S$ is a closed Jordan curve, k-sheetedly covering the circle $|z_1| = r$; the argument z_1 is a local parameter on it (Figure 5).

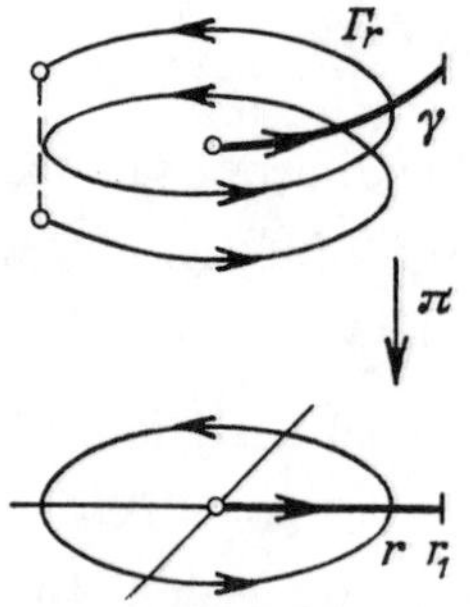

Figure 5.

Under a complete walk over this curve in the positive direction (with growth of the parameter), the increase in the argument z_1 is $2\pi \cdot k$, hence there is a unique parametrization γ_r: $[0,2\pi) \to \Gamma_r$ such that $\pi \circ \gamma_r(t) = re^{itk}$ and $\gamma_r(0) = \gamma(r)$. We define on S a continuous function $\zeta(z)$ by putting $\zeta(\gamma_r(t)) = (r/r_1)^{1/k} e^{it}$. It is clear that $\zeta(z)$ is equal to $(z_1/r_1)^{1/k}$ on $S \setminus \{0\}$ and therefore it is holomorphic on the manifold $S \setminus \{0\}$. By construction, the function ζ takes different values at distinct points of S, hence it realizes a one-to-one map of S onto the unit disk $|\zeta| < 1$ which is holomorphic on $S \setminus \{0\}$. We will call the function ζ a *local (normalizing) parameter* on S. Note that $\zeta(z)$ is not holomorphic on all of S, since 0 is a singular point. (E.g., for the semicubic parabola this function $\zeta(z) = z_2/z_1$ is only meromorphic in a neighborhood of S.) By Riemann's theorem on removable singularities the inverse map $z = z(\zeta)$ is holomorphic in the whole disk $|\zeta| < 1$; in particular, its components can be expanded in Taylor series: $z_j(\zeta) = \sum_1^\infty a_{j\nu}\zeta^\nu$. Thus, S is the holomorphic image of the unit disk under the map $\zeta \mapsto z(\zeta) = (z_1(\zeta), \ldots, z_n(\zeta))$; moreover, by construction $\mathrm{ord}_0 z_1(\zeta)$ is the number of sheets of the cover π: $S \to U_1$.

The parametric representation of S constructed above is sometimes written as $z_j = \sum_\nu a_{j\nu} z_1^{\nu/k}$, $j = 2, \ldots, n$. These series in fractional powers are called Puiseux series, hence the parametrization of S here described may be called a Puiseux parametrization.

So, at points of irreducibility a one-dimensional analytic set has the structure of a holomorphic disk; in particular, it is a topological manifold. At singular points it can not be smooth (C^1), according to the Levi-Civita theorem (A2.3). It does have some smoothness: if $\zeta(z)$ is the above-described holomorphic parameter in a neighborhood of 0 on S, then the function $\zeta(z) = (z_1/r_1)^{1/k}$ inverse to $z(\zeta)$

on S obviously satisfies on S the Hölder condition of order $1/k$, where k is the multiplicity of the cover S above $\mathbb{C}_1$ at 0. There is nothing similar for analytic sets of higher dimensions. In general, such an analytic set is not even a topological manifold in a neighborhood of a singular point of irreducibility.

Examples. (a) The analytic set $A: z_1^2 = z_3 z_2^2$ in $\mathbb{C}^3$ (cf. p.5.3) is irreducible at 0, but is reducible at all points $(0,0,z_3)$, $z_3 \neq 0$, along which it is the union of the two complex manifolds $z_1 = \pm\sqrt{z_3}\cdot z_2$. Hence locally it is not a topological manifold. (In this case the property of local reducibility wins.)

(b) A more delicate example is furnished by the analytic set $A: z_3^2 = z_1 z_2$ in $\mathbb{C}^3$, which is equivalent to the standard quadric $z_1^2 + z_2^2 + z_3^2 = 0$. It is locally and globally irreducible and has unique singular point 0. The set of regular points, $A \setminus \{0\}$, is the image of $\mathbb{C}^2 \setminus \{0\}$ under the map $(\zeta_1, \zeta_2) \mapsto (\zeta_1^2, \zeta_2^2, \zeta_1\zeta_2)$. Obviously, this is a locally biholomorphic two-sheeted cover $\mathbb{C}^2 \setminus \{0\} \to A \setminus \{0\}$, hence $A \setminus \{0\}$ cannot be simply connected, since there are no connected, one-sheeted covers above a simply connected manifold (the monodromy Theorem). The same is true in an arbitrary neighborhood $U \ni 0$: after removal of 0 the set $A \cap U$ becomes nonsimply connected. For this reason $A \cap U$ cannot be homeomorphic to a ball in $\mathbb{R}^4$, hence A is not a topological manifold in any neighborhood of 0, irrespective of its irreducibility at 0 and its regularity at all other points. In $\mathbb{P}_2$ the cone $z_3^2 = z_1 z_2$ corresponds to a smooth holomorphic curve (a one-dimensional analytic set without singularities), homeormophic to a torus, and in particular, nonsimply connected.

6.2. Normalization and uniformization. Apart from points of irreducibility, a pure one-dimensional analytic set A can also have points of "selfintersection". If such singularities are first "unglued" and the irreducible singularities are then "untwisted", as the result we obtain a one-dimensional complex manifold that almost bijectively parametrizes the initial analytic set. This process of removing one-dimensional singularities will be called normalization of the initial one-dimensional set. In concreto it is performed as follows.

The singular points $a_j \in A$ form in A a discrete subset. Construct pairwise disjoint coordinate neighborhoods $U_j \ni a_j$ (in the ambient manifold Ω) such that $A \cap U_j$ consists of a finite number of irreducible components $S_{j\nu}$, each of which can be holomorphically parametrized by a local normalizing parameter $\zeta_{j\nu}: S_{j\nu} \to \{|\zeta| < 1\}$ (the coordinates are chosen such that the projection of

$A \cap U_j$ to the $\mathbb{C}_1$-axis is proper in a neighborhood of a_j, subsequently we shrink U_j, as in p.6.1). Let $V_j \subset\subset U_j$ be a smaller neighborhood of a_j. Construct the abstract set $A^\star = (A \setminus \cup_i \overline{V}_i) \cup (\bigcup_{j,\nu} S_{j\nu})$, where $\bigcup$ is the disjunctive union (points of distinct $S_{j\nu}$ are regarded as distinct). It is obtained from A by "ungluing" the points of selfintersection (reducibility). Each point $z \in \operatorname{reg} A$ (see Figure 6) corresponds to exactly one point of $A^\star$, while each singular point a_j corresponds to as many points of $A^\star$ as there are irreducible components at it in the analytic set $A \cap U_j$. The inverse correspondence is determined by the natural projection $A^\star \to A$, which is bijective above reg A.

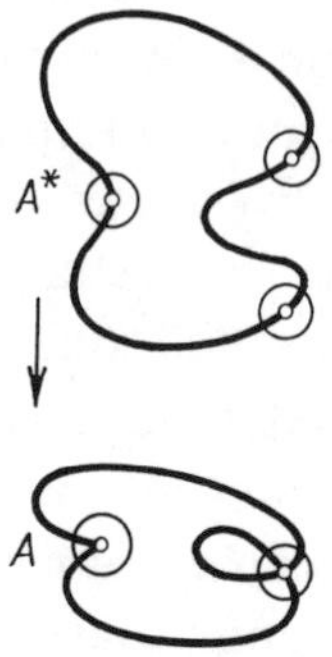

Figure 6.

The local holomorphic coordinates at regular points and the local normalizing parameters $\zeta_{j\nu}$ in neighborhoods of points $a_j \in \operatorname{sng} A$ define complex charts on the set $A^\star$. The charts on $A \setminus \cup \overline{V}_j$, as can easily be seen, are holomorphically compatible with all $(S_{j\nu}, \zeta_{j\nu})$, therefore all charts together define a complex structure on $A^\star$. The set $A^\star$ endowed with this complex structure is a one-dimensional complex manifold, called the normalization of the one-dimensional analytic set A. The projection $A^\star \to A$ is holomorphic with respect to this structure on $A^\star$, biholomorphic above reg A, and finite and proper in the large. Summed up:

P r o p o s i t i o n. *For any pure one-dimensional analytic set A there is a one-dimensional complex manifold $A^\star$ (the normalization of A) and a finite, proper, holomorphic map $A^\star \to A$ (projection), biholomorphic above* reg A.

So, a one-dimensional analytic set is a "slightly spoiled" one-dimensional complex manifold, "twisting" and selfintersecting on a finite set of points. If A is irreducible, A^* is connected, i.e. A^* is a Riemann surface. Thus we may say that an irreducible one-dimensional analytic set can be globally parametrized (up to inessential selfintersections) by the points of some Riemann surface. More on Riemann surfaces can be found in, e.g., [131], [155].

For an arbitrary complex manifold M there is uniquely defined (up to biholomorphisms) the so-called universal covering (manifold). This is a simply connected complex manifold $\tilde{M}$ together with a map (projection) $\pi: \tilde{M} \to M$ which is a local biholomorphic cover. $\tilde{M}$ is constructed is a standard manner, using the fundamental group of M (see, e.g., [155]). In case M is a Riemann surface, the universal covering manifold is one of the three manifolds: the unit disk, the complex plane, or the Riemann sphere (more details on universal covering manifolds are in [155]). Each function $f \in \mathcal{O}(M)$ can be lifted to a holomorphic function $f \circ \pi$ on $\tilde{M}$, hence if there are nonconstant bounded holomorphic functions on the Riemann surface M, its universal covering manifold must be the unit disk $\Delta \subset \mathbb{C}$. Turning to an irreducible one-dimensional analytic set A, we obtain a construct $A \to A^* \to \tilde{A}^*$ (from the set to its normalization, from the latter to the universal covering manifold) and corresponding projections $\tilde{A}^* \to A^* \to A$. The universal covering manifold of A^* will be called the *universal covering manifold* of the set A. Thus, if A is an irreducible one-dimensional analytic subset of a bounded domain in $\mathbb{C}^n$, then its universal covering manifold is the unit disk Δ; in particular, there is a holomorphic map $\pi: \Delta \to A$, $\pi(\Delta) = A$, that is locally biholomorphic above reg A.

6.3. Maximum principle. Related to the simple structure of one-dimensional analytic sets there is a natural desire to have a sufficiently large amount of such subsets on an analytic set of arbitrary dimension. Locally there are in fact a lot of such sets. Let A be a pure p-dimensional analytic set in $\mathbb{C}^n$, $0 \in A$, and suppose that coordinates have been chosen such that in some neighborhood $U = U' \times U'' \ni 0$ the projection $\pi: A \cap U \to U' \subset \mathbb{C}^p$ is proper; moreover, let $\pi^{-1}(0) \cap A \cap U = \{0\}$. For any complex line $L' \ni 0$ in $\mathbb{C}^p$ the analytic set $\pi^{-1}(L') \cap A \cap U$ is pure one-dimensional and is an analytic cover above $L' \cap U'$. If S is an irreducible component of this analytic set, then $\pi: S \to L' \cap U'$ is also an analytic cover. Since $\pi^{-1}(0) \cap A \cap U = \{0\}$, we have $0 \in S$. Thus we have obtained the following simple statement:

L e m m a. *Let A be an arbitrary analytic set and suppose* $\dim_a A \geq 1$. *Then there is a neighborhood $U \ni a$ such that $A \cap U$ can be represented as a union of irreducible one-dimensional analytic sets containing a.*

In general, these analytic sets are not holomorphic disks and can have common points other than a. The problem of the existence of one-dimensional layers under the condition that the layers intersect only at a and do not have singularities other than a has not been solved (cf. [148] questions 3.9 and 9.4). We use the Lemma in a proof of a maximum principle on analytic sets.

Let A be a pure one-dimensional analytic set and $\pi: A^* \to A$ its normalization. A function $u: A \to [-\infty, \infty)$ is called weakly subharmonic on A if $U \circ \pi$ is a subharmonic function on the complex manifold A^*. Let now A be an arbitrary analytic set. A function $u: A \to [-\infty, \infty)$ is called *weakly plurisubharmonic* on A if it is upper semicontinuous on A, if the set $\{z \in A : u(z) = -\infty\}$ is nowhere dense on A, and if for any irreducible one-dimensional analytic subset $A' \subset A$ the restriction $u|_{A'}$ is weakly subharmonic or $\equiv -\infty$. Such are, e.g., functions of the form $\tilde{u}|_A$ where $\tilde{u}$ is a plurisubharmonic function in a neighborhood of A on the ambient manifold Ω such that $\{\tilde{u} = -\infty\} \cap A$ is nowhere dense in A (this easily follows from the local parametrization of one-dimensional analytic sets and the preservation of plurisubharmonicity by holomorphic maps).

T h e o r e m (maximum principle). *Let A be a connected analytic set and u a weakly plurisubharmonic function on A. If $u(a) = \max_A u$ at some point $a \in A$, then $u \equiv u(a)$ on A.*

■ Let $E = \{z \in A: u(z) = u(a)\}$. Since $u(a)$ is the maximal value and u is upper semicontinuous, E is closed in A. On the other hand, if $b \in E$ then the Lemma implies that there is a neighborhood $U \ni b$ such that $A \cap U$ can be represented as a union of irreducible one-dimensional analytic sets $S \subset A$ containing b. If $\pi: S^* \to S$ is the normalization of S, then $u \circ \pi$ is a subharmonic function on the manifold S^* and it attains its maximum at the points of the set $\pi^{-1}(b) \subset S^*$. Since S^* is connected (by the irreducibility of S), the maximum principle on complex manifolds (which can be proved as for domains in $\mathbb{C}$) implies that $u \circ \pi \equiv$ const on S^*, hence $u \equiv$ const $= u(a)$ on S. Since $A \cap U$ is the union of such S, we have $u = u(a)$ on $A \cap U$, hence E is open in A. Since A is connected, $E = A$. ■

C o r o l l a r y. *Let A be an analytic set with compact closure $\overline{A}$ on a complex manifold, ϕ an upper bounded weakly plurisubharmonic function on A, and let the set $E \subset \partial A = \overline{A} \setminus A$ consist of those ζ at which $\lim_{z \to \zeta, z \in A} \phi(z) = -\infty$. Let u be an upper bounded weakly plurisubharmonic function on A. Then if $\overline{\lim}_{z \to \zeta, z \in A} u(z) \leqslant m$ at all $\zeta \in (\partial A) \setminus E$, then $u(z) \leqslant m$ on A.*

■ If $\phi \leqslant M$ on A and $\epsilon > 0$, then $u + \epsilon\phi \leqslant m + \epsilon M + \epsilon$ in a sufficiently small neighborhood of ∂A (this obviously follows from the definition of lim sup, the boundedness of u, and the definition of E). According to the Theorem, $u \leqslant m + \epsilon(M + 1 - \phi)$ everywhere on A. As $\epsilon \to 0$ we obtain in the limit $u \leqslant m$. ■

7. Algebraic sets

7.1. Chow's theorem. A set in $\mathbb{C}^n$ of the form $\{\zeta: p_1(\zeta) = \cdots = p_k(\zeta) = 0\}$ where p_j are polynomials in $(\zeta_1, \ldots, \zeta_n)$ is called an *affine algebraic set*. If $p(z_0, \ldots, z_n)$ is a homogeneous polynomial in $\mathbb{C}^{n+1}$ and if $z \sim w$ in $\mathbb{C}^{n+1}_*$, then $p(z) = 0$ if and only if $p(w) = 0$. Therefore, for points $[z] \in \mathbb{P}_n$ the property $p(z) = 0$ does not depend on the choice of homogeneous coordinates for $[z]$, and the notion of zeros of homogeneous polynomials in $\mathbb{P}_n$ makes sense. A set in $\mathbb{P}_n$ of the form $\{[z]: p_1(z) = \cdots = p_k(z) = 0\}$ where p_j are homogeneous polynomials in homogeneous coordinates is called a *projective algebraic set*. Both projective and affine algebraic sets are often simply called algebraic.

Consider the intersection of a projective algebraic set with $\mathbb{C}^n = \mathbb{P}_n \setminus H_0$. Suppose the set is defined by polynomials p_j. Then in $\mathbb{C}^n$ the equation $p_j(z) = 0$ is equivalent to $p_j(1, z_1/z_0, \ldots, z_n/z_0) = 0$, since $p_j(z) = z_0^{\deg p_j} p_j(1, z_1/z_0, \ldots, z_n/z_0)$. Thus, in $\mathbb{C}^n$ we obtain an affine algebraic set. Conversely, any polynomial in $(\zeta_1, \ldots, \zeta_n)$ can be obtained in this way: if we put $p^*(z_0, 'z) := z_0^s p('z/z_0)$, where s is the degree of p, then p^* is a homogeneous polynomial in $(z_0, \cdots, z_n)$; moreover, $p^*(1, \zeta) = p(\zeta)$. The polynomial p^* is called the *projectivization* (homogeneization) of p. Thus, every affine algebraic set can be represented as the intersection with $\mathbb{C}^n$ of some projective algebraic set, and any projective algebraic set is a locally affine, in particular an analytic, subset in $\mathbb{P}_n$. The local property of being analytic in $\mathbb{P}_n$ turns out, unexpectedly, to be equivalent to the global property of being algebraic - this is the assertion of a remarkable theorem of Chow.

T h e o r e m. *Every analytic subset A of $\mathbb{P}_n$ is a projective algebraic set.*

■ Let A be a pure p-dimensional analytic subset in $\mathbb{P}_n$ and let $\Pi\colon \mathbb{C}_*^{n+1} \to \mathbb{P}_n$ be the canonical projection. Then $\Pi^{-1}(A)$ is a pure $(p+1)$-dimensional analytic subset in $\mathbb{C}_*^{n+1}$ (p.2.5). The Remmert-Stein theorem (p.4.4) implies that the closure $\tilde{A} = \Pi^{-1}(A) \cup \{0\}$ of this set is an analytic subset in $\mathbb{C}^{n+1}$. Hence there exist functions $g_1, \ldots, g_k$, holomorphic in a ball with center at 0, such that $\tilde{A} \cap U$ is the set of their common zeros in U. Let $g_j = \sum_1^\infty (g_j)_m$ be the series expansions in homogeneous polynomials in z. If $z \in \tilde{A} \cap U$, then $tz \in \tilde{A} \cap U$ for all $t \in \mathbb{C}$, $|t| < 1$. Hence $g_j(tz) = \sum_1^\infty (g_j)_m(z) t^m \stackrel{t}{\equiv} 0$, implying that all $(g_j)_m(z) = 0$. Conversely, if all $(g_j)_m \equiv 0$ at a point $z \in \tilde{A} \cap U$ then all $g_j(z) = 0$, implying $z \in A \cap U$. Thus $A \cap U$ is the set of common zeros of the countable system of polynomials $(g_j)_m$, $j = 1, \ldots, k$; $m = 1, 2, \ldots$. Let $V \subset\subset U$ be a smaller neighborhood of 0. By p.5.7, the analytic set $\tilde{A} \cap V$ is defined by a finite tuple of these polynomials, which we will denote by $p_1, \ldots, p_N$. Since $\tilde{A}$ is a cone (i.e. $t\tilde{A} \subset \tilde{A}$ for all $t \in \mathbb{C}$), these same polynomials define all of $\tilde{A}$. Thus, $A = \{[z]\colon p_1(z) = \cdots = p_N(z) = 0\}$ is a projective algebraic set. ■

Chow's theorem has many beautiful corollaries, and is the basis for applying analytic methods in algebraic geometry.

R e m a r k. In the course of the proof we have proved the following:

If A is an analytic subset in $\mathbb{C}^n$ such that $tz \in A$ for all $z \in A$ and all $t \in \mathbb{C}$ (i.e. A is a cone with vertex at 0), then A is an affine algebraic set.

7.2. Closure of affine algebraic sets.

P r o p o s i t i o n 1. *The closure in $\mathbb{P}_n$ of an arbitrary algebraic subset $A \subset \mathbb{C}^n$ is a projective algebraic set.*

■ Let A be a pure p-dimensional set of common zeros of polynomials $p_1, \ldots, p_N$, and let A^- be the set of common zeros in $\mathbb{P}_n$ of the homogeneous polynomials p_j^* (the projectivizations of p_j), $j = 1, \ldots, N$. Since in $\mathbb{C}^n = \mathbb{P}_n \setminus H_0$ the zeros of p_j and p_j^* coincide, $A = A^- \setminus (A^- \cap H_0)$, hence, according to Corollary 5.1, the closure of A in $\mathbb{P}_n$ is an analytic subset in $\mathbb{P}_n$.

Chow's theorem implies that it is algebraic. ■

In a particular case this statement can be refined.

P r o p o s i t i o n 2. *The closure in* $\mathbb{P}_n$ *of an affine algebraic set* $A = \{\zeta \in \mathbb{C}^n : p(\zeta) = 0\}$, *where* p *is a polynomial of degree* s, *coincides with the projective algebraic set* $\{[z] \in \mathbb{P}_n : p^*(z) = 0\}$, *where* p^* *is the projectivization of* p. *In particular, the points at infinity of* $\bar{A}$ *form the set* $\{[0,'z] : (p)_s('z) = 0\}$, *where* $(p)_s$ *is the homogeneous polynomial composed of the monomials in* p *of maximum degree* s.

We stress that in this case A is defined by one equation, and that there is no such simple description of $\bar{A}$ in the general case.

■ The set $\{p^* = 0\}$ is closed in $\mathbb{P}_n$, hence contains $\bar{A}$. Since $p^*(z) = (p)_s('z) + z_0 q(z)$, where $q(z)$ is also a homogeneous polynomial, $p^*|_{H_0} \not\equiv 0$, hence the set $\{p^* = 0\} \cap H_0$ is pure $(n-2)$-dimensional (p.2.6). Since $\{p^* = 0\}$ is pure $(n-1)$-dimensional, in a neighborhood of each point $[0,'z] \in \{p^* = 0\} \cap H_0$ there are points of the set $\{p^* = 0\} \cap \mathbb{C}^n = A$, hence $\bar{A} \supset \{p^* = 0\}$. ■

7.3. Algebraic sets as analytic covers. First we prove the following simple statement.

L e m m a. *If* A *is a pure* p*-dimensional projective algebraic set in* $\mathbb{P}_n$, *then there is an* $(n-p-1)$*-dimensional complex plane* $L \subset \mathbb{P}_n$ *not intersecting* A.

■ Let $\tilde{A} = \Pi^{-1}(A) \cup \{0\} \subset \mathbb{C}^{n+1}$, where $\Pi: \mathbb{C}^{n+1}_* \to \mathbb{P}_n$ is the canonical projection. By the proof of Chow's theorem (p.7.1), $\tilde{A}$ is a pure $(p+1)$-dimensional analytic subset in $\mathbb{C}^{n+1}$. By p.3.5 there is a plane $\tilde{L} \subset \mathbb{C}^{n+1}$ of dimension $n-p$ such that 0 is an isolated point of the set $\tilde{A} \cap \tilde{L}$. Since $\tilde{A}$ and $\tilde{L}$ are cones, this implies that $\tilde{A} \cap \tilde{L} = \{0\}$, hence the set $A = \Pi(\tilde{A} \setminus \{0\})$ and the $(n-p-1)$-dimensional plane $L = \Pi(\tilde{L} \setminus \{0\})$ in $\mathbb{P}_n$ have no common points. ■

We now note that to each $(n-p-1)$-dimensional plane $L \subset \mathbb{P}_n$ is related a projection $\pi_L: \mathbb{P}_n \setminus L \to L^\perp$, corresponding to orthogonal projection onto $(\tilde{L})^\perp$ in $\mathbb{C}^{n+1}$ (here $L^\perp = \Pi(\tilde{L}^\perp)$ is the p-dimensional plane in $\mathbb{P}_n$ consisting of the points that are located diametrically opposite with respect to L), see A3.3. If A is a projective algebraic set in $\mathbb{P}_n$ not intersecting L, then the restriction $\pi_L|_A$ clearly is a proper holomorphic map (Figure 7). Let $\sigma \subset L^\perp$ be the critical set of

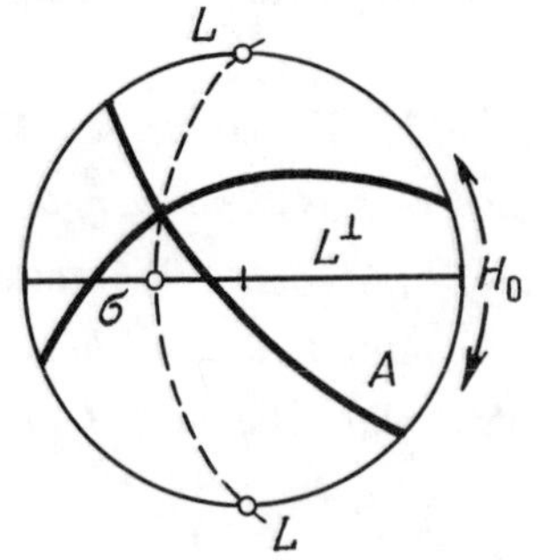

Figure 7.

this map, i.e. $\sigma = \pi_L(\mathrm{br}\, \pi_L|_A)$. If Λ is an arbitrary complex hyperplane in $\mathbb{P}_n$ containing L, then $\pi_L\colon A\setminus\Lambda \to L^{\perp}\setminus\Lambda$ is also a proper map. Since $L^{\perp}\setminus\Lambda$ is unitarily equivalent to $\mathbb{C}^p = \mathbb{P}_p\setminus H_0 \subset \mathbb{P}_n$, by p.4.3 the map $\pi_L\colon A\setminus\Lambda \to L^{\perp}\setminus\Lambda$ is an analytic cover with critical analytic set equal to $\sigma\setminus\Lambda$. Since Λ was arbitrary, this means that σ is an analytic subset of $L^{\perp}$ (algebraic by Chow's theorem); hence $\pi_L\colon A \to L^{\perp}$ is an analytic cover. Summing up:

P r o p o s i t i o n 1. *Let A be a pure p-dimensional projective algebraic set in $\mathbb{P}_n$ and let $L \subset \mathbb{P}_n$ be a complex $(n-p-1)$-dimensional plane not intersecting A (according to the Lemma, such L exist). Then the projection $\pi_L\colon A \to L^{\perp} \approx$ Ppubn is an analytic cover, and its critical set $\sigma \subset L^{\perp}$ is algebraic.*

If $L \subset H_0 \subset \mathbb{P}_n$ and L_1 is an $(n-p)$-dimensional plane containing L and the point $[1,0,\ldots,0]$ (the coordinate origin in $\mathbb{C}^n = \mathbb{P}_n\setminus H_0$), then in $\mathbb{C}^n$ the projection π_L is orthogonal projection onto $(L_1)^{\perp}$. If now A is an affine algebraic set in $\mathbb{C}^n$ whose closure in $\mathbb{P}_n$ does not intersect $L \subset H_0$, then, taking $\Lambda = H_0$ (and $\overline{A}$ instead of A) in the preceding reasoning, we obtain that $\pi_L\colon A \to L^{\perp} \cap \mathbb{C}^n$ is an analytic cover. Since $\overline{A} \cap H_0$ has dimension $p-1$ and $H_0 \approx \mathbb{P}_{n-1}$, such $(n-p-1)$-dimensional planes $L \subset H_0$ exist, by the Lemma. Thus the following Proposition holds:

P r o p o s i t i o n 2. *Let A be a pure p-dimensional affine algebraic set in $\mathbb{C}^n$. Then there is in $\mathbb{C}^n$ a p-dimensional subspace C such that orthogonal projection $\pi\colon A \to C$ is an analytic cover; moreover, this cover can be continued to an analytic cover of the closures $\overline{A} \to \overline{C}$. (For C we may take $L^{\perp} \cap \mathbb{C}^n$, where $L \subset H_0$ is an*

$(n-p-1)$-dimensional plane not intersecting $\overline{A}$.)

A more general statement can be found in p.17.2.

7.4. Some criteria for being algebraic. What conditions on an analytic subset $A \subset \mathbb{C}^n$ would guarantee that it is algebraic? The following geometric property of algebraic sets is very useful for answering this question (see, e.g., [42]).

Theorem 1. *A pure p-dimensional analytic subset $A \subset \mathbb{C}^n$ is algebraic if and only if its closure in $\mathbb{P}_n$ does not intersect some $(n-p-1)$-dimensional complex plane L located in the hyperplane at infinity $H_0 = \mathbb{P}_n \setminus \mathbb{C}^n$.*

■ Necessity was proved in Lemma 7.3. We prove sufficiency. Let $L \subset H_0$ be an $(n-p-1)$-dimensional complex plane not intersecting $\overline{A}$. Then projection $\pi_L: \overline{A} \to L^{\perp}$ is a proper map, and points at infinity (with respect to $\mathbb{C}^n$) are projected to points at infinity. Since $\dim L^{\perp} = p$ and $L^{\perp} \not\subset H_0$, we have $\dim L^{\perp} \cap H_0 = p-1$, thus $\pi_L(\overline{A} \setminus A) \subset H_0$ has Hausdorff dimension at most $2(p-1)$. Let Λ be an arbitrary complex hyperplane in $\mathbb{P}_n$ containing L. Then $\pi_L: \overline{A} \setminus \Lambda \to L^{\perp} \setminus \Lambda$ is also a proper holomorphic map of sets in $\mathbb{P}_n \setminus \Lambda \approx \mathbb{C}^n$. Since $\pi_L(\overline{A} \setminus A)$ is locally a removable subset in $L^{\perp}$ (see p.4.1 and A1.4), $\pi_L: \overline{A} \setminus \Lambda \to L^{\perp} \setminus \Lambda$ is a generalized analytic cover. By Theorem 4.3, $\overline{A} \setminus \Lambda$ is an analytic subset in $\mathbb{P}_n \setminus \Lambda$. Since Λ is arbitrary we obtain that $\overline{A}$ is an analytic subset in $\mathbb{P}_n$, hence, by Chow's theorem, $\overline{A}$ is a projective algebraic set, and A is affine. ■

For $p = n-1$ the plane L degenerates to a point and Theorem 1 can be regarded as a geometric analog of Sokhotskiĭ's theorem on essential singularities: either A is an algebraic set or the set of limit points of A at infinity contains the whole hyperplane at infinity H_0 (transcendency of analytic sets is analogous to essential singularities of holomorphic functions). The case $p < n-1$ can be regarded as a generalization of this "geometric Sokhotskiĭ theorem" to higher codimensions.

Theorem 1 is equivalent to the following criterion, which is formulated in terms relating to $\mathbb{C}^n$ only.

Theorem 2. *A pure p-dimensional analytic subset $A \subset \mathbb{C}^n$ is algebraic if and only if it is contained, after some unitary change of coordinates, in the union of a ball*

B: $|z| < R$ *and a cone K*: $|z''| < C|z'|$, *where* $z = (z', z'')$, $z' = (z_1, \ldots, z_p)$.

■ The limit points of the cone K in H_0 form the set $H_0 \cap \{|z''| \leqslant C|z'|\}$, which does not intersect the complex plane $L = \{z' = 0\} \cap H_0$ of dimension $n-p-1$. In other words, the exterior of $\overline{B} \cup \overline{K}$ is a neighborhood of L in $\mathbb{P}_n$. On the other hand, any neighborhood of L in $\mathbb{P}_n$ obviously contains the complement of some set of the form $\overline{B} \cup \overline{K}$. Therefore the statement of the Theorem follows from Theorem 1. ■

The following criterion of Rudin (cf. [116]) is, like Theorem 2, in statement analogous to Liouville's theorem.

Theorem 3. *A pure p-dimensional analytic subset* $A \subset \mathbb{C}^n$ *is algebraic if and only if it is contained, after some unitary change of coordinates, in a domain* D: $|z''| < C(1+|z'|)^s$, *where* $z = (z', z'')$, $z' = (z_1, \ldots, z_p)$, *and* C, s *are certain constants.*

■ If A is algebraic then, by Theorem 2, A is contained in a unitary image of such a domain D with $s = 1$ and a certain constant $C > 0$. Conversely, let $A \subset D$; then the projection $\pi: A \to \mathbb{C}_{1 \ldots p}$ is proper, hence A is a k-sheeted analytic cover above $\mathbb{C}_{1 \ldots p}$. By Theorem 4.3, A is the set of common zeros of canonically defining functions $\Phi_I(z', z'')$, $|I| = k$, which are holomorphic in $\mathbb{C}^n$ and are polynomials in z'', $\Phi_I = \sum_{|J| \leqslant k} \phi_{IJ}(z')(z'')^J$. Since $A \subset D$, Viète's generalized formula (cf. p.4.2 and p.4.3) implies that $|\phi_{IJ}(z')| \leqslant C(1+|z'|)^{ks}$ for all $z' \in \mathbb{C}_{1 \ldots p}$. Liouville's theorem (A1.1) implies that the holomorphic functions ϕ_{IJ} are also polynomials. Hence the Φ_I are polynomials in z, and A is an algebraic set. ■

Further properties of algebraic sets, including some other criteria for being algebraic, can be found in p.11.3 and p.17.2.

Chapter 2

TANGENT CONES AND INTERSECTION THEORY

8. Tangent cones

8.1. Definitions and simplest properties. Let E be an arbitrary set in $\mathbb{R}^N$. A vector $v \in \mathbb{R}^N$ is called tangent to E at a point $a \in \overline{E}$ if there exist a sequence of points $a^j \in E$ and numbers $t_j > 0$ such that $a^j \to a$ and $t_j(a^j - a) \to v$ as $j \to \infty$. The set of all such tangent vectors is denoted by $C(E,a)$ and is called the *tangent cone* to E at a. This really is a cone with vertex 0 since if $v \in C(E,a)$, then tv also belongs to $C(E,a)$ for all $t \geqslant 0$. Geometrically the cone $C(E,a)$ is the set of limit positions of secants of E passing through a; it is the set of limit points of the family of sets $t(E-a) = \{t(x-a):x \in E\}$ as $t \to \infty$. If $a \notin \overline{E}$ then, by definition, the set $C(E,a)$ is empty.

We note some obvious properties of tangent cones:

1) monotonicity: if $E_1 \subset E_2$, then $C(E_1,a) \subset C(E_2,a)$;

2) additivity: $C(E_1 \cup E_2,a) = C(E_1,a) \cup C(E_2,a)$;

3) $C(E_1 \cap E_2,a) \subset C(E_1,a) \cap C(E_2,a)$;

4) $C(E_1 \times E_2,(a_1,a_2)) = C(E_1,a_1) \times C(E_2,a_2)$;

5) if E is a manifold of class C^1 and $a \in E$, then $C(E,a)$ coincides with the tangent space to E at a.

Of interest to us are tangent cones to analytic sets in $\mathbb{C}^n \approx \mathbb{R}^{2n}$; to some extent they must replace tangent planes at the most interesting points of an analytic set, its singularities. An almost ideal situation is obtained in case the analytic set is one-dimensional.

P r o p o s i t i o n. *The tangent vectors at 0 to an analytic set $A \subset \mathbb{C}^n$ which is irreducible and one-dimensional at 0, form a complex line. If this line is the $\mathbb{C}_1$-axis, there are a neighborhood $U = U' \times U'' \ni 0$ and a holomorphic one-to-one map $z = z(\zeta)$ of the unit disk $|\zeta| < 1$ in $\mathbb{C}_\zeta$ onto $A \cap U$, such that $z_1(\zeta) = c\zeta^k$, where k is the number of sheets of $A \cap U$ above U_1; moreover,* $\operatorname{ord}_0 z_j(\zeta) > k$ *for all $j > 1$.*

■ Let v be some unit tangent vector to A at 0. By a unitary change of coordinates, $v = (1,0,\ldots,0)$. Since A is irreducible at 0, $A \cap \mathbb{C}_{2\ldots n}$ is zero-dimensional at 0, hence there is a neighborhood $U = U_1 \times U'' \ni 0$ such that the projection $\pi: A \cap U \to U_1$ is an analytic cover; moreover $\pi^{-1}(0) \cap A \cap U = \{0\}$. By p.6.1, there is a holomorphic one-to-one map $z = z(\zeta)$ of the unit disk $|\zeta| < 1$ onto $A \cap U$ such that $z_1(\zeta) = c\zeta^k$, where k is the multiplicity of the cover. By the definition of tangent vector there is a sequence $\zeta_i \to 0$ such that

$$\frac{z(\zeta_i)}{|z(\zeta_i)|} \to v = (1,0,\ldots,0).$$

If $\operatorname{ord}_0 z_j(\zeta) = k_j$, then $z_j(\zeta) = (c_j + o(1))\zeta^{k_j}$, and since $z_j(\zeta_i) = o(z_1(\zeta_i))$, $j > 1$, all $k_j > k$ for $j > 1$. But then $z(\zeta) = z_1(\zeta)(v + o(1))$ as $\zeta \to 0$, and all tangent vectors to A at 0 are complexly proportional to v. Since $\pi(A \cap U)$ contains a full neighborhood of 0 in $\mathbb{C}_1$, the tangent vectors to A at 0 fill the whole complex line $\mathbb{C}_1 = \mathbb{C}\cdot v$. ■

C o r o l l a r y. *The tangent cone to a pure one-dimensional analytic set A at an arbitrary point $a \in A$ is a finite union of complex lines (their number does not exceed the number of irreducible components of A at a).*

In higher dimensions the situation is much richer and much more complicated. E.g., if $p(z)$ is a homogeneous polynomial in z in $\mathbb{C}^n$, then its zero set is itself a cone with vertex at 0, and for $n \geqslant 3$ this cone has a very complicated structure (in particular, it does not split into complex hyperplanes in general). In the general situation there is also no local parametrization, and other methods of investigation must be applied.

8.2. The tangent cone and maps. Let $f: U \to V$ be a differentiable map of neighborhoods of 0 in $\mathbb{R}^N$ and $\mathbb{R}^m$, $f(0) = 0$, and let f' be the linear map $\mathbb{R}^N \to \mathbb{R}^m$ tangent at 0. Let E be a set in $\mathbb{R}^N$ for which 0 is a limit point, and suppose $v \in C(E, 0)$, $|v| = 1$. Then there is a sequence $x^j \to 0$, $x^j \in E$, such that $(x^j / |x^j|) \to v$. The points $y^j = f(x^j)$ belong to $f(E)$, and $y^j = f'(x^j) + o(|x^j|)$. Hence the sequence

$$\frac{y^j}{|x^j|} = f'\left[\frac{x^j}{|x^j|}\right] + o(1)$$

has a limit, equal to $f'(v)$. Thus, $f'(v) \in C(f(E),0)$, and we have proved

P r o p o s i t i o n 1. *Let $f: U \to V$ be a map between neighborhoods of 0 in $\mathbb{R}^N$ and $\mathbb{R}^m$, differentiable at 0, let $f(0) = 0$, and let f' be the linear map $\mathbb{R}^N \to \mathbb{R}^m$ tangent at 0. Then, for any set $E \subset \mathbb{R}^N$ for which 0 is a limit point, the inclusion $f'(C(E, 0)) \subset C(f(E),0)$ holds.*

The converse inclusion is, in general, not valid, because of possible degeneracy of f' (e.g., if $f' \equiv 0$ at 0 and 0 is a limit point of $f(E)$). However, if the kernel of f' is transversal to E at 0, then equality holds:

P r o p o s i t i o n 2. *Let $f: U \to V$ be a map between neighborhoods of 0 in $\mathbb{R}^N$ and $\mathbb{R}^m$, differentiable at 0, let $f(0) = 0$, let f' be the linear map $\mathbb{R}^N \to \mathbb{R}^m$ tangent at 0, and let $Z = \{x: f'(x) = 0\}$ be the kernel of f'. Let E be a set in $\mathbb{R}^N$ for which 0 is a limit point and such that $f'(0) \cap \bar{E} = \{0\}$ and $Z \cap C(E, 0) = \{0\}$. Then $f^{-1}(C(E, 0)) = C(f(E),0)$.*

■ It suffices to prove the inclusion $\supset$. Let $w \in C(f(E),0)$, $|w| = 1$, and let a sequence $y^j \in f(E)$ be such that $y^j / |y^i| \to w$ as $j \to \infty$. Since $f^{-1}(0) \cap \bar{E} = \{0\}$, there is a sequence of points $x^j \in E$ such that $f(x^j) = y^j$ and $x^j \to 0$ as $j \to \infty$. By passing to a subsequence we may assume that $(x^j / |x^j|) \to v \in C(E, 0)$. Since $y^j = f'(x^j) + o(|x^j|)$, this implies that $(y^j / |x^j|) \to f'(v)$ as $j \to \infty$. But $|v| = 1$ and $Z \cap C(E, 0) = \{0\}$, hence the vector $\tilde{w} = f'(v) \neq 0$. Since $(y^j / |y^j|) \to w$, this implies that $w = t\tilde{w}$ for some

$t > 0$; hence

$$w = tf'(v) = f'(tv) \in f'(C(E, 0)).\blacksquare$$

The conditions of the Proposition are fulfilled if, e.g., f satisfies an inequality $|f(x)| \geqslant \theta|x|$ for all $x \in E$ with some constant $\theta > 0$; in particular, if f is a diffeomorphism in a neighborhood of 0. We note a particular instance of Proposition 2 of interest to us.

C o r o l l a r y. *Let A be an analytic set in $\mathbb{C}^n$, $0 \in A$, and let $\pi: \mathbb{C}^n \to \mathbb{C}^p$ be orthogonal projection. Assume that $\pi: A \to U' \subset \mathbb{C}^p$ is a proper map and that $\pi^{-1}(0) \cap A = \pi^{-1}(0) \cap C(A, 0) = \{0\}$. Then $\pi(C(A, 0)) = C(\pi(A), 0)$.*

8.3. The tangent cone and the σ-process. We remind the reader that the incidence manifold $I(1,n)$ in $\mathbb{C}^n \times \mathbb{P}_{n-1}$ consists of the pairs $(z, [w])$ such that the vectors $z, w \in \mathbb{C}^n$ are complexly proportional. The projection π of the manifold $I(1,n)$ onto $\mathbb{C}^n$ is one-to-one above $\mathbb{C}^n_*$, and the fiber above 0 is all of $\mathbb{P}_{n-1}$. Hence $I(1,n)$ can be regarded as $\mathbb{C}^n$ to which $\mathbb{P}_{n-1}$ is "glued" at 0 as a complex submanifold (cf. A3.5). The transition from $\mathbb{C}^n$ to $I(1,n)$ is called the σ-process at 0.

A set $A \subset \mathbb{C}^n$ can be naturally lifted to $I(1,n)$ as follows: we first lift $A \setminus \{0\}$, putting $\tilde{A}_0 = \tilde{A} \setminus (\{0\} \times \mathbb{P}_{n-1}) := \pi^{-1}(A \setminus \{0\}) \cap I(1,n)$, and the complete lift $\tilde{A}$ is, by definition, the closure of $\tilde{A}_0$ in $A \times \mathbb{P}_{n-1}$. Thus, if A is closed in a domain D in $\mathbb{C}^n$, then $\tilde{A}$ is closed in $D \times \mathbb{P}_{n-1}$ and $\tilde{A}_0$ is dense in $\tilde{A}$. The set $\tilde{A} \cap (\{0\} \times \mathbb{P}_{n-1})$ reflects in the following way the tangent structure of the set A at 0.

P r o p o s i t i o n 1. *Let $A \ni 0$ be a pure p-dimensional analytic subset of a domain D in $\mathbb{C}^n$, let $\tilde{A}$ be its lift to the incidence manifold $I(1,n)$, and let*

$$\tilde{A} \cap (\{0\} \times \mathbb{P}_{n-1}) =: \{0\} \times C^\star(A, 0).$$

Then $\tilde{A}$ is an analytic subset in $D \times \mathbb{P}_{n-1}$, $C^\star(A, 0)$ is a pure $(p-1)$-dimensional algebraic subset in $\mathbb{P}_{n-1}$, and

$$C(A, 0) \setminus \{0\} = \{v: [v] \in C^\star(A, 0)\}.$$

■ Let $A' = (A \times \mathbb{P}_{n-1}) \cap I(1,n)$ be the trivial lift of A to $I(1,n)$. Obviously,

A' is an analytic subset in $D \times \mathbb{P}_{n-1}$ and $\tilde{A}_0 = A' \setminus (\{0\} \times \mathbb{P}_{n-1})$. By Corollary 5.1, $\tilde{A}$ is also an analytic subset in $D \times \mathbb{P}_{n-1}$, hence $\{0\} \times C^*(A, 0) := \tilde{A} \cap (\{0\} \times \mathbb{P}_{n-1})$ is an analytic subset in $\{0\} \times \mathbb{P}_{n-1}$, and thus, by Chow's theorem, $C^*(A, 0)$ is an algebraic subset in $\mathbb{P}_{n-1}$. Concerning dimensions: since $\tilde{A}_0$ is pure p-dimensional and everywhere dense in $\tilde{A}$, the set $\tilde{A}$ is also pure p-dimensional, and $\{0\} \times C^*(A, 0)$ has dimension $< p$. On the other hand, $\{0\} \times C^*(A, 0) = \tilde{A} \cap (\{0\} \times \mathbb{P}_{n-1})$ on the n-dimensional manifold $I(1,n)$. Hence, by p.3.5, $\dim_{[v]} C^*(A, 0) \geqslant p - 1$ at every point $[v] \in C^*(A, 0)$.

Finally, let $v \in C(A, 0)$, $|v| = 1$, and $a^j \in A \setminus \{0\}$, $a^j \to 0$, be such that $a^j / |a^j| \to v$. Then as $j \to \infty$ the points $(a^j,[a^j]) = (a^j,[a^j / |a^j|]) \in I(1,n)$ tend to $(0,[v])$, hence $[v] \in C^*(A, 0)$. Conversely, suppose $v' \in C^*(A, 0)$ and let a sequence $(a^j,[w^j]) \in \tilde{A}_0$ converge to $(0,[v'])$ as $j \to \infty$. Since $(a^j,[w^j]) \in I(1,n)$, we have $w^j = \lambda_j a^j$ for certain $\lambda_j \in \mathbb{C}$, hence $[w^j] = [a^j / |a^j|]$. If v is the limit vector for the sequence $a^j / |a^j|$, then $[v] = [v']$ because $(a^j,[a^j / |a^j|]) \to (0,[v'])$. But $v \in C(A, 0)$, and thus we have proved that every point of the set $C^*(A, 0)$ has the form $[v]$ with $v \in C(A, 0)$. ■

C o r o l l a r y. *The cone $C(A, 0)$ is a pure p-dimensional algebraic subset in $\mathbb{C}^n$, complexly homogeneous in the sense that for any $v \in C(A, 0)$ the vectors λv, $\lambda \in \mathbb{C}$, also belong to $C(A, 0)$.*

The set $C^*(A, 0) \subset \mathbb{P}_{n-1}$ will be called the *projective tangent cone* to A at 0. As was proved, $C^*(A, 0) = \Pi(C(A, 0) \setminus \{0\})$, where $\Pi\colon \mathbb{C}^n_* \to \mathbb{P}_{n-1}$ is the canonical projection.

8.4. Analytic description. What equations define $C(A, 0)$ if the equations defining A are known? The simplest answer is obtained in the case of a hypersurface.

P r o p o s i t i o n 1. *Let $A\colon f = 0$ be a principal analytic set in a neighborhood of 0 in $\mathbb{C}^n$, let $\mu \geqslant 1$ be the multiplicity of the zero of f at 0, and let $f(z) = \sum_{\mu}^{\infty} (f)_m(z)$ be the expansion in homogeneous polynomials. Then $C(A, 0) = \{z\colon (f)_\mu(z) = 0\}$.*

■ Suppose f is holomorphic in the closure of a neighborhood $U = U' \times U_n \ni 0$, and assume $(f)_\mu(0, \ldots, 0, 1) \neq 0$ (this assumption does not restrict the generality). By shrinking U_n we may assume that in U_n the function

$f(0,z_n)$ has no zeros distinct from $z_n = 0$. By shrinking U' we may assume that f and $(f)_\mu$ do not have zeros on $U' \times \partial U_n$. Finally, by shrinking U also, we may assume that $\sum_{m>\mu} |(f)_m(z)| < |(f)_\mu(z)|$ for all $z \in U' \times \partial U_n$. The set $(tA) \cap U$ is defined by the functions

$$f_t(z) := t^\mu f(z/t) = (f)_\mu(z) + \sum_{m>\mu} (f)_m(z) t^{\mu-m};$$

as $t \to \infty$, these functions tend to $(f)_\mu$. The construction implies that they are all zero free on $U' \times \partial U_n$. By Rouché's theorem, for each fixed $z' = U'$ the zeros of $f_t(z)$ tend to zeros of the polynomial $(f)_\mu$, hence $\{(f)_\mu = 0\} \subset C(A, 0)$. On the other hand, on any compact set of the form $\{|(f)_\mu| \geqslant \epsilon\} \cap \overline{U}$, $\epsilon > 0$, from some $t(\epsilon)$ onwards all $|f_t| \geqslant \epsilon/2$, hence all limit points of $(tA) \cap U$ belong to the zero set of $(f)_\mu$. ■

Thus, in the case of codimension 1 the tangent cone is given by the initial homogeneous polynomial from the expansion of the holomorphic function defining the analytic set itself. For higher codimensions the situation is more complicated. If, in a neighborhood of 0, the set A is given by the system of equations $f_1 = \cdots = f_k = 0$ and if $(f_j)^*$ are the initial homogeneous polynomials of the functions f_j at 0, then, in general, the set of solutions of the system $(f_1)^* = \cdots = (f_k)^* = 0$ does not coincide with $C(A, 0)$. E.g., if $\operatorname{ord}_0 f_1 < \operatorname{ord}_0 f_j$, $j > 1$, then the functions $f_1, f_1 + f_2, \ldots, f_1 + f_k$ define the same A, but the initial polynomials of these functions at 0 all coincide with $(f_1)^*$, and the set of their common zeros may be too large in order to be $C(A, 0)$.

Suppose A is pure p-dimensional and $0 \in A$. Then $C(A, 0)$ is also a pure p-dimensional analytic set and hence there is an $(n-p)$-dimensional plane $L \subset \mathbb{C}^n$ such that 0 is an isolated point of $C(A, 0) \cap L$ (see p.3.5). Since $C(A, 0)$ and L both are cones it follows that $C(A, 0) \cap L = \{0\}$. Such planes L will be called *transversal* to A at 0. This notion plays an important role in the study of infinitesimal properties of analytic sets (see, e.g., p.11.2 and p.12.5).

P r o p o s i t i o n 2. *Let A be a pure p-dimensional analytic set in $\mathbb{C}^n$, $0 \in A$, and suppose that the $\mathbb{C}_{(p+1)\cdots n}$-plane is transversal to A at 0. Then there is a neighborhood $U = U' \times U'' \ni 0$, $U' \subset \mathbb{C}^p$, such that the projection $\pi: A \cap U \to U'$ is proper and $\pi^{-1}(0) \cap A \cap U = \{0\}$. Let Φ_I be canonical defining functions of the analytic cover $\pi|_{A \cap U}$ and Φ_I^* their initial homogeneous* *polynomials at 0. Then*

$$C(A,0) = \{z\colon \Phi_I^*(z) = 0 \ \text{ for all } \ I\}.$$

■ The definitions of tangent cone and transversality obviously imply that in a small neighborhood of 0 the set A lies inside a cone $|z''| \leqslant C|z'|$ with some constant C. By definition, for each $z' \in U'$ the function $\Phi_I(z',z'')$ is the coefficient at w^I of the polynomial

$$P(z,w) = <w,z'' - \alpha^1(z')> \cdots <w,z'' - \alpha^k(z')>, \quad w \in \mathbb{C}^{n-p},$$

where $\alpha^1(z'), \ldots, \alpha^k(z')$ are the z''-coordinates of the points of the fiber $\pi^{-1}(z') \cap A \cap U$ (counted with multiplicities in case z' is a critical value for $\pi_{|A \cap U}$, see p.4.3.). The set $A \cap U$ is defined by the condition $P(z,w) \overset{w}{\equiv} 0$, hence $(tA) \cap U$, $t \geqslant 1$, is defined by $P(z/t,w) \overset{w}{\equiv} 0$, or, equivalently, by $t^k P(z/t,w) \overset{w}{\equiv} 0$. Since

$$t^k P(z/t,w) = \sum_{|I|=k} t^k \Phi_I(z/t) w^I = \prod_1^k <w,z'' - t\alpha^j(z'/t)>$$

and $|\alpha^j(z'/t)| \leqslant c|z'/t|$ in view of the transversality, we have $\operatorname{ord}_0 t^k \Phi_I(z/t) = k$ for all $t \geqslant 1$ and all I, $|I| = k$, and the holomorphic functions $t^k \Phi_I(z/t)$ have Taylor coefficients that are uniformly bounded in $t \geqslant 1$ (the coefficients at $(z'')^J$, $|J| = k$, do not depend on t at all). The expansion of these functions in homogeneous polynomials implies that

$$\lim_{t\to\infty} t^k \Phi_I(z/t) = \Phi_I^*(z)$$

uniformly for $z \in U$. For each fixed $z' \in U'$ we can choose a subsequence in the sequence $\{t\alpha^1(z'/t), \ldots, t\alpha^k(z'/t)\}$ that is convergent, say to $\{\alpha^1, \ldots, \alpha^k\}$. Since

$$\prod <w,z'' - t\alpha^j(z'/t)> \to \prod <w,z'' - \alpha^j>$$

for this subsequence in t, the polynomials $\Phi_I^*(z',z'')$ represent canonical defining functions for the system $\{\alpha^1, \ldots, \alpha^k\} \subset \mathbb{C}_{z''}^{n-p}$. Since all $(z',\alpha^j) \in C(A,0)$, this implies that the set of common zeros of the polynomials Φ_I^* belongs to $C(A,0)$.

Conversely, if $v = t_j a^j + \epsilon_j$, where $a^j \in A$, $a^j \to 0$, $\epsilon_j \to 0$, then

$$0 = t_j^k \Phi_I(a^j) = t_j^k \Phi_I\left[\frac{v - \epsilon_j}{t_j}\right] = \Phi_j^*(v - \epsilon_j) + O\left(\frac{1}{t_j}\right) \to \Phi_I^*(v),$$

hence all $\Phi_I^\star$ vanish on $C(A, 0)$. ■

A weaker (in the case of pure dimension) but better known version of Proposition 2 is the following:

Let $I(A, 0)$ be the ideal in the ring $\mathcal{O}_0$ of holomorphic functions at 0 consisting of the functions that vanish on A (each in its own neighborhood of 0). Then

$$C(A, 0) = \{v\colon (f)^\star(v) = 0, \quad f \in I(A, 0)\}.$$

(This statement also holds for analytic sets that are not of pure dimension.)

The basic inclusion $\supset$ was proved in Proposition 2. In essence, the inclusion $\subset$ is proved similarly: if $f \in I(A, 0)$, $v \in C(A, 0) \setminus \{0\}$, and $v = t_j a^j + \epsilon_j$, where $a^j \in A$, a^j, $\epsilon_j \to 0$ as $j \to \infty$, and if $k = \operatorname{ord}_0 f$, then

$$0 = t_j^k f(a^j) = t_j^k f\left[\frac{v - \epsilon_j}{t_j}\right] = (f)^\star(v - \epsilon_j) + O\left[\frac{1}{t_j}\right] \to (f)^\star(v)$$

hence $(f)^\star \mid_{C(A, 0)} = 0$.

8.5. Tangent vectors and one-dimensional sections. As was proved in p.8.1, the tangent cones to a one-dimensional analytic set have a very simple structure: they are finite unions of complex lines. It is therefore natural to ask whether there exists on a given analytic set $A \ni 0$ a one-dimensional analytic subset with tangent cone spanned on a given vector $v \in C(A, 0)$. The positive answer to this question readily follows from the analytic description of tangent cones.

P r o p o s i t i o n 1. *Let A be a pure p-dimensional analytic set in $\mathbb{C}^n$, $0 \in A$, and let $v \in C(A, 0)$, $v \neq 0$. Let coordinates be chosen such that the plane $\mathbb{C}_{(p+1)\cdots n}$ is transversal to A at 0, and let v' be the projection of v to $\mathbb{C}_{1\cdots p}$. Then v is a tangent vector to a one-dimensional analytic set $A \cap \{|z| < r,\ z' \in \mathbb{C}v'\}$ and hence a tangent vector to an irreducible component of this set; if r is sufficiently small, the full tangent cone to this component at 0 is equal to $\mathbb{C}v$.*

■ Let $U = U' \times U''$ be a neighborhood of 0 in $\mathbb{C}^n$ such that $\pi\colon A \cap U \to U'$ is an analytic cover; moreover, assume $A \cap U \cap \mathbb{C}_{(p+1)\cdots n} = \{0\}$. If Φ_I are canonical defining functions of this cover, then $\Phi_I(\lambda v', z'')$ are canonical defining functions of the one-dimensional analytic set

$$A_v = A \cap U \cap \{z' \in \mathbb{C}v'\}$$

in $(\mathbb{C}v') \times \mathbb{C}_{(p+1)\cdots n} \approx \mathbb{C}^{n-p+1}_{\lambda,z''}$. The polynomials $\Phi_I^*(\lambda v', z'')$, which are homogeneous in (λ, z''), define in $(\mathbb{C}v) \times \mathbb{C}_{(p+1)\cdots n}$ the tangent cone to A_v (p.8.4), hence

$$C(A_v, 0) = C(A, 0) \cap \{z' \in \mathbb{C}v'\}$$

and, in particular, $v \in C(A_v, 0)$. The remainder obviously follows from the definition of irreducibility (global and at a point), see p.5.3. ■

C o r o l l a r y. *Under the same assumptions there is a one-dimensional analytic subset $A' \subset A$, irreducible at 0, such that $v \in C(A', 0)$ and $A' \cap \operatorname{br}(\pi|_A) \subset \{0\}$.*

■ Since tangent cones are invariant under biholomorphic transformations (p.8.2) we may assume that the critical set σ of the analytic cover $\pi|_{A \cap U}$ intersects every line $\mathbb{C}v'$, $v' \in \mathbb{C}^p_*$, in a zero-dimensional set. If $v' = \pi(v)$, we obtain, by shrinking U', that $\sigma \cap \mathbb{C}v' \cap U \subset \{0\}$ hence the zero-dimensional analytic set $\pi^{-1}(\mathbb{C}v') \cap A \cap U$ can intersect $\operatorname{br} \pi|_A$ only at 0. As was proved, for sufficiently small U' this set has a component A' irreducible at 0, such that $C(A', 0) = \mathbb{C}v$. ■

P r o p o s i t i o n 2. *Let A be an analytic set in $\mathbb{C}^n$, $0 \in A$, and let $v \in C(A, 0)$, $v \neq 0$. Then there is a smooth (C^1), piecewise analytic Jordan arc $\gamma: (-r, r) \to A$ such that $\gamma(0) = 0$ and $\gamma'(0) = \lim_{t \to 0} \gamma(t)/t = v$.*

■ By Proposition 1 it suffices to consider the case in which A is a one-dimensional analytic set, irreducible at 0. Then $C(A, 0)$ is a complex line, which we may assume to be the $\mathbb{C}_1$-axis. Since $v \in \mathbb{C}_1$ we may assume $v = (1, 0, \ldots, 0)$. By p.6.1, in a small neighborhood of 0 the set A is the one-to-one image of the unit disk $|\zeta| < 1$ in $\mathbb{C}$ under a holomorphic map $z = z(\zeta)$; moreover, $z_1(\zeta) = r\zeta^k$ where k is the multiplicity of the cover A above $\mathbb{C}_1$ in a neighborhood of 0. The $\mathbb{R}$-analytic set $A \cap \{\operatorname{Im} z_1 = 0\}$ consists of $2k$ arcs, each of which can be projected onto $(-r, 0]$ or onto $[0, r)$ and is tangent to the $(x_1 = \operatorname{Re} z_1)$-axis at 0. The required arc γ is obtained by gluing two such arcs, one above $(-r, 0]$ and another above $[0, r)$, with total parametrization interval $(-r, r)$. ■

It is easy to see that the smoothness of γ at 0 is at least $1 + 1/k$, where k is the multiplicity of the cover used in the proof (from the preceding Section it is clear it is at least equal to the deviation of A from $C(A,0)$ at 0). As the simplest examples (say the semicubic parabola) indicate, in general there is no better smoothness.

8.6. Deviation. At a regular point a of an analytic set A the distance between the tangent plane and the set itself tends to zero like $o(r)$ in an r-neighborhood of a. An analogous property holds in the general case - for tangent cones (this, mainly, justifies the term 'tangent cone'). We remind the reader that the Hausdorff distance dist (E, E') between two sets E, E' in a metric space is the infimum of all numbers $\delta > 0$ such that E and E' lie in δ-neighborhoods of each other. For $E \subset \mathbb{C}^n$ we denote by E_r, $r > 0$, the set $E \cap \{|z| < r\}$.

Proposition. *Let A be a pure p-dimensional analytic set in $\mathbb{C}^n$, $0 \in A$. Then there are positive constants ϵ, C, and r_0 such that*

$$\operatorname{dist}(A_r, C(A,0)_r) \leqslant Cr^{1+\epsilon}$$

for all $r < r_0$.

■ Represent A in a neighborhood $U \ni 0$ as an analytic cover above a domain $U' \subset \mathbb{C}^p_{z'}$ such that

$$A \cap U \cap C^{n-p}_{z''} = C(A,0) \cap \mathbb{C}^{n-p}_{z''} = \{0\}.$$

Let Φ_I be canonical defining functions of this cover. By p.8.4, $C(A,0)$ is the set of common zeros of the system of initial homogeneous polynomials Φ_I^* of the functions Φ_I at 0; moreover $\Phi_I(\cdot, z'')$ and $\Phi_I^*(\cdot, z'')$ are canonical defining functions of systems of k (not necessarily distinct) points in $\mathbb{C}^{n-p}_{z''}$. Since $A \cap U \cap \mathbb{C}^{n-p}_{z''} = \{0\}$ and the plane $\mathbb{C}^{n-p}_{z''}$ is transversal to A at 0, we find that $C(A,0)$ and $A \cap U$ lie in some cone $|z''| \leqslant C_1 |z'|$. Since $\operatorname{ord}_0 \Phi_I = k$ for all I (see the proof of Proposition 2, p.8.4), Schwarz' lemma (A1.1) implies that

$$|\Phi_I(z) - \Phi_I^*(z)| \leqslant C_2 |z|^{k+1} \leqslant C_3 |z'|^{k+1}$$

in U. By Corollary 4.2 this means that for each fixed $z' \in U'$ the Hausdorff distance between the zero sets of the systems $\{\Phi_I\}$ and $\{\Phi_I^*\}$ does not exceed $C'|z'|^{1+1/k}$. Choose $r < 1$ small such that $2C'r^{1/k} < 1$ and such that the ball $|z| < 2C'r$ belongs to U. Then, by what has already been proved, there is for each point $z = (z', z'') \in A$, a point $\tilde{z} = (z', \tilde{z}'') \in C(A, 0)$ such that $|z - \tilde{z}| < C' r^{1+1/k}$. Since $C(A, 0)$ is a cone, the point $z^* = \tilde{z}/(1 + C' r^{1/k})$ belongs to $C(A, 0)$, $|z^*| < r$, and

$$|z - z^*| \leqslant C' r^{1+1/k} + C_4 |\tilde{z}| r^{1/k} \leqslant C'' r^{1+1/k}.$$

Conversely, if $z^* \in C(A, 0)_r$, then $\tilde{z} = z^* C' r^{1/k}$ belongs to $C(A, 0) \cap U$, and $(\tilde{z})' \in U'$. By what has already been proved, there is a point $z \in A \cap U$ such that $|z - \tilde{z}| < C' r^{1+1/k}$, hence $|z - z^*| < C''' r^{1+1/k}$. Since $|z| \leqslant 2C' r^{1+1/k} < r$, we find $z \in A_r$. Thus we have proved that A_r and $C(A, 0)_r$ lie in δ-neighborhoods of each other, where

$$\delta \leqslant \max(C', C'') \cdot r^{1+1/k}. \quad \blacksquare$$

The following inequality is more convenient in certain applications:

If $\pi: A \cap U \to U'$ is an analytic cover and

$$\pi^{-1}(0) \cap A \cap U = \pi^{-1}(0) \cap C(A, 0) = \{0\},$$

then there are constants ϵ, C, and r_0 such that the distance between the sets $A \cap U \cap \{|z'| < r\}$ and $C(A, 0) \cap \{|z'| < r\}$ does not exceed $Cr^{1+\epsilon}$ for all $r < r_0$.

C o r o l l a r y 1. *Let U be an arbitrary neighborhood of a projective tangent cone $C^*(A, 0)$ in $\mathbb{P}_{n-1}$, and let $\tilde{U} = \Pi^{-1}(U)$ be the corresponding open cone in $\mathbb{C}^n_*$. Then for all sufficiently small r the set A_r belongs to $\tilde{U} \cup \{0\}$.*

C o r o l l a r y 2. *If an analytic set A is irreducible at $0 \in A$, then the set $C^*(A, 0)$ is connected.*

$\blacksquare$ Assume that $C^*(A, 0) = C_1 \cup C_2$, where C_1, C_2 are closed nonintersecting subsets in $\mathbb{P}_{n-1}$. Then they have nonintersecting neighborhoods $U_1 \supset C_1$, $U_2 \supset C_2$ in $\mathbb{P}_{n-1}$. By Corollary 1, for sufficiently small r the set $A_r \setminus \{0\}$ belongs to the union of the nonintersecting open cones $\tilde{U}_j = \Pi^{-1}(U_j)$. This, obviously, implies that $A' = A_r \cap \tilde{U}_1$ and $A'' = A_r \cap \tilde{U}_2$ are analytic subsets of the domain $0 < |z| < r$. By the theorem of Remmert-Stein (p.4.4), the closure of each of them in the ball $|z| < r$ is an analytic subset of this ball, hence the set

$$A_r = (A' \cup \{0\}) \cup (A'' \cup \{0\})$$

is reducible. Since r is arbitrarily small, this implies that A is reducible at 0. ■

The converse statement is, of course, not true. E.g., for the analytic set $A: z_2(z_2 - z_1^2) = 0$ in $\mathbb{C}^2$, which is reducible at 0, the tangent cone at 0 is the $\mathbb{C}_1$-axis, and $C^*(A, 0)$ is the single point $[1,0]$ in $\mathbb{P}_1$. The stronger statement that $C^*(A, 0)$ is irreducible is also not true (for sets of dimension > 1). As an example we may take the hypersurface $A: z_3^3 = z_1 z_2$ in $\mathbb{C}^3$. It is irreducible at 0 (the cover of A above $\mathbb{C}^2 \setminus \{z_1 z_2 = 0\}$ is locally biholomorphic and connected); however, by p.8.4 the tangent cone to A at 0 is the set $z_1 z_2 = 0$ in $\mathbb{C}^3$; in $\mathbb{P}_2$ it corresponds to the set $C^*(A, 0)$, consisting of two hyperplanes.

D e f i n i t i o n. The *order of deviation* (or, simply, the *deviation*) of an analytic set A from its tangent cone $C(A, 0)$ is the number

$$\kappa = \underline{\lim}_{r \to 0} \frac{\log \operatorname{dist}(A_r, C(A, 0)_r)}{\log r}.$$

The proof of the Proposition shows that in it we may take $1/k$ as an ϵ, where k is the multiplicity of the transversal cover $A \cap U \to U'$ (i.e. the multiplicity of the analytic set A at 0, see p.11.2). Thus, always $\kappa \geqslant 1 + 1/k > 1$. For one-dimensional analytic sets the deviation can be calculated most simple.

L e m m a. *Suppose an analytic set $A \subset \mathbb{C}^n$ is the one-to-one image of the disk $|\zeta| < 1$ under a holomorphic map $z = z(\zeta)$, $z(0) = 0$, where, moreover,*

$$\operatorname{ord}_0 z_1(\zeta) = k < \operatorname{ord}_0 z_j(\zeta), \quad j > 1.$$

Let k' be the minimal order at 0 of the functions $z_2(\zeta), \ldots, z_n(\zeta)$. Then the (order of) deviation of A from $C(A, 0)$ is equal to $\kappa = k'/k$; moreover, there is a constant $C > 0$ such that

$$\operatorname{dist}(A_r, C(A, 0)_r) \sim C r^{\kappa}$$

(the quotient tends to 1 as $r \to 0$).

■ By the condition, in a sufficiently small neighborhood $U \ni 0$ of the form $U_1 \times U''$, $U_1 \subset \mathbb{C}_1$, the set $A \cap U$ is an analytic cover above U_1 that is transversal to $\mathbb{C}_{z''}^{n-1}$. The tangent cone to A at 0 clearly coincides with the $\mathbb{C}_1$-

axis. As noted above, it suffices to estimate the order of deviation of A and $\mathbb{C}_1$ along the variable z'' as $z_1 \to 0$. For a fixed $z_1 \in U_1$ there are k parameter values $\zeta^{(1)}, \ldots, \zeta^{(k)}$. Since $\operatorname{ord}_0 z_1(\zeta) = k$, Taylor expansion obviously implies that $|\zeta^{(i)}| \sim C_1 |z_1|^{1/k}$, $i = 1, \ldots, k$, with some constant $C_1 > 0$ depending on the series of $z_1(\zeta)$ only. Hence Taylor's formula for $z_j(\zeta)$ implies that

$$\lim_{\zeta \to 0} \frac{|z''(\zeta)|}{|z_1(\zeta)|^{k'/k}} = C \in (0, \infty).$$

Thus, for each fixed $z_1 \in U_1$ the estimate $|z''| \sim C|z_1|^{k'/k}$ holds for points $(z_1, z'') \in A \cap U$. This obviously implies that

$$\operatorname{dist}(A_r, C(A, 0)_r) \sim Cr^{k'/k}. \quad \blacksquare$$

The pair k', k for such a one-dimensional analytic set A that is irreducible at 0, does not depend on the choice of a holomorphic parametrization: k is the number of sheets of A above its tangent plane (in a neighborhood of 0), and k'/k is the deviation of A from $C(A, 0)$. These numbers are called the *Puiseux indices* of the holomorphic curve A at 0.

Finding the order of deviation in the general case, especially in dimensions exceeding 1, is a very difficult problem that has been almost not investigated up till now.

9. Whitney cones

9.1. Definitions and simplest properties. Besides the definition of tangent cone to an analytic set in $\mathbb{C}^n$ given above, there are also other definitions possible, which give one and the same tangent space at regular points. In Whitney's paper [148] the most natural definitions of this notion, and their relations, have been studied. In the sequel such cones will be called Whitney cones. We give the definitions of these cones $C_i(A,a)$, $i = 1, \ldots, 6$, in terms of the vectors forming them.

D e f i n i t i o n. Let A be an analytic set in $\mathbb{C}^n$ and suppose $a \in A$.

1) A vector v in $\mathbb{C}^n$ belongs to $C_1(A,a)$ if there are a neighborhood $U \ni a$ in $\mathbb{C}^n$ and a holomorphic vectorfield $v(z)$ in U (i.e. a holomorphic map of U into $\mathbb{C}^n$) such that $v(a) = v$ and $v(z) \in T_z(A)$ for all $z \in (\operatorname{reg} A) \cap U$.

2) $v \in C_2(A,a)$ if for every $\epsilon > 0$ there is a $\delta > 0$ such that for $z \in \operatorname{reg} A$ with $|z-a| < \delta$ we have $|v - v'| < \epsilon$ for some $v' \in T_z A$.

3) $C_3(A,a) = C(A,a)$ is the tangent cone defined in p.8.1.

4) $v \in C_4(A,a)$ if there are sequences of points $z^j \in \operatorname{reg} A$ and vectors $v^j \in T_{z^j} A$ such that $z^j \to a$ and $v^j \to v$ as $j \to \infty$.

5) $v \in C_5(A,a)$ if there are sequences of points $z^j, w^j \in A$ and numbers $\lambda_j \in \mathbf{C}$ such that $z^j \to a$, $w^j \to a$, and $\lambda_j(z^j - w^j) \to v$ as $j \to \infty$.

6) $v \in C_6(A,a)$ if $(df)_a(v) = 0$ for every function $f \in I(A,a)$, i.e. holomorphic in a neighborhood of a and vanishing on A (within this neighborhood). This set is also denoted by $T(A,a)$ and is called the *tangent space* to A at a.

It is clear from the definitions that the sets $C_i(A,a) \subset \mathbf{C}^n$ are closed and that they really are cones with vertices at 0; more precisely, if $v \in C_i(A,a)$ then $\lambda v \in C_i(A,a)$ also for all $\lambda \in \mathbf{C}$ (for C_3 this statement was proved in p.8.3). Moreover, the cones C_1, C_2, and C_6 are, obviously, $\mathbf{C}$-linear subspaces in $\mathbf{C}^n$. We note some properties of Whitney cones, which directly follow from the definitions.

1. If $a \in \operatorname{reg} A$, then all $C_i(A,a) = T_a A$.

2. The Whitney cones are invariant under biholomorphic transformations:

$$C_i(f(A), f(a)) = f'_a(C_i(A,a)), \quad i = 1, \ldots, 6,$$

where f'_a is the linear map tangent at a to the biholomorphism f, which is defined in a neighborhood of a.

■ For $\mathbf{C}$-linear maps this property is obvious, and in the general case it follows from the expansion $\Delta f = f'_a(\Delta z) + o(\Delta z)$. ■

3. $C_i(A' \cup A'', a) = C_i(A',a) \cup C(A'',a)$, $i = 3,4$; for $i = 5,6$ only the inclusion $\supset$ is valid; for $i = 1,2$ in general neither one holds.

As examples we give a table (Table 1) of the Whitney cones at 0 for certain most simple algebraic sets:

$$A_1\colon z_1 z_2 = 0, \quad A_2\colon z_2^2 = z_1^3 \text{ in } \mathbf{C}^2,$$

$$A_3\colon z_3^2 = z_1 z_2, \quad A_4^2\colon z_3^2 = z_1 z_2^2, \quad A_5 = \mathbf{C}_1 \cup \mathbf{C}_2 \cup \mathbf{C}_3,$$

$$A_6 = A_2 \times \mathbf{C}_3 \text{ in } \mathbf{C}^3.$$

	C_1	C_2	C_3	C_4	C_5	C_6
A_1	0	0	A_1	A_1	$\mathbb{C}^2$	$\mathbb{C}^2$
A_2	0	$\mathbb{C}_1$	$\mathbb{C}_1$	$\mathbb{C}_1$	$\mathbb{C}^2$	$\mathbb{C}^2$
A_3	0	0	A_3	$\mathbb{C}^3$	$\mathbb{C}^3$	$\mathbb{C}^3$
A_4	0	$\mathbb{C}_1$	$\mathbb{C}_{12}$	$\mathbb{C}^3$	$\mathbb{C}^3$	$\mathbb{C}^3$
A_5	0	0	A_5	A_5	$z_1z_2z_3=0$	$\mathbb{C}^3$
A_6	$\mathbb{C}_3$	$\mathbb{C}_{13}$	$\mathbb{C}_{13}$	$\mathbb{C}_{13}$	$\mathbb{C}^3$	$\mathbb{C}^3$

Table 1.

Note that the cones C_1, C_2 seldom occur (most often they are trivial). We will not use them, and we will consider only the cones C_i with $i \geqslant 3$ in the sequel.

9.2. Hierarchy and analyticity. We prove the inclusions $C_3 \subset \cdots \subset C_6$ for the Whitney cones at an arbitrary point a of an analytic set A in $\mathbb{C}^n$ (the inclusions $C_1 \subset C_2 \subset C_3$ are also valid, see [148]).

$C_3 \subset C_4$. ■ We first note that if A is one-dimensional and irreducible at a, then local holomorphic parametrization (p.6.1) makes it clear that the planes T_zA tend to the cone $C(A,a)$ as $z \to a$, $z \in \operatorname{reg} A$. This, of course, implies that in this situation $C_3(A,a) = C_4(A,a)$. In the general situation let $v \in C(A,a)$. By p.8.5 there is a one-dimensional analytic subset $A' \subset A$, irreducible at a, such that $A' \cap \operatorname{sng} A \subset \{a\}$ and $v \in C(A',a)$. If $z^j \in A' \setminus \{a\}$ and $z^j \to a$, then, as proved above, $T_{z^j}A' \to C(A',a) \ni v$. Since $z^j \in \operatorname{reg} A$ and $T_{z^j}A' \subset T_{z^j}A$, the vector v is the limit vector of some sequence of vectors $v^j \in T_{z^j}A$. ■

$C_4 \subset C_5$. ■ Let $v \in C_4(A,a)$; suppose $z^j \in \operatorname{reg} A$, $v^j \in T_{z^j}A$ are such that $z^j \to a$, $v^j \to v$ as $j \to \infty$. The definition of v^j implies that there are points $w^j \in A$, $|w^j - z^j| < |z^j - a|$, and numbers $t_j \to 0$ such that $|t_j(w^j - z^j) - v^j| < \epsilon_j \to 0$ as $j \to \infty$. Since $v^j \to v$, we find $v = \lim_{j\to\infty} t_j(w^j - z^j)$, i.e. $v \in C_5(A,a)$. ■

The inclusion $C_5 \subset C_6$ is trivial.

Analyticity of the cones C_4, C_5 can be derived from the following stronger statement (see [148]); analyticity of C_3 was proved in p.8.3.

P r o p o s i t i o n. *Let A be an analytic subset of a domain D in $\mathbb{C}^n$ and put*

$$\mathbb{C}_i(A) := \{(z,v): z \in A, \quad v \in \mathbb{C}_i(A,z)\}, \quad i = 4,5.$$

Then $C_i(A)$ are analytic subsets in $D \times \mathbb{C}^n$, and C_4 is the smallest closed subset in $D \times \mathbb{C}^n$ containing the tangent bundle $\{(z,v) := z \in \operatorname{reg} A,\ v \in T_zA\}$ to the manifold $\operatorname{reg} A$. *Moreover,* $\dim C_4(A) = \dim C_5(A) = 2 \dim A$.

■ Without loss of generality we may assume that A is pure p-dimensional, that it is defined in a neighborhood $U = U' \times U''$ in $\mathbb{C}^n$, $U' \subset \mathbb{C}^p$, and that the projection $\pi: A \to U'$ is a k-sheeted analytic cover. Let Φ_I, $|I| = k$, be canonical defining functions of this cover, and let $\sigma \subset U$ be the critical set of this cover.

$i = 4$. Put $\Sigma = \pi^{-1}(\sigma)$ and $C_4^0 = C_4(A) \cap ((D \setminus \Sigma) \times \mathbb{C}^n)$. For each point $z \in \operatorname{reg} A$ such that $z' \notin \sigma$ the tangent space T_zA consists of the vectors v for which

$$<d\Phi_I(z),v> := \sum_j \frac{\partial \Phi_I}{\partial z_j}(z) v_j = 0, \quad |I| = k$$

(see p.8.4). Therefore, inside $(D \setminus \Sigma) \times \mathbb{C}^n$ the set C_4^0 is defined by this system of holomorphic equations. But this system is defined and holomorphic everywhere in $D \times \mathbb{C}^n$, hence C_4^0 is part of the analytic subset

$$\tilde{C}_4 = \{(z,v): <d\Phi_I(z),v> = 0, \quad |I| = k\}$$

in $D \times \mathbb{C}^n$, and $\tilde{C}_4 \setminus C_4^0 = \tilde{C}_4 \cap (\Sigma \times \mathbb{C}^n)$ is an analytic subset in $D \times \mathbb{C}^n$. By Corollary 5.1, the closure of C_4^0 in $\tilde{C}_4$ is an analytic subset in $D \times \mathbb{C}^n$. Since C_4^0 is dense in C_4 (by the definition of C_4) and since $C_4(A)$ is closed in $D \times \mathbb{C}^n$, we have $C_4(A) = \overline{C_4^0} \cap (D \times \mathbb{C}^n)$. Since $z \mapsto T_zA$ is continuous on $\operatorname{reg} A$, above $\operatorname{reg} A$ the set $C_4(A)$ coincides with the tangent bundle to $\operatorname{reg} A$. Since clearly $\dim C_4^0 = 2p$, and since C_4^0 is dense in $C_4(A)$, we have $\dim C_4(A) = 2p$.

$i = 5$. Consider in $D \times D \times \mathbb{C}^n$ the analytic set

$$\tilde{A} = \{(z,\zeta,v): z \in A, \quad \zeta \in A, \quad (z-\zeta) \wedge v = 0\}.$$

Let $\Delta = \{(z,z): z \in A\}$ be the diagonal in $A \times A$, and let $\tilde{\pi}: (z,\zeta,v) \mapsto (z,\zeta)$. It is easily seen that $\pi: \tilde{A} \to A \times A$ is a map *onto* and that above each point $(z,\zeta) \in (A \times A) \setminus \Delta$ the fiber is a complex line (consisting of the vectors that are complexly proportional to $z - \zeta$). As above, the closure of $\tilde{A} \setminus \tilde{\pi}^{-1}(\Delta)$ in

$D \times D \times \mathbb{C}^n$ is an analytic subset (obviously, of dimension $2p+1$). The set of its limit points on $\tilde{\pi}^{-1}(\Delta)$, which will be denoted by $\tilde{C}$, is also an analytic subset in $D \times D \times \mathbb{C}^n$, and $\dim \tilde{C} \leq 2p$ since $\tilde{C}$ is nowhere dense in the closure of $\tilde{A} \setminus \tilde{\pi}^{-1}(\Delta)$. Let π': $(z,\zeta,v) \mapsto (z,v)$. The definition of C_5 readily implies that the projection π': $\tilde{C} \to C_5(A)$ is one-to-one, in particular proper. Thus, $C_5(A)$ is an analytic subset in $D \times \mathbb{C}^n$ of dimension $\leq 2p$. Since $C_4 \subset C_5$, we have $\dim C_5(A) = 2p$. ■

C o r o l l a r y 1. *The cones $C_4(A,a)$ and $C_5(A,a)$ are algebraic subsets in $\mathbb{C}^n$.*

■ Analyticity of these cones follows from the Proposition, since $C_i(A,a)$ is the fiber of $C_i(A)$ above a. Since the Whitney cones are complexly homogeneous, they are algebraic according to the Remark following Chow's theorem (p.7.1). ■

C o r o l l a r y 2. *The cones $C_i(A,a)$, $i = 4,5$, depend upper semicontinuously on $z \in A$; more precisely, if $z \to a \in A$, then the limit set of the family of cones $C_i(A,z)$ belongs to $C_i(A,a)$.*

■ The set $C_i(A)$ is closed, and $C_i(A,z)$ is the intersection of it by the plane $\{z\} \times \mathbb{C}^n$. ■

This property also holds for $C_6(A,z)$, but for $C_3(A,z)$ it does not hold: if $C_3(A,a) \neq C_4(A,a)$, then there is a sequence $z^j \in \operatorname{reg} A$ such that the limit set of the family $T_{z^j}A$ does not belong to $C_3(A,a)$.

9.3. Tangent space. The cone $C_6(A,a) = T(A,a)$ is a $\mathbb{C}$-linear subspace in $\mathbb{C}^n$ since differentials of holomorphic functions are $\mathbb{C}$-linear (see the definition of C_6). The dimension of the tangent space $T(A,a)$ turns out to be equal to the minimal dimension of a complex manifold into which A can be embedded in a neighborhood of the point a.

P r o p o s i t i o n. *Let A be an analytic set in $\mathbb{C}^n$, $a \in A$, and let $\dim T(A,a) = d$. Then there are a neighborhood $U \ni a$ and a d-dimensional complex manifold $M \subset U$ such that $A \cap U \subset M$. A manifold of lower dimension and with this property does not exist.*

■ The last statement obviously follows from the fact that if $A \subset M$ then $T(A,a) \subset T_aM$. We now assume that $d < n$ (otherwise the first statement of the

Proposition is trivial). Then among the linear functions $(df)_a$, $f \in I(A,a)$, defining $T(A,a)$ we can find $n-d$ linearly independent ones, say $(df_1)_a, \ldots, (df_{n-d})_a$. By the implicit function Theorem there is a neighborhood $U \ni a$ such that the set

$$M = \{z \in U: f_1(z) = \cdots = f_{n-d}(z) = 0\}$$

is a d-dimensional complex submanifold in U. Since all $f_j|_{A \cap U} \equiv 0$ for U sufficiently small, we have $A \cap U \subset M$. ■

We say that a set A is minimally embedded in $\mathbb{C}^n$ at a point $a \in A$ if $\dim T(A,a) = n$. If this does not hold, then by the Proposition there is a biholomorphic transformation of a neighborhood of a mapping A into an analytic set located in a neighborhood of 0 of a space $\mathbb{C}^d \subset \mathbb{C}^n$.

E x a m p l e. Let A: $z_1 z_2 = 0$, $z_3 = z_1^2$ in $\mathbb{C}^3$. Since M: $z_3 = z_1^2$ is a submanifold in $\mathbb{C}^3$ with tangent space $\mathbb{C}_{12}$ at 0 and since $A \subset M$, we have $T(A,0) \subset \mathbb{C}_{12}$. On the other hand, the vectors (1,0,0) and (0,1,0) are obviously tangent to A at 0 (i.e. belong to $C(A,0) \subset T(A,0)$). Hence $T(A,0) = \mathbb{C}_{12}$.

9.4. Whitney cones and projections. We give some simple properties of analytic covers related to the cones C_4 and C_5. A more refined discussion of similar results can be found in Stutz [132], [133], from which we have taken the statements given below.

P r o p o s i t i o n 1. *Let $A \ni 0$ be a pure p-dimensional analytic set in $\mathbb{C}^n$ such that $C_4(A,0) \cap \mathbb{C}_{(p+1)\cdots n} = \{0\}$. Then there is a neighborhood $U \ni 0$ in $\mathbb{C}^n$ such that*

$$(\operatorname{sng} A) \cap U = \operatorname{br} \pi|_{A \cap U}, \quad \text{where } \pi: z \mapsto (z_1, \ldots, z_p).$$

■ Recall that $\operatorname{br} \pi|_{\operatorname{reg} A}$ consists of the points at which the tangent space intersects $\mathbb{C}_{(p+1)\cdots n}$ along at least one line. Hence if our statement is not true, there are sequences of points $z^j \in \operatorname{reg} A$ and vectors $v^j \in T_{z^j}A$, $|v^j| = 1$, such that $z^j \to 0$ and $v^j \to v \in \mathbb{C}_{(p+1)\cdots n}$ as $j \to \infty$. But then $v \neq 0$ and $v \in C_4(A,0)$, by the definition of C_4, contradicting the conditions. ■

In order that the conditions be fulfilled it is, of course, necessary that $\dim C_4(A,0) \leqslant p$; if this property holds, then almost all $(n-p)$-dimensional planes $L \ni 0$ intersect $C_4(A,0)$ only at 0, hence, by what has been proved, the projections along them have, in a neighborhood of 0, critical points on sng A only.

P r o p o s i t i o n 2. *Let $A \ni 0$ be a pure p-dimensional analytic set in $\mathbb{C}^n$ such that $C_5(A,0) \cap \mathbb{C}_{(p+2)\cdots n} = \{0\}$, and let $\tilde{\pi}: z \mapsto (z_1, \ldots, z_{p+1})$. Then there is a neighborhood $U \ni 0$ in $\mathbb{C}^n$ such that the almost single-sheeted projection $\tilde{\pi}|_{A \cap U}$ is a homeomorphism of $A \cap U$ onto some hypersurface in $U \cup \mathbb{C}_{1\cdots(p+1)}$.*

■ Assume that the statement is not true in any neighborhood of 0. Then we can find two sequences $\{z^j\}$, $\{w^j\}$ in A, $z^j \neq w^j$, z^j, $w^j \to 0$, such that $\tilde{\pi}(z^j) = \tilde{\pi}(w^j)$, i.e. the vectors $w^j - z^j$ are parallel to $\mathbb{C}_{(p+2)\cdots n}$. By taking subsequences we may assume that $(w^j - z^j)/|w^j - z^j| \to v \in \mathbb{C}_{(p+2)\cdots n}$, $|v| = 1$. But by the definition of C_5 the vector v must belong to $C_5(A,0)$, contradicting the conditions. ■

Again, in order that the conditions be fulfilled it is necessary that $\dim C_5(A,0) \leqslant p+1$. If this is so, then almost all $(n-p-1)$-dimensional planes $L \ni 0$ intersect $C_5(A,0)$ at 0 only, hence the projections along them to A are single-sheeted in a neighborhood of 0. But what in case $\dim C_5 = p$?

C o r o l l a r y. *If $\dim C_5(A,0) = p$, then $0 \in \operatorname{reg} A$.*

■ Since $C_5(A,0)$ is an algebraic cone, after a unitary change of coordinates the condition $C_5(A,0) \cap \mathbb{C}_{(p+1)\cdots n} = \{0\}$ is fulfilled. Put $\pi: z \mapsto (z_1, \ldots, z_p)$. As in the proof of Proposition 2 we obtain that in a neighborhood of 0 the projection $\pi|_A$ is one-to-one, hence in a neighborhood of 0 the set A is a p-dimensional complex manifold, by p.3.3. ■

In relation to this role of the dimension of C_4 and C_5 it is expedient to clarify on which subsets in A the dimension of these cones is too large in order to obtain good results.

L e m m a. *Let A be a pure p-dimensional analytic subset of a domain D in $\mathbb{C}^n$. Then*

$$\{z \in A : \dim C_4(A,z) > p\} \text{ and } \{z \in A : \dim C_5(A,z) > p+1\}$$

are also analytic subsets in D, of dimension not exceeding $p-2$.

■ Denote these sets by Σ_4 and Σ_5. It being not necessary to prove analyticity (see [132]), we confine ourselves to estimating the Hausdorff dimensions. Since the set $C_4(A)$ from p.9.2 has dimension $2p$ and since $C_4(\operatorname{reg} A)$ is everywhere dense in $C_4(A)$, while $\Sigma_4 \subset \operatorname{sng} A$, the dimension of the set $\{(z,v) \in C_4(A): z \in \Sigma_4\}$ is strictly less than $2p$. Each fiber of $C_4(A,z)$, for $z \in \Sigma_4$, of this set has dimension $\geqslant p+1$ by the requirement. Hence, by A6.3 the Hausdorff dimension of Σ_4 does not exceed $2(p-2)$ (hence $\dim \Sigma_4 \leqslant p-2$). The case Σ_5 is analogous; the dimension of the set $\{(z,v) \in C_5(A): z \in \Sigma_5\}$ does not exceed that of $C_5(A)$, which equals $2p$, and all fibers above Σ_5 have at least dimension $p+2$. ■

9.5. Singularities of codimension 1. Puiseux normalization.

P r o p o s i t i o n 1. *Let A be a pure p-dimensional analytic set in $\mathbb{C}^n$, irreducible at $0 \in \operatorname{reg}(\operatorname{sng} A)$. Assume that $\dim_0(\operatorname{sng} A) = p-1$ and $\dim C_4(A,0) = p$. Then there are a neighborhood $U \ni 0$ and a one-to-one holomorphic map $z = z(\zeta)$ of the polydisk Δ^p: $|\zeta_j| < 1$, $j = 1, \ldots, p$, in $\mathbb{C}^p$ onto the set $A \cap U$, such that $(\operatorname{sng} A) \cap U = z(\Delta^p \cap \{\zeta_1=0\})$ and such that the map $\Delta^p \setminus \{\zeta_1=0\} \to (\operatorname{reg} A) \cap U$ is biholomorphic. If in a neighborhood of 0 coordinates have been chosen such that $\operatorname{sng} A \subset \mathbb{C}_{2\cdots p}$ and $C_4(A,0) \cap \mathbb{C}_{(p+1)\cdots n} = \{0\}$ (this can be achieved by a biholomorphic transformation), then one such parametrization has the form*

$$z(\zeta) = (r\zeta_1^k, r\zeta_2, \ldots, r\zeta_p, z_{p+1}(\zeta), \ldots, z_n(\zeta)),$$

where $r > 0$ is a constant, k is the multiplicity of the cover A over $\mathbb{C}_{1\cdots p}$ in a neighborhood of 0, and $z_j(\zeta)$ are holomorphic functions of the form $\zeta_1^k \cdot \phi_j(\zeta)$, where ϕ_j are also holomorphic in Δ^p, $j = p+1, \ldots, n$.

By analogy with one-dimensional local parametrization, the last representation may be called a *Puiseux parametrization* (normalization) for the analytic set A in a neighborhood of 0.

■ Let, in a neighborhood of 0, $\operatorname{sng} A = \mathbb{C}_{2\cdots p}$. Since by assumption $\operatorname{sng} A$ is a $(p-1)$-dimensional complex manifold, this can be achieved by a biholomorphic change of coordinates in a neighborhood of 0 in $\mathbb{C}^n$. The dimension of C_4 does not change under biholomorphic transformations, hence we may assume also that

$$C_4(A,0) \cap \mathbb{C}_{(p+1)\cdots n} = \{0\}.$$

Since $C_3 \subset C_4$, the plane $\mathbb{C}_{(p+1)\cdots n}$ is transversal to A at 0, hence there is a small neighborhood $U = U' \times U'' \ni 0$ such that U' is a polydisk

$\{|z_j| < r, j = 1, \ldots, p\}$ in $\mathbf{C}_{1 \cdots p}$, the projection $\pi: A \cap U \to U'$ is proper, and $\pi^{-1}(0) \cap A \cap U = \{0\}$. For $r > 0$ sufficiently small we have by p.9.4 that $\operatorname{br} \pi|_{A \cap U} = (\operatorname{sng} A) \cap U = \mathbf{C}_{2 \cdots p} \cap U = U' \cap \{z_1 = 0\}$, and, moreover, this coincides with the critical set σ of the cover $\pi|_{A \cap U}$. Since the cones $C_4(A,z)$ are semicontinuous in z (Corollary 2, p.9.3), we may also assume that $C_4(A,z) \cap \mathbf{C}_{(p+1) \cdots n} \cap \{0\}$ for all $z \in A \cap U$.

Let k be the multiplicity of the cover $\pi: A \cap U \to U'$. Since A is irreducible at 0, the sets $(\operatorname{reg} A) \cap U$ and $M = A \cap U \setminus \pi^{-1}(\sigma)$ are connected, and the projection $\pi: M \to U' \setminus \sigma$ is a locally biholomorphic k-sheeted cover. Since $U' \setminus \sigma = \Delta_* \times r\Delta^{p-1}$, where $\Delta_*: 0 < |z_1| < r$ is a punctured disk, the holomorphic function $\zeta_1(z) = (z_1/r)^{1/k}$ is defined on M (such that $\zeta_1(z)^k = z_1/r$). Indeed, any closed path on M can be continuously deformed into a path lying above $(U' \setminus \sigma) \cap \mathbf{C}_1 = \Delta_* \times \{0\}$. In particular, this implies that the manifold $M \cap \mathbf{C}_{1(p+1) \cdots n}$ is connected, hence $A \cap U \cap \mathbf{C}_{1(p+1) \cdots n}$ is irreducible. By p.6.1, the required continuous root $(z_1/r)^{1/k}$ exists on the irreducible one-dimensional analytic set $A \cap U \cap \mathbf{C}_{1(p+1) \cdots n}$, hence for every closed path on M the increment of the argument z_1 is an integral multiple of $2\pi k$. Hence, by determining $(z_1/r)^{1/k}$ positively on one of the sheets above $U' \cap \{z_1 > 0\}$, we can continue it analytically to all of M, and as a result we obtain a single-valued, hence holomorphic, function $\zeta_1(z)$. On distinct sheets of the cover $\pi: M \to U' \setminus \sigma$ (for fixed z') it takes distinct values (since the number of sheets is also equal to k). Hence the map $z \mapsto \zeta(z) = (\zeta_1(z), z_2/r, \ldots, z_p/r)$ biholomorphically transforms M onto $\Delta^p \setminus \{\zeta_1 = 0\}$. Let $z = z(\zeta)$ be the inverse map. Then

$$z(\zeta) = (r\zeta_1^k, r\zeta_2, \ldots, r\zeta_p, z_{p+1}(\zeta), \ldots, z_n(\zeta)).$$

Since $z_j(\zeta)$ is holomorphic and bounded on $\Delta^p \setminus \{\zeta_1 = 0\}$, it holomorphically continues to all of Δ^p. Since M is dense in $A \cap U$, the image of Δ^p under this continuation $z = z(\zeta)$ will be all of $A \cap U$. Since $U \cap \mathbf{C}_{2 \cdots p}$ is the image of $\Delta^p \cap \{\zeta_1 = 0\}$ (both sets are irreducible and belong to $A \cap U$), we have that $z_j(0, \zeta_2, \ldots, \zeta_p) \equiv 0$ for all $j > p$. Thus, in passing we have proved that $A \cap U \cap \{z_1 = 0\} = \mathbf{C}_{2 \cdots p} \cap U = (\operatorname{sng} A) \cap U$. Since $C_4(A,z) \cap \mathbf{C}_{(p+1) \cdots n} = \{0\}$ for all $z \in A \cap U$, for each point $(0, 'a, 0, \ldots, 0) \in U$, where $'z = (z_2, \ldots, z_p)$, there is a constant $C('a)$ such that the inequalities $|z_j| \leqslant C('a)|z_1|$ for all $j > p$ are fulfilled on the set $A \cap U \cap \{'z = 'a\}$. Thus, the functions $\phi_j(\zeta) = rz_j(\zeta)/z_1(\zeta)$, which are holomorphic in $\Delta^p \setminus \{\zeta_1 = 0\}$, are uniformly bounded on the one-dimensional sections $(\Delta^p \setminus \{\zeta_1 = 0\}) \cap \{'\zeta = 'a\}$. The Theorem on removable singularities implies that they can be holomorphically continued to Δ^p in the variable ζ_1, while by Lemma 2, A1.1, these continuations are holomorphic in Δ^p in all variables. Thus, $z_j(\zeta) = \zeta_1^k \phi_j(\zeta)$, where $\phi_j \in \mathcal{O}(\Delta^p)$, $j > p$. ■

R e m a r k. In the Puiseux representation we may, of course, assume that the functions ϕ_j are holomorphic in a neighborhood of $\bar{\Delta}^p$. If $\phi_j(\zeta) = c_j('\zeta) + \Sigma_1^\infty c_{j\nu}('\zeta)\zeta_1^\nu$ are the Taylor expansions in ζ_1, then by the additional biholomorphic change of coordinates

$$z \mapsto \left[z_1, \ldots, z_p, z_{p+1} - \frac{1}{r} c_{p+1}('z) z_1, \ldots, z_n - \frac{1}{r} c_n('z) z_1\right]$$

we obtain an additional property of the Puiseux parametrization: $|\phi_j(\zeta)| \leq C|\zeta_1|$, hence $\operatorname{ord}_{(0,'\zeta)} z_j(\zeta) > k$ for all $j > p$ and $'\zeta \in \Delta^{p-1}$. In terms of the (new) variable z this means that the uniform estimate $|z''| = o(|z_1|)$ as $z_1 \to 0$ holds on $A \cap U$. This implies that the tangent planes to A at points $z \in (\operatorname{reg} A) \cap U$ are uniformly close to $\mathbb{C}_{1 \ldots p}$, i.e. uniformly tend to $\mathbb{C}_{1 \ldots p}$ as $r \to 0$.

This Remark readily implies

C o r o l l a r y 1. *Let A be a pure p-dimensional analytic set in $\mathbb{C}^n$, irreducible at $0 \in \operatorname{reg}(\operatorname{sng} A)$, and let $T_0(\operatorname{sng} A) = \mathbb{C}_{1 \ldots (p-1)}$. Then if* $\dim C_4(A, 0) = p$,

1) $C_3(A, 0)$ coincides with the p-dimensional complex plane $T_0(\operatorname{sng} A) \times C(A \cap \mathbb{C}_{p \ldots n}, 0)$; and

2) $C_3(A, 0) = C_4(A, 0)$.

■ Since, obviously, $C_3(A, 0)$ is contained in the p-dimensional plane indicated and $C_3 \subset C_4$, it suffices to prove that $C_4(A, 0)$ coincides with this p-dimensional plane. By the Remark, all planes $T_z A$, $z \in (\operatorname{reg} A) \cap U$, tend to $\mathbb{C}_{1 \ldots p}$ as $z \to 0$ (with respect to the coordinates chosen in this Remark), hence $C_4(A, 0)$ is this limiting p-dimensional plane. On the other hand, since $A \cap U \cap \mathbb{C}_{p \ldots n}$ is parametrically given by the system $z_1 = \cdots = z_{p-1} = 0$, $z_p = r\zeta_p^k$, $z_{p+1} = o(\zeta_p^k)$, the tangent cone at 0 to this irreducible one-dimensional analytic set is the complex line $\mathbb{C}_p$. Thus, $C_4(A, 0) = \mathbb{C}_{1 \cdots (p-1)} \times \mathbb{C}_p = T_0(\operatorname{sng} A) \times C(A \cap \mathbb{C}_{p \cdots n}, 0)$. ■

C o r o l l a r y 2. *Let A be a pure p-dimensional analytic set in $\mathbb{C}^n$, and suppose that the point $0 \in \operatorname{reg}(\operatorname{sng} A)$ is such that $\dim_0(\operatorname{sng} A) = p - 1$ and $\dim C_4(A, 0) = p$. Then there is a neighborhood $U \ni 0$ such that for all $z \in A \cap U$ the tangent cones $C_3(A,z) = C_4(A,z)$ are finite unions of p-dimensional complex planes, and, moreover, these cones change continuously on $\operatorname{sng} A \cap U$.*

■ Indeed, if A' is an irreducible component of A in a small neighborhood of 0, then A' is irreducible at 0 and $\dim C_4(A',0) = p$ also. By Proposition 1, p.9.4, $\operatorname{sng} A' = \operatorname{br} \pi|_{A'}$ in a neighborhood of 0. Since A' is irreducible at 0, Proposition 4.5 implies that $\dim_0(\operatorname{sng} A') = p - 1$ (or $\operatorname{sng} A'$ is empty). Finally, since $\operatorname{sng} A' \subset \operatorname{sng} A$ and since in a neighborhood of 0 the set $\operatorname{sng} A$ is a $(p-1)$-dimensional complex manifold, 0 is a regular point of $\operatorname{sng} A'$. Thus A' satisfies the conditions of Corollary 1, hence $C_1(A',0) = C_4(A',0)$ is a p-dimensional complex plane. The same conditions hold at all points $z \in \operatorname{sng} A'$ sufficiently close to 0. The remaining follows from Corollary 1 and from the additivity property of the cones C_3, C_4 (p.9.1). ■

In conclusion we mention that the conditions $\dim_0 \operatorname{sng} A = p - 1$ and $\dim C_4(A, 0) = p$ are not independent:

P r o p o s i t i o n 2. *Let A be a pure p-dimensional analytic set in $\mathbb{C}^n$, irreducible at $0 \in \operatorname{sng} A$. If $\dim_0 \operatorname{sng} A < p - 1$, then $\dim C_4(A, 0) > p$.*

■ Assume the contrary. Then, after a suitable unitary transformation, $C_4(A, 0) \cap \mathbb{C}_{(p+1) \cdots n} = \{0\}$. Hence there is a neighborhood $U = U' \times U'' \ni 0$ such that $A \cap U \cap \mathbb{C}_{(p+1) \cdots n} = \{0\}$ and $\pi: A \cap U \to U' \subset \mathbb{C}_{1 \cdots p}$ is an analytic cover. Let σ be the critical analytic set of this cover. Since $0 \in \operatorname{sng} A$, $\sigma \ni 0$. Since A is irreducible at 0, $\dim_0 \sigma = p - 1$ (see Proposition 4.5). Since $\dim_0 \operatorname{sng} A < p - 1$ this implies that there exists a sequence of points $a^j \in (\operatorname{reg} A) \cap \operatorname{br} \pi|_A$ converging to 0. By the definition of critical points, $T_{a^j}A$ contains a vector $v^j \in \mathbb{C}_{(p+1) \cdots n}$, $|v^j| = 1$. The definition of $C_4(A, 0)$ implies that $C_4(A, 0)$ also contains such a vector v, contradicting the assumption that $C_4(A, 0) \cap \mathbb{C}_{(p+1) \cdots n} = \{0\}$. ■

Thus, in Proposition 1 and Corollary 2 the condition $\dim_0 \operatorname{sng} A = p - 1$ is superfluous; it automatically follows from the other condition $\dim C_4(A, 0) = p$. Proposition 2 also implies that the set $\{z \in A\colon \dim C_4(A, 0) > p\}$ contains all singular points of A at which $\dim_z \operatorname{sng} A < p - 1$; in particular, it contains all isolated singular points of A if $p > 1$.

10. Multiplicities of holomorphic maps

10.1. Multiplicity of projections. Let A be a pure p-dimensional analytic set in $\mathbb{C}^n$, $a \in A$, and L an $(n-p)$-dimensional complex subspace in $\mathbb{C}^n$ and such that a is an isolated point of the set $A \cap (a+L)$. Then, as we know, there is a domain $U \ni a$ in $\mathbb{C}^n$ such that $A \cap U \cap (a+L) = \{a\}$ such that the projection $\pi_L\colon A \cap U \to U'_L \subset L^\perp$ along L is a k-sheeted analytic cover, for some $k \in \mathbb{N}$. The critical analytic set σ of this cover does not partition the domain U'_L and is nowhere dense in it, therefore the number of sheets of this cover does not change when shrinking U (if z' is the projection of z in $L^\perp$ and $z' \in U'_L \setminus \sigma$, then

$$\sharp A \cap U \cap (z+L) = k$$

and all k points of the fiber above z' tend to a as $z' \to a'$). This number is called the *multiplicity of the projection* $\pi_L|_A$ at a, and is denoted by $\mu_a(\pi_L|_A)$ (see also p.4.3); it is also called the intersection index of A and the plane $a + L$ at a, and is then denoted by $i_a(A, a+L)$, see p.12.1. At points $z \in A$ for which the dimension of the fiber of $\pi_L|_A$ is positive we put $\mu_z(\pi_L|_A) = +\infty$, by definition; for points $z \notin A$ we conveniently set $\mu_z(\pi_L|_A) = 0$.

For any point z in the above-indicated small neighborhood $U \ni a$ the number of sheets of the cover $A \cap U \to U'_L$ does not exceed k in a neighborhood of z (it may be less), hence the function $\mu_z(\pi_L|_A)$ is upper semicontinuous on A; it is also upper semicontinuous in any domain containing A as a closed subset. More detailed information about the level sets of this function is contained in the following statement.

L e m m a 1. *Let A be a pure p-dimensional analytic subset of a domain $U = U' \times U''$ in $\mathbb{C}^n$ such that the projection $\pi\colon A \to U'$ is an analytic cover. Then for each natural number m the set*

$$\{z \in A: \mu_z(\pi|_A) \geqslant m\}$$

is analytic.

■ Denote this set by A_m^π. We first investigate the case $p = n - 1$. Let k be the number of sheets of the cover $\pi|_A$, let $\sigma \subset U'$ be its critical set, and let $F(z',z_n)$ be the Weierstrass polynomial of degree k in z_n corresponding to this cover (see p.2.8). For each fixed $z' \in U' \setminus \sigma$ all zeros of $F(z',z_n)$ in z_n are simple. If now $a \in A_m^\pi$, then there are a neighborhood $V \ni a$ such that $\pi^{-1}(a') \cap A \cap \overline{V} = \{a\}$ and a sequence of points $z_j' \in U' \setminus \sigma$ such that $z_j' \to a'$ and $\sharp \pi^{-1}(z_j') \cap A \cap V \geqslant m$ for all j. Thus, at least m distinct simple zeros of the polynomial $F(z'_j, z_n)$ tend to a_n as $j \to \infty$. Rouché's theorem implies that the order of the zero of $F(a',z_n)$ at $z_n = a_n$ is greater than or equal to m. Analytically this is described by the system of equations $(\partial^i F / \partial z_n^i)(a) = 0$, $i = 0, \ldots, m-1$, hence in case $p = n - 1$ the sets A_m^π are analytic.

The general case is reduced to the previous case by using almost single-sheeted projections (see p.3.6). Let $\lambda(z'')$ be a linear function such that the map $\pi_\lambda: z = (z',z'') \mapsto z_\lambda = (z', \lambda(z''))$ almost single-sheetedly transforms A onto an analytic set A_λ of codimension 1 in $U_\lambda = \pi_\lambda(U) \subset \mathbf{C}^{p+1}$. Let $\pi': (z',\lambda) \mapsto z'$. Then $\pi = \pi' \circ \pi_\lambda$, hence $\pi': A_\lambda \to U'$ is also an analytic cover (see p.3.1) with critical analytic set $\sigma_\lambda \subset U'$. Since $\operatorname{sng} A_\lambda \subset (\pi')^{-1}(\sigma_\lambda)$ and $\pi_\lambda|_A$ is a single-sheeted map above $\operatorname{reg} A_\lambda$, the numbers of sheets of the covers $\pi: A \to U'$ and $\pi': A_\lambda \to U'$ coincide, and $\mu_z(\pi|_A) \leqslant \mu_{z_\lambda}(\pi'|_{A_\lambda})$, in particular $\pi_\lambda(A_m^\pi) \subset (A_\lambda)_m^{\pi'}$. According to the case of codimension 1 already investigated, $(A_\lambda)_m^{\pi'}$ is an analytic subset in U_λ. Thus, A_m^π is contained in the analytic set $C_m^\lambda = (\pi_\lambda)^{-1}((A_\lambda)_m^{\pi'}) \cap A$, and hence also in the intersection of the C_m^λ for all $\lambda \in (\mathbf{C}^{n-p})^*$ such that $\pi_\lambda|_A$ is an almost single-sheeted projection. By p.3.6 such λ form an everywhere dense subset in $(\mathbf{C}^{n-p})^*$, hence for a fixed $a \in A$ the function λ can be chosen such that it takes different values at distinct points of the fiber $\pi^{-1}(a') \cap A$. This additional property and the almost single-sheetedness of $\pi_\lambda|_A$ imply that the map $(\pi|_A) = \mu_{a_\lambda}(\pi'|_{A_\lambda})$. If $a \notin A_m^\pi$ then for this λ also $a_\lambda \notin (A_\lambda)_m^{\pi'}$, hence $a \notin C_m^\lambda$. Thus, $A_m^\pi = \cap C_m^\lambda$ is an analytic subset in U. ■

As a Corollary we note the dependence on a parameter of the multiplicity of projections.

L e m m a 2. *Let A be a pure $(p+q)$-dimensional analytic subset of a domain*

$U \times V$ in $\mathbb{C}^n_z \times \mathbb{C}^q_t$, let $M \subset A$ be a complex submanifold, $M = \{(z,t) : z = z(t),\ t \in V\}$, and let $A_t = \{z: (z,t) \in A\}$. Assume that for each $t \in V$ the set A_t is pure p-dimensional, $p \geqslant 1$, and that the restriction onto each A_t of the projection $\pi: z \mapsto z' \in \mathbb{C}_{1\dots p}$ is an analytic cover above a domain $U' \subset \mathbb{C}_{1\dots p}$ (not depending on t). Let $\tilde{\pi}: (z,t) \mapsto (z',t)$. Then there are a natural number m and a proper (possibly empty) analytic subset $\Sigma \subset V$ such that

$$\mu_{z(t)}(\pi|_{A_t}) = \mu_{(z(t),t)}(\tilde{\pi}|_A) = m \quad \text{for } \textit{all } t \in V \setminus \Sigma.$$

■ By the conditions $\tilde{\pi}|_A$ is an analytic cover above a domain $U \times V$. Since $M \subset A = \bigcup_1^\infty A_j^{\tilde{\pi}}$ and the $A_j^{\tilde{\pi}}$ are becoming smaller as j grows while their intersection over all j is empty, there is an m such that $M \subset A_m^{\tilde{\pi}}$ but M is not contained in $A_{m+1}^{\tilde{\pi}}$. Then $\mu_{(z,t)}(\tilde{\pi}|_A) = m$ for all (z,t) from $M \setminus A_{m+1}^{\tilde{\pi}}$, and this set is nonempty. By Lemma 1, $M \cap A_{m+1}^{\tilde{\pi}}$ is a proper analytic subset in M, hence its projection Σ_1 in V is also a proper analytic subset, in V. By construction, $\mu_{(z(t),t)}(\tilde{\pi}|_A) = m$ for all $t \in V \setminus \Sigma_1$.

Further, let σ be the critical analytic set of the cover $\tilde{\pi}|_A$ and let

$$\Sigma_2 = \{t \in V: (z',t) \in \tilde{\sigma} \text{ for all } z' \in U'\}.$$

Then Σ_2 is a proper analytic subset in V (it is equal to $\cap\ \{t: (z',t) \in \tilde{\sigma}\}$ over all $z' \in U'$). Put $\Sigma = \Sigma_1 \cup \Sigma_2$. Let $t \in V \setminus \Sigma$, and let W be a neighborhood of the point $(z(t),t)$ such that $\tilde{\pi}|_{A \cap W}$ is an m-sheeted analytic cover (by construction such a neighborhood does exist, since $t \notin \Sigma_1$). Since $t \notin \Sigma_2$ there is a sequence of points $z_j' \in U'$ such that

$$z_j' \to z(t)' \text{ and } \#\tilde{\pi}^{-1}((z'_j,t)) \cap A \cap W = m$$

for all j. Since $\tilde{\pi}^{-1}((z_j',t)) = (\pi^{-1}(z_j'),t)$, the number of points of the fiber $\pi^{-1}(z_j') \cap A_t$ in the neighborhood $W_t = \{z: (z,t) \in W\}$ of $z(t)$ is also equal to m for all j, and cannot be larger. Thus, $\mu_{z(t)}(\pi|_{A_t}) = m$. ■

Note the particular case in which Σ is empty:

C o r o l l a r y. *If for all $t \in V$ the following conditions are fulfilled: 1) $\pi^{-1}(z(t)') \cap A_t = \{z(t)\}$; and 2) $\operatorname{br} \tilde{\pi}|_A$ is nowhere dense in $A_t \times \{t\}$, then $\mu_{z(t)}(\pi|_{A_t}) = \mu_{(z(t),t)}(\tilde{\pi}|_A) = m$ for all $t \in V$.*

■ Indeed, 1) implies that $\mu_{(z(t),t)}(\tilde{\pi}|_A)$ coincides with the multiplicity of the whole cover $\tilde{\pi}|_A$; if it equals m, then the set Σ_1 from Lemma 2 is empty. Since $\tilde{\sigma}$ is the projection of br $\tilde{\pi}|_A$, the second condition implies that Σ_2 is also empty. ■

10.2. Multiplicity of maps. Multiplicity of projections is a particular instance of the more general notion of multiplicity of holomorphic maps. Let A be a pure p-dimensional analytic set on a complex manifold and let $f: A \to Y$ be a holomorphic map into another complex manifold Y (see p.2.7). Assume that $a \in A$ is an isolated point in the fiber $f^{-1}(f(a))$ and that $\dim Y = p$. Then (p.3.1) there are coordinate neighborhoods $U \ni a$ and $V \ni f(a)$ such that $f: A \cap U \to V$ is a proper map and, moreover, $f^{-1}(f(a)) \cap A \cap \overline{U} = \{a\}$. By p.4.3 it is an analytic cover. Its multiplicity (number of sheets) does not depend on the choice of U, V with the properties indicated; it is called *multiplicity of the map f* at a, and is denoted by $\mu_a(f|_A)$ or simply by $\mu_a(f)$. For each point $w \in V \setminus \sigma$, where σ is the critical analytic set of the cover given above, the number of pre-images, $\# f^{-1}(w) \cap A \cap U$, is exactly equal to $\mu_a(f)$, all these points on A are regular, f has maximal rank p at them (the Jacobian with respect to local coordinates is distinct for zero), and is locally biholomorphic. For points $w \in \sigma$ the number of pre-images on $A \cap U$ is strictly less then $\mu_a(f)$ (see p.3.3), hence

$$\mu_a(f) = \overline{\lim_{w \to f(a)}} \# f^{-1}(w) \cap A \cap U.$$

This equation allows us to define the notion of multiplicity of a map also in case $\dim Y > p$. This definition requires U only to be such that

$$f^{-1}(f(a)) \cap A \cap \overline{U} = \{a\}.$$

If $\dim Y = p$, the definition implies that $\mu_a(f) = 1$ for $a \notin \operatorname{br} f$ and $\mu_a(f) > 1$ for $a \in \operatorname{br} f$; in particular, $\mu_a(f) > 1$ at all singular points of the set A (at the points $a \in A$ that are not isolated in their fiber $f^{-1}(f(a))$ we put $\mu_a(f) = +\infty$).

The action of an arbitrary holomorphic map leads to projection of its graph. If $\Gamma_f = \{(z, f(z)) : z \in A\} \subset A \times Y$ is the graph of a map $f: A \to Y$, and if $\pi: A \times Y \to Y$ is the natural projection, then, obviously,

$$\mu_a(f|_A) = \mu_{(a, f(a))}(\pi|_{\Gamma_f}).$$

Since multiplicity is a local notion, the proofs of a lot of statements concerning multiplicities lead to the standard situation of p.10.1, in which A is an analytic subset of a domain $U = U' \times U''$ in $\mathbb{C}^n$ and $f = \pi|_{A \cap U}$ is a proper projection into U'.

If $A = \cup_1^k A_j$ with all A_j pure p-dimensional, and if $\dim A_j \cap A_i < p$ for $i \neq j$, then, obviously,

$$\mu_a(f|_{\cup A_j}) = \sum \mu_a(f|_{A_j})$$

(*additivity* of multiplicity). Since $A \cap U$ decomposes into irreducible components and since these components are irreducible at a if U is a small neighborhood of a (p.5.4), we often restrict ourselves to analytic sets that are irreducible at a point.

Further, if $A \to^f Y \to^g Z$ are holomorphic maps with discrete fibers, then it is also obvious that

$$\mu_a(g \circ f|_A) = \mu_a(f|_A) \cdot \mu_{f(a)}(g|_{f(A)})$$

(*multiplicativity* of multiplicity). Hence, in particular, we obtain that *multiplicities of maps are invariant under biholomorphic transformations, where the multiplicities are, of course, taken at the image and pre-image.*

If $f: A \to Y$, $g: B \to Z$, then the map $(f,g): (z,w) \mapsto (f(z), f(w))$ can be represented as the superposition $(z,w) \mapsto (f(z), w) \mapsto (f(z), g(w))$. The multiplicity at a of the first map is obviously equal to $\mu_a(f|_A)$, the multiplicity at $(f(a), b)$ of the second map is equal to $\mu_b(g|_B)$, hence

$$\mu_{(a,b)}((f,g)|_{A \times B}) = \mu_a(f|_A) \cdot \mu_b(g|_B).$$

The notion of multiplicity of a map is meaningful not only on analytic sets but also on complex manifolds themselves. In case $A = \operatorname{reg} A$ is a complex manifold and $\dim Y = \dim A$, the set of points at which the multiplicity of a map $f: A \to Y$ exceeds 1, coincides with $\operatorname{br} f$ (locally it is the zero set of the Jacobian with respect to local coordinates).

We consider some examples of calculating multiplicities.

E x a m p l e s. (a) If $\dim A = 1$ and $Y = \mathbb{C}$, a holomorphic map $f: A \to Y$ is a holomorphic function on A. Rouché's theorem readily implies that at regular points $a \in A$ we have $\mu_a(f) = \operatorname{ord}_a(f - f(a))$ (the order of the zero is considered with respect to local coordinates on $\operatorname{reg} A$).

(b) Let A be one-dimensional and irreducible at a point $a \in \operatorname{sng} A$. By p.6.1 there are a neighborhood $U \ni a$ and a holomorphic parametrization $z = z(\zeta)$ of the set $A \cap U$, $z(0) = a$. For any holomorphic function $f\colon A \to \mathbb{C}$, $f(a) = 0$, the function $f(z(\zeta))$ is holomorphic in the disk $|\zeta| < 1$. Since the map $z = z(\zeta)$ is one-to-one, Example (a) and multiplicativity of multiplicity imply $\mu_a(f) = \operatorname{ord}_0 f(z(\zeta))$.

(c) Let A be a pure p-dimensional, $f\colon A \to U' \subset \mathbb{C}^p$ a proper map, σ its critical analytic set, $a \in A$, and $f^{-1}(f(a)) = \{a\}$. Let $S' \subset U'$ be an arbitrary one-dimensional complex manifold containing $f(a)$ and intersecting with σ in a zero-dimensional set. Then the set $f^{-1}(S') \cap A = S$ is pure one-dimensional and $f: S \to S'$ is an analytic cover. Since the numbers of sheets of the covers $f|_A$ and $f|_S$ coincide, $\mu_a(f|_A) = \mu_a(f|_S)$.

(d) Let $f\colon (z_1,z_2) \mapsto (z_1^k, z_2^l)$ be a map from $\mathbb{C}^2$ into $\mathbb{C}^2$. Each point $w = (w_1,w_2)$ with $w_1 w_2 \neq 0$ has number of pre-images equal to kl. As $w \to 0$ all these tend to 0, hence $\mu_{(0,0)}(f) = kl$. If $w = (a_1^k, w_2)$, $a_1 \neq 0$ and $w_2 \to 0$, then to each point $(a_1 e^{2\pi\nu i/k}, 0)$, $\nu = 1,\ldots,k$, tend exactly l pre-images of w, hence $\mu_{(a_1,0)}(f) = l$. Similarly, $\mu_{(0,a_2)}(f) = k$ if $a_2 \neq 0$. Since $\operatorname{br} f \subset \{z_1 z_2 = 0\}$, we have $\mu_a(f) = 1$ at all other points.

We will investigate one effective method allowing us to calculate multiplicities of holomorphic maps in the simples cases (cf. [205]). First we formulate a generalization of the important Example (b).

L e m m a 1. *Let A be a one-dimensional analytic subset of a domain $U = U_1 \times U''$ in $\mathbb{C}^n$ such that the projection $\pi\colon A \to U_1$ is a k-sheeted analytic cover with unique critical point a (or without critical points at all); moreover, suppose $\pi^{-1}(a) \cap A = \{a\}$. Let $f: A \to \mathbb{C}$ be a holomorphic function, $f(a) = 0$. Then:*

(1) if A is irreducible at a and $z = z(\zeta)$ is its Puiseux parametrization in a neighborhood of $a = z(0)$, then

$$\mu_a(f) = \operatorname{ord}_0 f(z(\zeta));$$

(2) if $\pi^{-1}(z_1) \cap A = \{(z_1, \alpha^j(z_1)),\ j = 1,\ldots,k\}$, $z_1 \neq 0$, and

$$g(z_1) = \prod_1^k f(z_1, \alpha^j(z_1)),\quad g(a_1) = 0,$$

then $\mu_a(f) = \operatorname{ord}_{a_1} g(z_1)$.

■ The first situation is investigated in Example (b), we prove the second statement. Let $a = 0$. By shrinking U we may assume that $A = \bigcup_1^l A_j$, $l \leq k$, where all A_j are one-dimensional, irreducible at 0, and also satisfy the conditions of the Lemma. The function g for A is equal to $g_1 \cdots g_l$, where g_j corresponds to A_j. Hence it suffices to consider the case in which A is irreducible at 0 and is the bijective image of a disk under a holomorphic map $z = z(\zeta) = (\zeta^k, z''(\zeta))$. Using a suitable numeration, $\alpha^j(z_1) = z''(z_1^{1/k} e^{2\pi ji/k})$, $j = 1, \ldots, k$ ($z_1^{1/k}$ is some fixed value of the root), hence the function

$$g(z_1(\zeta)) = \prod_1^k f(\zeta^k, z''(\zeta e^{2j\pi i/k}))$$

is holomorphic in a disk $|\zeta| < r$. Since all functions

$$f(\zeta^k, z''(\zeta e^{2j\pi i/k})) = f(z(\zeta e^{2j\pi i/k}))$$

have at $\zeta = 0$ identical order of the zero, equal to $\mu_a(f)$ (Example (b)), $\mu_0(g(z_1(\zeta))) = k\mu_0(f)$. Since $z_1(\zeta) = \zeta^k$, in view of the multiplicativity of multiplicity $\operatorname{ord}_0 g(z_1(\zeta)) = k \operatorname{ord}_0 g(z_1)$, hence $\operatorname{ord}_0 g(z_1) = \mu_a(f)$. ■

The following Lemma indicates how we can diminish the dimension of a set being mapped; in particular, how to reduce holomorphic maps in $\mathbf{C}^m$, $m > 1$, to holomorphic functions.

L e m m a 2. *Let A be a pure m-dimensional analytic set, and let $f = (f_1, \ldots, f_m)\colon A \to \mathbf{C}^m$ be a holomorphic map such that $f(a) = 0$ and such that a is an isolated point of the fiber $f^{-1}(0)$. Assume that the p-dimensional analytic set $'A = \{z \in A\colon f_{p+1}(z) = \cdots = f_m(z) = 0\}$ intersects* $\operatorname{br} f|_A$ *in a set of dimension $< p$. Then $\mu_a(f|_A) = \mu_a('f|_{'A})$, where $'f = (f_1, \ldots, f_p)$.*

■ The statement being local we assume that A is an analytic set in $\mathbf{C}^n$. Let $\tilde{A} = \{(z, f(z))\colon z \in A\}$ be the graph of $f|_A$ in $\mathbf{C}^n \times \mathbf{C}^m$, and let $\tilde{\pi}\colon (z, w) \mapsto w$. Since a is isolated in $f^{-1}(0)$ on A, there is a domain $\tilde{U} = V \times U \ni (a, 0)$ in $\mathbf{C}^n \times \mathbf{C}^m$ such that $\tilde{\pi}\colon \tilde{A} \cap \tilde{U} \to U$ is a proper map with $\tilde{\pi}^{-1}(0) \cap \tilde{A} \cap \tilde{U} = \{(a, 0)\}$. By construction and the definition of multiplicity, the number of sheets of this cover is $\mu_a(f|_A) =: \mu$.

The set $'A$ corresponds on $\tilde{A}$ to the set

$$'\tilde{A} = \{(z, w) \in \tilde{A}\colon w_{p+1} = \cdots = w_n = 0\}.$$

If $\tilde{\sigma} \subset U$ is the critical analytic set of the cover $\tilde{\pi}|_{\tilde{A} \cap \tilde{U}}$, then $\tilde{\sigma} = \tilde{\pi}(\operatorname{br} \tilde{\pi}|_{\tilde{A}}) = f(\operatorname{br} f|_A)$. The condition implies that

$$\tilde{\sigma} \cap \{w_{p+1} = \cdots = w_n = 0\} = \tilde{\sigma} \cap \tilde{\pi}('\tilde{A})$$

is nowhere dense in $\tilde{\pi}('\tilde{A})$, hence the cover $\tilde{\pi}: '\tilde{A} \to U \cap \{w_{p+1} = \cdots = w_n = 0\}$ is also μ-sheeted. Since $\tilde{\pi}^{-1}(0) \cap '\tilde{A} \cap U = \{(a, 0)\}$, we find $\mu_0('f|_{'A}) = \mu_{(a,0)}(\tilde{\pi}|_{'\tilde{A}}) = \mu$. ■

C o r o l l a r y. *Let D be a domain in $\mathbf{C}^n$, and let $f: D \to \mathbf{C}^n$ be a holomorphic map such that $f(a) = 0$ and such that a is an isolated point in $f^{-1}(0)$. Let $S = \{z \in D: f_2(z) = \cdots = f_n(z) = 0\}$, and suppose that the rank of the matrix $(\partial f_j / \partial z_k)$, $j \geqslant 2$, $k \geqslant 1$, is equal to $n-1$ on an everywhere dense subset in S. Then $\mu_a(f) = \mu_a(f_1|_S)$.*

■ At all points where this rank equals $n-1$ the analytic set S is one-dimensional. Since this set, $'S$, is dense in S, S is pure-one-dimensional. Since a is isolated in $f^{-1}(0)$, in a neighborhood of a the fibers of $f_1|_S$ are zero-dimensional, hence $df_1|_S \neq 0$ on an everywhere dense open subset $S'' \subset S \cap U$, where U is a sufficiently small neighborhood of a. But then the rank of the matrix $\partial f / \partial z$ is equal to n at the points of the set $S' \cap S'' \subset S$, i.e. all these points lie outside $\operatorname{br} f$. Since $S' \cap S''$ is everywhere dense in S, this implies that the set $S \cap (\operatorname{br} f)$ is zero-dimensional, hence the conditions of Lemma 2 are fulfilled. ■

The condition on the rank of the Jacobi matrix (as well as the condition on the intersection of $\operatorname{br} f$ and $'A$ in Lemma 2) can be dropped, but it is then necessary to regard $'A$ with the corresponding multiplicities as a holomorphic chain (see p.11.5 and §12).

E x a m p l e s. (a) Let $f: (z_1, z_2) \mapsto (z_1^{k_1} z_2^{k_2}, z_2^2 - z_1^3)$ and put $a = (0,0)$. Then $S: z_2^2 = z_1^3$ is the semicubic parabola, and the conditions of the Corollary are obviously fulfilled. The set S is parametrized by the map $\zeta \mapsto (\zeta^2, \zeta^3)$, hence $\mu_a(f) = \operatorname{ord}_0 f_1(z(\zeta)) = 2k_1 + 3k_2$.

(b) For the map $f: (z_1, z_2) \mapsto (z_1^{k_1} - z_1 z_2,\ z_2^{k_2} - z_1 z_2)$ the conditions of the Corollary are also fulfilled, and for $k_2 > 1$ the set S is the union of the manifolds $S_1: z_2 = 0$ and $S_2: z_1 = z_2^{k_2 - 1}$. Since $f_1|_{S_1} = z^{k_1}$, we have $\mu_0(f_1|_{S_1}) = k_1$. Since $f_1|_{S_2} = z_2^{k_1(k_2-1)} - z_2^{k_2}$ and z_2 is a local coordinate on S_2, we have $\mu_0(f_1|_{S_2}) = k_2$ $(k_1 > 1)$. Thus, $\mu_0(f) = k_1 + k_2$ if $k_1, k_2 > 1$.

(c) Let

$$g(z) = z_1^4 + z_2^4 + z_3^4 - z_1 z_2 z_3$$

and

$$f\colon (z_1,z_2,z_3) \mapsto \nabla g = (4z_1^3 - z_2 z_3, 4z_2^3 - z_1 z_3, 4z_3^3 - z_1 z_2).$$

It is easy to convince ourselves that the rank of the Jacobi matrix for (f_2,f_3) is equal to 2 on all of the set $S\colon f_2 = f_3 = 0$, except at $z = 0$, i.e. the conditions of the Corollary are fulfilled. The set S is an analytic cover above the $\mathbb{C}_1$-axis (the unique limit point of S at infinity in $\mathbb{P}_3$ is $[0,1,0,0]$), and $S \cap \mathbb{C}_{23} = \{0\}$. By Bezout's theorem (p.10.4) the number of sheets of S above $\mathbb{C}_1$ is equal to 9. It is easily seen that for $z \in S$ the values of z_2 and z_3 are proportional to $\sqrt{z_1}$ hence, at the points of the fiber above z_1 distinct from $(z_1,0,0)$, f_1 is proportional to z_1, and $f_1(z_1,0,0) = 4z_1^3$. Hence the function $g(z_1)$ from Lemma 1 is equal to Cz_1^{8+3}, so $\mu_0(f) = 11$.

10.3. Multiplicities and initial polynomials. In order to calculate multiplicities of maps Rouché's theorem, which is well known in the one-dimensional case, is very useful.

T h e o r e m 1. *Let D be a domain with compact closure on an n-dimensional complex manifold Ω, let f,g be holomorphic maps from D into $\mathbb{C}^n$, continuous on $\overline{D}$, and such $|f - g| < |f|$ on ∂D. Then the numbers of zeros (pre-images of the coordinate origin $0 \in \mathbb{C}^n$) of f and g in D, counted with multiplicities, are equal:*

$$\sum_{f(a)=0} \mu_a(f|_D) = \sum_{g(b)=0} \mu_b(g|_D).$$

■ Let $f_t = f + t(g - f)$, $t \in \mathbb{C}$, $|t| < 1 + \epsilon$, where $\epsilon > 0$ is small such that

$$\min_{\partial D} |f_t| \geqslant \min_{\partial D}(|f| - (1+\epsilon)|g - f|) = r > 0.$$

Define in $\Omega \times \mathbb{C}$ the domain $\tilde{D} = \{(z,t)\colon z \in D, |f_t(z)| < r\}$. Then $F\colon (z,t) \mapsto (f_t(z),t)$ is a proper map, hence an analytic cover of $\tilde{D}$ above the domain $W = \{|w| < r, \quad |t| < 1 + \epsilon\} \subset \mathbb{C}^{n+1}$. Since $\det(\partial F / \partial(z,t)) = \det(\partial f_t / \partial z)$, the critical set σ of this cover is $\{(w,t)\colon |t| < 1 + \epsilon, w \in \sigma_t\}$, where σ_t is the critical analytic set of the cover $f_t\colon \{|f_t(z)| < r\} \cap D \to \{|w| < r\}$. Each point $(w,t) \notin \sigma$ has the same number of pre-images in $\tilde{D}$ under F, say equal to k. Therefore, for fixed t, $|t| < 1+\epsilon$, the number of pre-images of $w \notin \sigma_t$, $|w| < r$, in D under f_t is also equal to k. Since $f = f_0$ and $g = f_1$, we have $\sharp f^{-1}(w) \cap D = \sharp g^{-1}(w) \cap D$ if $w \notin \sigma_0 \cup \sigma_1$, $|w| < r$. Since $\sigma_0 \cup \sigma_1$ is nowhere dense in the ball $|w| < r$, this equality (already counted with multiplicities) holds for all w, $|w| < r$, in particular for $w = 0$. ■

R e m a r k. The only fact that was used essentially in the proof is the existence of a family of holomorphic maps f_t: $D \to \mathbb{C}^n$, holomorphically depending on the parameter t in a domain in the plane $\mathbb{C}$ containing 0 and 1, with the condition $f_0 = f$, $f_1 = g$, all f_t being continuous on $\overline{D}$ and zero free on ∂D. Hence the condition $|f - g| < |f|$ in Rouché's theorem can be varied in various directions, e.g. it suffices to require that only $|f_1 - g_2| < |f_1|$ on ∂D, etc. The main corollary of this Theorem is that under holomorphic variation of a map the number of zeros does not change (until the zeros appear on the boundary), and that zero sets change continuously under such variations (in the Hausdorff metric of deviation, see p.8.6). We will repeatedly use these properties in the sequel.

C o r o l l a r y 1. *Let $f = (f_1, \ldots, f_n)$ be a holomorphic map defined in a neighborhood of 0 in $\mathbb{C}^n$, let $(f_j)_*$ be the initial homogeneous polynomials in the expansion of f at 0, and let $(f)_* = ((f_1)_*, \ldots, (f_n)_*)$. Now, if 0 is the only zero of the system $(f)_*$, then $\mu_0(f) = \mu_0((f)_*)$.*

■ Taylor's formula implies that

$$f_j(tz) = t^{k_j}[(f_j)_*(z) + tg_j(z,t)],$$

where g_j are holomorphic in a neighborhood of 0 in $\mathbb{C}^n_z \times \mathbb{C}_t$. Since by requirement $\min_{|z|=1} |(f)_*(z)| = m > 0$, there is an $r > 0$ such that $f(tz)$ is defined and $|tg(z,t)| < m$ for $|z| \leqslant 1$, $|t| \leqslant r$. Thus, for $|z| = r$ we have $|f - (f)_*| < |(f)_*|$, hence, by Rouché's theorem, f and $(f)_*$ have the same number of zeros, counted with multiplicities, in the ball $|z| < r$. Since r can be taken arbitrarily small and since 0 is the only zero of $(f)_*$, we have $\mu_0(f) = \mu_0((f)_*)$. ■

The following statement is a natural generalization of the Theorem on the tangent cone to a hypersurface (p.8.4).

C o r o l l a r y 2. *Let $f_1, \ldots, f_q$ be holomorphic functions in a neighborhood of 0 in $\mathbf{C}^n$, let $A\colon f_1 = \cdots = f_q = 0$, let $(f_j)_*$ be the initial homogeneous polynomial of f_j at 0, and let $(f)_* = ((f_1)_*, \ldots, (f_q)_*)$. If the dimension of the set $A_* = \{z\colon (f)_*(z) = 0\}$ is equal to $n-q$, then*

$$C(A,0) = A_* = \cap_{j=1}^{q} C(\{f_j = 0\}, 0).$$

■ If $\dim A_* = n-q$, there is a q-dimensional subspace $L \subset \mathbf{C}^n$ such that $L \cap A_* = \{0\}$. Rotation gives $L = \mathbf{C}^q_{z''}$, where $z = (z', z'')$, $z'' = (z_{n-q+1}, \ldots, z_n)$. Let $v \in A_*$, $|v| = 1$. Then $v' \neq 0$ and v is an isolated point of the set $\{z' = v'\} \cap A_*$ (projection of A_* to $\mathbf{C}^{n-q}_{z'}$ is proper). Therefore, for every $\epsilon > 0$ there is a neighborhood $U''_\epsilon \ni v''$ in the ball $B(v'', \epsilon) \subset \mathbf{C}^q_{z''}$ such that $(v' \times \partial U''_\epsilon) \cap A_*$ is empty, i.e. $|(f)_*(v', z'')| \geqslant c_\epsilon > 0$ for all $z'' \in \partial U''_\epsilon$. Since

$$f_j(tv', tz'') = t^{k_j}[(f_j)_*(v', z'') + t g_j(v', z'', t)],$$

there is an $r_\epsilon > 0$ such that

$$|(f - (f)_*)(tv', tz'')| < |(f)_*(tv', tz'')|$$

for all $z'' \in \partial U''_\epsilon$ and $0 < |t| < r_\epsilon$. By Rouché's theorem, the system of equations $f(tv', tz'') = 0$ has a solution $z''_t \in U''_\epsilon$ (i.e. $|z''_t - v''| < \epsilon$). Setting $z^t = (tv', tz''_t)$ we obtain $z^t \in A$, $z^t \to 0$ as $t \to 0$, and $|t^{-1} z^t - v| = |(v', z''_t) - v| < \epsilon$. Since $\epsilon > 0$ was arbitrary, we find $v \in C(A, 0)$, i.e. we have proved that $A_* \subset C(A, 0)$. The converse inclusion is trivial (see p.8.4). ■

The following Theorem of A.K. Tsikh and A.P. Yuzhakov, [205], [3], essentially supplements the statements proved, and can be very useful when calculating multiplicities.

T h e o r e m 2. *Let $f = (f_1, \ldots, f_n)$ be a holomorphic map of a neighborhood of 0 in $\mathbf{C}^n$ such that $f(0) = 0$ and $f^{-1}(0) = \{0\}$. Let $(f_j)_*$ be the initial homogeneous polynomial of f_j at 0, and let $(f)_* = ((f_1)_*, \ldots, (f_n)_*)$.*

Now if $(f)_*^{-1}(0) = \{0\}$ (i.e. the $(f_j)_*$ have unique common zero $z = 0$), then

$$\mu_0(f) = \prod_1^n \operatorname{ord}_0 f_j = \prod_1^n \deg (f_j)_*.$$

In the opposite case $\mu_0(f) > \prod_1^n \mathrm{ord}_0 f_j$.

■ For $n = 1$ only the first situation occurs, and this situation is well known from a standard course on the theory of functions of a complex variable.

$n-1 \Rightarrow n$. By the condition there are neighborhoods V, $W \ni 0$ such that $f\colon V \to W$ is proper. Let $\sigma \subset W$ be the critical analytic set of this cover, and let $L \ni 0$ be a complex line such that 0 is an isolated point in $L \cap \sigma$. Put $S = f^{-1}(L \cap W)$. Then S is a one-dimensional analytic subset in V and $\mu_0(f) = \mu_0(f|_S)$ (see Example 10.2(c)). Multiplicity does not change under unitary changes of coordinates in the image, hence we may assume that $L = \mathbb{C}_1$, i.e. $S\colon f_2 = \cdots = f_n = 0$. Now we change the coordinates in the pre-image such that the one-dimensional cone $C(S,0)$ is properly projected into the $\mathbb{C}_1$-axis. Then 0 is an isolated point in $S \cap \mathbb{C}_{2\ldots n}$, hence there is a neighborhood $U = U_1 \times U'' \subset V$ such that $\pi\colon S \cap U \to U_1 \subset \mathbb{C}_1$ is a k-sheeted analytic cover with unique singular point 0 or without singular points at all, and with $\pi^{-1}(0) \cap S \cap U = \{0\}$. What is k equal to?

Since 0 is an isolated point in $L \cap \sigma$, the rank of the Jacobian matrix $\partial f''/\partial z$, where $f'' = (f_2, \ldots, f_n)$, at points $z \in S \setminus \{0\}$ is equal to $n-1$. Since the plane $z_1 = \text{const}$ is transversal to $S \cap U$ if U is sufficiently small, the rank of $\partial f''/\partial z''$ at the points of $S \cap U \setminus \{0\}$ is also equal to n^{-1}. Thus we may assume that for each fixed $z_1 = t \in U_1 \setminus \{0\}$ the system

$$f''(t,z'') = (f_2(t,z''), \cdots, f_n(t,z''))$$

has only simple zeros (i.e. zeros of multiplicity 1) in $U'' \subset \mathbb{C}^{n-1}$, and that their number is k. The maps $f''(t,z'')$ holomorphically depend on $t \in U_1$, hence (see Rouché's theorem) the map $f''(0,z'')$ also has k zeros in U'', counted with multiplicities. Since $z'' = 0$ is the only zero of the latter system, we have proved that $k = \mu_0(f''(0,z''))$. By the induction hypothesis,

$$k \geqslant \prod_2^n \mathrm{ord}_0 f_j(0,z'') \geqslant \prod_2^n \mathrm{ord}_0 f_j(z).$$

If $(f_j)_*(0,z'') \equiv 0$, then for this j we have $\mathrm{ord}_0 f_j(z) < \mathrm{ord}_0 f_j(0,z'')$, hence $k > \prod_2^n \mathrm{ord}_0 f_j$. If all $(f_j)_*(0,z'') \not\equiv 0$ but have a common zero $z'' \neq 0$, then $k > \prod_2^n \mathrm{ord}_0 f_j(0,z'')$ by the induction hypothesis. Thus $k = \prod_2^n \mathrm{ord}_0 f_j(z)$ if and only if $S_* \cap \mathbb{C}^{n-1}_{z''} = \{0\}$, where $S_*\colon (f_2)_* = \cdots = (f_n)_* = 0$ (in this case, of course dim $S_\star = 1$).

If $z_1 \in U_1 \setminus \{0\}$, and $\alpha^j(z_1)$, $j = 1, \ldots, k$, are the z''-coordinates of points in the fiber $\pi^{-1}(z_1) \cap S \cap U$, then $|\alpha^j(z_1)| \leqslant C_1 |z_1|$ in view of the transversality of S to $\mathbf{C}_{2\ldots n}$. Hence $|f_1(z_1, \alpha^j(z_1))| \leqslant C_2 |z_1|^{k_1}$, where $k_1 = \operatorname{ord}_0 f_1(z)$, and thus, by Lemma 1, p.10.2,

$$\mu_0(f) = \mu_0(f_1|_S) = \operatorname{ord}_0 \prod_1^k f_1(z_1, \alpha^j(z_1)) \geqslant k k_1 \geqslant \prod_1^n \operatorname{ord}_0 f_j(z).$$

As already been shown, for equality it is necessary that $S_\star$ be one-dimensional. Suppose this condition is fulfilled. Then, by Corollary 2, $S_\star = C(S, 0)$. If $(f_1)_\star(v) = 0$ for some vector $v \in S_\star \setminus \{0\}$, and if $S' \subset S$ is the irreducible component of S with tangent cone $\mathbf{C}v$, then $|f_1(z)| = o(|z_1|^{k_1})$ for $z \in S'$, $z \to 0$, and $|\prod_1^k f_1(z_1, \alpha^j(z_1))| = o(|z_1|^{kk_1})$ as $z_1 \to 0$. Thus, $\mu_0(f) > \prod_1^n \operatorname{ord}_0 f_j$ in this case. If now $S_\star \cap \{(f_1)_\star = 0\} = \{0\}$, then on $S \cap U$ we have the estimate

$$|f_1(z)| = |(f_1)_\star(z)|(1 + o(1)) = C_3 |z_1|^{k_1}(1 + o(1))$$

for some constant $C_3 > 0$. It implies that

$$\operatorname{ord}_0 \prod_1^k f_1(z_1, \alpha^j(z_1)) = k \cdot k_1 = \prod_1^n \operatorname{ord}_0 f_j(z). \quad \blacksquare$$

In applications we more often use a simple generalization of this Theorem, in which the homogeneous polynomials are replaced by quasihomogeneous polynomials (see, e.g., [3]). Recall that a polynomial $p(z)$ is called *weighted homogeneous* (or *quasihomogeneous*) with respect to a weight $(a_1, \ldots, a_n)$, where $a_i > 0$ are rational numbers, if all monomials $z_1^{i_1} \cdots z_n^{i_n}$ (with nonzero coefficients) have one and the same weighted degree $d = i_1 a_1 + \cdots + i_n a_n$.

T h e o r e m 2'. *Let $f = (f_1, \ldots, f_n)$ be a holomorphic map of a neighborhood of 0 in $\mathbf{C}^n$ such that $f(0) = 0$, $f^{-1}(0) = \{0\}$. Assume that $f_j = p_j + \epsilon_j$, where p_j is a weighted homogeneous polynomial of degree d_j with respect to a weight $(a_1, \ldots, a_n)$, and ϵ_j is a series of polynomials having (with respect to this weight) degrees larger than d_j, $j = 1, \ldots, n$. Then*

$$\mu_0(f) \geqslant \frac{d_1}{a_1} \cdots \frac{d_n}{a_n},$$

and equality holds if and only if $z = 0$ is an isolated zero of the system $(p_1, \ldots, p_n)$.

■ Let $m \in \mathbb{N}$ be such that all numbers ma_j are integers. Make the change of variables $z_j = \zeta_j^{ma_j}$, $j = 1, \ldots, n$. Then $z_1^{i_1} \cdots z_n^{i_n}$ is transformed to a monomial in ζ of degree $m(i_1 a_1 + \cdots + i_n a_n)$, and thus

$$F_j(\zeta) := f_j(z(\zeta)) = P_j(\zeta) + \delta_j(\zeta),$$

where P_j is a homogeneous polynomial of degree md_j and $\delta_j(\zeta)$ is a series of polynomials of degrees larger than md_j. By Theorem 2, $\mu_0(F) \geqslant md_1 \cdots md_n$, with equality if and only if $\zeta = 0$ is an isolated zero of the system $(P_1, \ldots, P_n)$. Since, obviously, the map $z = z(\zeta)$ has multiplicity $ma_1 \cdots ma_n$ at 0 and is finitely-sheeted, multiplicativity of multiplicity implies that

$$\mu_0(F(\zeta)) = \mu_0(f(z)) \cdot \mu_0(z(\zeta)) = \mu_0(f) \cdot ma_1 \cdots ma_n,$$

i.e. $\mu_0(f) \geqslant (d_1 \cdots d_n) / (a_1 \cdots a_n)$, with equality if and only if 0 is an isolated zero of the system $(p_1, \ldots, p_n)$. ■

E x a m p l e. Let $f_1(z) = z_1^2 + z_2^4$, $f_2(z) = z_1^3 + z_2^5$. By Theorem 2, we can only say that $\mu_0(f) \geqslant 6$. However, with respect to the weight $a_1 = 2$, $a_2 = 1$ the first function is quasihomogeneous of degree 4, while for f_2 the initial quasihomogeneous polynomial is z_2^5 (of weighted degree 5). Since f_1 and z_2^5 have unique common zero at $z = 0$, Theorem 2' implies $\mu_0(f) = 10$.

10.4. Bezout's theorem. Using Theorem 2, p.10.3, and Rouché's theorem we can easily prove Bezout's remarkable theorem concerning the number of solutions of a system of polynomial equations. Let $p = (p_1, \ldots, p_n)$ be a system of homogeneous polynomials in $(z_0, \ldots, z_n)$, and let $a \in \mathbb{P}_n$ be an isolated zero of this system (i.e. an isolated point of the set of common zeros in $\mathbb{P}_n$ of the polynomials p_j). Let $(a_0, \ldots, a_n)$ be the homogeneous coordinates of a, with $a_i = 1$ for a certain i. The multiplicity $\mu_a(p)$ of the system p at a is by definition the multiplicity of the zero of the system of functions $p_j(z_0, \ldots, z_{i-1}, 1, z_{i+1}, \ldots, z_n)$, $j = 1, \ldots, n$, at $(a_0, \ldots, a_{i-1}, a_{i+1}, \ldots, a_n) \in \mathbb{C}^n$. Since multiplicity of holomorphic maps is invariant under biholomorphic changes of coordinates, this definition does not depend on the choice of an affine chart in $\mathbb{P}_n$ containing the point a.

T h e o r e m. *Let $p = (p_1, \ldots, p_n)$ be a system of homogeneous polynomials in* *homogeneous coordinates in* $\mathbb{P}_n$ *with finite set of common zeros in* $\mathbb{P}_n$. *Then the number of these zeros, counted with multiplicities, is equal to the product of the degrees of the polynomials:*

$$\sum_{p(a)=0} \mu_a(p) = \prod_{j=1}^{n} \deg p_j.$$

■ After a suitable unitary transformation in $\mathbb{P}_n$ we may assume that all zero of the system p lie in $\mathbb{C}^n = \mathbb{P}_n \setminus \{z_0=0\}$. Then these zeros are points $a^k = [1,\alpha^k]$, where $\alpha^k \in \mathbb{C}^n$, $k = 1, \ldots, s$. Consider in $\mathbb{C}^n$ the system of polynomials $p_j(t,z_1, \ldots, z_n)$, $j = 1, \ldots, n$, which holomorphically depend on the parameter $t \in \mathbb{C}$. For fixed t, $|t| < 2$, all zeros of this system lie in some ball $|z| < R$ (independent of t). By Rouché's theorem, the number of zeros in this ball of the system $p(t,z)$ (counted with multiplicities) does not depend on t, $|t| < 2$. For $t = 1$ the zeros of the system are the points α^k, $k = 1, \ldots, s$, with multiplicities $\mu_{a^k}(p) = \mu_{\alpha^k}(p(1,z_2, \ldots, z_n))$ (according to the definition given above). For $t = 0$ the polynomials $p_j(0,z)$ are homogeneous in $z = (z_1, \ldots, z_n)$. Since no $p_j(z_0, \ldots, z_n)$ can be divided by z_0 (otherwise they would have a common zero on the hyperplane $z_0 = 0$, and we have kicked them out of it), then

$$\deg p_j(0,z_1, \ldots, z_n) = \deg p_j(z_0, \ldots, z_n).$$

Since $z = 0$ is an isolated zero of the system $p(0,z)$ in $\mathbb{C}^n$, Theorem 2, p.10.3 implies that

$$\mu_0(p(0,z)) = \prod_1^n \deg p_j.$$

But $z = 0$ is the only zero of the system $p(0,z)$ if $|z| < R$, hence by Rouché's theorem $\sum_{k=1}^{s} \mu_{\alpha^k}(p(1,z)) = \mu_0(p(0,z))$. As was proved the lefthand side of this equation is $\sum_{p(a)=0} \mu_a(p)$, and the righthand side is $\prod_1^n \deg p_j$. ■

C o r o l l a r y. *Let* $p = (p_1, \ldots, p_n)$ *be a system of polynomials in* $\mathbb{C}^n$, *and let* q_j *consist of the monomials in* p_j *of maximal degree* $\deg p_j$. *Assume that* $z = 0$ *is the only zero of the system* $(q_1, \ldots, q_n)$. *Then the number of zeros in* $\mathbb{C}^n$ *of the system* p *(counted with multiplicities) is equal to* $\prod_1^n \deg p_j$.

■ To a polynomial p_j corresponds its projectivization, the homogeneous polynomial

$$p_j^*(z_0,z) = z_0^{d_j} p_j \left(\frac{z}{z_0} \right) = q_j(z) + z_0 r_j(z_0,z),$$

where $d_j = \deg p_j$ and r_j is also a homogeneous polynomial in (z_0,z) (see p.7.1). Since $(q_1, \ldots, q_n)$ has the unique zero $z = 0$ in $\mathbb{C}^n$, and since $p_j^*(0,z) = q_j(z)$, all zeros of the system $p^* = (p_1^*, \ldots, p_n^*)$ lie in $\mathbb{C}^n = \mathbb{P}_n \setminus \{z_n = 0\}$ and form a compact set in it. By p.3.3, this set is finite. Further, by the definition of multiplicities in $\mathbb{P}_n$, the systems p and p^* have, within $\mathbb{C}^n$, identical zeros with identical multiplicities. As already said, on $\{z_0 = 0\}$ there are no zeros of p^*, i.e. all its zeros lie in $\mathbb{C}^n$. Bezout's theorem implies that their number is $\prod_1^n \deg p_j$. ■

10.5. Milnor numbers. Let f be a holomorphic function in a neighborhood of a point a in $\mathbb{C}^n$, with a an isolated critical point of f, i.e. an isolated zero of the system

$$df := \left(\frac{\partial f}{\partial z_1}, \ldots, \frac{\partial f}{\partial z_n} \right).$$

The *Milnor number* of f at a is the multiplicity of the zero of this system at $z = a$ (cf. [87]); it is denoted by $\mu^{(n)}(f,a)$. If a is a nonisolated critical point of f, it is convenient to put $\mu^{(n)}(f,a) = +\infty$. The Milnor number is closely related to the topology of the sets $f = \text{const}$ in a small neighborhood of a (see [87], [72]). Regrettably, shortage of space does not allow us to present the results in question. Since, after all, this notion is extensively used in modern investigations of singularities of analytic sets, we prove some simple facts concerning Milnor numbers.

Rouché's theorem (p.10.3) obviously implies the following property (independence on a holomorphically varying parameter):

L e m m a. *Let F be a holomorphic function in a domain $U \times V$ in $\mathbb{C}_z^n \times \mathbb{C}_t^m$ such that the zero set of the system $(\partial F / \partial z_1, \ldots, \partial F / \partial z_n)$ is $\{0\} \times V$. Then the Milnor number $\mu^{(n)}(f_t,0)$ of the function $f_t(z) = F(z,t)$ does not depend on $t \in V$.*

P r o p o s i t i o n 1. *The Milnor number is invariant under biholomorphic transformations: if ϕ: $U \to V$ is a biholomorphic map and if $f \in \mathcal{O}(U)$, then*

$$\mu^{(n)}(f,a) = \mu^{(n)}(f\circ\phi^{-1},\phi(a)).$$

■ We will assume that $a = \phi(a) = 0$. Under linear transformations of coordinates in $\mathbf{C}^n$ the vector df is multiplied by a constant nondegenerate matrix, hence (in view of the multiplicativity of multiplicity) the multiplicity does not change. Thus we may assume that $\phi^{-1}(z) = z + \epsilon(z)$, where $\epsilon = (\epsilon_1, \ldots, \epsilon_n)$ and $\operatorname{ord}_0 \epsilon_j > 1$ for all j. Put $f_t(z) = f(z + \epsilon(tz)/t)$, where $t \in \mathbf{C}$, $|t| < 2$. Then

$$\frac{\partial f_t}{\partial z_j}(z) = \frac{\partial f}{\partial z_j}(z^t) + \sum_{k=1}^{n} \frac{\partial f}{\partial z_k}(z^t)\cdot\frac{\partial \epsilon_k}{\partial z_j}(tz),$$

where $z^t = z + \epsilon(tz)/t$, or, abbreviated, $df_t(z) = df(z^t)\cdot(E + (\partial\epsilon/\partial z)(zt))$, where E is the identity matrix. Since $\operatorname{ord}_0 \epsilon_k > 1$, in a sufficiently small neighborhood $U_0 \ni 0$ the matrix $E + (\partial\epsilon/\partial z)(zt)$ is invertible for any t, $|t| < 2$. If the system $df(z)$ has unique zero $z = 0$ in a domain $W \ni 0$, and if $U_0 \subset W$ is sufficiently small, then for each $z \in U_0$ the points z^t, $|t| < 2$, belong to W, hence $df(z^t) \neq 0$ if $z \neq 0$. Thus, there is a neighborhood $U \ni 0$ such that for any t, $|t| < 2$, the system $df_t(z)$ has only the zero $z = 0$ in U. By the Lemma, $\mu^{(n)}(f_t, 0)$ does not depend on t. It remains to remark that $f_0 = f$ and $f_1 = f\circ\phi^{-1}$. ■

P r o p o s i t i o n 2. *Let a function f be holomorphic in a domain $U \ni 0$ in $\mathbf{C}^n$, and suppose $z = 0$ is the only critical point of f in U. Then for each k, $1 \leqslant k < n$, there is a compact set $\Sigma_f(k,n) \subset G(k,n)$ of Hausdorff codimension $\geqslant 2$ such that for each k-dimensional plane $L \ni 0$, $L \notin \Sigma_f(k,n)$, the point 0 is an isolated singular point of the function f: $L \to \mathbf{C}$, and the Milnor numbers $\mu^{(k)}(f|_L, 0)$ for all such $L \notin \Sigma_f(k,n)$ are identical.*

We immediately note that $\mu^{(k)}(f|_L, 0)$ does not depend on the choice of coordinates on L (if only by Proposition 1).

■ A point $z \in L$ is critical for $f|_L$ if and only if the gradient of f, the vector $\overline{df(z)}$, is orthogonal to L in $\mathbf{C}^n$. Consider in $U \times G(k,n)$ the analytic set $\tilde{A}$ consisting of the points (z, L) for which $\overline{df(z)}$ is orthogonal to L (in local coordinates, e.g. in $U \times U_{1\ldots k}$, see A.3.4, this condition can be written in matrix form as $(E, W)\cdot df(z) = 0$). If $df(z) \neq 0$, then all such planes $L \subset \mathbf{C}^n$ lie in $\overline{df(z)}^{\perp} \approx \mathbf{C}^{n-1}$. If moreover $L \ni z$, then L is uniquely determined by the $(k-1)$-dimensional plane $L \cap z^{\perp} \cap \overline{df(z)}^{\perp}$ in $z^{\perp} \cap \overline{df(z)}^{\perp} \approx \mathbf{C}^{n-2}$. Thus, if a point $z \in U \setminus \{0\}$ is fixed, the set of planes $L \ni z$ orthogonal to $\overline{df(z)}$ form a submanifold in $G(k,n)$, biholomorphic to $G(k-1,n-2)$, i.e. of dimension $(n-k-1)(k-1)$. But if $z \in L$ and L is orthogonal to $\overline{df(z)}$, then $z \perp \overline{df(z)}$, i.e. z lies on the hypersurface defined by the equations

$$\sum_1^n z_j \frac{\partial f}{\partial z_j}(z) = 0 \text{ in } U.$$

Thus, the analytic set

$$A = \tilde{A} \cap I(k,n) = \{(z,L) : z \in L \cap U,\ L \subset \overline{df(z)}^{\perp}\}$$

has dimension

$$(n-1) + (n-k-1)(k-1) = (n-k)k = \dim G(k,n).$$

Let $\pi_1 \colon I(k,n) \to \mathbf{C}^n$ and $\pi_2 \colon I(k,n) \to G(k,n)$ be the natural projections (see A3.5). Then $A_L := \pi_1(\pi_2^{-1}(L) \cap A) = \{z \in L \cap U \colon L \subset \overline{df(z)}^{\perp}\}$. By p.3.8 there is a compact set $\Sigma_1 \subset G(k,n)$ of Hausdorff codimension $\geqslant 2$ such that $\dim_{(0,L)} \pi_2^{-1}(L) \cap A = \dim_0 A_L = 0$ for all $L \notin \Sigma_1$. Thus, for all $L \notin \Sigma_1$ the set of critical points of $f|_L$ on $L \cap U$ is discrete.

The set A contains the fiber $\{0\} \times G(k,n)$ as an irreducible component. By p.5.1, the closure A_1 of $A \setminus (\{0\} \times G(k,n))$ in A is an analytic subset in A, and the dimension of $A_1 \cap (\{0\} \times G(k,n))$ is strictly less than that of $G(k,n)$. Let $\Sigma_2 := \pi_2(A_1 \cap (\{0\} \times G(k,n)))$. Then $\Sigma = \Sigma_1 \cup \Sigma_2$ is a compact set in $G(k,n)$ of Hausdorff codimension $\geqslant 2$. By construction of Σ_1 and Σ_2, for each $\Lambda \notin \Sigma$ there are neighborhoods $V \ni 0$ in $\mathbf{C}^n$ and $W \ni \Lambda$ in $G(k,n)$ such that for any $L \in W$ the point $z = 0$ is the only critical point of $f|_L$ in $L \cap V$. By the Lemma (applied to the holomorphic family $f|_L$ in the domain $V \times W$), the number $\mu^{(k)}(f|_L, 0)$ is constant for $L \in W$. Since the complement to Σ is connected (A6.1), this implies that $\mu^{(k)}(f|_L, 0)$ is constant in $G(k,n) \setminus \Sigma$. ■

This value of $\mu^{(k)}(f|_L, 0)$, common to almost all k-dimensional planes $L \ni 0$, is called the *k-th Milnor number* of f at 0; it is denoted by $\mu^{(k)}(f, 0)$. Note that the above-described critical sets in $G(k,n)$ are in fact algebraic (for the proof it is necessary to strengthen Corollary 1, p.3.8 in a corresponding way; we will not do so).

11. Multiplicities of analytic sets

11.1. Multiplicity of an analytic set at a point. Let A be a pure p-dimensional analytic set in $\mathbf{C}^n$ and let $a \in A$. For every $(n-p)$-dimensional plane $L \ni 0$ such that a is an isolated point in $A \cap (a+L)$, the multiplicity of the projection, $\mu_a(\pi_L|_A)$, is finite. The minimum of these numbers over all $(n-p)$-dimensional

$$\mu_a(A) := \min \{\mu_a(\pi_L|_A): L \in G(n-p,n)\}.$$

If a is a regular point of A, then obviously $\mu_a(A) = 1$. Conversely, if $\mu_a(A) = 1$, there are an $(n-p)$-dimensional plane L and a neighborhood $U \ni a$ such that the projection $\pi_L: A \cap U \to U' \subset L^\perp$ is one-to-one; by Proposition 3, p.3.3, in this situation a is a regular point of A.

On A the function $\mu_a(A)$ is upper semicontinuous (this easily follows from the semicontinuity of $\mu_z(\pi_L|_A)$). If we put $\mu_z(A) = 0$ for $z \notin A$, this function becomes upper semicontinuous in any domain in which A is closed.

The definition of multiplicity of an analytic set at a point, given above, uses the geometry of $\mathbf{C}^n$ in an essential way. The following Proposition, stating the invariance of multiplicity under biholomorphic transformations, allows us to generalize this notion to arbitrary complex manifolds.

P r o p o s i t i o n. *Let A be a p-dimensional analytic subset of a domain D in $\mathbf{C}^n$, and let $\phi: D \to G \subset \mathbf{C}^n$ be a biholomorphic map. Then $\mu_a(A) = \mu_{\phi(a)}(\phi(A))$ for every $a \in A$.*

■ The proof is similar to that of Rouché's theorem. The statement being local we may assume that A is defined in D by holomorphic functions $f_1, \ldots, f_N$. Let $a = \phi(a) = 0$. By an additional linear transformation (under which $\mu_0(\pi_L|_A)$, and hence also $\mu_0(A)$, are invariant, see p.10.2) we can represent $\phi^{-1}(z)$ in the form $z + \epsilon(z)$, where $\epsilon(z) = o(|z|)$ as $z \to 0$. In the domain $U \times \{|t|<2\}$, $t \in \mathbf{C}$, we consider the analytic set $\tilde{A} = \{(z,t): f_j(z+\epsilon(tz)/t) = 0, j = 1, \ldots, N\}$ and we put $A_t = \{z: (z,t) \in \tilde{A}\}$ (the domain U is chosen small such that $z + \epsilon(tz)/t \in D$ for all $z \in U$ and all t, $|t| \leq 2$). Then $A_0 = A$ and $A_1 = \phi(A)$. Let 0 be an isolated point of $A \cap \mathbf{C}_{(p+1)\cdots n}$, and let $\mu_0(A) = \mu_0(\pi|_A)$, where $\pi: z \mapsto z' = (z_1, \ldots, z_p)$. Since the family of biholomorphisms $z \mapsto z + \epsilon(tz)/t$, $|t| \leq 2$, is compact and since their linear parts equal z, without loss of generality we may assume that $U = U' \times U''$, $U' \subset \mathbf{C}^p$, that all projections $\pi\colon A_t \cap U \to U'$ are proper, and that $\pi^{-1}(0) \cap A_t \cap U = \{0\}$ (this all is true if U is sufficiently small).

Let $\tilde{\pi}\colon (z,t) \mapsto (z',t)$. Then $\tilde{\pi}\colon \tilde{A} \to U' \times \{|t|<2\}$ is also an analytic cover, and $\operatorname{br} \tilde{\pi}|_{\tilde{A}}$ is obviously equal to $\cup_{|t|<2}((\operatorname{br} \pi|_{A_t}) \times \{t\})$, hence $\operatorname{br} \tilde{\pi}|_{\tilde{A}}$ is nowhere dense in each $A_t \times \{t\}$. Thus, the analytic set $\tilde{A}$ (the family A_t) satisfies the conditions of Lemma 2 and Corollary 10.1, according to which $\mu_0(\pi|_{A_t})$ does not depend on t and is equal to $\mu_0(\pi|_A) = \mu_0(A)$. Thus, $\mu_0(\pi|_{\phi(A)}) = \mu_0(A)$, and thus $\mu_0(\phi(A)) \leqslant \mu_0(A)$.

Applying the same reasoning to ϕ^{-1} instead of ϕ (and $\phi(A)$ instead of A), we obtain the converse inequality. ■

Hence the notion of multiplicity of an analytic set at a point is defined for analytic sets on an arbitrary complex manifold (using local coordinates; as in $\mathbf{C}^n$).

11.2. Multiplicities and the tangent cone. For which planes L is the minimum of the numbers $\mu_a(\pi_L|_A)$ attained, where A is a given pure p-dimensional analytic set in $\mathbf{C}^n$ and $a \in A$? It turns out that this is completely determined by the position of L relative to the tangent cone $C(A,0)$.

Proposition 1. *Let $A\colon f = 0$ be a principal analytic set in a neighborhood of 0 in $\mathbf{C}^n$, with f the minimal defining function for A. Then for any complex line $L \ni 0$,*

$$\mu_0(\pi_L|_A) = \operatorname{ord}_0 f\,|_L \geqslant \operatorname{ord}_0 f;$$

equality holds if and only if $L \cap C(A,0) = \{0\}$.

■ Let 0 be an isolated point in $A \cap L$. By a unitary transformation L becomes the $\mathbf{C}_n$-axis, after which we may assume, by the Weierstrass preparation theorem, that

$$f(z',z_n) = z_n^k + c_1(z')z_n^{k-1} + \cdots + c_k(z'),$$

where $k = \operatorname{ord}_0 f|_{\mathbf{C}_n} = \mu_0(\pi_L|_A)$. Since $\operatorname{ord}_0 f|_L \geqslant \operatorname{ord}_0 f$ the inequality is proved. If $\mathbf{C}_n \not\subset C(A,0)$, there is a constant C such that in a small neighborhood of 0 the set A belongs to the cone $|z_n| \leqslant C|z'|$. The construction of the Weierstrass polynomials and Viète's formula imply that in this case $\operatorname{ord}_0 c_j \geqslant j$, and thus $\text{ord}_0 f = k$.

Conversely, assume that $\text{ord}_0 f = k$. Then $\text{ord}_0 c_j \geqslant j$, since the monomials in f have different degrees in z_n and hence their initial homogeneous polynomials at 0 cannot cancel each other when added. Fix a small z', write the roots of $f(z',z_n)$ in order of decreasing modulus: $z_{n\nu} = z_{n\nu}(z')$, $\nu = 1,\ldots,k$, and divide $f(z',z_{n1}) = 0$ by z_{n1}^k. We obtain $1 = -\sum_1^k c_j(z')/z_{n1}^j$, which obviously implies $k\cdot\max_j |c_j(z')/z_{n1}^j| \geqslant 1$, hence

$$|z_{nj}| \leqslant |z_{n1}| \leqslant \max_j (k|c_j(z')|)^{1/j} \leqslant C|z'|,$$

since $\text{ord}_0 c_j \geqslant j$. Thus, in a neighborhood of 0 all of A belongs to the cone $|z_n| \leqslant C|z'|$, hence $\mathbb{C}_n \cap C(A,0) = \{0\}$. ■

C o r o l l a r y. *If f is the minimal defining function for the set $A: f = 0$, then* $\mu_0(A) = \text{ord}_0 f$.

The same role is played by the tangent cone in the general situation.

P r o p o s i t i o n 2. *Let A be a pure p-dimensional analytic set in a neighborhood of 0 in $\mathbb{C}^n$, and let $L \in G(n-p,n)$. The equality $\mu_0(\pi_L|_A) = \mu_0(A)$ holds if and only if the plane L is transversal to A at 0, i.e. $L \cap C(A,0) = \{0\}$.*

Note that the planes $L \in G(n-p,n)$ intersecting the p-dimensional cone $C(A,0)$ in a set of dimension $\geqslant 1$ form in $G(n-p,n)$ a closed (as a matter of fact, an algebraic) subset of Hausdorff codimension $\geqslant 2$ (see Corollary 2, p.3.8). Hence, by A6.1,

The planes L transversal to a pure p-dimensional analytic set $A \ni 0$ at 0 in $\mathbb{C}^n$ form a domain that is everywhere dense in $G(n-p,n)$.

We turn to the proof of Proposition 2.

■ We will assume that $L = \mathbb{C}_{z''}^{n-p}$, where $z = (z',z'')$, $z' \in \mathbb{C}^p$. Let a neighborhood $U' \times U'' = U \ni 0$ be such that $(0' \times U'') \cap A = \{0\}$ and $\pi: A \cap U \to U'$ is a k-sheeted analytic cover. Then, by definition, $k = \mu_0(\pi_L|_A)$. By p.3.6 there is a linear function $\lambda(z'')$ such that the projection $\pi_\lambda: z \mapsto (z',\lambda(z'')) \in \mathbb{C}^{p+1}$ is almost single-sheeted on $A \cap U$; in particular, if $\pi = \pi' \circ \pi_\lambda$, then $\pi': A_\lambda \to U'$ is also a k-sheeted analytic cover ($A_\lambda = \pi_\lambda(A \cap U)$). If $\mathbb{C}_{z''}^{n-p} \cap C(A,0)$ contains a vector $v \neq 0$, then λ can be chosen such that $\lambda(v'') \neq 0$. Then, obviously, $(0', \lambda(v'')) \in \mathbb{C}_\lambda \cap C(A_\lambda, 0)$, hence, by Proposition 1, $\mu_0(A_\lambda) < k$. Since the projection $\pi_\lambda|_{A \cap U}$ is almost single-sheeted, $\mu_0(A_\lambda) \geqslant \mu_0(A)$; hence we have proved that $\mu_0(\pi_L|_A) > \mu_0(A)$ in the non-transversal situation.

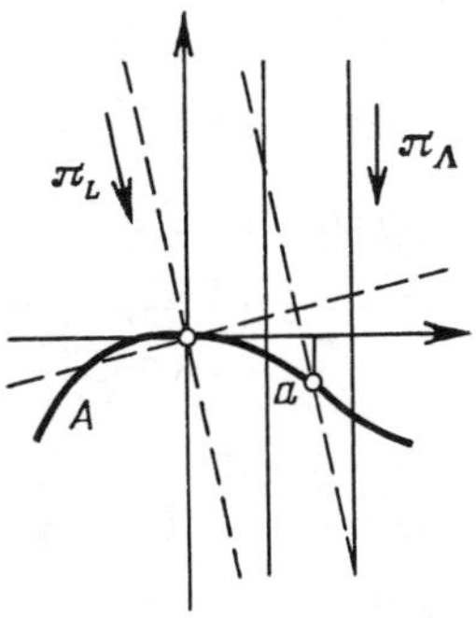

Figure 8.

By what was proved above, the planes $L \in G(n-p,n)$ that are transversal to A at 0 form a domain G_0 that is everywhere dense in $G(n-p,n)$. If $\Lambda \subset G_0$, there is a connected neighborhood $U_\Lambda \ni 0$ in $\mathbb{C}^n$ of the form $U'_\Lambda \times U''_\Lambda$, where $U'_\Lambda \subset \Lambda^\perp$, $U''_\Lambda = U_\Lambda \cap \Lambda$, such that $A \cap \Lambda \cap \overline{U}_\Lambda = \{0\}$, such that the projection $\pi_\Lambda \colon A \cap U_\Lambda \to U'_\Lambda$ is proper, and such that A does not intersect the compact set $\overline{U}'_\Lambda \times \partial U''_\Lambda$. If the plane $L \in G_0$ is sufficiently close to Λ, then $A \cap L \cap \overline{U}_\Lambda = \{0\}$ (otherwise we would have $\Lambda \cap C(A,0) \neq \{0\}$). Let $a \in A \cap U_\Lambda$ be such that $\pi_\Lambda(a) \in U'_\Lambda$ is not a critical value for $\pi_\Lambda|_{A \cap U_\Lambda}$ (Figure 8). Then there are neighborhoods $W' \subset\subset V' \subset U'_\Lambda$ of $\pi_\Lambda(a)$ such that $\pi_\Lambda^{-1}(V') \cap A \cap U_\Lambda$ is the union of k graphs V^j of holomorphic functions, each of which is projected by π_Λ onto V', while the tangent planes to all V^j at all points form with $\Lambda^\perp$ angles $\leqslant \alpha < \pi/2$. Put $W^j = V^j \cap \pi_\Lambda^{-1}(W')$. Then if the plane $L \in G_0$ is sufficiently close to Λ, the maps $\pi_L|_{W^j}$ and $\pi_\Lambda|_{W^j}$ are single-sheeted, $\pi_L(a)$ is a noncritical value for $\pi_L|_{A \cap U_\Lambda}$, and $\#\pi_L^{-1}(\pi_L(a)) \cap A \cap U_\Lambda = k$. Finally, if L is sufficiently close to Λ, then there is a domain $U'_L \subset L^\perp$, containing 0 and $\pi_L(a)$, such that the restriction of π_L onto $A \cap U_\Lambda \cap \pi_L^{-1}(U'_L)$ is a proper map. From all this we conclude that if L is sufficiently close to Λ, then $\mu_0(\pi_L|_A) = \mu_0(\pi_\Lambda|_A)$. Since G_0 is connected, this implies $\mu_0(\pi_L|_A) \equiv \text{const}$ in G_0, and since $\min \mu_0(\pi_L|_A)$ is attained in G_0,

$\mu_0(\pi_L|_A) \equiv \mu_0(A)$ for all $L \in G_0$. ■

C o r o l l a r y 1. *Let* $U = U' \times U'' \ni 0$ *be such that the projection* $\pi: A \cap U \to U' \in \mathbb{C}^p_{z'}$ *is proper,* $A \cap U \cap \mathbb{C}^{n-p}_{z''} = C(A,0) \cap \mathbb{C}^{n-p}_{z''} = \{0\}$, *and the restriction onto* $A \cap U$ *of the projection* $\tilde{\pi}: z \mapsto (z', z_{p+1})$ *is almost single-sheeted. Let* $\tilde{A} = \tilde{\pi}(A)$, *and let* $f(z_1, \ldots, z_{p+1})$ *be the minimal defining function of this hypersurface in a neighborhood of* 0 *in* $\mathbb{C}^{p+1}$. *Then* $\mu_0(A) = \mu_0(\tilde{A}) = \operatorname{ord}_0 f$.

■ By p.8.2, the set $\tilde{A}$ is transversal to the $\mathbb{C}_{p+1}$-axis at 0. By Proposition 1, $\mu_0(\tilde{A}) = \mu_0(\pi'|_{\tilde{A}}) = \operatorname{ord}_0 f$, where $\pi': (z', z_{p+1}) \mapsto z'$. Since $\pi' \circ \tilde{\pi} = \pi: z \mapsto z'$ and $\tilde{\pi}|_{A \cap U}$ is one-to-one above reg $\tilde{A}$ (see p.3.6), $\mu_0(\pi'|_{\tilde{A}}) = \mu_0(\pi|_A)$. ■

C o r o l l a r y 2. *Let, in a domain* $U = U' \times \mathbb{C}^{n-p}_{z''} \ni 0$ *in* $\mathbb{C}^n$, $f_j(z', z_{p+j}) = z_{p+j}^{d_j} + c_{j1}(z') z_{p+j}^{d_j - 1} + \cdots + c_{jd_j}(z')$ *be Weierstrass polynomials in* z_{p+j}, $j = 1, \ldots, n-p$, *without multiple factors, and such that* $\operatorname{ord}_0 c_{j\nu} \geqslant \nu$ *for all* j *and* ν. *Let* $A = \{z \in U: f_j(z) = 0,\ j = 1, \ldots, n-p\}$. *Then* $\mu_0(A) = d_1 \cdots d_{n-p}$.

■ The condition on the coefficients is equivalent to the $\mathbb{C}_{(p+1)\cdots n}$-plane being transversal to A at 0 (one implication follows obviously from Viètè's formula, the other from the proof of Proposition 1). Hence if $\pi: (z', z'') \mapsto z'$, then $\mu_0(A) = \mu_0(\pi|_A)$. For the same reason the multiplicity of the set A_j: $f_j = 0$ in $\mathbb{C}_{1 \cdots p(p+j)}$ at 0 is equal to the multiplicity of its projection into U'. Since f_j has no multiple irreducible factors, f_j is the minimal defining function for A_j, hence, by Proposition 1, $\mu_0(A_j) = d_j$. Since $A = \cap\, \pi_j^{-1}(A_j)$ where $\pi_j: z \mapsto (z', z_{p+j})$, we have $\mu_0(\pi|_A) = d_1 \cdots d_{n-p}$. ■

Yet two more simple Corollaries from Proposition 2. The first expresses additivity of multiplicity:

C o r o l l a r y 3. *If* $A = \bigcup_1^N A_j$ *where all* A_j *are pure p-dimensional, then* $\mu_a(A) \leqslant \sum_1^N \mu_a(A_j)$. *Equality holds if and only if* $\dim_a (A_i \cap A_j) < p$ *for* $A_j \neq A_i$.

The *proof* follows in an obvious manner from the existence of a unique $(n-p)$-dimensional plane transversal to all $A_j \ni a$ at a.

The following statement is called multiplicativity of multiplicity.

C o r o l l a r y 4. $\mu_{(a', a'')}(A' \times A'') = \mu_{a'}(A') \cdot \mu_{a''}(A'')$.

■ Let $a' = a'' = 0$, $\dim_0 A' = p$, $\dim_0 A'' = q$, and suppose that the planes L', of dimension $n - p$, and L'', of dimension $n - q$, are transversal to A' and A'', respectively, at 0. Since $C(A' \times A'', (0,0)) = C(A',0) \times C(A'',0)$, the plane $L = L' \times L''$ is transversal to $A' \times A''$ at $(0,0)$, hence $\mu_{(0,0)}(\pi_L|_{A' \times A''}) = \mu_{(0,0)}(A' \times A'')$. Since $\pi_L(z,w) = (\pi_{L'}(z), \pi_{L''}(w))$ and the multiplicity is the number of pre-images of a generic point under a local analytic cover, $\mu_{(0,0)}(A' \times A'') = \mu_0(\pi_{L'}|_{A'}) \cdot \mu_0(\pi_{L''}|_{A''}) = \mu_0(A') \cdot \mu_0(A'')$. ■

11.3. Degree of an algebraic set. In case the analytic set A is itself a cone with vertex at 0 (i.e. $A = C(A, 0)$), the transversality condition simply means that $A \cap L = \{0\}$. In this situation $\mu_0(\pi_L|_A) = \mu_0(A)$ for all L such that 0 is an isolated point in $A \cap L$. This common multiplicity is called the *degree* of the cone A, and is denoted by $\deg A$. This name is motivated by the following example.

If A is the zero set of a homogeneous polynomial p and if p does not have multiple factors (which is equivalent to $dp \neq 0$ on an everywhere dense subset of A, see p.2.9), then by a unitary transformation we may assume that p contains the monomial $z_n^{\deg p}$ with nonzero coefficient, hence projection of A to the plane $z_n = 0$ is an analytic cover. Since $dp \neq 0$ on $\operatorname{reg} A$, for a fixed z' outside the critical analytic set of this cover all zeros of $p(z', z_n)$ in z_n are simple, hence the multiplicity of this cover (equal, by definition, to $\deg A$) is equal to $\deg p$.

If A is a projective algebraic set in $\mathbb{P}_n$, it has a corresponding cone $\tilde{A} = \Pi^{-1}(A) \cup \{0\}$ in $\mathbb{C}^{n+1}$, where $\Pi\colon \mathbb{C}_*^{n+1} \to \mathbb{P}_n$ is the canonical projection. By definition, the degree of A is taken to the degree of the cone $\tilde{A}$; i.e. $\deg A := \mu_0(\tilde{A})$.

We will clarify the geometrical meaning of $\deg A$ in terms of $\mathbb{P}_n$. If A is pure p-dimensional, by p.7.3 there is an $(n - p - 1)$-dimensional complex plane in $\mathbb{P}_n$ not intersecting A. For each such plane L a projection $\pi_L\colon \mathbb{P}_n \setminus L \to L^{\perp}$ is defined, whose restriction onto A is an analytic cover above the p-dimensional plane $L^{\perp}$ in $\mathbb{P}_n$ (see p.7.3). A plane L has a corresponding $(n - p)$-dimensional subspace $\tilde{L} = \Pi^{-1}(L) \cup \{0\}$ in $\mathbb{C}^{n+1}$, and π_L is the projection $\pi_{\tilde{L}}$ onto the $(p+1)$-dimensional plane $\tilde{L}^{\perp}$ in $\mathbb{C}^{n+1}$ (in the sense that $\Pi \circ \pi_{\tilde{L}} = \pi_L \circ \Pi$, see A3.2 and p.7.3). The multiplicities of the analytic covers $\pi_L\colon A \to L^{\perp}$ and $\pi_{\tilde{L}}\colon \tilde{A} \to \tilde{L}^{\perp}$ coincide, since for any fixed $z \in \tilde{L}^{\perp}$, $z \neq 0$, the points in the fiber $\pi_{\tilde{A}}^{-1}(z)$ are pairwise noncollinear and thus pairwise distinct points in $\mathbb{P}_n$

correspond to them. Since, $\mu_0(\pi_{\tilde{L}}|_{\tilde{A}}) = \mu_0(\tilde{A})$ we obtain

P r o p o s i t i o n 1. *Let A be a pure p-dimensional algebraic subset in $\mathbb{P}_n$, and let L be an arbitrary $(n-p-1)$-dimensional plane in $\mathbb{P}_n$ not intersecting A. Then the multiplicity of the analytic cover $\pi_L\colon A \to L^{\perp}$ is equal to* $\deg A$.

Let now Λ be an arbitrary $(n-p)$-dimensional plane in $\mathbb{P}_n$ containing L. Then π_L maps the set $A \cap \Lambda$ to the point $\Lambda \cap L^{\perp}$, i.e. $A \cap \Lambda$ is the fiber of $\pi_L|_A$. The number of points of this fiber can be less than the geometric multiplicity of $\pi_L|_A$ (equal to $\deg A$), but if the points are counted with the multiplicities of the cover we do obtain $\deg A$. Thus, $\sum_{a\in A\cap\Lambda}\mu_a(\pi_L|_A) = \deg A$. The number $\mu_0(\pi_L|_A)$ is also called the intersection index of Λ and A at a (see p.12.1), and is denoted by the symbol $i_a(A,\Lambda)$. Since the number $\mu_a(\pi_L|_A)$ continuously depends on $L \subset \Lambda\setminus A$ and is integer-valued (while $A \cap \Lambda$ is finite), this quantity really depends on A, Λ, and a only. The number $\sum_{a\in A\cap\Lambda} i_a(A,\Lambda)$ is called the complete intersection index of the analytic sets A and Λ, and is denoted by the symbol $i(A,\Lambda)$. Thus we have

P r o p o s i t i o n 2. *Let A be a pure p-dimensional algebraic subset in $\mathbb{P}_n$, and let Λ be an arbitrary $(n-p)$-dimensional plane in $\mathbb{P}_n$ such that the set $A \cap \Lambda$ is zero-dimensional. Then*

$$i(A,\Lambda) := \sum_{a\in A\cap\Lambda} i_a(A,\Lambda) = \deg A.$$

By definition, the degree of an affine algebraic set $A \subset \mathbb{C}^n$ is equal to the degree of its closure in $\mathbb{P}_n \supset \mathbb{C}^n$.

C o r o l l a r y 1. *Let A be a pure p-dimensional algebraic set in $\mathbb{C}^n$, and let L be an arbitrary $(n-p)$-dimensional plane in $\mathbb{C}^n$ such that the closures of A and L in $\mathbb{P}_n$ do not have points at infinity (i.e. in $\mathbb{P}_n\setminus\mathbb{C}^n$) in common. Then $\pi_L\colon A\to L^{\perp}\approx\mathbb{C}^p$ is an analytic cover with number of sheets equal to* $\deg A$.

C o r o l l a r y 2. *Let $A_j \subset \mathbb{C}^{n_j}$, $j = 1,\ldots,k$, be affine algebraic sets. Then*

$$\deg(A_1\times\cdots\times A_k) = \prod_1^k \deg A_j.$$

■ Let L_j be a plane of dimension $n_j - \dim A_j$ in $\mathbb{C}^{n_j}$ such that the projection $\pi_{L_j}\colon A_j \to L_j^{\perp}$ has $\deg A_j$ sheets. Then $L = \prod L_j$ is a plane of dimension $(\sum n_j) - (\sum \dim A_j)$ in $\mathbb{C}^N$, $N = \sum n_j$, such that $\pi_L\colon \prod A_j \to L^{\perp}$ is an analytic cover with $\prod \deg A_j$ sheets. Since the closures of A_j and L_j in $\mathbb{P}_n$ do not have points at infinity in common, A_j lies in some domain of the form $D_j\colon \{|z| \leqslant R\} \cup K_j$, where K_j is a closed cone intersecting L at its vertex only (see p.7.4). Thus, $\prod A_j \subset \prod D_j$, and the closures of $\prod A_j$ and L in $\mathbb{P}_n$ do not have points at infinity in common. By Corollary 1, the number of sheets of the cover $\pi_L\colon \prod A_j \to L^{\perp}$ is equal to $\deg \prod A_j$. ■

11.4. Multiplicity sets. We begin with a theorem of Whitney [150] concerning analytic level sets of the function $\mu_z(A)$.

T h e o r e m. *Let A be a pure p-dimensional analytic set in $\mathbb{C}^n$, and put $A^{(m)} := \{z \in A\colon \mu_z(A) \geqslant m\}$. Then $A^{(m)}$, $m = 1,2,\ldots$, are analytic subsets in A, and $A^{(1)} = A$, $A^{(2)} = \operatorname{sng} A$.*

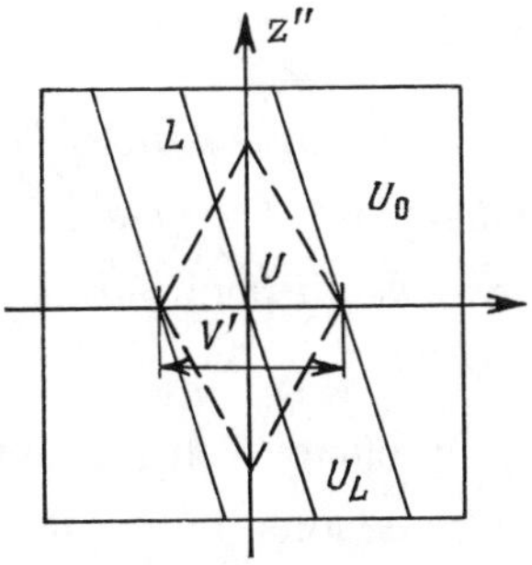

Figure 9.

■ The statement being local we may assume that $0 \in A^{(m)}$ and prove the analyticity of $A^{(m)}$ in a neighborhood of 0. We choose coordinates such that the plane $L_0 = \mathbb{C}_{(p+1)\cdots n}$ is transversal to A at 0. Then there is a neighborhood $W \ni L_0$ in $G(n-p,p)$ such that all planes $L \in W$ are transversal to A at 0. Let $U_0 = U_0' \times U_0''$ be a neighborhood of 0 in $\mathbb{C}^n$ such that the projection $\pi\colon A \cap U_0 \to U_0'$ is an analytic cover with number of sheets equal to $\mu_0(\pi_{L_0}|_A) = \mu_0(A)$, and let $V' \subset\subset U_0'$ be a smaller neighborhood of 0 in $\mathbb{C}^n$

(see Figure 9). If the plane L is sufficiently close to L_0, and if $z' \in V'$, then the set $(z'+L) \cap \overline{U}_0$ does not intersect $(\partial \overline{U}_0') \times \overline{U}_0''$. Hence, if W is sufficiently small, and if $U_L := \cup_{z' \in V'}(z'+L) \cap U_0$, the projection $\pi_L|_{A \cap U_L}$ is also proper. Put $U = \cap_{L \in W} U_L$. By Lemma 1, p.10.1, the set $A_L^m = \{z \in A \cap U_L: \mu_z(\pi_L|_A) \geq m\}$ is analytic. Since $\mu_z(A) \leq \mu_z(\pi_L|_A)$ for all L, we have $A^{(m)} \cap U \subset \cap_{L \in W}(A_L^m \cap U)$. On the other hand, let $a \in (A \cap U) \setminus A^{(m)}$. By p.11.2, the planes L such that $a + L$ are transversal to A at a form in $G(n-p,n)$ an everywhere dense domain; in particular, this set contains an $L \in W$. By Proposition 2, p.11.2, $\mu_a(\pi_L|_A) = \mu_a(A) < m$, hence $a \notin \cap_{L \in W}(A_L^m \cap U)$. Thus we have proved that $A^{(m)} \cap U$ coincides with the analytic set $\cap_{L \in W}(A_L^m \cap U)$. The last statement in the Theorem clearly follows from: $\mu_z(A) = 1$ if and only if $z \in \operatorname{reg} A$ (see p.11.1). ■

Let now A' be an arbitrary irreducible analytic set belonging to A. Since A' is the union of the decreasing sequence of sets $A' \cap A^{(m)}$ with empty intersection, there is a maximal number $m \in \mathbb{N}$ such that $A' \subset A^{(m)}$. Since $A' \not\subset A^{(m+1)}$ and A' is irreducible, $A'' = A' \cap A^{(m+1)}$ is nowhere dense in A' and $\dim A'' < \dim A'$. At the points of $A^{(m)} \setminus A^{(m+1)}$ the multiplicity of A is m, hence we have proved

P r o p o s i t i o n 1. *Let A be a pure p-dimensional analytic set in $\mathbb{C}^n$, and let $A' \subset A$ be an irreducible analytic set. Then there is a natural number m such that $\mu_z(A) = m$ for all $z \in A'$ except, possibly, for the points of an analytic subset $A'' \subset A'$ of dimension $< \dim A'$.*

This m will be called the multiplicity of the analytic set A along the irreducible analytic set $A' \subset A$, and will be denoted by $\mu(A,A')$. For the relation to multiplicities of sections of A by planes transversal to A' see below, p.11.7.

At points of the same multiplicity an analytic set has a certain homogeneity with respect to projections.

P r o p o s i t i o n 2. *Let A be a pure p-dimensional analytic set in $\mathbb{C}^n$, $0 \in A$, $\mu_0(A) = m$, and let $L \ni 0$ be a plane of dimension $n-p$ and transversal to A at 0. Let a domain $U \ni 0$ be such that $\pi_L|_{A \cap U}$ is an m-sheeted analytic cover (above a certain domain in $L^{\perp}$). Then the restriction $\pi_L|_{A^{(m)} \cap U}$ is one-to-one, and for each point $z \in A^{(m)} \cap U$ the plane $z + L$ is transversal to A at z.*

■ Without loss of generality we may assume that $L = \mathbb{C}_{(p+1)\cdots n}$. By requirement, the projection $\pi: A \cap U \to U' \subset \mathbb{C}_{1\cdots p}$ is m-sheeted, i.e. the number of points in each fiber of $\pi|_{A\cap U}$, counted with the multiplicities of $\pi|_A$, is equal to m. Since $m \geqslant \mu_z(\pi|_A) \geqslant \mu_z(A) = m$ for $z \in A^{(m)} \cap U$, equality holds everywhere, and there are no other points than z in the fiber $\pi^{-1}(z') \cap A \cap U$, i.e. $\pi|_{A^{(m)}\cap U}$ is one-to-one. Finally, since $\mu_z(\pi|_A) = \mu_z(A)$ for $z \in A^{(m)} \cap U$, by Proposition 2, p.11.2, the plane $z + L$ is transversal to A at z. ■

In conclusion we note a relation between the multiplicities of an analytic set and those of its plane sections.

P r o p o s i t i o n 3. *Let A be a pure p-dimensional analytic set in $\mathbb{C}^n$, $0 \in A$, and let $m > n - p$. Then there is a compact set Σ of Hausdorff codimension $\geqslant 2$ in $G(m,n)$ such that $\mu_0(A \cap L) = \mu_0(A)$ for all $L \in G(m,n)\setminus\Sigma$.*

■ Let $I(m,n)$ be the incidence manifold in $\mathbb{C}^n \times G(m,n)$ (see A3.5), and put

$$\tilde{A} = \{(z,L) \colon z \in A \cap L\} = \pi_1^{-1}(A) \cap I(m,n),$$

where π_1 denotes projection to $\mathbb{C}_1$. If π_2 is projection to $G(m,n)$, then

$$A \cap L = \pi_1(\pi_2^{-1}(L) \cap \tilde{A}).$$

If $a \in (\operatorname{reg} A) \cap L$ but $\dim (T_a A) \cap L > (m+p) - n$, then (a,L) is a critical point for $\pi_2|_{\tilde{A}}$, and conversely. Corollary 1, p.3.8, implies that the planes L for which $\dim_z A \cap L > (m+p) - n$ at a certain point z, $|z| \leqslant r$ (where r is sufficiently small), form a compact set Σ_1 of Hausdorff codimension $\geqslant 2$. If $\operatorname{br} \pi_2|_{\tilde{A}}$ is somewhere dense on $\tilde{A} \cap \pi_2^{-1}(L) \cap \{|z| \leqslant r\}$, by the rank Theorem this L belongs to Σ_1. Thus, if $L \notin \Sigma_1$, then $\dim_0 A \cap L = (m+p) - n$, and $\dim (T_z A) \cap L = (m+p) - n$ for almost all $z \in \operatorname{reg} A$, $|z| \leqslant r$.

On the other hand, by Corollary 2, p.3.8, the planes L for which

$$\dim_0 L \cap C(A,0) > (m+p) - n,$$

also form a compact set, Σ_2, of Hausdorff codimension 2 in $G(m,n)$. Put $\Sigma = \Sigma_1 \cup \Sigma_2$.

Let $L \in G(m,n)\setminus\Sigma$. Then there is an $(n-p)$-dimensional plane $\Lambda \subset L$ transversal to A at 0 (since $L \notin \Sigma_2$). Let U be a neighborhood of 0 such that $\pi_\Lambda \colon A \cap U \to U' \subset \Lambda^\perp$ is an analytic cover with number of sheets equal to

$\mu_0(A)$ (by Proposition 2, p.11.2, such a U exists). Since $L \notin \Sigma_1$, we have that $\pi_\Lambda(A \cap L \cap U)$ does not belong to the critical analytic set of this cover; hence the number of sheets of the cover $\pi_\Lambda\colon A \cap U \to L \cap U'$ is also equal to $\mu_0(A)$. On the other hand, since Λ is transversal to $A \cap L$ at 0, this number is equal to $\mu_0(A)$ (by Proposition 2, p.11.2). ■

11.5. Holomorphic chains. Not once but already several times, although in different situations, analytic sets with multiplicities occurred. These objects can be included in a general notion. A *holomorphic chain* on a complex manifold Ω is a formal, locally finite sum $T = \sum k_j A_j$, where A_j are pairwise distinct irreducible analytic subsets in Ω and $k_j \neq 0$ are integers. Here, locally finite means that for every compact set $K \subset \Omega$ there are only finitely many indices j for which $K \cap A_j$ is nonempty. (For an informal definition of holomorphic chains, in terms of currents, see below, p.16.1.) The analytic set $\bigcup A_j$ is called the *support* of the chain T, and is denoted by supp T or $|T|$. The sets A_j are called the *components* of T, and the numbers k_j are called the multiplicities of T (or of the components of T). To every formal, locally finite sum $\sum k_j A_j$, where the A_j are analytic subsets in Ω and $k_j \in \mathbb{Z}$, uniquely corresponds a holomorphic chain, constructed as follows: split A_j into irreducible components A_{ji}, replace in the sum A_j by the chain $\sum A_{ji}$, reduce similar sets (i.e. add the coefficients of equal A_{ji}), and leave only those A_{ji} which have obtained nonzero coefficients. It may happen, of course, that all coefficients turn out to be equal to zero, therefore we add the zero element $T = 0$ with empty support to the set of holomorphic chains. After this, holomorphic chains can be multiplied by integers, added, and subtracted: $k(\sum k_j A_j) = \sum k k_j A_{ji}$; by definition, the sum $(\sum k_j A_j) + (\sum k_i' A_i')$ is the holomorphic chain corresponding to the formal sum $\sum_{i,j}(k_j A_j + k_i' A_i')$; and $T - T' := T + (-1)T'$.

If all components of a chain T are pure p-dimensional, T is called a *holomorphic p-chain* (or, simple, a p-chain). We will deal mainly with such chains.

A chain is called *positive* if all its coefficients are positive. Up till now we have only encountered positive chains (divisors of holomorphic functions, multiplicity sets, etc.). Negative chains are needed too, and not only for the algebraic formalism. As an example we may give divisors of meromorphic functions, which are defined analogously to divisors of holomorphic functions, but the irreducible components of polar sets are conveniently ascribed the corresponding negative multiplicities.

Holomorphic chains are the natural object arising when solving systems of holomorphic equations. It is clear that a solution must be taken with corresponding multiplicities, and this is no other than a holomorphic chain. In the course of solving a system, say, with isolated zeros (a 0-chain corresponds to it) intermediate holomorphic chains of positive dimension arise, as, e.g., in p.10.2 (see also p.18.6).

The multiplicity of a holomorphic map f defined on the support of a chain $T = \sum k_j A_j$ is defined by additivity:

$$\mu_0(f|_T) := \sum k_j \mu_a(f|_{A_j});$$

analogously, $\mu_a(T) := \sum k_j \mu_a(A_j)$.

Let $T = \sum k_j A_j$ be a holomorphic p-chain in a domain $U = U' \times U''$ in $\mathbb{C}^n$ such that its support $A = \cup A_j$ is an analytic cover above $U' \subset \mathbb{C}^p$ when projecting $\pi: (z',z'') \mapsto z'$ (in such a case we also say that T is an analytic cover above U', and we write $\pi: T \to U'$. Let $\sigma \subset U'$ be the critical analytic set of the cover $\pi: A \to U'$ (it is critical also for $\pi: T \to U'$, by definition). Since $\pi: A_j \to U'$ are also analytic covers (for all j), for each point $z' \in U' \setminus \sigma$ the fiber $\pi^{-1}(z') \cap A_j$ consists of m_j distinct points, whose z'-coordinates will be denoted by $\alpha_{j1}(z'), \ldots, \alpha_{jm_j}(z')$. The number $k = \sum k_j m_j = \sum_{\pi(z)=z'} \mu_z(\pi|_T)$ will be called the multiplicity of the cover $\pi: T \to U'$, and for $z' \in U' \setminus \sigma$ we form the system of k points $\alpha_l(z'), \ldots, \alpha_k(z')$ in $\mathbb{C}^{n-p}_{z''}$, in which each $\alpha_{ji}(z')$ is repeated m_j times. The definition of multiplicity of projections implies that if the points $\alpha_{ji}(z') \in \mathbb{C}^{n-p}_{z''}$ corresponding to a $z' \in \sigma$ are taken as many times as the multiplicity of $\pi|_T$ at $(z',\alpha_{ji}(z'))$, we also obtain a system of k points; moreover, the systems constructed vary continuously as z' varies in U' (see p.4.3). The canonical defining functions of these systems of k points, $\Phi_I(z',z'')$, $|I| = k$, as in the case of analytic sets, are holomorphic in $U' \times \mathbb{C}^{n-p}_{z''}$ (they are polynomials in z'' of degree k with coefficients from $\mathcal{O}(U')$). They are called canonical defining functions of the cover $\pi: T \to U'$. Thus, the terminology of analytic covers can be completely transferred to holomorphic chains, and we will freely use it in the sequel.

11.6. The tangent cone as chain. Holomorphic chains naturally arise also in processes involving limit transitions (e.g. as in the multiplicity of a map at a point). We will demonstrate this by the example of the tangent cone.

Let $A \ni 0$ be a pure p-dimensional analytic set in $\mathbb{C}^n$ such that in a domain $U = U' \times U'' \ni 0$ the projection $\pi: A \cap U \to U'$ is a transversal analytic cover with canonical defining functions Φ_I, $|I| = k$. We will take U small such that $\pi^{-1}(0) \cap A \cap U = \{0\}$. (The transversality condition means that for all $z \in A \cap U$ an estimate $|z''| \leq C|z'|$ holds, with some constant C, see p.8.6.)

In order to define the tangent cone as a chain we consider the family of analytic sets $A/t = \{\zeta/t : \zeta \in A \cap U\}$, $t \in \mathbb{C}$, $|t| < 1$. If $z \in A/t$, then $tz \in A \cap U$, i.e. $\Phi_I(tz) = 0$, $|I| = k$. Since $\mathrm{ord}_0 \Phi_I = k$ (see p.8.4), it is more convenient to take the function $t^{-k}\Phi_I(tz)$, instead of $\Phi_I(tz)$. This function is also holomorphic in all variables in the domain $U \times \{|t| < 1\} \subset \mathbb{C}^{n+1}$ (at $t = 0$ it is defined by continuity). Thus, the family of sets A/t (which give the tangent cone $C(A, 0)$ as $t \to 0$) defines an analytic set $\tilde{A} = \{(z,t): |t| < 1, t^{-k}\Phi_I(tz) = 0, |I| = k\}$ in $\mathbb{C}^{n+1}$. On the section $t = 0$ only the initial homogeneous polynomials $\Phi_I^*(z)$ of degree k in z remain of the $t^{-k}\Phi_I(tz)$, and the set $\tilde{A} \cap \{t=0\}$ is $C(A, 0) \times \{0\}$, see p.8.4. For each $z' \in U'$ the functions $\Phi_I^*(z', \cdot)$, $|I| = k$, are canonical defining functions for the systems of k points in $\mathbb{C}_{z''}^{n-p}$, and the projection $\pi: C(A, 0) \to \mathbb{C}^p$ can have fewer than k sheets; thus, the functions Φ_I^*, $|I| = k$, determine at the points of the tangent cone certain multiplicities. We investigate this in more detail.

Let σ be the critical analytic set of the cover $\pi: C(A, 0) \to \mathbb{C}^p$, let C_j be the irreducible components of $C(A, 0)$, and let a simply connected domain $V' \subset\subset \mathbb{C}^p \setminus \sigma$. Then $\pi^{-1}(V') \cap C(A, 0)$ splits into a finite number of pairwise disjoint manifolds $S_{ji} \subset C_j$. Let V_{ji} be arbitrary small pairwise disjoint neighborhoods of the S_{ji} intersected with $\pi^{-1}(V')$. Then, for sufficiently small t the set $(A/t) \cap \pi^{-1}(V')$ belongs to the union of the V_{ji}. In each V_{ji} we have the analytic set $(A/t) \cap V_{ji}$, which is clearly an analytic cover above V' with number of sheets equal to, say, k_{ji}. This number is independent of t ($|t| < \epsilon$) in view of the continuity, and all these k_{ji} sheets "stick together" in the single sheet S_{ji} as $t \to 0$. It obviously follows from the construction above that the multiplicity determined at the points of $C(A, 0)$ by the functions Φ_I^*, $|I| = k$, is equal to k_{ji} everywhere on S_{ji} (we can say that $(A/t) \cap V_{ji} \to k_{ji}S_{ji}$ in the sense of holomorphic chains, as $t \to 0$; a correct definition of this convergence in terms of currents can be found in p.16.1). Since $\mathbb{C}^p \setminus \sigma$ is connected, and since, obviously, the multiplicity found is locally constant on $C(A, 0) \setminus \pi^{-1}(\sigma)$, the numbers k_{ji} in fact only depend on j, i.e. the chains $k_{ji}S_{ji}$ stick together (and continue across $\pi^{-1}(\sigma)$) in the single holomorphic chain $\sum k_j C_j$. This chain is called the *tangent cone* to A at 0 *in the sense of holomorphic chains*; it will be denoted by $C([A],0)$.

For general holomorphic chains the tangent cone is naturally defined by additivity: $C(\sum k_j A_j, 0)$ is the holomorphic chain corresponding to $\sum k_j C([A_j],0)$. (In the general case it is necessary to reduce to similar terms; moreover, certain coefficients may turn out to be zero, hence in general the support of $C(\sum k_j A_j, 0)$ can be smaller than that of $\cup C(A_j, 0)$.) The degree of a cone-chain is also defined by additivity: $\deg \sum k_j C_j := \sum k_j \deg C_j$.

As an example we consider the holomorphic curve $A: z_2^k = z_1^l$ in $\mathbb{C}^2$, where $k, l \in \mathbb{N}$, $k < l$. The set A/t is defined by the equation $z_2^k = t^{l-k} z_1^l$. As $t \to 0$ we obtain the tangent cone $C(A, 0) = \mathbb{C}_1$, and the equation $z_2^k = 0$, ascribing multiplicity k to the points of $\mathbb{C}_1$. On the other hand, every A/t is a k-sheeted cover (above any domain in $\mathbb{C}_1 \setminus \{0\}$), and as $t \to 0$ all k sheets tend to the $\mathbb{C}_1$-plane. Thus, the multiplicity defined above is equal to k, and the chain $C([A],0) = k\mathbb{C}_1$. The fact that the multiplicity of the set A itself figures here is no coincidence, in the general situation the following holds:

P r o p o s i t i o n. *Let T be a holomorphic p-chain in a neighborhood of 0 in $\mathbb{C}^n$. Then $\mu_0(T) = \deg C(T, 0)$. In particular, if $A \ni 0$ is a pure p-dimensional analytic set, then $\mu_0(A) = \deg C([A],0)$.*

■ The proof follows from the definitions. If $\pi: A \cap U \to U'$ is a transversal cover with number of sheets $k = \mu_0(A)$, then the definition of $C([A],0)$ implies that, obviously, $\pi: C([A],0) \to \mathbb{C}^p$ is also k-sheeted (counted with multiplicities). If $C([A],0) = \sum k_j C_j$, then $k = \sum k_j \mu_0(C_j) = \sum k_j \deg C_j =: \deg C([A],0)$. The general case is obtained by additivity. ■

11.7. Dependence of the tangent cone on parameters. In general, the tangent cone at the points of an analytic set A in $\mathbb{C}^n$ is not even a "semicontinuous" function of these points: it is easy to construct an example in which $C(A, 0)$ does not contain the limit set of a family $C(A,z)$ as $z \to 0$, $z \in A$. Nevertheless, such discontinuities are encountered relatively seldom, and the following holds:

P r o p o s i t i o n 1. *Let A be a pure p-dimensional analytic set in $\mathbb{C}^n$ and let $A^{(m)} \subset A$ consist of the points at which $\mu_z(A) \geqslant m$. Then on $A^{(m)} \setminus A^{(m+1)}$ the tangent cones $C(A,z)$ vary continuously, both as sets (in the Hausdorff metric the compact sets $C(A,z) \cap \{|z| \leqslant 1\}$ vary continuously, see p.8.6) and as holomorphic chains (see p.12.2 and p.16.1).*

■ Let $0 \in A^{(m)} \setminus A^{(m+1)}$ and let $U = U' \times U''$ be a neighborhood of 0 such that the projection $\pi: A \cap U \to U'$ is exactly m-sheeted. Then (p.11.4) the restriction $\pi: A^{(m)} \cap U \to \sigma'$ is one-to-one for some analytic subset $\sigma' \subset U'$, i.e. $A^{(m)} \cap U$ is the graph of a vector function $z'' = \phi(z')$ on σ', such this function is moreover continuous since $A^{(m)} \cap U$ is closed in U (on the set reg σ' the function ϕ is even holomorphic by Proposition 3, p.3.3). Let Φ_I, $|I| = m$, be canonical defining functions of the cover $\pi: A \cap U \to U'$, and let $\Phi_I(z)_a^\star$ be the initial homogeneous polynomial of degree m in z in the expansion of $\Phi_I(z)$ in powers of $z - a$ (i.e. $\Phi_I(z) = \Phi_I(z)_a^\star + o(|z-a|^m)$). Since the plane $a + \mathbb{C}_{(p+1)\cdots n}$ is transversal to A at the points $a \in A^{(m)} \cap U$ (p.11.4), by Proposition 2, p.8.4, $C(A,a)$ is the set of common zeros of the polynomials $\Phi_I(z)_a^\star$, $|I| = m$. Since $\Phi_I(z) = \sum_{|J| \leqslant m} \phi_{IJ}(z')(z'')^J = \sum_{|J| \leqslant m} \phi_{IJ}(a'+(z-a)')\ (z''-a''+\phi(a'))^J$, the coefficients of these $\Phi_I(z)_a^\star$ continuously depend on a, both in the Hausdorff metric and when counted with multiplicities. ■

P r o p o s i t i o n 2. *For any compact set $K \subset A^{(m)} \setminus A^{(m+1)}$ there are positive constants ϵ, C, and r_0 such that for every point $a \in K$ the Hausdorff distance between $A \cap \{|z-a| \leqslant r\}$ and $a + C(A,a) \cap \{|z| \leqslant r\}$ does not exceed $Cr^{1+\epsilon}$ for $r \leqslant r_0$.*

■ The constants ϵ, C, and r_0 characterizing the deviation of the analytic set A from its tangent cone $C(A,a)$ can be estimated in terms of the coefficients of canonical defining functions of the transversal cover A in some neighborhood of a (see p.8.6). Since for $a \in K$ these coefficients vary continuously (see the proof of Proposition 1), $\epsilon(a)$, $C(a)$, and $r_0(a)$ can be chosen locally constant. Now use the compactness of K. ■

C o r o l l a r y. *Let in a neighborhood of 0 the set $A^{(m)} \setminus A^{(m+1)}$ coincide with*

$\mathbb{C}_{1 \cdots q}$, *and suppose that the* $\mathbb{C}_{(p+1) \cdots n}$*-plane is transversal to A at* 0. *Then there are a neighborhood* $'U \times U' \times U'' \ni 0$, *where* $'U \subset \mathbb{C}_{1 \cdots q}$, $U' \subset \mathbb{C}_{(q+1) \cdots p}$, $U'' \subset \mathbb{C}_{(p+1) \cdots n}$, *and a constant C such that the set* $A \cap U$ *lies within the "wedge"* $|z''| \leqslant C|z'|$, *where* $z' = (z_{q+1}, \ldots, z_p)$ *and* $z'' = (z_{p+1}, \ldots, z_n)$.

Finally we consider how the tangent cone changes under holomorphic variation of defining functions.

P r o p o s i t i o n 3. *Let A be a pure* $(p+q)$*-dimensional analytic subset of a domain* $U \times V$ *in* $\mathbb{C}^n_z \times \mathbb{C}^q_t$, *and let* $A_t = \{z: (z,t) \in A\}$. *Assume that* $U \times \{0\}$ *is entirely contained in* $A^{(m)} \setminus A^{(m+1)}$ *and that the restriction onto A of the projection* $\tilde{\pi}: (z,t) \mapsto (z',t) \in \mathbb{C}_{1 \cdots p} \times \mathbb{C}^q$ *is an m-sheeted analytic cover. Then* $C(A_t, 0) = \{z: \Phi_I(z)^*_t = 0, |I| = m\}$, *where* $\Phi_I(z)^*_t$ *are the initial homogeneous polynomials of degree m in the Taylor expansion with respect to z of canonical defining functions* $\Phi_I(z',t,z'')$ *of the cover* $\pi|_A$ *(the variable t is regarded here as a parameter). In particular, the tangent cones to* A_t *at* 0 *vary continuously (in the Hausdorff metric).*

■ In essence there is nothing to prove. For each fixed $t \in V$ the functions $\Phi_I(z',t,z'')$, $|I| = m$, form canonical defining functions for the holomorphic chain with support A_t (and multiplicities $\mu_{(z,t)}(\tilde{\pi}|_A)$, which are constant everywhere on each irreducible component of A_t). By p.11.6, the tangent cone to this chain is defined by the functions $\Phi_I(z)^*_t$ (both as a set and as a chain). ■

C o r o l l a r y. *Let A be a pure p-dimensional analytic set in* $\mathbb{C}^n$, *let* $M \subset A$ *be a q-dimensional complex manifold, and let* $L \ni 0$ *be a plane of dimension* $n-q$ *such that for each* $z \in M$ *the plane* $z + L$ *is transversal to M at z while the set* $A_z = A \cap (z+L)$ *is pure* $(p+q)$*-dimensional. Then the cones* $C(A_z, z)$ *continuously depend on* $z \in M \setminus A^{(m+1)}$, *where* $m = \mu(A,M)$.

■ The statement is local, hence after a suitable biholomorphic change of coordinates we may assume that M is a domain on the $\mathbb{C}_{1 \cdots q}$-plane. By Corollary 1, for each $a \in M \setminus A^{(m+1)}$ there is an $(n-p)$-dimensional plan $L' \subset L$ such that $a + L'$ is transversal to A at a. Therefore, by Proposition 3, the tangent cones $C(A_z, z)$ for $z \in M \setminus A^{(m+1)}$ vary continuously. ■

12. Intersection indices

12.1. The case of complementary codimensions. We will immediately consider the intersection of several analytic sets, since the formalism for two sets is not simpler than that in the general situation. Let $A_1, \ldots, A_k$ be analytic subsets of a domain $D \subset \mathbf{C}^n$ of pure dimensions $p_1, \ldots, p_k$, respectively. By p.3.5, the codimension of the analytic set $A_1 \cap \cdots \cap A_k$ at each point does not exceed the sum of the codimensions of the A_j. In this and in the next section we will assume that

1. the sum of the codimensions of the A_j is equal to n (i.e. $\sum p_j = (k-1)n$);
2. $\cap_1^k A_j$ is zero-dimensional (the minimal possible dimension).

Under these assumptions we will define the multiplicities with which the sets A_j intersect at points, and in this way we will construct a corresponding 0-chain in D. At first glance the definition given below may seem somewhat fanciful, but technically it is very convenient, since we can prove at once that this definition is well compatible with the multiplicities from §10 and §11.

The set $A_1 \times \cdots \times A_k =: \prod A_j$ in $\mathbf{C}^{kn} = \mathbf{C}_{z^1}^{n} \times \cdots \times \mathbf{C}_{z^k}^{n}$ has dimension $\Sigma p_j = (k-1)n$ (by assumption 1), and the diagonal $\Delta = \{z^1 = \cdots = z^k\} \subset \mathbf{C}^{kn}$ has complementary dimension n. If a is isolated in $\cap A_j$ (assumption 2), then $(a)^k := (a, \ldots, a) \in \mathbf{C}^{kn}$ is isolated in $(\prod A_j) \cap \Delta$, hence the projection $\pi_\Delta|_{\prod A_j}$ along Δ on $\Delta^\perp$ (in $\mathbf{C}^{kn}$) is an analytic cover in some neighborhood of $(a)^k$.

D e f i n i t i o n. The *intersection index* of the sets $A_1, \ldots, A_k$ at a point $a \in \cap A_j$ is (under the assumptions 1 and 2) the multiplicity of the projection $\pi_\Delta|_{\prod A_j}$ at $(a)^k \in \mathbf{C}^{kn}$:

$$i_a(A_1, \ldots, A_k) := \mu_{(a)^k}(\pi_\Delta|_{\prod A_j}).$$

For $a \in D \setminus \cap A_j$ it is convenient to put $i_a(A_1, \ldots, A_k) = 0$.

It is clear from the definition that:

The intersection index is a symmetric function of the sets, i.e.

$i_a(A_{j_1}, \ldots, A_{j_k}) = i_a(A_1, \ldots, A_k)$ *for any permutation* $(1, \ldots, k) \mapsto (j_1, \ldots, j_k)$.

The following Proposition discloses the geometrical meaning of the intersection index in terms of the space $\mathbf{C}^n$ itself: under small shifts of the sets A_j the point a splits into exactly $i_a(A_1, \ldots, A_k)$ distinct points of simple intersection, see Figure 10. (This property may be taken as definition of the index, but technically it is less convenient.)

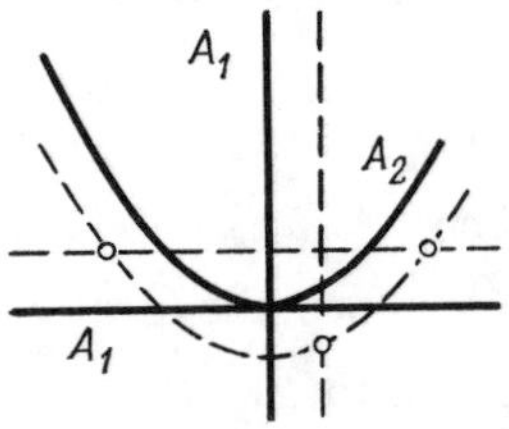

Figure 10.

Proposition. *Let* $A_, \ldots, A_k$ *be analytic sets in* $\mathbf{C}^n$ *of pure complementary codimensions (i.e.* $\sum \operatorname{codim} A_j = n$*), and let* a *be an isolated point of the set* $\cap A_j$*. Then there are a neighborhood* $U \ni a$ *and a number* $\epsilon > 0$ *such that for almost all* $c = (c^1, \ldots, c^k)$ *with* $|c^j| < \epsilon$ *the number of points of the set* $(\cap(c^j + A_j)) \cap U$ *is equal to* $i_a(A_1, \ldots, A_k)$*. For such* c *all points of the set* $(\cap(c^j + A_j)) \cap U$ *are regular on each* $c^j + A_j$*, and the tangent planes to the* $c^j + A_j$ *at these points intersect transversally (i.e. at a point). The exceptional values* c *from the domain* $\{|c^j| < \epsilon: j = 1, \ldots, k\} \subset \mathbf{C}^{kn}$*, i.e. for which this property does not hold, form in this domain a proper analytic subset, and for them*

$$\#(\cap(c^j + A_j)) \cap U < i_a(A_1, \ldots, A_k).$$

■ Since $(a)^k$ is isolated in $(\prod A_j) \cap \Delta$ in $\mathbf{C}^{kn}$, there is a neighborhood $\tilde{U} \ni (a)^k$ such that $\tilde{A} = (\prod A_j) \cap \tilde{U}$ is an analytic cover above a certain domain $\tilde{U}'$ in $\Delta^\perp$ (with respect to the projection π_Δ); moreover, $\tilde{A} \cap \Delta = \{(a)^k\}$, and the number of sheets of this cover is equal to $m = \mu_{(a)^k}(\pi_\Delta|\tilde{A})$. Let $\tilde{\sigma} \subset \tilde{U}'$ be its critical analytic set. Now, if $-\pi_\Delta(c) \in \tilde{U}' \setminus \tilde{\sigma}$, then the plane $\Delta - c$

intersects $\tilde{A}$ at exactly m distinct points; all these points are regular on $\tilde{A}$, and the intersection of $\Delta - c$ and the manifold reg $\tilde{A}$ at them is transversal. We also note that $\pi_\Delta((a)^k) = 0$, i.e. U' is a neighborhood of 0 in $\Delta^\perp$.

Since $\tilde{A} \cap \Delta = \{(a)^k\}$, here are neighborhood $U \supset V \ni a$ and a number $\epsilon > 0$ such that the ϵ-neighborhoods of V belongs to U, the Cartesian product of the ϵ-neighborhoods of U belongs to $\tilde{U}$, and $\tilde{A} \cap (\Delta - c) \subset V \times \cdots \times V$ if $c = (c^1, \ldots, c^k)$ and all $|c^j| < \epsilon$. Fix such a c, under the condition $-c' = -\pi_\Delta(c) \notin \tilde{\sigma}$.

Let $b = (b^1, \ldots, b^k) \in \tilde{A} \cap (\Delta - c)$. Then $b^j \in A_j \cap V$, $j = 1, \ldots, k$, $b + c \in \Delta$, i.e. $b^1 + c^1 = \cdots = b^k + c^k \in U$ is one and the same point. By construction, this point belongs to all $(c^j + A_j) \cap U$. Conversely, if $b^0 \in (\cap(c^j + A_j)) \cap U$, then $b_j = b^0 - c^j \in A_j \cap (U - c^j)$, hence $b = (b^1, \ldots, b^k) \in \tilde{A} \cap (\Delta - c)$. Since the projection of $\Delta - c$ to each $\mathbb{C}^{n_j}_z$ is one-to-one, we have thereby proved that $(\cap(c^j + A_j)) \cap U$ and $\tilde{A} \cap (\Delta - c)$ have the same number of points, i.e. m. This number is maximal possible (the cover $\pi_\Delta|_{\tilde{A}}$ is m-sheeted), hence for exceptional values c, forming the analytic set

$$\pi_\Delta^{-1}(-\tilde{\sigma}) \cap \{|c^j| < \epsilon : j = 1, \ldots, k\},$$

the number of points in $(\cap(c^j + A_j)) \cap U$ is strictly less than m.

As proved above, if $-\pi_\Delta(c) \notin \tilde{\sigma}$, then each point $b \in \tilde{A} \cap (\Delta - c)$ is regular on $\tilde{A}$, i.e. $b \in \operatorname{reg} \tilde{A}$, hence

$$b^0 = b^j + c^j \in \operatorname{reg}(c^j + A_j)$$

for all j. It remains to remark that the tangent space to reg $\tilde{A}$ at b is the Cartesian product of the tangent spaces T_j to the sets reg $(c^j + A_j)$ at b^0. Transversality of intersection of $\tilde{A}$ and $\Delta - c$ at b means that $(T_1 \times \cdots \times T_k) \cap \Delta = \{0\}$, which is clearly equivalent to the condition $T_1 \cap \cdots \cap T_k = \{0\}$, i.e. transversality of the T_j at b^0. ■

R e m a r k. The set of exceptional values c is not only nowhere dense in the domain $\{|c^j| < \epsilon : j = 1, \ldots, k\}$ for a sufficiently small $\epsilon > 0$, but also on the intersection with each plane $c^j = 0$, since, say,

$$\#\cap(c^j + A_j) \cap U = \#(c^j - c^1 + A_j) \cap U;$$

i.e. the number of points in the intersection does not change when all sets are shifted by one and the same vector $c^1 \in \mathbb{C}^n$.

C o r o l l a r y. *If A is a pure p-dimensional analytic set in $\mathbb{C}^n$, $A \ni 0$, and if an $(n-p)$-dimensional plane L is such that 0 is an isolated point in $A \cap L$, then $i_0(A,L) = \mu_0(\pi_L|_A)$.*

For holomorphic chains the intersection indices are defined by additivity: if

$$T_j = \sum c_{jl} A_{jl}, \quad j = 1, \ldots, k,$$

are holomorphic chains of pure dimensions p_j, respectively, and if $\sum p_j = (k-1)n$, then at each isolated point $a \in \cap |T_j|$ the intersection index of these chains is defined as

$$i_a(T_1, \ldots, T_k) = \sum c_{1l_1} \cdots c_{kl_k} \cdot i_a(A_{1l_1}, \ldots, A_{kl_k}).$$

Here it may happen that at certain points $a \in \cap |T_j|$ the intersection index is equal to zero (because of the presence of multiplicities of opposite sign). If the whole set $\cap |T_j|$ is zero-dimensional in D, then by ascribing to each point $a \in \cap |T_j|$ its intersection index we obtain a holomorphic 0-chain in D:

$$T_1 \wedge \cdots \wedge T_k := \sum_{a \in \cap |T_j|} i_a(T_1, \ldots, T_k) \cdot \{a\}.$$

If the T_j are positive chains, the Proposition implies that this intersection chain varies continuously (in subdomains $G \subset D$ not containing points of $\cap |T_j|$ on the boundary) under small shifts of the chains T_j. Moreover, if $T_j = A_j$ (i.e. the multiplicities are equal to 1), then for almost all small $c = (c^1, \ldots, c^k)$ all multiplicities of the points of the 0-chain $((c^1 + A_1) \cap G) \wedge \cdots \wedge ((c^k + A_k) \cap G)$ are also equal to 1.

Addition of positive holomorphic chains corresponds to union of analytic sets. The definition of intersection indices of holomorphic chains is compatible with the definition for analytic sets, also taking into account these operations. Indeed, if A_1', A_1'' are pure p_1-dimensional analytic sets and $\dim A_1' \cap A_1'' < p_1$, then

$$\#(A_1' \cup A_1'') \cap \left[\bigcap_2^k (c^j + A_j) \right] =$$

$$= \# A_1' \cap \left[\bigcap_2^k (c^j + A_j) \right] + \# A_1'' \cap \left[\bigcap_2^k (c^j + A_j) \right]$$

for almost all $(c^2, \ldots, c^k)$, $|c^j| < \epsilon$. By the Proposition and the Remark this means that

$$i_a(A_1' \cup A_1'', A_2, \ldots, A_k) = i_a(A_1', A_2, \ldots, A_k) + i_a(A_1'', A_2, \ldots, A_k).$$

The property of multilinearity and the symmetry of the function $i_a(T_1, \ldots, T_k)$ with respect to T_j is included in the definition.

12.2. Some properties of indices. First of all we note a relation between intersection indices and multiplicities of holomorphic maps. If a function f is holomorphic in a domain D, if Z_f is its divisor (its zero set, counted with multiplicities), and if $\operatorname{supp} Z_f = \cup S_j$ is the decomposition into irreducible components, then the multiplicity of the zero of f at almost all points of S_j is one and the same, say k_j (see p.1.5). Thus, the divisor of f can be naturally regarded as a holomorphic $(n-1)$-chain in D: $Z_f = \sum k_j S_j$.

P r o p o s i t i o n 1. *Let $f = (f_1, \ldots, f_n)$: $D \to \mathbb{C}^n$ be a holomorphic map of a domain $D \subset \mathbb{C}^n$, and let $a \in D$ be an isolated point of the zero set of f (which is equal to $\cap |Z_{f_j}|$). Then*

$$\mu_a(f) = i_a(Z_{f_1}, \ldots, Z_{f_n}),$$

where Z_{f_j} is the divisor of f_j.

■ If at a all supports of the chains Z_{f_j} are regular and intersect transversally, the proof is simple: in a neighborhood of a the functions f_j have the form $h_j^{k_j}$, where $h_j \in \mathcal{O}_a$, $dh_j(a) \neq 0$, hence $\mu_a(f) = k_1 \cdots k_n$ (see p.10.2). On the other hand, the index is also equal to $k_1 \cdots k_n \cdot i_a(Z_{h_1}, \ldots, Z_{h_n}) = k_1 \cdots k_n$ since the Z_{h_j} are mutually transversal complex manifolds with multiplicities 1 (by the proof of Proposition 12.1, the intersection index of such manifolds is 1).

The chain Z_{f_j} shifted by a vector c^j is the divisor of the holomorphic function $f_j(z - c^j) =: g_j(z)$ (everything considered in a neighborhood of $a \in \cap |Z_{f_j}|$). By Proposition 12.1 there are arbitrarily small $c^1, \ldots, c^n$ such that in a fixed neighborhood $U \ni a$ the supports of the chains Z_{g_j} intersect at regular points only, and then only transversally. Let $a^1, \ldots, a^\mu$ be these intersection points $(c^1, \ldots, c^n$ being fixed). By what has already been proved, $\mu_{a^j}(g) = i_{a^j}(Z_{g_1}, \ldots, Z_{g_n})$. By Rouché's theorem (p.10.3), $\sum \mu_{a^j}(g) = \mu_a(f)$ if U

and c^j are sufficiently small. On the other hand, under the same assumptions $\sum i_{a^j}(Z_{g_1}, \ldots, Z_{g_n}) = i_a(Z_{f_1}, \ldots, Z_{f_n})$ by Proposition 12.1. ■

The *complete intersection index* of holomorphic chains $T_1, \ldots, T_k$ of complementary pure codimensions *in a domain* D is the number $i_D(T_1, \ldots, T_k) := \sum i_a(T_1, \ldots, T_k)$, where the summation extends over all points in the set $\cap \, |T_j|$, under the assumption that this set is finite.

We say that a sequence of holomorphic p-chains T^ν in a domain $D \subset \mathbf{C}^n$ converges to a p-chain T in D if

1) the limit set of the family $|T^\nu|$ in D (i.e. the set of all limit points for sequences of the form $\{z^\nu\}$, $z^\nu \in |T^\nu|$) belongs to $|T|$;
2) for each point $a \in \operatorname{reg} |T|$ and each $(n-p)$-dimensional plane $L \ni a$ transversal to $|T|$ at a there are a neighborhood $U \ni a$ and an index ν_0 such that $|T| \cap L \cap U = \{a\}$ and such that the total multiplicity of each 0-chain $T^\nu \wedge L$ in U (i.e. the sum of the multiplicities of $T^\nu \wedge L$ over all points in $|T^\nu| \cap L \cap U$) is equal to the multiplicity of T at a for all $\nu > \nu_0$.

This can be briefly stated thus: $T^\nu \to T$ if both the supports and the multiplicities converge (this definition is stronger than that of convergence in the sense of currents; coincidence holds for positive chains only, see p.16.1).

The following Proposition establishes the continuous dependence of the intersection index of intersecting chains.

P r o p o s i t i o n 2. *Let $T_1, \ldots, T_k$ be holomorphic chains of complementary pure codimensions, defined is a neighborhood of the closure of a bounded domain D in $\mathbf{C}^n$, and such that $\cap \, |T_j| \cap \overline{D}$ is a finite set completely belonging to D. Let $\{T_j^\nu\}$ be sequences of holomorphic chains converging to T_j, respectively, in a neighborhood of $\overline{D}$. Then, from some index ν_0 onwards, we have*

$$i_D(T_1^\nu, \ldots, T_k^\nu) = i_D(T_1, \ldots, T_k).$$

■ The definitions of convergence and direct product of chains obviously imply that the chain $T_j^\nu = T_1^\nu \times \cdots \times T_k^\nu$ converges, as $\nu \to \infty$, to the chain $\prod T_j$ in a neighborhood of $(\bar{D})^k$ in $\mathbb{C}^{kn}$. Since $|\prod T_j| \cap \Delta$ is a zero-dimensional set, there are neighborhoods $\mathfrak{U} \supset \Delta$, and $\mathfrak{U}' \ni 0$ in $\Delta^\perp$, such that the projection $\pi_\Delta: (\prod T_j) \cap \mathfrak{U} \to \mathfrak{U}'$ is an analytic cover, with number of sheets equal to $i_D(T_1, \ldots, T_k)$ (counted with the multiplicities of $\prod T_j$). The planes $c + \Delta$, $c \in \Delta^\perp$, not intersecting the critical analytic set of this cover, do intersect $(\prod T_j) \cap \mathfrak{U}$ at regular points only, and then only transversally. Hence the definition of convergence $\prod T_j^\nu \to \prod T_j$ implies that from some ν_0 onwards the maps $\pi_\Delta: (\prod T_j^\nu) \cap \mathfrak{U} \to \mathfrak{U}'$ also define analytic covers, and the numbers of sheets, counted with the multiplicities of these covers (by definition equal to $i_D(T_1^\nu, \ldots, T_k^\nu)$) are also equal to $i_D(T_1, \ldots, T_k)$ for $\nu > \nu_0$. ■

C o r o l l a r y. *If $c^j \in \mathbb{C}$ are arbitrary, sufficiently small vectors ($|c^j| < \epsilon$), then* $i_D(c^1 + T_1, \ldots, c^k + T_k) = i_D(T_1, \ldots, T_k)$.

We stress that the c^j are arbitrary here, in distinction to Proposition 12.1 (in which intersection points were counted without multiplicities).

The image of a holomorphic chain under a biholomorphic map ϕ is defined by additivity: $\phi(\sum k_j A_j) = \sum k_j \phi(A_j)$. The invariance of intersection indices under biholomorphic maps can be easily derived from the definition of index and the invariance of multiplicities of holomorphic maps.

P r o p o s i t i o n 3. *Let $T_1, \ldots, T_k$ be holomorphic chains in some domain in $\mathbb{C}^n$, let a be an isolated point in the intersection of their supports, and let ϕ be a biholomorphic map defined in a neighborhood of a. Then*

$$i(T_1, \ldots, T_k) = i_{\phi(a)}(\phi(T_1), \ldots, \phi(T_k)).$$

■ Suppose $a = \phi(a) = 0$ and $\phi(A_j) = C_j$. By definition, $i_0(C_1, \ldots, C_k) = \mu_0(\pi_\Delta|_{\prod C_j})$. The set $\prod C_j = C_1 \times \cdots \times C_k$ is obtained from $\prod A_j$ by the biholomorphic transformation $(z^1, \ldots, z^k) \mapsto (\phi(z^1), \ldots, \phi(z^k))$ in a neighborhood of 0 in $\mathbb{C}^{kn}$. Since multiplicities of projections are invariant under such transformations (see p.10.2), $\mu_0(\pi_\Delta|_{\prod C_j}) = \mu_0(\pi_\Delta|_{\prod A_j}) = i_0(A_1, \ldots, A_k)$. ■

Thus, the notion of intersection index of holomorphic chains using local coordinates is well-defined on an arbitrary complex manifold. The following geometrical analog of Bezout's theorem is of interest in this respect.

Proposition 4. *Let $T_1, \ldots, T_k$ be positive holomorphic chains of complementary pure dimensions in $\mathbb{P}_n$, with algebraic supports and zero-dimensional intersection. Then*

$$i(T_1, \ldots, T_k) := \sum_{a \in \cap |T_j|} i_a(T_1, \ldots, T_k) = \prod_1^k \deg T_j.$$

(Recall that $\deg (\sum k_i A_i) := \sum k_i \deg A_i$.)

■ The definitions of index and degree obviously imply that it suffices to investigate the case when $T_j = A_j$ are algebraic subsets in $\mathbb{P}_n$ (i.e. the multiplicities are equal to 1). Since $\cap A_j$ consists of finitely many points, we may assume that $\cap A_j \subset \mathbb{C}^n = \mathbb{P}_n \setminus \{z_n = 0\}$. Put $A_j^0 = A_j \cap \mathbb{C}^n$. The definition of index implies that $\sum i_a(A_1^0, \ldots, A_k^0)$ is equal to the multiplicity of the analytic cover $\pi_\Delta: \prod A_j \to \Delta^\perp$ in $\mathbb{C}^{kn}$. From $\cap A_j \subset \mathbb{C}^n$ and the fact that this also holds for small displacements of A_j in $\mathbb{P}_n$ we readily see that the closures of $\prod A_j$ and Δ in $\mathbb{P}_{kn}$ do not have points at infinity in common. By Corollaries 1 and 2, p.11.3, the multiplicity of the projection $\pi_\Delta: \prod A_j^0 \to \Delta^\perp$ is equal to $\deg (\prod A_j) = \prod \deg A_j$. ■

12.3. Intersections of holomorphic chains. We now consider the situation in which the intersection of analytic sets need not be zero-dimensional. In this situation it is especially convenient to give all expositions in terms of holomorphic chains.

Let $T_1, \ldots, T_k$ be holomorphic chains in some domain D in $\mathbb{C}^n$, with supports $|T_j|$ of pure dimensions p_j, respectively. We say that these chains *intersect properly at a point* $a \in \cap |T_j|$ if the dimension of $\cap |T_j|$ at this point is the theoretically minimal possible value, i.e. equals $(\sum p_j) - (k-1)n$ (see p.3.5). If this property holds at all points of $\cap |T_j|$ (in D), we say that $T_1, \ldots, T_k$ *intersect properly* (in the large, in D). Note that if $T_1, \ldots, T_k$ intersect properly at a, then every subsystem $T_{j_1}, \ldots, T_{j_l}$ also intersects properly at this point. This obviously follows from p.3.5: if the dimension of $\cap_{j \in J} |T_j|$ at a is not the minimal possible value, $J = (j_1, \ldots, j_l)$, then

$$\dim_a \bigcap_1^k |T_j| \geqslant \dim_a \bigcap_{j \in J} |T_j| + \sum_{j \notin J} p_j - (k-l)n > (\sum_1^k p_j) - (k-1)n.$$

In the large this property does not hold; e.g., the sets $A_1: z_1(z_2 z_3 - 1) = 0$,

A_2: $z_2(z_2z_3-1)=0$, and A_3: $z_3=0$ in $\mathbb{C}^3$ intersect properly (the only common point is the coordinate origin), but $A_1 \cap A_2$ contains the hypersurface $z_2z_3=1$. In this and the following Section we will assume that the chains intersect properly at all points under consideration.

The notion of intersection index of chains in case $\cap\, |T_j|$ is not zero-dimensional naturally reduces to the situation of zero-dimensional intersection considered above. We add to the family $T_1,\ldots,T_k$ a plane L of complementary dimension $n-\dim_a \cap\, |T_j|$, located "in general position" with respect to $\{T_j\}$ at a. The nature of such planes will become clear in the following Proposition.

Proposition. *Let $T_1,\ldots,T_k$ be holomorphic chains of pure dimensions in some domain in $\mathbb{C}^n$, properly intersecting at a point a, and let G be the set of all planes of dimension $p=n-\dim_a \cap\, |T_j|>0$, transversal to $\cap\, |T_j|$ at a. Then $i_a(T_1,\ldots,T_k,L)$ is independent of $L\in G$.*

■ Suppose $a=0$. By Corollary 2, p.3.8, the set G is an everywhere dense subdomain of the Grassmannian $G(p,n)$. By Proposition 2, p.12.2, $i_a(T_1,\ldots,T_k,L)$ does not change under small displacements of L, i.e. on G this function is locally constant. Since G is connected, $i_a(T_1,\ldots,T_k,L)$ is one and the same number for all $L\in G$. ■

Definition 1. The number $i_a(T_1,\ldots,T_k):=i_a(T_1,\ldots,T_k,L)$, $L\in G$, is called the *intersection index of the chains $T_1,\ldots,T_k$ at the point a.*

Corollary 1. *$i_a(T_1,\ldots,T_k)$ is a biholomorphic invariant: if ϕ is biholomorphic in a neighborhood of a, then*

$$i_{\phi(a)}(\phi(T_1),\ldots,\phi(T_k))=i_a(T_1,\ldots,T_k).$$

■ Suppose $a=\phi(a)=0$. After a suitable linear change of coordinates, $\phi(z)=z+\epsilon(z)$, where $\epsilon(z)=o(|z|)$. Put $\phi_t(z)=z+t^{-1}\epsilon(tz)$. Then ϕ_t is also a biholomorphism in some neighborhood $U\ni 0$, independent of $t\in(0,1]$. By the definition and Proposition 2, p.12.2, the function $i_0(\phi_t(T_1),\ldots,\phi_t(T_k))$ is locally constant in t, hence constant in the large. For $t=0$ the arguments are the initial chains $T_1,\ldots,T_k$, and for $t=1$ the chains $\phi(T_1),\ldots,\phi(T_k)$. ■

C o r o l l a r y 2. *If M is a complex manifold of dimension* $n - \dim_a \cap |T_j|$, *transversal to* $\cap |T_j|$ *at a, then* $i_a(T_1, \ldots, T_k) = i_a(T_1, \ldots, T_k, M)$.

■ After a biholomorphic transformation ϕ of a neighborhood of a mapping $M \cap U$ into a plane L, transversality is preserved; hence, by Corollary 1, $i_a(T_1, \ldots, T_k, M) = i_{\phi(a)}(\phi(T_1), \ldots, \phi(T_k), L) = i_{\phi(a)}(\phi(T_1), \ldots, \phi(T_k)) = i_a(T_1, \ldots, T_k)$. ■

C o r o l l a r y 3. *If* $a \in \operatorname{reg} \cap |T_j|$, *then there is a neighborhood* $U \ni a$ *such that* $i_z(T_1, \ldots, T_k)$ *does not depend on* $z \in \cap |T_j| \cap U$.

■ If $a + L$ is transversal to $\cap |T_j|$ at a, then $z + L$ is transversal to $\cap |T_j|$ at z, for all $z \in \cap |T_j|$ sufficiently close to a. Hence the statement follows from the definition of index and the continuity of the latter (Proposition 2, p.12.2). ■

Thus, on connected components of $\cap |T_j|$ the intersection index $i_z(T_1, \ldots, T_k)$ is constant.

D e f i n i t i o n 2. Let S be an irreducible component of $\cap |T_j|$. The common value $i_z(T_1, \ldots, T_k)$ at the points of the manifold

$$S \cap \operatorname{reg} \cap |T_j|,$$

which is everywhere dense on S, is called the *intersection index of the chains* $T_1, \ldots, T_k$ *along* S, and is denoted by $i_S(T_1, \ldots, T_k)$. If $\cap |T_j| = \bigcup S_j$ is the decomposition into irreducible components and $k_j = i_{S_j}(T_1, \ldots, T_k)$, then the chain $\sum k_j S_j$ is called the *intersection* (*chain*) of the holomorphic chains $T_1, \ldots, T_k$ (in their common domain of definition); it is denoted by $T_1 \wedge \cdots \wedge T_k$.

The definition and Proposition 2, p.12.2, obviously imply that the intersection chain depends continuously on the initial chains:

C o r o l l a r y 4. *Let* $T_1^\nu, \ldots, T_k^\nu$ *be holomorphic chains of pure dimensions, properly intersecting in a domain D,* $\nu = 1, 2, \ldots$, *and converging as* $\nu \to \infty$ *to chains* $T_1, \ldots, T_k$ *which also intersect properly in D. Assume also that the limit set of the family* $|T_1^\nu \wedge \cdots \wedge T_k^\nu|$ *in D belongs to* $|T_1 \wedge \cdots \wedge T_k|$. *Then* $T_1^\nu \wedge \cdots \wedge T_k^\nu \to T_1 \wedge \cdots \wedge T_k$ *in D as* $\nu \to \infty$.

12.4. Properties of intersection chains. We start with the important property of being associative (commutativity and multilinearity of $T_1 \wedge \cdots \wedge T_k$ in the "variables" $T_1, \cdots, T_k$ clearly follow from the definition).

P r o p o s i t i o n 1. *Let T_1, T_2, T_3 be chains of pure dimensions, properly intersecting in a domain D. Then*

$$T_1 \wedge T_2 \wedge T_3 = (T_1 \wedge T_2) \wedge T_3.$$

R e m a r k. In general the chains T_1, T_2 do not intersect properly in D (see, e.g., the example at the beginning of p.12.3). However, at all points of the set $|T_1| \cap |T_2| \cap |T_3|$ their intersection is proper, hence the chain $T_1 \wedge T_2$ is defined in a neighborhood of the set $|T_1| \cap |T_2| \cap |T_3|$. The equality stated must be understood as equality of chains in a neighborhood of their (common) support, in the sequel we will not mention this anymore.

■ In view of the additivity it suffices to consider the case when $T_j = A_j$ are analytic subsets in D. Let $A_1 \wedge A_2 = \sum k_j S_j$, where $k_j = i_{S_j}(A_1, A_2)$, and let C be an arbitrary irreducible component of $A_1 \cap A_2 \cap A_3$ in D. Our statement is equivalent to: the equality $i_C(A_1, A_2, A_3) = \sum k_j i_C(S_j, A_3)$ holds for each component C. Suppose $0 \in \operatorname{reg} C$, and let L be a plane of dimension $n - \dim C$ transversal to C at 0. Then $i_C(A_1, A_2, A_3) = i_0(A_1, A_2, A_3, L)$, and $i_C(S_j, A_3) = i_0(S_j, A_3, L)$ for all j. By Proposition 12.1 there are a neighborhood $U \ni 0$ and arbitrarily small vectors c^3, $c^4 \in \mathbf{C}^n$ such that for each j the analytic sets S_j, $c^3 + A_3$, $c^4 + L$ intersect within U only at regular points $a_{j\nu}$; moreover, the intersection is transversal, and $i_0(S_j, A_3, L)$ is the number of points $a_{j\nu}$:

$$i_0(S_j, A_3, L) = \sharp S_j \cap (c^3 + A_3) \cap (c^4 + L) \cap U.$$

Denote by $U_{j\nu}$ pairwise disjoint neighborhoods of $a_{j\nu}$ in U, and put $V_j = \cup_\nu U_{j\nu}$, $V = \cup_j V_j$. Choose $U_{j\nu}$ small such that $S_j \cap V_j$, $(c^3 + A_3) \cap V$, and $M := (c^3 + A_3) \cap (c^4 + L) \cap V$ is a complex manifold (the latter is possible in view of the transversality of the intersection of the manifolds $c^3 + A_3$ and $c^4 + L$ at all points $a_{j\nu}$). By Corollary 2, p.12.3, for each ν the equality $k_j := i_{S_j}(A_1, A_2) = i_{a_{j\nu}}(A_1, A_2, M)$ holds, hence

$$\sum_j k_j i_C(S_j, A_3) = \sum_j k_j i_0(S_j, A_3, L) =$$

$$= \sum_j k_j(\sharp S_j \cap M) = \sum_{j\nu} i_{a_{j\nu}}(A_1, A_2, M).$$

On the other hand,

$$i_{a_{j\nu}}(A_1, A_2, M) = \sharp(c^1 + A_1) \cap (c^2 + A_2) \cap M \cap U_{j\nu}$$

for almost all sufficiently small $(c^1, \cdots, c^2)$, hence

$$\sum_j k_j i_C(S_j, A_3) = \sharp(c^1 + A_1) \cap (c^2 + A_2) \cap M =$$

$$= \sharp(c^1 + A_1) \cap (c^2 + A_2) \cap (c^3 + A_3) \cap (c^4 + L) \cap U$$

for almost all sufficiently small (c^1, c^4). By Proposition 12.1 the last expression is equal to $i_0(A_1, A_2, A_3, L)$, i.e. to $i_C(A_1, A_2, A_3)$. ■

Corollary 1. *The operation $(T_1, \ldots, T_k) \mapsto T_1 \wedge \cdots \wedge T_k$ on holomorphic chains T_j in a domain D is associative, under the condition that the chains intersect properly.*

Corollary 2. *For each point $a \in \cap |T_j|$ the equality $i_a(T_1, \ldots, T_k) = \mu_a(T_1 \wedge \cdots \wedge T_k)$ holds.*

■ The multiplicity of the chain $T_1 \wedge \cdots \wedge T_k$ at a is, by definition, equal to the intersection index at a of this chain with an arbitrary plane $L \ni a$ of complementary dimension and transversal to $\cap |T_j|$ at a. By Proposition 12.1 there are a neighborhood $U \ni a$ and a vector $c \in \mathbb{C}^n$ such that $|T_1 \wedge \cdots \wedge T_k|$ and $c + L$ intersect at regular points a_ν only, and then only transversally, and, moreover, $\mu_a(T_1 \wedge \cdots \wedge T_k) = i_a(T_1 \wedge \cdots \wedge T_k, L) = \sum_\nu i_{a_\nu}(T_1 \wedge \cdots \wedge T_k, c + L)$. But at regular points $a_\nu \in \cap |T_j|$, in view of the associativity, $i_{a_\nu}(T_1 \wedge \cdots \wedge T_k, c + L) = i_{a_\nu}(T_1, \ldots, T_k, c + L)$. By Proposition 2, p.12.2, $\sum i_{a_\nu}(T_1, \ldots, T_k, c + L) = i_a(T_1, \ldots, T_k, L) = i_a(T_1, \ldots, T_k)$ for sufficiently small c (the last equality is by definition). ■

Thus, the chain $T_1 \wedge \cdots \wedge T_k$, which was at first defined only in terms of intersection indices at its regular points, contains complete information on the intersection indices of the chains $T_1, \ldots, T_k$ at all points $a \in \cap |T_j|$.

Corollary 3. *Let T be a holomorphic p-chain in a neighborhood of 0 in $\mathbb{C}^n$,* *and let L be a plane of dimension q and such that* $\dim_0 |T| \cap L = (p+q)-n$ *(proper intersection). Let* $L' \subset L$ *be a plane of dimension* $n-p$*, transversal to* $|T| \cap L$ *at 0 on the plane L. Then* $i_0(T,L) = i_0(T,L')$.

■ Let $\Lambda = L' + L^{\perp}$. Then Λ is transversal to $|T| \cap L$ at 0, hence $i_0(T,L) = i_0(T,L,\Lambda)$. By Proposition 1, $T \wedge L \wedge \Lambda = T \wedge (L \wedge \Lambda) = T \wedge L'$. Hence, by Corollary 2, $i_0(T,L,\Lambda) = i_0(T,L')$. ■

The general situation of intersection of several chains can be reduced to intersection of a chain with a plane by using direct products (as in p.12.1). For chains $T_j = \sum c_{ji} A_{ji}$ in $\mathbb{C}^{n_j}$, $j = 1, \ldots, k$, the direct product is defined as

$$\prod T_j := \sum c_{1i_1} \cdots c_{ki_k} A_{1i_1} \times \cdots \times A_{ki_k} \quad \text{in } \mathbb{C}^{n_1} \times \cdots \times \mathbb{C}^{n_k}.$$

Suppose $n_1 = \cdots = n_k = n$, suppose all $T_j \subset \mathbb{C}^n$ intersect properly at a point a, and let $L \ni a$ be a plane of dimension complementary to the dimension of $\cap |T_j|$ and such that a is isolated in $(\cap |T_j|) \cap L$. Put

$$(L)^k = (L \times \cdots \times L) \cap \Delta = \{(z, \ldots, z) \in \mathbb{C}^{kn} : z \in L\}.$$

Then $\prod T_j$ and $(L)^k$ intersect properly at $(a)^k \in \mathbb{C}^{kn}$, since $\dim (L)^k = \dim L = \operatorname{codim}_a \cap |T_j| = \sum \operatorname{codim}_a |T_j| = \operatorname{codim}_{(a)^k} |\prod T_j|$.

L e m m a 1. $i_a(T_1, \ldots, T_k, L) = i_{(a)^k}(\prod T_j, (L)^k)$.

■ In view of the additivity with respect to T_j of both sides, it suffices to investigate the situation when $T_j = A_j$ are analytic sets. If U is the neighborhood of a from Proposition 12.1 and $b \in (c^j + A_j) \cap L \cap U$, then $b - c^j$ belongs to A_j and $(L - c^j) \cap U$ at the same time, i.e. $(b-c^1, \ldots, b-c^k) \in (\prod A_j) \cap ((L)^k - c)$, where $c = (c^1, \ldots, c^k)$. It is clear that in this way a one-to-one correspondence between $\cap (c^j + A_j) \cap L \cap U$ and $(\prod A_j) \cap ((L)^k - c) \cap \mathfrak{U}$ is established, where $\mathfrak{U}$ is a suitable neighborhood of $(a)^k$, sufficiently small if U is sufficiently small. Since $c^1, \ldots, c^k$ are arbitrary, the statement is obtained from the Definition, Proposition, and Corollary in p.12.1. ■

If L is transversal to $\cap |T_j|$ at a, Lemma 1 implies

$$i_a(T_1, \ldots, T_k) = i_{(a)^k}(\prod T_j, (L)^k).$$

L e m m a 2. $i_a(T_1, \ldots, T_k) = i_{(a)^k}(\prod T_j, \Delta)$.

■ Let $L \ni a$ be a plane of dimension $n - \dim_a \cap |T_j|$ transversal to $\cap |T_j|$ at a. By Lemma 1, $i_a(T_1, \ldots, T_k) = i_{(a)^k}(\prod T_j, (L)^k)$. On the other hand, the plane $(L)^k \subset \Delta$ is clearly transversal to $|\prod T_j| \cap \Delta$ at $(a)^k$ on the plane Δ, hence, by Corollary 3, $i_{(a)^k}(\prod T_j, (L)^k) = i_{(a)^k}(\prod T_j, \Delta)$. ■

The projection $\pi\colon (z)^k \to z$ of the diagonal $\Delta \subset \mathbf{C}^{kn}$ onto $\mathbf{C}^n$ is one-to-one, hence for any holomorphic chain $T = \sum k_j S_j$ in $\mathbf{C}^{kn}$ with support on Δ the projection $\pi_* T := \sum k_j \pi(S_j)$ is well-defined; it is a holomorphic chain in (the corresponding domain) in $\mathbf{C}^n$.

P r o p o s i t i o n 2. $T_1 \wedge \cdots \wedge T_k = \pi_*((\prod T_j) \wedge \Delta)$.

■ If $a \in \operatorname{reg} \cap |T_j|$, the equality of Lemma 2 contains at the righthand side the multiplicity of the chain $(\prod T_j) \wedge \Delta$ at $(a)^k$, and at the lefthand side the multiplicity of the chain $T_1 \wedge \cdots \wedge T_k$ at $a = \pi((a)^k)$. ■

C o r o l l a r y 1. *Let T_1, T_2 be holomorphic chains of pure dimensions in a domain D in $\mathbf{C}^n$, intersecting property in D, and let T_1', T_2' be similar chains in a domain G in $\mathbf{C}^m$. Then in $D \times G$ the equality*

$$(T_1 \times T_1') \wedge (T_2 \times T_2') = (T_1 \wedge T_2) \times (T_1' \wedge T_2')$$

holds.

■ In view of the additivity of both sides it suffices to investigate the situation when T_1, T_2 and T_1', T_2' are analytic sets (this implies, in particular, that the supports of the chains being investigated coincide). Let $a \in |T_1 \wedge T_2|$, $a' \in |T_1' \wedge T_2'|$, and let L, L' be planes transversal to $|T_1 \wedge T_2|$ and $|T_1' \wedge T_2'|$, respectively, at these points. By Lemma 1 and Corollary 2,

$$\mu_a(T_1 \wedge T_2) = i_{(a)^2}(T_1 \times T_2, (L)^2), \quad \mu_{a'}(T_1' \wedge T_2') = i_{(a')^2}(T_1' \times T_2', (L')^2)$$

hence

$$\mu_{(a,a')}((T_1 \wedge T_2) \times (T_1' \wedge T_2')) = i_{(a)^2}(T_1 \times T_2, (L)^2) \cdot i_{(a')^2}(T_1' \times T_2', (L')^2).$$

Proposition 12.1 and the equality $(A \cap C) \times (B \cap D) = (A \times B) \cap (C \times D)$ for arbitrary sets imply that the righthand side of the last equation is equal to $i_b(T_1 \times T_2 \times T_1' \times T_2', (L)^2 \times (L')^2)$, where $b = (a,a,a',a')$. In turn, this is clearly equal to

$$i_c(T_1 \times T_1' \times T_2 \times T_2', (L \times L')^2) = \mu_{(a,a')}((T_1 \times T_1') \wedge (T_2 \times T_2')),$$

where $c = (a, a', a, a')$. Thus, we have proved that the chains being investigated have equal multiplicities at all points. ■

Corollary 2. *If chains $T_1, \ldots T_k$ in $\mathbb{C}^m$ and chains $T_1', \ldots, T_k'$ in $\mathbb{C}^n$ intersect properly (in domains D and G, respectively), then*

$$(T_1 \times T_1') \wedge \cdots \wedge (T_k \times T_k') = (T_1 \wedge \cdots \wedge T_k) \times (T_1' \wedge \cdots \wedge T_k'),$$

$$i_{(a,b)}(T_1 \times T_1', \ldots, T_k \times T_k') = i_a(T_1, \ldots, T_k) \cdot i_b(T_1', \ldots, T_k').$$

■ The first equality can obviously be obtained by induction from Corollaries 1 and 4:

$$((T_1 \times T_1') \wedge \cdots \wedge (T_{k-1} \times T_{k-1}')) \wedge (T_k \times T_k') =$$

$$= ((T_1 \wedge \cdots \wedge T_{k-1}) \times (T_1' \wedge \cdots \wedge T_{k-1}')) \wedge (T_k \times T_k') =$$

$$= (T_1 \wedge \cdots \wedge T_{k-1} \wedge T_k) \times (T_1' \wedge \cdots \wedge T_{k-1}' \wedge T_k').$$

The second equality follows from the first one and Corollary 2, since multiplicities are multiplied when taking a direct product (p.11.2). ■

We combine the main statements proved above:

Theorem. *Let $T_1, \ldots, T_k$ be holomorphic chains of pure dimensions in a domain $D \subset \mathbb{C}^n$, intersecting properly in D. Then in D is defined the holomorphic chain $T_1 \wedge \cdots \wedge T_k$ of dimension $(\sum \dim |T_j|) - (k-1)n$, with support belonging to $\cap |T_j|$ and whose multiplicity on each irreducible component C of $\cap |T_j|$ is equal to $i_C(T_k, \ldots, T_k)$. This chain is the image of the chain $(\prod T_j) \cap \Delta$ in $(D)^k \subset \mathbb{C}^{kn}$ under the natural projection $(z, \ldots, z) \mapsto z$. At each point $a \in \cap |T_j|$ the multiplicity of $T_1 \wedge \cdots \wedge T_k$ is equal to $i_a(T_1, \ldots, T_k)$. The operation $(T_1, \ldots, T_k) \mapsto T_1 \wedge \cdots \wedge T_k$ is multilinear over the ring $\mathbb{Z}$, is commutative, associative, invariant under biholomorphic transformations of D, and continuous.*

12.5. Multiplicities and transversality. How do the multiplicities of the chain $T_1 \wedge \cdots \wedge T_k$ relate to those of the initial chains $T_1, \ldots, T_k$? In order to answer this question it is expedient to introduce the notion of transversality, which was already successfully used in the case of intersections of analytic sets with planes of complementary dimensions (p.12.2). We say that the holomorphic chains $T_1, \ldots, T_k$ of pure dimensions $p_1, \ldots, p_k$, respectively, in a domain D in $\mathbf{C}^n$ intersect *transversally* at a point $a \in \cap |T_j|$ if the tangent cones to the supports of these chains at a intersect properly at the coordinate origin, i.e. if $\dim_0 \cap_1^k C(|T_j|, a) = (\sum_1^k p_j) - (k-1)n$. If this property is fulfilled at all points $a \in \cap |T_j|$, we say that $T_1, \ldots, T_k$ intersect transversally in D. Transversality of an intersection (both at a point and globally) implies that the intersection is proper. This clearly follows from the property that an analytic set is close to its tangent cone (p.8.6). For this natural property, in the definition of transversality we must take the tangent cones to the supports of the chains, and note to the chains: if the multiplicities of T_j are of different signs, $|C(T_j,a)| \subset C(|T_j|,a)$, and equality need not hold. Hence proper intersection of the cones $C(T_j,a)$ does in general not imply proper intersection of the T_j themselves. We yet not that our definition of transversality is wider than that generally adopted: usually transversality is defined at regular points only, and this regularity is assumed in the very notion of transversality.

The importance of this notion can be seen from the following assertion.

T h e o r e m. *Let $T_1, \ldots, T_k$ be positive holomorphic chains of pure dimensions in a domain $D \subset \mathbf{C}^n$, intersecting properly in D, and let $a \in \cap |T_j|$. Then*

$$i_a(T_1, \ldots, T_k) = \mu_a(T_1 \wedge \cdots \wedge T_k) \geqslant \mu_a(T_1) \cdots \mu_a(T_k),$$

and equality holds if and only if the chains $T_1, \ldots, T_k$ intersect transversally at a.

(For nonpositive chains the Theorem, obviously, does not hold, since the left-hand side may turn out to be equal to, e.g., zero.)

■ The proof is relatively simple. The chain $T_1 \wedge \cdots \wedge T_k$ is the projection of $(\prod T_j) \cap \Delta$, which lies in $\mathbf{C}^{kn}$, and $i_a(T_1, \ldots, T_k) = i_{(a)^k}(\prod T_j, \Delta)$, see p.12.4. If the plane $L \ni a$ of dimension $n - \dim_a \cap |T_j|$ is transversal to $\cap |T_j|$ at a, then $i_a(T_1, \ldots, T_k) = i_{(a)^k}(\prod T_j, (L)^k)$ (Lemma 1, p.12.4). The definition of multiplicities of analytic sets and multiplicativity of these multiplicities (p.11.1 and p.11.2) imply

$$i_a(T_1, \ldots, T_k) = i_{(a)^k}(\prod T_j, (L)^k) =$$

$$= \mu_{(a)^k}(\pi_{(L)^k} | \prod T_j) \geqslant \mu_{(a)^k}(\prod T_j) = \prod \mu_a(T_j).$$

By Proposition 2, p.11.2, equality holds if and only if L can be chosen such that $(L)^k$ is transversal to $\prod T_j$ at $(a)^k$ (the initial and last terms in this inequality are independent of L). By p.8.1, $C(|\prod T_j|,(a)^k) = \prod C(|T_j|,a)$, hence $(L)^k$ is transversal to $\prod T_j$ (more precisely, $(L)^k - (a)^k$ intersects $C(|\prod T_j|,(a)^k)$ at $0 \in \mathbf{C}^{kn}$ only) if and only if $L - a$ intersects $\cap C(|T_j|,a)$ at 0 only. But such an L of dimension $n - \dim_a \cap |T_j|$ exists if and only if $\dim_0 \cap C(|T_j|,a) = \dim_a \cap |T_j|$, i.e. if the T_j intersect transversally at a. ■

Note the following simple Corollary for algebraic sets.

C o r o l l a r y. *Let $T_1, \ldots, T_k$ be positive chains of pure dimensions with algebraic supports, intersecting properly in $\mathbf{P}_n$. Then* $\deg T_1 \wedge \cdots \wedge T_k = \prod_1^k \deg T_j$.

■ Let $\dim \cap |T_j| = p$, and let L be a complex plane in $\mathbf{P}_n$ of dimension $n-p$ such that $\cap |T_j| \cap L$ is zero-dimensional. Then $\deg T_1 \wedge \cdots \wedge T_k = i(T_1 \wedge \cdots \wedge T_k, L) = i(T_1, \ldots, T_k, L)$ (in view of the associativity of intersections, see p.12.4). By Proposition 4, p.12.2, the last expression is equal to $\prod_1^k \deg T_j$, since $\deg L = 1$. ■

12.6. Multiplicities of fibers of holomorphic maps. If in Theorem 12.5 we take for T_j the divisors of holomorphic functions f_j in D and if $k = n$, then by Proposition 1, p.12.2, we obtain the Tsikh-Yuzhakov theorem (p.10.3) concerning multiplicities of zeros of holomorphic maps. Theorem 12.5 generalizes such results in two directions: first, instead of divisors we may take, e.g., fibers of holomorphic maps with appropriate multiplicities (see below); secondly, instead of maps between complex manifolds we may consider maps of analytic sets (in the final result we have in mind multiplicities of points of these sets). We will explain this in more detail.

Let X be an n-dimensional complex manifold, and $f\colon X \to Y$ a holomorphic map into an m-dimensional manifold, $n \geqslant m$, such that all fibers $f^{-1}(f(x))$, $x \in X$, have dimension $n - m$ (which is minimal possible). Let $a \in X$ and let M be a complex manifold in X of dimension m, passing through a transversally to the fiber $f^{-1}(b)$, where $b = f(a)$. The multiplicity of the map f at a is defined by the formula $\mu_a(f) := \mu_a(f|_M)$, where $\mu_a(f|_M)$ is defined in p.10.2 (as the multiplicity of the analytic cover $f|_M$ at a). This definition does not depend on the choice of M, which easily follows from the invariance of multiplicity under biholomorphisms and Rouché's theorem (see p.10.2 and p.10.3). (In other terms this number may be defined as the intersection index of the graph Γ_f of f with the manifold $w = f(z)$ at $(a, f(a)) \in X \times Y$.) Obviously, at regular points of the fiber $f^{-1}(b)$ the multiplicity of f is locally constant, hence if $f^{-1}(b) = \cup S_j$ is the decomposition into irreducible components, $\mu_z(f) \equiv k_j \in \mathbb{Z}_+$ on $S_j \cap \operatorname{reg} f^{-1}(b)$. The chain $\sum k_j S_j$ is called the fiber of the map f counted with multiplicities (or the *fiber in the sense of holomorphic chains*); it is denoted by $f^*[b]$. If $f: X \to \mathbb{C}$ and $b = 0$, this chain coincides with the divisor of f.

Let now A be a pure p-dimensional analytic set in X such that all fibers of $f|_A$ have, in a neighborhood of a, dimension $p - m$. Then the multiplicity of $f|_A$ at a is, by definition, equal to $\mu_a(f|_A) := i_a(A, f^*[f(a)])$, and the fiber of $f|_A$ above a, counted with multiplicities, is the holomorphic $(p-m)$-chain $A \wedge f^*[f(a)]$ (of course, instead of A we may consider any p-chain). For $p = m$ this multiplicity of $f|_A$ coincides with that defined in p.10.2. Indeed, we may assume that $f: A \to Y$ is a k-sheeted analytic cover and that $f^{-1}(f(a)) \cap A = \{a\}$. Then $k = \mu_a(f|_A)$, according to the definition in p.10.2. On the other hand, by p.3.8, for almost all $w \in f(X)$ the fibers $f^{-1}(w)$ have in X multiplicity 1, i.e. do not belong to $\operatorname{br} f$ (under the assumption that all fibers are $(n-m)$-dimensional). Moreover, the rank Theorem implies that almost all these fibers intersect A at regular points only, and then only transversally, i.e. $\sum_{f(z)=w} i_z(A, f^*[w]) = k$ for them. The remaining follows from the continuity of the index (p.12.2) and from the obvious fact that $f^*[w] \to f^*[f(a)]$ in the sense of holomorphic chains (see p.12.2) as $w \to f(a)$.

CHAPTER 3

METRICAL PROPERTIES OF ANALYTIC SETS

13. The fundamental form and volume forms

13.1. Hermitian manifolds. A complex manifold Ω is called Hermitian if in the fibers $T_z\Omega$ of the tangent space $T\Omega$ a positive definite Hermitian quadratic form $H_z(v,v')$ is defined that is an infinitely differentiable function of z (i.e. if $v(z)$, $v'(z)$ are vector fields of class C^∞, then $H_z(v(z),v'(z))$ is a function of class C^∞). The notion of Hermiticity requires, of course, a complex structure in the fibers of $T\Omega$. It is introduced in a standard manner as follows: if z are the local complex coordinates in a neighborhood of $a \in \Omega$ and $z = x + iy$, then the complex structure operator J in $T_a\Omega$ acts according to $J(\partial / \partial x_k) = \partial / \partial y_k$, $J(\partial / \partial y_k) = -\partial / \partial x_k$, and is further $\mathbb{R}$-linear. The space $T_a\Omega$ is naturally isomorphic to $(T_{1,0}\Omega)_a$ ($\sum\{v_j(\partial / \partial z_j)+\overline{v}_j(\partial / \partial \overline{z}_j)\} \leftrightarrow \sum v_j(\partial / \partial z_j)$), and under this isomorphism the operator J corresponds to ordinary multiplication by i in the $\mathbb{C}$-linear space $(T_{1,0}\Omega)_a$ (see A4.2). After the introduction of the complex structure we naturally define in $T_a\Omega$ the operation of multiplication by a complex number $((\alpha+i\beta)v=\alpha v+\beta Jv)$, and complex conjugation $\sum\{\alpha_k(\partial / \partial x_k)+\beta_k(\partial / \partial y_k)\} \mapsto \sum\{\alpha_k(\partial / \partial x_k)-\beta_k(\partial / \partial y_k)\}$ (in $T\Omega$ the operator $\sum v_j(\partial / \partial z_j) \mapsto \sum \overline{v}_j(\partial / \partial z_j)$ corresponds to it; do not confuse it with the corresponding operation in $\mathbb{C}T\Omega$!). Hermiticity of $H_a(v,v')$ means $\mathbb{C}$-linearity in the first argument v, and Hermitian symmetry: $H_a(v,v') = \overline{H_a(v',v)}$ for any $v,v' \in T_a\Omega$. In local coordinates the form H can be represented as $\sum h_{jk} dz_j \otimes d\overline{z}_k$, where $h_{jk} = \overline{h_{kj}}$ are functions and $(dz_j \otimes d\overline{z}_k)(v,v') := dz_j(v) \cdot d\overline{z}_k(v')$.

We introduce the notations $S = \operatorname{Re} H$, $\omega = -\operatorname{Im} H$. The form S is symmetric on T_z: $S(v,v') = S(v',v)$, $\mathbb{R}$-bilinear, and also positive definite. Such a form on a differentiable manifold is called a Riemannian metric (it is usually denoted by the symbol ds^2). Thus, every Hermitian manifold (Ω,H) is automatically a Riemannian manifold, with metric $\operatorname{Re} H$. We yet note that $S(iv,iv') = S(v,v')$, and that the vectors v, iv are orthogonal with respect to S, i.e. $S(v,iv) = 0$ for all $v \in T_z\Omega$ (this property clearly follows from the definition). If $v^1, \ldots, v^n$ is an orthogonal basis in $T_z\Omega$ with respect to H (i.e. $H(v^j,v^k) = \delta_{jk}$), then $v^1, iv^1, \ldots, v^n, iv^n$ is an $\mathbb{R}$-linear basis, orthonormal with respect to S. Under permutations of the v^k the pairs v^k, iv^k in the second basis permute, i.e. we obtain even permutations. Thus, on an arbitrary complex manifold a canonical orientation is defined: in the fibers $T_z\Omega$ bases of the form $v^1, iv^1, \ldots, v^n, iv^n$ are considered as being positively oriented (this definition does, obviously, not depend on the Hermitian structure, and is determined by the complex structure on the manifold only).

Clearly the form $\omega = -\operatorname{Im} H$ is antisymmetric: $w(v,v') = -\operatorname{Im} H(v,v') = -w(v',v)$, and $\mathbb{R}$-bilinear, i.e. ω is a real differential form of degree 2 (see A4.2). If

$$H = \sum h_{jk} dz_j \otimes d\bar{z}_k,$$

$$v = \sum \left[v_j \frac{\partial}{\partial z_j} + \bar{v}_j \frac{\partial}{\partial \bar{z}_j} \right], \quad v' = \sum \left[v_j' \frac{\partial}{\partial z_j} + \bar{v}_j' \frac{\partial}{\partial \bar{z}_j} \right],$$

then $H(v,v') = \sum h_{jk} v_j \bar{v}_k'$, hence

$$\omega(v,v') = -\frac{1}{2i} \sum h_{jk} (v_j \bar{v}_k' - v_j' \bar{v}_k) = \left[\frac{i}{2} \sum h_{jk} dz_j \wedge d\bar{z}_k \right] (v,v').$$

Thus, in local coordinates $\omega = i(\sum h_{jk} dz_j \wedge d\bar{z}_k)/2$. We yet note that $\omega(iv,iv') = \omega(v,v')$, and $\omega(v,iv) = H(v,v) > 0$ if $v \neq 0$. The form H can be uniquely recovered from ω; more precisely, $H(v,v') = \omega(v,iv') - i\omega(v,v')$. This differential form ω is called the *fundamental form* of the Hermitian manifold Ω. If ω is closed, i.e. $d\omega = 0$, then the manifold (Ω,H) is called a *Kähler manifold.* We investigate some important examples.

a) In $\mathbb{C}^n$ there is a natural Hermitian structure: $H = \sum_1^n dz_j \otimes d\bar{z}_j$. The Riemannian metric corresponding to it is the Euclidean metric

$$ds^2 = \sum |dz_j|^2 = \operatorname{Re} \sum dz_j \otimes d\bar{z}_j.$$

The fundamental form ω can in this case be represented as

$$\omega = \frac{i}{2}\sum dz_j \wedge d\bar{z}_j = \frac{i}{2}\partial\bar{\partial}|z|^2 = \frac{1}{4}dd^c|z|^2.$$

With this standard Hermitian structure the space $\mathbf{C}^n$ is a Kähler manifold. If $z_j = x_j + iy_j$, then $dz_j \wedge d\bar{z}_j = -2i\, dx_j \wedge dy_j$, hence $\omega = \sum dx_j \wedge dy_j$ in real coordinates. The representation $\omega = (dd^c|z|^2)/4$ implies the simple, but important, fact that ω is invariant under unitary transformations of $\mathbf{C}^n$ (under which $|z|$ and the operators d, d^c are invariant): if z' are the coordinates obtained from z by a unitary transformation, then $\omega = (dd^c|z'|^2)/4$.

In a neighborhood of every point a of the Hermitian manifold (Ω, H) there are local coordinates that are orthogonal at a with respect to H (this clearly follows from the possibility of bringing a Hermitian form to diagonal form). In such coordinates

$$H_a = \sum (dz_j)_a \otimes (d\bar{z}_j)_a, \quad (\omega)_a = \frac{i}{2}\sum (dz_j)_a \wedge (d\bar{z}_j)_a,$$

i.e. in a neighborhood of a the form ω is "Euclidean", up to infinitesimals.

b) If M is a complex manifold embedded in the Hermitian manifold (Ω, H), then the Hermitian structure $H|_M$ is naturally induced on it. The corresponding Riemannian metric is $S|_M$, and the fundamental form coincides with $\omega|_M$.

c) The form $\tilde{\omega}_0 = (dd^c \log|z|^2)/4$ in $\mathbf{C}^{n+1}_*$ is invariant under dilatations, hence it corresponds in $\mathbf{P}_n$ to a form ω_0 such that $\tilde{\omega}_0$ is the pre-image of ω_0 under the canonical map $\Pi: \mathbf{C}^{n+1}_* \to \mathbf{P}_n$ (this condition determines ω_0 uniquely). It is constructed as follows. In the coordinate neighborhood U_j: $z_j \neq 0$ we put

$$\omega_0 := \frac{1}{4}dd^c \log\left[1 + \sum_{k \neq j}\left|\frac{z_k}{z_j}\right|^2\right].$$

In intersections $U_i \cap U_j$ the functions under the sign of the logarithm differ by the factor $|z_i/z_j|^2$. Since the function z_i/z_j is holomorphic and zero free in $U_i \cap U_j$, we have $dd^c \log|z_i/z_j|^2 = 2i\partial\bar{\partial}\log|z_i/z_j|^2 = 0$, hence the definitions given above coincide in $U_i \cap U_j$. The form ω_0 in $\mathbf{P}_n$ thus constructed is called the *Fubini-Study form* (one often writes $\omega_0 = (dd^c \log|z|^2)/4$, where z are the homogeneous coordinates in $\mathbf{P}_n$, but this notation is incorrect, the form ω_0 in $\mathbf{P}_n$ is not d-exact). Since the function $\log(1+|\zeta|^2)$ is strictly plurisubharmonic in $\mathbb{C}^n_\zeta$ (which can be checked directly), ω_0 is positive; more precisely, $\omega_0(v,iv)>0$ for any tangent vector $v \in T_a\mathbb{P}_n$, $v \neq 0$. Since ω_0 is also invariant under multiplication by i in $T\mathbb{P}_n$ (which is obvious from the local representations), a unique Hermitian structure H_0 on $\mathbb{P}_n$ such that $\omega_0 = -\mathrm{Im}\, H_0$ corresponds to it. The corresponding Riemannian metric $S_0 = \mathrm{Re}\, H_0$ in $\mathbb{P}_n$ is called the *Fubini-Study metric.* We can prove that the distance in $\mathbb{P}_n$ determined by it is proportional to the distance defined in A3.4. Since $d\omega_0 = 0$, with this Hermitian structure $\mathbb{P}_n$ is a Kähler manifold, hence every complex manifold embedded in it is also a Kähler manifold.

The form ω_0 is invariant under unitary transformations $l^*: \mathbb{P}_n \to \mathbb{P}_n$, i.e. $l^*(\omega_0) = \omega_0$. This clearly follows from the facts that $\Pi^*(\omega_0) = \tilde{\omega}_0$ and that $\tilde{\omega}_0$ is invariant under unitary transformations of $\mathbb{C}^{n+1}$.

In $\mathbb{C}^n = \mathbb{P}_n \setminus \{z_0=0\}$ with affine coordinates $\zeta_j = z_j / z_0$ the projective fundamental Fubini-Study form has the representation

$$\omega_0 = \frac{i}{2}\partial\bar{\partial}\log(1+|\zeta|^2) = \frac{i}{2}\left[\frac{\partial\bar{\partial}|\zeta|^2}{1+|\zeta|^2} - \frac{\partial|\zeta|^2\wedge\bar{\partial}|\zeta|^2}{(1+|\zeta|^2)^2}\right].$$

For $n = 1$ this is equal to

$$\frac{i}{2}\,\frac{d\zeta\wedge d\bar{\zeta}}{(1+|\zeta|^2)^2}$$

and the corresponding metric $|d\zeta|^2/(1+|\zeta|^2)^2$ is the classical spherical metric in $\mathbb{C}$. For each fixed point $a \in \mathbb{P}_n$ we can change the homogeneous coordinates by a unitary transformation so that $a = [1,0,\ldots,0]$ in the new coordinates. In the corresponding affine coordinates a will be the coordinate origin, and the Fubini-Study metric can, in a neighborhood of a, be written as $(dd^c|\zeta|^2)/4 + O(|\zeta|^2)$, i.e. is almost Euclidean.

We denote the form $\tilde{\omega}_0$ in $\mathbb{C}^{n+1}_*$ also by ω_0 in the sequel; it will always be clear from the context which form is meant.

13.2. Volume forms. In $\mathbb{C}^n$ the Euclidean volume form dV is equal to $dx_1\wedge dy_1\wedge\cdots\wedge dx_n\wedge dy_n$. In complex coordinates it may be written as

$$dV = \frac{i}{2}dz_1\wedge d\bar{z}_1\wedge\cdots\wedge\frac{i}{2}dz_n\wedge d\bar{z}_n = \frac{1}{n!}\omega^n,$$

where $\omega = i(\partial\bar{\partial}|z|^2)/2 = (dd^c|z|^2)/4$ is the Euclidean fundamental form. If L is a linear subspace in $\mathbb{C}^n$ of dimension p, we can map it onto $\mathbb{C}^p \subset \mathbb{C}^n$ by a unitary change of coordinates. The restriction of ω to $\mathbb{C}^p$ is $i(\sum_1^p dz_j \wedge d\bar{z}_j)/2$, i.e. coincides with the Euclidean fundamental form in $\mathbb{C}^p$. Hence the Euclidean volume form in $\mathbb{C}^p \subset \mathbb{C}^n$ is equal to $(\omega|_{\mathbb{C}^p})^n/p! = (\omega^p|_{\mathbb{C}^p})/p!$. Since ω and its powers are invariant under unitary transformations, we thus obtain that on any p-dimensional subspace $L \subset \mathbb{C}^n$ the Euclidean volume form is equal to $(\omega^p|_L)/p!$.

Let now M be an arbitrary p-dimensional complex manifold in $\mathbb{C}^n$. If $a \in M$, then, as has just been proved, the value of the form $\omega^p/p!$ on an orthonormal frame with canonical orientation in T_aM is equal to 1, hence $dV_M = (\omega^p|_M)/p!$

Finally, consider a general Hermitian manifold (Ω, H) with fundamental form ω. As was noted in p.13.1, for each point $a \in \Omega$ there are local coordinates $z_1, \ldots, z_n$ in which $(\omega)_a = i(\sum dz_j \wedge d\bar{z}_j)_a/2$, i.e. on $T_a\Omega$ we are in the Euclidean situation. As has been proved, the value of $\omega^p/p!$ on any canonically oriented and orthonormal frame in any p-dimensional complex subspace in $T_a\Omega$ is equal to 1. Hence, for any p-dimensional complex manifold M in Ω the restriction $(\omega^p|_M)/p!$ coincides with the volume form on M. Hence we have proved the following important statement:

W i r t i n g e r ' s t h e o r e m (infinitesimal version). *Let Ω be a Hermitian complex manifold with fundamental form ω. Then for any p-dimensional complex manifold M embedded in Ω the restriction of $\omega^p/p!$ to M coincides with the volume form on M in the induced metric. In short,*

$$dV_M = \frac{1}{p!}\omega^p|_M.$$

Traditionally the volumes of subsets of a k-dimensional real manifold are called k-dimensional volumes, and are denoted by vol_k. Hence the equality obtained may be written as:

$$\mathrm{vol}_{2p} M = \frac{1}{p!}\int_M \omega^p$$

for any p-dimensional complex manifold M embedded in Ω.

The presence of one differential form in Ω measuring the volumes of all embedded complex manifolds of the same dimension is an essential advantage of Hermitian differential geometry in comparison to Riemannian geometry. In Wirtinger's theorem a lot of information concerning the structure of complex manifolds is hidden (especially in the compact case), and if we bring together even only those few consequences of this, in general very simple, Theorem, that are used in the book, a sufficiently impressive picture would be obtained.

The fundamental form is useful also in the computation of volumes of real manifolds embedded in a Hermitian manifold. We restrict ourselves to one important example.

L e m m a. *Let* ρ *be a real-valued function of class* C^1 *on an n-dimensional Hermitian manifold* (Ω, H) *with fundamental form* ω, *and let the real hypersurface* $\Gamma = \{z \in \Omega\colon \rho(z)=0,\ (d\rho)_z \neq 0\}$ *have the same orientation as the boundary of the set* $\rho < 0$ *(with the canonical orientation on* Ω*). Then on* Γ,

$$d^c\rho \wedge \omega^{n-1} = (n-1)!\,|d\rho|\,dV_\Gamma.$$

Here $|d\rho|(a)$ is the norm of the covector $(d\rho)_a$ as a linear functional on the space $T_a\Omega$ with Hermitian scalar product H_a. In $\mathbf{C}^n$ (and also with respect to local coordinates at a that are orthonormal with respect to H_a), $|d\rho|(a) = (\sum |\partial\rho/\partial x_j|^2 + |\partial\rho/\partial y_j|^2))^{1/2}(a)$ is the length of the gradient of ρ. We yet remark that $|d^c\rho| = |d\rho|$ (this can be easily verified in suitable local coordinates, see the proof).

■ Fix a point $a \in \Gamma$ and choose local coordinates in a neighborhood of a such that they are orthonormal at a, $dx_1|_{T_a\Gamma} = 0$, and $(\partial\rho/\partial x_1)(a) > 0$. Then $dz_1|_{T_a^c\Gamma} = 0$, hence at a the form dV_Γ is equal to $dy_1 \wedge \prod_2^n \{i(dz_j \wedge d\bar{z}_j)/2\}$ (the orientation is taken into account by the choice of the sign of $(\partial\rho/\partial x_1)(a)$). Since in $T_a\Omega$ the linear forms $(dx_1)_a$ and $(d\rho)_a$ have identical zero sets, they are proportional, hence $(d\rho)_a = |d\rho|(a)\cdot(dx_1)_a$, since $|(dx_1)_a| = 1$. Because $dy_1 = d^c x_1$, this implies $dy_1 = (d^c\rho)/|d\rho|$ at a. Furthermore, the form $\omega^{n-1}/(n-1)!$ is, in our coordinates, $dz_1 \wedge d\bar{z}_1 \wedge \alpha + \prod_2^n \{i(dz_j \wedge d\bar{z}_j)/2\}$. Since $dz_1 \wedge d\bar{z}_1 = -2i\, dx_1 \wedge dy_1$, after multiplication by dy_1 the first term gives zero. Hence at a we have $dV_\Gamma = dy_1 \wedge \prod_2^n \{i(dz_j \wedge d\bar{z}_j)/2\} = (d^c\rho/|d\rho|) \wedge \omega^{n-1}/(n-1)!$. ■

13.3. Wirtinger's inequality. In general the restriction of $\omega^p/p!$ to a $2p$-dimensional real manifold in Ω does not give a volume (e.g., in $\mathbb{C}^2$ the restriction of ω to $\mathbb{R}^2$ or to the plane $z_2 = \bar{z}_1$ vanishes). But there is a useful one-sided estimate in this situation.

Remember that the value of an antisymmetric multilinear s-form ϕ on an s-vector $v^1\wedge \cdots \wedge v^s$ is, by definition, equal to $\phi(v^1, \ldots, v^s)$, and that the norm $|\cdot|$ in the space of s-vectors in $\mathbb{R}^N$ is defined as the Euclidean norm with respect to the basis $\{e^{j_1}\wedge\ldots\wedge e^{j_s}\}$, where $e^1, \ldots, e^N$ form a canonical basis in $\mathbb{R}^N$. In $\mathbb{C}^n \approx \mathbb{R}^{2n}$ a canonical real basis is the basis $e^1, ie^1, \ldots, e^n, ie^n$, where $e^1, \ldots, e^n$ form a standard $\mathbb{C}$-basis in $\mathbb{C}^n$ (see A4.1).

Theorem. *Let $v = v^1\wedge \cdots \wedge v^{2p}$ be an arbitrary $2p$-dimensional vector in $\mathbb{C}^n \approx \mathbb{C}^{2n}$, and let $\omega = i(\sum z_j\wedge \bar{z}_j)/2$. Then*

$$\omega^p(v) \leqslant p!\,|v|,$$

and equality holds if and only if v is a positive $2p$-vector, i.e. $v = v^1\wedge iv^1\wedge \cdots \wedge v^p\wedge iv^p$ for some $v^1, \ldots, v^p \in \mathbb{C}^n$.

By replacing v with $-v$ we obtain that also $|\omega^p(v)| \leqslant p!\,|v|$.

■ Let L be the $2p$-dimensional real subspace in $\mathbb{C}^n \approx \mathbb{R}^{2n}$ spanned on the vectors $v^1, \ldots, v^{2p}$. Without loss of generality we may assume that v^j are mutually orthogonal and normalized.

For $p = 1$ we may assume that v^1 is directed along the x_1-axis. Since $\omega = \sum x_j\wedge y_j$, we have $\omega(v^1, v^2) = (x_1\wedge y_1)(v^1, v^2)$. Put $v^2 = \lambda i v^1 + v'$, where $\lambda = \mathrm{Re}(v^2, iv^1) \leqslant 1$ and $v' \in \mathbb{C}_{2\cdots n}$. Since $x_1(v') = y_1(v') = 0$, by bilinearity we have $(x_1\wedge y_1)(v^1, v^2) = \lambda(x_1\wedge y_2)(v^1, iv^1) = \lambda$. Hence $\omega(v) \leqslant 1$, and $\omega(v) = 1$ if and only if $\lambda = 1$, i.e. $v^2 = iv^1$.

In the general case we use induction with respect to p. Since $\mathrm{Re}(v^1, iv^1) = 0$, the projection of iv^1 to L is an $\mathbb{R}$-linear combination of $v^2, \ldots, v^{2p}$. By an orthonormal change of the system of vectors $v^2, \ldots, v^{2p}$ (such that, moreover, $v^1\wedge \cdots \wedge v^{2p}$ does not change), we may assume that the projection of iv^1 to L is proportional to v^2, i.e. $v^3, \ldots, v^{2p}$ are complex orthogonal to v^1. A unitary transformation of $\mathbb{C}^n$ gives v^1 the direction of the x_1-axis, and then $v^3, \ldots, v^{2p}$ will lie in $\mathbb{C}_{2\cdots n}$. In view of the anticommutativity of exterior multiplication, $\omega^p = p\cdot x_1\wedge y_1\wedge \omega('z)^{p-1} + (\omega('z))^p$, where $\omega('z) = i(\sum_2^n z_j\wedge \bar{z}_j)/2$. Since

$v^1 \in \mathbf{C}_1$, we have $(\omega('z))^p(v) = 0$, hence $\omega^p(v) = p(x_1 \wedge y_1 \wedge (\omega('z))^{p-1})(v)$. Put $v^2 = \lambda i v^1 + 'v^2$, where $\lambda = \mathrm{Re}(v^2, iv^1)$, $|\lambda| \leqslant 1$, and put $v' = v^3 \wedge \cdots \wedge v^{2p}$. Since $y_1|_{\mathbf{C}_{2\cdots n}} = 0$, we have

$$(x_1 \wedge y_1 \wedge \omega^{p-1})(v^1 \wedge 'v^2 \wedge 'v) = 0,$$

hence, in view of the multilinearity,

$$\omega^p(v) = \lambda \omega^p(v^1 \wedge iv^1 \wedge 'v) = \lambda p \cdot (x_1 \wedge y_1)(v^1, iv^1) \cdot (\omega('z))^{p-1}('v).$$

According to the induction hypothesis, $|(\omega('z))^{p-1}| \leqslant (p-1)!$, with equality only if $'v$ or $-'v$ is positive. Thus, $|\omega^p(v)| \leqslant |\lambda| p! \leqslant p!$, with equality if and only if $|\lambda| = 1$ (i.e. $v^2 = \pm iv^1$) and $'v$ or $-'v$ positive. It remains to note that ω^p is positive on positive $2p$-vectors, since it is unitarily invariant and its restriction to $\mathbf{C}^p \subset \mathbf{C}^n$ is

$$p! dx_1 \wedge dy_1 \wedge \cdots \wedge dx_p \wedge dy_p. \quad \blacksquare$$

C o r o l l a r y. *Let M be an arbitrary oriented 2p-dimensional manifold of class C^1, embedded in a Hermitian manifold Ω with fundamental form ω. Then*

$$\frac{1}{p!} \int_M \omega^p \leqslant \mathrm{vol}_{2p} M,$$

and in the case of finite volume equality holds if and only if M is a complex manifold in Ω with the canonical orientation.

This estimate is called the *Wirtinger inequality*, and reflects a minimality property of complex manifolds (see p.14.4).

$\blacksquare$ By A4.3 and the above-mentioned Theorem,

$$\frac{1}{p!} \int_M \omega^p = \int_M \left< \frac{1}{p!} \omega^p, \tau_M \right> d\mathcal{H}_{2p} \leqslant \int_M 1 \cdot d\mathcal{H}_{2p} = \mathrm{vol}_{2p} M,$$

and (by the Theorem) equality holds if and only if for almost all $z \in M$ the unit $2p$-vectors $\tau_M(z)$, corresponding to $T_z M$, are positive, i.e. the planes $T_z M$ are complex. But then, by continuity, all $T_z M$ are complex, hence, by Theorem A2.3, M is a complex manifold. $\blacksquare$

In $\mathbf{C}^n$ with $\omega = i(\sum dz_j \wedge d\bar{z}_j)/2$ the volume form of an embedded p-dimensional complex manifold M is equal to $(\omega^p|_M)/p!$ (by p.13.2). Since now

$\omega^p / p! = \sum'_{\# I = p} dV_I$, where $dV_I = \prod_2^p (dx_{i_\nu} \wedge dy_{i_\nu})$ are the volume forms on the corresponding coordinate planes $\mathbb{C}_I$, according to the Lemma we have

$$\mathrm{vol}_{2p} M = \sum_{\# I = p}{}' \int_M dV_I = \sum_{\# I = p}{}' \int_{\pi_I(M)} \#(\pi_I^{-1}(z_I) \cap M) dV_I,$$

where π_I is projection into $\mathbb{C}_I$. The integral over $\pi_I(M)$ is the area of the projection of M in $\mathbb{C}_I$, counted with the multiplicities of $\pi_I|_M$. Hence the following holds:

W i r t i n g e r ' s t h e o r e m (global version). *Let M be a p-dimensional complex manifold in $\mathbb{C}^n$. Then its 2p-dimensional volume is equal to the sum of the volumes of the projections of M on all p-dimensional coordinate planes, counted with the multiplicities of these projections.*

For one-dimensional manifolds this statement can be regarded as the complex analog of Pythagoras' theorem - both in contents and, yes, in merits.

13.4. Integration in $\mathbb{P}_n$. The volume form in $\mathbb{P}_n$ with the Fubini-Study metric is equal to $dV_{\mathbb{P}} := \omega_0^n / n!$ (by Theorem 13.2). In affine coordinates in $\mathbb{C}^n = \mathbb{P}_n \setminus \{z_0 = 0\}$ it has the representation

$$dV_{\mathbb{P}} = \frac{1}{n!}\left[\frac{i}{2}\right]^n \left[\frac{(\partial\bar\partial|\zeta|^2)^n}{(1+|\zeta|^2)^n} - n\frac{(\partial\bar\partial|\zeta|^2)^{n-1} \wedge \partial|\zeta|^2 \wedge \bar\partial|\zeta|^2}{(1+|\zeta|^2)^{n+1}}\right].$$

It is easy to check that

$$n(\partial\bar\partial|\zeta|^2)^{n-1} \wedge \partial|\zeta|^2 \wedge \bar\partial|\zeta|^2 = |\zeta|^2 (\partial\bar\partial|\zeta|^2)^n$$

(both forms are homogeneous, unitarily invariant in $\mathbb{C}_\zeta^n$, and equality obviously holds at $(1,0,\ldots,0) \in \mathbb{C}_\zeta^n$). Hence $dV_{\mathbb{P}} = (1+|\zeta|^2)^{-n-1} dV_\zeta$, where dV_ζ is the Euclidean volume form in $\mathbb{C}_\zeta^n$. The volume of all of $\mathbb{P}_n$ can be computed in $\mathbb{C}^n$ using polar coordinates, with respect to which $dV_\zeta = r^{2n-1} dr \wedge \sigma$, where σ is the Euclidean volume form on the unit sphere in $\mathbb{C}^n$. Since the volume of this sphere is $2\pi^n / (n-1)!$, and since $\int_0^\infty r^{2n-1} / (1+r^2)^{n+1} dr = 1/2n$, the volume of $\mathbb{P}_n$ in the Fubini-Study metric is $\pi^n / n!$. The form ω_0 is invariant under unitary transformations of $\mathbb{P}_n$, and its restriction to $\mathbb{P}_p = \{z_{p+1} = \cdots = z_n = 0\} \subset \mathbb{P}_n$ obviously is the Fubini-Study form in the space $\mathbb{P}_p$. Hence the following holds:

P r o p o s i t i o n. *The volume of any p-dimensional complex plane in* $\mathbb{P}_n$, $0 < p \leqslant n$, *with respect to the Fubini-Study metric, is equal to* $\pi^p / p!$.

Integrals in $\mathbb{C}^{n+1}$ can be conveniently expressed as repated integrals: first over complex rays, then averaging over $\mathbb{P}_n$. We give two simple versions of Fubini's theorem, related to the canonical fibering Π: $\mathbb{C}_*^{n+1} \to \mathbb{P}_n$.

L e m m a 1. *Let a function f be integrable on the sphere* $S^{2n+1}: |z| = 1$ *in* $\mathbb{C}^{n+1}$, *and let* $f_*([z]) := (1/2\pi)\int_0^{2\pi} f(e^{i\phi}z)d\phi$, $z \in S^{2n+1}$. *Then* $\int_{2n+1} f\sigma = 2\pi \int_{\mathbb{P}_n} f_* dV_{\mathbb{P}}$, *where* σ *is the Euclidean volume form on* S^{2n+1}.

■ By Fubini's theorem (A4.4),

$$\int f\sigma = \int_{\mathbb{P}_n} \left[\int_{\gamma[z]} f \cdot \left| \frac{\sigma}{dV_{\mathbb{P}}} \right| d\phi \right] dV_{\mathbb{P}}.$$

where $\gamma[z] = \{e^{i\phi}z: 0 \leqslant \phi < 2\pi\}$ is the fiber of $\Pi|_{2n+1}$ above the point $[z] \in \mathbb{P}_n$ (on it the volume form is clearly equal to $d\phi$). In view of the unitarily invariance, the weight $|\sigma / dV_{\mathbb{P}}|$ is constant. Putting $f \equiv 1$ and using the Proposition we find that it equals 2π. ■

L e m m a 2. *Let a function f be integrable in a ball* B_R: $|z| < R$ *in* $\mathbb{C}^{n+1}$. *Then*

$$\int_{B_R} f dV = \int_{\mathbb{P}_n} \left[\int_{|\lambda| < R} |\lambda|^{2n} f\left[\lambda \frac{z}{|z|} \right] \frac{i}{2} d\lambda \wedge d\bar{\lambda} \right] dV_{\mathbb{P}}.$$

■ The pre-image of $dV_{\mathbb{P}}$ in $\mathbb{C}_*^{n+1}$ is $\omega_0^n / n!$, where $\omega_0 = (dd^c \log |z|^2)/4$. The Euclidean volume is represented in $\mathbb{C}_*^{n+1}$ by $d|z| \wedge d^c|z| \wedge \omega_0^n / n! = |z|^{2n} d|z| \wedge d^c |z| \wedge \omega^n / n!$. Hence $|dV / dV_{\mathbb{P}}|(z) = |z|^{2n}$. It remains to substitute this in Fubini's formula. ■

13.5. Integration over incidence manifolds. Crofton's formulas. The Grassmannians $G(k,n)$ are projective algebraic manifolds. Under the canonical Plücker embedding of $G(k,n)$ into $\mathbb{P}_{\binom{n}{k}-1}$ a Hermitian structure and metric are induced on it, which will be called the Fubini-Study metric in $G(k,n)$. By Proposition A3.6, this metric is invariant under unitary automorphisms of $G(k,n)$. By representing the elements of the Grassmannian by nonsingular $(k \times n)$-matrices (z_j^i), the fundamental Fubini-Study form can be written as

$$\omega_0 = \frac{1}{4} dd^c \log |z^1 \wedge \cdots \wedge z^k|^2.$$

In the coordinate neighborhood $U_{1 \cdots k} \approx \mathbb{C}^{k(n-k)}$, whose points can be uniquely represented by matrices (E, W), the w_j^i, $1 \leqslant i \leqslant k$, $k+1 \leqslant j \leqslant n$, serve as local affine coordinates. At the coordinate origin, corresponding to the matrix $(E, 0)$, the Fubini-Study form coincides with the Euclidean form in $\mathbb{C}^{k(n-k)}$, since for the matrix $Z = (E, W)$ we have $|z^1 \wedge \cdots \wedge z^k|^2 = 1 + \sum_{i,j} |w_j^i|^2 + o(|w|^2)$. The volume form in $G(k,n)$ with respect to the Fubini-Study metric will be denoted by dV_G. This form can conveniently be normalized, by multiplying by a constant, such that the volume of all of $G(k,n)$ be 1. This normalized volume form will be denoted by dL_G, or simply by dL.

The complex planes in $\mathbb{P}_n$ of dimension k form the Grassmannian $\tilde{G}(k,n)$, which will be naturally identified with $G(k+1, n+1)$ using the canonical projection Π: $\mathbb{C}_*^{n+1} \to \mathbb{P}_n$. Under this identification, the form dL_G on $G(k+1, n+1)$ corresponds to $dL_{\tilde{G}}$ (we will denote it also by dL), is invariant under unitary transformations, and is such that the volume of all of $\tilde{G}(k,n)$ is equal to 1. Put $\tilde{G}(k,n) = \tilde{G}$.

Introduce in $\mathbb{P}_n \times \tilde{G}$ the Hermitian structure with fundamental form $\omega_0 = \omega_0(z) + \omega_0(L)$. Let $\tilde{I} \subset \mathbb{P}_n \times \tilde{G}$ be the projective incidence manifold corresponding to the Grassmannian $\tilde{G}$, i.e. the complex submanifold of $\mathbb{P}_n \times \tilde{G}$ consisting of the points (z, L) with $z \in L$ (see A3.5). It is the image of the incidence manifold $I(k+1, n+1)$ under the map from

$$\mathbb{C}_*^{n+1} \times G(k+1, n+1)$$

to $\mathbb{P}_n \times \tilde{G}$ which is induced by the canonical projection Π: $\mathbb{C}_*^{n+1} \to \mathbb{P}_n$.

Let π_1: $(z, L) \mapsto z$ and π_2: $(z, L) \mapsto L$ be the natural projections onto $\mathbb{P}_n$ and $\tilde{G}$. For an arbitrary complex manifold M in $\mathbb{P}_n$ we denote by $\tilde{M}$ the submanifold in $\tilde{I}$ consisting of the points (z, L) with $z \in M$. The fiber $\tilde{I}_z = \pi_1^{-1}(z) \cap \tilde{I}$ is a complex manifold, isomorphic to $G(k,n)$; clearly, all $\tilde{I}_z$ are unitarily equivalent. The fibers in the other direction, $\pi_2^{-1}(L) \cap \tilde{M}$, are projected in $\mathbb{P}_n$ onto the corresponding set $M \cap L$. These fibers may be of very different natures. We will further assume that $\dim M = p \geqslant n-k$. Then $\tilde{M}$ is a complex manifold of dimension $p + k(n-k)$, and $\operatorname{br} \pi_2|_{\tilde{M}}$ is an analytic subset of $\tilde{M}$ of smaller dimension. Since $\tilde{M} = (M \times G) \cap \tilde{I}$, we have

$$T_{(z,L)}\tilde{M} = (T_z M \times T_L \tilde{G}) \cap T_{(z,L)}\tilde{I}.$$

Since $\pi^{-1}(L) \cap \tilde{I} = L \times \{L\}$, the tangent space at (z,L) to this fiber is equal to $T_z L \times \{0\} \subset T_{(z,L)}\tilde{I}$. Thus, the tangent plane to $\tilde{M} \cap \pi_z^{-1}(L)$ is equal to $((T_z M) \cap (T_z L)) \times \{0\}$, hence $\tilde{M}_0 = \tilde{M} \setminus (\mathrm{br}\, \pi_2 \,|\, _{\tilde{M}})$ consists of the pairs $(z,L) \in \tilde{M}$ at which

$$\dim T_z M \cap T_z L = p + k - n$$

(transversal intersection of M and L at z). At all points of $\tilde{M}$ the projection $\pi_2 \,|\, _{\tilde{M}}$ has maximum possible rank, equal to $\dim \tilde{G}$; in particular, the fibers of $\pi_2 \,|\, _{\tilde{M}_0}$ are pure $(p+k-n)$-dimensional.

The fibering $\pi_1 \colon \tilde{I} \to \mathbb{P}_n$ naturally determines on $\tilde{I}$ a real analytic differential form $dV_{\tilde{I}} / dV_{\mathbb{P}}$ of degree $2k(n-k)$, which, at each point $(z,L) \in \tilde{I}$, is tangent to $\tilde{I}_z$ and coincides with the volume form on $\tilde{I}_z$ (see A4.4). This form will be denoted by $dV_{\tilde{I}_z}$. Similarly, the fibering $\pi_2 \colon \tilde{M}_0 \to \tilde{G}$ determines on $\tilde{M}_0$ a form of degree $2(p+k-n)$, tangent to the fibers of $\pi_2 \,|\, _{\tilde{M}_0}$ and coinciding with the volume form on a fiber. This $2(p+k-n)$-form on $\tilde{M}_0$ will be denoted by $dV_{M \cap L}$. On the other hand, we can naturally lift to $\tilde{M}$ the p-dimensional Fubini-Study volume form on $M \subset \mathbb{P}_n$ (using π_1) and the Fubini-Study volume form on $\tilde{G}$ (using π_2). These lifts to $\tilde{M}$ are denoted by dV_M and $dV_{\tilde{G}}$, respectively. As a result, two differential forms of maximal degree are obtained on $\tilde{M}_0$: $dV_{\tilde{I}_z} \wedge dV_M$ and $dV_{M \cap L} \wedge dV_{\tilde{G}}$.

We want to compute the proportionality coefficient of these forms on $\tilde{M}_0$. In order to do this we introduce the notion of angle between the manifold M and a plane L at a point z belonging to their transversal intersection. The tangent spaces $T_z M$ and $T_z L$ lie in the $\mathbb{C}$-linear space $T_z \mathbb{P}_n$. Let L' be the orthogonal complement to $T_z M \cap T_z L$ in $T_z M$, and let L'' be the orthogonal complement to $T_z L$ in $T_z \mathbb{P}_n$. Then $\dim L' = \dim L'' = n-k$. In $\wedge_{\mathbb{C}}^{n-1}(T_z \mathbb{P}_n)$ to them correspond $(n-k)$-vectors $v' = v_1' \wedge \cdots \wedge v_{n-k}'$ and $v'' = v_1'' \wedge \cdots \wedge v_{n-k}''$, where $\{v'_j\}, \{v_j''\}$ are basis in L', L'', respectively. We define $\sin(M,L,z)$ to be the cosine of the angle between these $(n-k)$-vectors; more precisely, it is $|(v',v'')| \,/\, |v'| \cdot |v''|$ (scalar product and norm in $\wedge_{\mathbb{C}}^{n-k}(T_z \mathbb{P}_n)$ are naturally induced by the Hermitian structure in $T_z \mathbb{P}_n$).

L e m m a. *At all points* $(z,L) \in \tilde{M}_0$ *the equality*

$$dV_{M\cap L}\wedge dV_{\tilde{G}} = \sin^2(M,L,z)\cdot dV_M \wedge dV_{\tilde{I}_z}$$

holds, regarded as forms on $\tilde{M}_0$.

■ Fix a point $(a,\Lambda) \in \tilde{M}_0$. Since the quantities given do not change under unitary transformations of $\mathbb{P}_n$, we may assume that $a = [1,0,\ldots,0]$. Introduce in $\mathbb{C}^n = \mathbb{P}_n \setminus \{z_0=0\}$ the standard affine coordinates with center at a. A unitary transformation maps $(T_aM) \cap \Lambda$ onto the $\mathbb{C}_{1\cdots(p+k-n)}$-plane, and $\Lambda \cap \mathbb{C}^n$ onto the $\mathbb{C}_{1\ldots k}$-plane. Put $z' = (z_1,\ldots,z_{p+k-n})$, $(z','z) = (z_1,\ldots,z_k)$, $z'' = (z_{k+1},\ldots,z_n)$, supplement z' to orthonormal coordinates in T_aM, by appending to it linear functions ζ', and then take also orthonormal coordinates ζ'' in $(T_aM)^\perp$. Then $('z,z'')$ can be unitarily expressed in terms of (ζ',ζ''), and T_aM is defined by the system $\zeta''=0$. In these coordinates

$$dV_{M\cap\Lambda} = dV(z') \text{ and } dV_M = dV(z')\wedge dV(\zeta')$$

(here and below, $dV(\eta)$ denotes the Euclidean volume form in the space of variables η).

Let $v_1',\ldots,v_{n-k}'$ be an orthonormal $\mathbb{C}$-basis in $(T_aM) \cap (\Lambda\cap T_aM)^\perp$, and put $v' = v_1'\wedge\cdots\wedge v_{n-k}'$. Similarly, let $v'' = v_1''\wedge\cdots\wedge v_{n-k}''$ correspond to $\Lambda^\perp$. Represent v' as $v' = (v',v'')v''+\tilde{v}$, where $\tilde{v}$ is orthogonal to v'' in $\wedge_{\mathbb{C}}^{n-k}(T_a\mathbb{P}_n)$, and put $dz'' = dz_{k+1}\wedge\cdots\wedge dz_n$. Since z'' are coordinates in $\Lambda^\perp$ corresponding to v'', we have $dz''(v'')=1$, and $dz''(\tilde{v})=0$; hence $dz''(v') = (v',v'')$. Since the forms $(i/2)^{n-k}\, dz''\wedge d\bar{z}''$, $dV(z'')$, and $dx''\wedge dy''$ can differ only in sign, A2.1 implies that

$$|dV(z'')(v'\wedge iv')| = |dz''(v')|^2 = |(v',v'')|^2 = \sin^2(M,\Lambda,a).$$

Since $dz_{k+j} = 0$ on $\Lambda\cap T_aM$, we have simultaneously proved that

$$dV(z'')|_{T_aM} = \sin^2(M,\Lambda,a)\, dV(\zeta')$$

(remember that ζ' are coordinates in $(T_aM) \cap (\Lambda\cap T_aM)^\perp$).

Now we linearize the situation in another direction. Introduce in $U_{0\ldots k} \subset \tilde{G}$ affine coordinates with center at Λ, corresponding to the matrix representation $L\sim(E,W)$ (the plane Λ corresponds to the matrix $(E,0)$). The incidence condition $z\sim[1,z] \in L\sim(E,W)$ can be written as

$$z_j = 1\cdot w_j^0 + z_1 w_j^1 + \cdots + z_k w_j^k, \ j = k+1, \ldots, n$$

(see A3.5). With respect to the coordinates chosen in $\mathbf{C}_z^n \times \mathbf{C}_w^{(k+1)(n-k)} \subset \mathbf{P}_n \times \tilde{G}$, the tangent plane to $\tilde{I}$ at (a, Λ) is given by the system $z_j = w_j^0$, $j = k+1, \ldots, n$, or, for short, by $z'' = w'$, where $w' := (w_{k+1}^0, \ldots, w_n^0)$. The fiber $\tilde{I}_a$ consists of the points (a, L) for which $L \ni a = [1, 0, \ldots, 0]$. In matrix form this means that the first column of (E, W) is $(1, 0, \ldots, 0)$, i.e. $w' = 0$. Thus, $\tilde{I}_a$ is defined by $w' = 0$, hence $dV_{\tilde{I}_a} = dV(w'')$. Finally, $dV_{\tilde{G}} = dV(w') \wedge dV(w'')$ (all at the point (a, Λ)). Since $T_{(a,\Lambda)}\tilde{M} = \pi_1^{-1}(T_a M) \cap T_{(a,\Lambda)}\tilde{I}$, we have that $T_{(a,\Lambda)}\tilde{M}$ is defined by the system $\zeta'' = 0$, $z'' = w'$. Thus, on $T_{(a,\Lambda)}M$ we have

$$\begin{aligned} dV_{M \cap \Lambda} \wedge dV_{\tilde{G}} &= dV(z') \wedge dV(w') \wedge dV(w'') = \\ &= dV(z') \wedge dV(z'') \wedge dV(w_n'') = \\ &= \sin^2(M, \Lambda, a)\, dV(z') \wedge dV(\zeta') \wedge dV(w'') = \\ &= \sin^2(M, \Lambda, a)\, dV_M \wedge dV_{\tilde{I}_a}. \ \blacksquare \end{aligned}$$

Since $\tilde{M} \setminus \tilde{M}_0$ is an analytic subset of dimension $< \dim \tilde{M}$, the following Proposition clearly follows from the Lemma and Fubini's theorem (A4.4).

P r o p o s i t i o n. *Let M be a p-dimensional complex manifold in $\mathbf{P}_n$, let $k \geqslant n-p$, and let $\tilde{M}$ be the "lift" of M to the incidence manifold in $\mathbf{P}_n \times \tilde{G}(k,n)$. Then for any function $f(z, L)$ continuous and integrable on $\tilde{M}$ we have*

$$\int_M \left[\int_{\tilde{I}_z} f(z, L) \sin^2(M, L, z) dV_{\tilde{I}_z}(L) \right] dV_M(z) = \qquad (\star)$$

$$= \int_{\tilde{G}(k,n)} \left[\int_{M \cap L} f(z, L)\, dV_{M \cap L}(z) \right] dV_{\tilde{G}}(L).$$

Integrals over $M \cap L$ are defined here only for almost all L; more precisely, for all L such that $\dim M \cap L = p + k - n$. Moreover, by definition an integral over $M \cap L$ is equal to the integral over reg $(M \cap L)$.)

The case when f is independent either of z or of L is of interest. If f is independent of z, the second expression in $(\star)$ equals

$$\int_{\tilde{G}(k,n)} f(L) \operatorname{vol}_{2(p+k-n)}(M \cap L)\, dV_{\tilde{G}}(L)$$

and is the averaged (with weight f) volume of the sections of M by k-dimensional complex planes $L \subset \mathbb{P}_n$. Equation (⋆) with f independent of L (i.e. in fact given on M only) leads to generalizations of the so-called Crofton formula from integral geometry. In this case the first expression in (⋆) is

$$\int_M f(z) \left[\int_{J_z} \sin^2(M,L,z)\, dV_{J_z}(L) \right] dV_M(z).$$

Here the inner integral does not change under unitary transformations of $\mathbb{P}_n$. For a fixed point $z \in M$ we can choose such a transformation for which $z = a = [1,0, \ldots, 0]$ and $T_z M = T_a \mathbb{P}_p$. Hence the inner integral is constant, not dependent on z and M, i.e. the formula

$$\int_M f dV_M = c(p,k,n) \int_{\tilde{G}(k,n)} \left[\int_{M \cap L} f dV_{M \cap L} \right] dL$$

holds. In order to compute the constant we take $M = \mathbb{P}_p \subset \mathbb{P}_n$ and $f \equiv 1$. Then, taking into account Proposition 13.4, this formula leads to $\pi^p / p! = c(p,k,n)\, \pi^{p+k-n} / (p+k-n)!$, i.e. $c(p,k,n) = \pi^{n-k}(p+k-n)!/p!$.

For $f \equiv 1$ we obtain the well-known *Crofton formula* for volumes of complex manifolds embedded in $\mathbb{P}_n$:

$$\mathrm{vol}_{2p} M = \pi^{n-k} \frac{(p+k-n)!}{p!} \int_{\tilde{G}(k,n)} \mathrm{vol}_{2(p+k-n)}(M \cap L)\, dL.$$

13.6. Relation between projective and affine volumes. If M is a complex manifold in $\mathbb{C}^n \subset \mathbb{P}_n$ and if $0 \notin M$, then next to the family of all k-dimensional planes in $\mathbb{C}^n$, forming an open dense subset of the Grassmannian $\tilde{G}(k,n)$, it is natural also to consider the family of k-dimensional planes passing through the coordinate origin 0, and forming the ordinary Grassmannian $G(k,n)$. Using the canonical projection $\Pi: \mathbb{C}^n_* \to \mathbb{P}_{n-1}$ the problem on the relation between the volume of M and that of its sections by planes $L \in G(k,n)$ can be reduced to Proposition 13.5. The following statement is, perhaps, most simple.

P r o p o s i t i o n 1. *Let M be a p-dimensional complex manifold in $\mathbb{C}^n$ such that $0 \notin M$ and such that the rank of the map $\Pi: M \to \mathbb{P}_{n-1}$ equals p almost everywhere on M. Let $\omega_0 = (dd^c \log |z|^2)/4$, and let f be a continuous function on M,* *integrable with respect to* $\omega_0^p|_M$. *Then we have for any* $k \geqslant n-p$,

$$\int_M f\omega_0^p = \pi^{n-k} \int_{G(k,n)} \left[\int_{M \cap L} f\omega_0^{p+k-n} \right] dL. \tag{**}$$

If $k = n-p$, *the inner integral at the righthand side is, by definition, equal to* $\sum_{z \in M \cap L} f(z)$.

(As in p.13.5, the inner integrals are defined for almost all L and are understood as integrals over reg $(M \cap L)$.)

■ The set br $\Pi|_M$, consisting of all points at which the rank of $\Pi|_M$ is less than p, is an analytic subset in M of dimension less than p. By p.3.8, the planes $L \in G(k,n)$ whose intersections with br $\Pi|_M$ have dimension $\geqslant p+k-n$ form in $G(k,n)$ a set of measure zero. Thus, without loss of generality we may assume that rank $\Pi|_M \equiv p$. Then $\Pi|_M$ is a local embedding, i.e. there are an open covering $\mathfrak{U} = \{U_j\}$ of the set M and complex manifolds M_j in $\mathbb{P}_n$ such that all $\Pi\colon M \cap U_j \to M_j$ are biholomorphic maps. Let $\{\lambda_j\}$ be a partition of unity on M subordinate to $\mathfrak{U}$, and put $f_j = (\lambda_j f)\circ\Pi^{-1}|_{M_j}$. By Proposition 13.5,

$$\int_{M_j} f_j\omega_0^p = \pi^{n-k} \int_{\tilde{G}} \left[\int_{M_j \cap \tilde{L}} f_j\omega_0^{p+k-n} \right] d\tilde{L},$$

where $\tilde{G} = \tilde{G}(k-1, n-1)$, $\tilde{L} = \Pi(L \setminus \{0\})$, and ω_0 is the Fubini-Study form in $\mathbb{P}_{n-1}$ (whose pre-image in $\mathbb{C}_*^n$ is $(dd^c \log |z|^2)/4$). Passing to pre-images and taking into account that $d\tilde{L}$ is mapped to dL, we obtain

$$\int_{M \cap U_j} \lambda_j f\omega_0^p = \pi^{n-k} \int_{G(k,n)} \left[\int_{M \cap L} \lambda_j f\omega_0^{p+k-n} \right] dL,$$

which has to be summed over j only. ■

In order to replace projective volume by Euclidean volume we may use the following Lemma.

L e m m a. *Let* M *be a* p*-dimensional complex manifold in* $\mathbb{C}^n$, $p > 0$, *and let* $z \in M \setminus \{0\}$. *Then*

$$\omega_0^p|_{T_zM} = \frac{\sin^2(z, M)}{|z|^{2p}} \omega^p|_{T_zM},$$

here $\omega_0 = (dd^c \log |z|^2)/4$, $\omega = (dd^c |z|^2)/4$, *and* $\sin^2(z,M)$ *is the sine of the angle between the vector* z *and the manifold* M *(i.e. its tangent plane) at* z *in* $\mathbb{C}^n \approx \mathbb{R}^{2n}$.

■ By a unitary change of coordinates in $\mathbb{C}^n$ the geometric tangent plane to M at z can be made parallel to the $\mathbb{C}_{1\ldots p}$-plane. The new coordinates ζ are partitioned into ζ', ζ'', where $\zeta' = (\zeta_1, \ldots, \zeta_p)$. Differentiation with respect to ζ and taking into account that $dd^c|a'+\zeta'|^2 = dd^c|\zeta'|^2$ for any $a' \in \mathbb{C}^p$ gives

$$\omega_0|_{T_zM} = \frac{i}{2}\partial\bar\partial \log(|z'+\zeta'|^2 + |z''|^2)|_{\zeta=0} =$$

$$= \frac{i}{2}\left[\frac{\partial\bar\partial|\zeta'|^2}{|z|^2} - \sum_{j,k=1}^{p} \frac{\bar z_j z_k}{|z|^4} d\zeta_j \wedge d\bar\zeta_k\right],$$

since $d\zeta''|_{T_zM} = 0$. Hence

$$\omega_0^p|_{T_zM} = \frac{\omega(\zeta')^{p-1}}{|z|^{2p}} \wedge \left[\omega(\zeta') - p\frac{i}{2}\sum_{j,k=1}^{p} \frac{\bar z_j z_k}{|z|^2} d\zeta_j \wedge d\bar\zeta_k\right].$$

Since $(\omega(\zeta'))^{p-1} = (p-1)!\sum_{j=1}^{p}\prod_{\nu\neq j}^{p} i(d\zeta_\nu \wedge d\bar\zeta_\nu)/2$, we have

$$\omega_0^p|_{T_zM} = \frac{(p-1)!}{|z|^{2p}}\sum_{j=1}^{p}\left[dV(\zeta') - p\frac{\bar z_j z_j}{|z|^2}dV(\zeta')\right] =$$

$$= \frac{p!}{|z|^{2p}}\left[1 - \frac{|z'|^2}{|z|^2}\right]dV(\zeta).$$

It remains to note that $p!dV(\zeta') = \omega^p|_{T_zM}$, that z' is the projection of z to T_zM, and that hence $1 - (|z'|^2/|z|^{2)} = \sin^2(z,M)$. ■

When replacing projective volume by Euclidean volume, in (**) appear weights which cannot be simultaneously removed from both sides of this equation. Hence the formulation with Euclidean volume is less transparent. Here is one of the statements clearly following from Proposition 1 (in view of the arbitrariness of f) and the Lemma.

P r o p o s i t i o n 2. *Let* M *be a* p*-dimensional complex manifold in* $\mathbb{C}^n$ *such that* $0 \notin M$ *and such that the rank of the map* $\Pi: M \to \mathbb{P}_{n-1}$ *is* p *almost everywhere on*

M. Suppose a function f is continuous on M and integrable with respect to the

$$\int_M f dV_M = \pi^{n-k}\frac{(p+k-n)}{p!} \times$$

$$\times \int_{G(k,n)} \left[\int_{M\cap L} f(z)|z|^{2(n-k)} \frac{\sin^2(z, M\cap L)}{\sin^2(z,M)} dV_{M\cap L} \right] dL,$$

where $dV_{M\cap L}$ is the Euclidean volume on $M \cap L$. In case $k = n-p$ we must take $\sin^2(z, M\cap L) = 1$, and the inner integral at the righthand side must then be replaced by $\sum_{z\in M\cap L} f(z)\cdot|z|^{2p} / \sin^2(z,M)$.

14. Integration over analytic sets

14.1. Lelong's theorem. If A is a pure p-dimensional analytic subset of a complex manifold Ω, then reg A is a $2p$-dimensional complex manifold embedded in Ω, and we can integrate $2p$-forms over it. The singularities of A form a very thin set, hence we may hope that improper integrals over reg A of forms of class $\mathcal{D}^{2p}(\Omega)$ converge. In order to prove this it suffices to prove that $\mathcal{H}_{2p}(A\cap K) < \infty$ for any compact set $K \subset \Omega$ (in p.3.7 it was proved only that $\mathcal{H}_{2p+\epsilon}(A) = 0$ for any $\epsilon > 0$). The latter statement is local, so we may assume that Ω is a domain in $\mathbf{C}^n$ (the local coordinate maps satisfy a Lipschitz condition in both directions, hence do not change the boundedness of $\mathcal{H}_{2p}$). The form $\omega^p/p!$, measuring the volumes of p-dimensional complex manifolds in $\mathbf{C}^n$ can written as $\mathbf{C}^n$, can be written as

$$\frac{1}{p!}\omega^p = \sum_{\#I=p}' dV_I,$$

where $I = (i_1, \ldots, i_p)$, and dV_I denotes the volume form $\prod_1^p dx_{i_\nu} \wedge dy_{i_\nu}$ in the coordinate plane $\mathbf{C}_I$. If $0 \in A$, by Lemma 2, p.3.4, we may assume that in a neighborhood of 0, A can be properly projected into all $\mathbf{C}_I$, $\#I = p$; more precisely, there are bounded neighborhoods $U_I \ni 0$ such that all projections $\pi_I\colon A \cap U_I \to U_I' = \mathbf{C}_I \cap U_I$ are proper. Let $\sigma_I \subset U_I'$ be the critical analytic set and k_I the number of sheets of the cover $\pi_I|_{A\cap U_I}$ above $U_I' \setminus \sigma_I$. Since $\pi_I^{-1}(\sigma_I) \cap A \cap U_I$ is an analytic set of dimension $<p$, it has no influence on the integral over (reg A) $\cap U_I$ with respect to dV_I. Since above $U_I' \setminus \sigma_I$ the projection

$\pi_I|_{A\cap U_I}$ is locally biholomorphic, the definition of integral with respect to a form clearly implies that

$$\int_{\operatorname{reg} A\cap U_I} dV_I = k_I \int_{U_I'\setminus\sigma_I} dV_I = k_I \operatorname{vol}_{2p} U_I'.$$

Let $U = \cap U_I$. Since the dV_I are positive, we have

$$\int_{\operatorname{reg} A\cap U} \frac{1}{p!}\omega^p = \sum_{\# I=p}{}' \int_{\operatorname{reg} A\cap U} dV_1 \leqslant \sum_{\# I=p}{}' k_I \operatorname{vol}_{2p} U_I' < \infty.$$

Since $\mathcal{H}_{2p}(\operatorname{sng} A) = 0$, the lefthand side is equal to $\mathcal{H}_{2p}(A\cap U)$. Thus we have proved the following Theorem (Lelong [77]).

T h e o r e m. *A pure p-dimensional analytic subset A of a Hermitian complex manifold Ω has locally finite 2p-measure, i.e. $\mathcal{H}_{2p}(A\cap K) < \infty$ for any compact set $K\subset\Omega$. If A is an analytic set in $\mathbb{C}^n$, its 2p-measure is equal to the sum of the volumes of the projections of A on the p-dimensional coordinate planes $\mathbb{C}_I$, counted with the multiplicities of the projections, i.e.*

$$\operatorname{vol}_{2p} A = \sum_{\# I=p}{}' \int_{\pi_I(A)} \#(\pi_I^{-1}(z_I)\cap A)\, dV_I.$$

(We will call the last statement *Wirtinger's theorem* for analytic sets, since for reg A this is Wirtinger's theorem 13.3.)

We can introduce a Hermitian structure on any complex manifold Ω (e.g. locally, as in $\mathbb{C}^n$, and then gluing using a partition of unity, see [151]). After this, for any pure p-dimensional analytic subset $A\subset\Omega$ and for any continuous, compactly supported $2p$-form ϕ in Ω the improper integral

$$\int_{\operatorname{reg} A}\phi = \int_{\operatorname{reg} A} \langle\phi,\tau_A\rangle\, d\mathcal{H}_{2p}$$

exists (and does not depend on the Hermitian structure chosen, since the integral at the lefthand side does not depend on it). It clearly depends continuously on ϕ, hence

C o r o l l a r y. *Let A be a pure p-dimensional analytic subset of a complex manifold Ω. Then integration of 2p-forms over* reg A *defines in Ω a current* $[A]$ *of measure type and of dimension 2p:*

$$<[A],\phi> := \int_{\operatorname{reg} A} \phi, \quad \phi \in \mathcal{D}^{2p}(\Omega).$$

In the sequel we will assume $\int_A = \int_{\operatorname{reg} A}$, by definition.

14.2. Properties of integrals over analytic sets. Lemma 1, A4.2, clearly implies:

P r o p o s i t i o n 1. *The current $[A]$ given by integration over a pure p-dimensional analytic subset A of a complex manifold has bidimension (p,p), i.e. $\int_A \phi^{r,s} = 0$ for any continuous compactly supported form $\phi^{r,s}$ of bidegree $(r,s) \neq (p,p)$, $r+s = 2p$.*

The restriction of a positive (p,p)-form in Ω to the manifold $\operatorname{reg} A$ has the representation $\lambda\, dV_A$, $\lambda \geqslant 0$, hence we have

P r o p o s i t i o n 2. *The current $[A]$ is positive, i.e. $\int_A \phi \geqslant 0$ for all positive integrable forms ϕ on* reg *A of bidegree (p,p).*

For the proof of the next property we first find a convenient way to represent forms of degree $2p-1$ in $\mathbf{C}^n$. Note that forms of bidegree (r,s) with $r>p$ or $s>p$ vanish when restricted to an arbitrary p-dimensional complex manifold in $\mathbf{C}^n$ (A4.2), hence for our purposes it suffices to consider forms of bidegree $(p,p-1)$ and $(p-1,p)$. A form ϕ of bidegree $(p,p-1)$ can be written as $\sum dz_k \wedge \psi_k$, where the ψ_k are forms of bidegree $(p-1,p-1)$ whose coefficients coincide, up to sign, with the corresponding coefficients of ϕ. Expanding ψ_k in a basis $\{\alpha_j\}$ of principal positive forms (cf. A4.5), we obtain that every form ϕ of bidegree $(p,p-1)$ in a domain in $\mathbf{C}^n$ can be represented as a linear combination of forms $dz_k \wedge \alpha_j$ with coefficients that are $\mathbf{C}$-linearly expressible in terms of the coefficients ϕ_{Jk} of ϕ. Similarly, a $(p-1,p)$-form is obtained this way from forms $K\bar{z}_k \wedge \alpha_j$.

P r o p o s i t i o n 3. *Let A be a pure p-dimensional analytic subset of a complex manifold Ω. Then the current given by integration over A is d-closed, i.e.*

$$\int_A d\phi = 0, \quad \phi \in \mathcal{D}^{2p-1}(\Omega).$$

■ Representing $\phi = \sum(\lambda_j \phi)$, where $\{\lambda_j\}$ is a partition of unity on Ω, we see that it suffices to prove the corresponding local statement. Hence we will assume that A lies in $\mathbb{C}^n$, $0 \in A$, and we will prove that $\int_A d\phi = 0$ for all smooth ϕ supported in a small neighborhood of 0.

First we consider the case $p = 1$. The set A can be taken irreducible at 0 (see p.5.4). By p.6.1, in some neighborhood $U \ni 0$ the analytic set $A \cap U$ is the image of the unit disk $\Delta \subset \mathbb{C}_\zeta$ under a one-to-one holomorphic map $z = f(\zeta)$, $f: \Delta \to A \cap U$; moreover $f^{-1}((\operatorname{reg} A) \cap U) \supset \Delta \setminus \{0\}$. Hence, if $\phi \in \mathfrak{D}^1(U)$, then

$$\int_A d\phi = \int_{\Delta\setminus\{0\}} f^*(d\phi) = \int_{\Delta\setminus\{0\}} d(f^*\phi) = \int_\Delta d(f^*\phi) = 0$$

by Stokes' theorem, since $f^*\phi \in \mathfrak{D}^1(\Delta)$.

Let now $p > 1$, and let $\{\alpha_j\}$ be a basis of principal positive forms of bidegree $(p-1, p-1)$. By definition, $\alpha_j = \prod_\nu i(dl_{j\nu} \wedge d\overline{l}_{j\nu})/2$, where $l_{j\nu}(z) = (z, a^{j\nu})$ are certain linear functions in $\mathbb{C}^n$. Denote by L_{jk} the complex plane spanned by $\{a^{j\nu}\}_{\nu=1}^{p-1}$ and the unit vector $e_k \in \mathbb{C}_k$. By Lemma 2, p.3.4, after a suitable unitary transformation of $\mathbb{C}^n$ there are neighborhoods $U_{jk} \ni 0$ such that the orthogonal projections $A \cap U_{jk} \to L_{jk} \cap U_{jk}$ are proper for all indices j,k for which $\dim L_{jk} = p$. Denote by U the intersection of all these U_{jk}. We will show that $\int_A d\phi = 0$ for all $\phi \in \mathfrak{D}^{2p-1}(U)$. By what was proved above the Proposition, it suffices to consider forms

$$\phi = (\lambda\, dz_k + \mu\, d\overline{z}_k) \wedge \alpha_j, \quad \lambda, \mu \in \mathfrak{D}(U).$$

Fix j,k; after a unitary transformation we have $L_{jk} = \mathbb{C}^p$ and $\alpha_j = dV'$, where dV' is the volume form in $\mathbb{C}_{1\ldots p}$. Then $\phi = \gamma \wedge \alpha_j$, where $\gamma = \alpha\, dz_p + \beta\, d\overline{z}_p \in \mathfrak{D}^1(U)$, and $d\phi = d''\gamma \wedge dV'$, where d'' is the differential with respect to $z'' = (z_p, \ldots, z_n)$ only. By Fubini's theorem (A4.4),

$$\int_A d\phi = \int_{\mathbb{C}^{p-1}} \left[\int_{\pi^{-1}(z')\cap A} d''\gamma \right] dV',$$

where π is the projection $z \mapsto z' = (z_1, \ldots, z_{p-1})$. Since γ is supported in $U \subset U_{jk}$ and all fibers $\pi^{-1}(z) \cap A \cap U_{jk}$ are one-dimensional analytic subsets in $\pi^{-1}(z') \cap U_j \subset \{z'\} \times \mathbb{C}_{p\ldots n}$, we find, using the case $p = 1$, that all inner integrals vanish. ■

14.3. Stokes' theorem. The last Proposition is a particular instance of Stokes' theorem on analytic sets. For a statement of the latter we must introduce the notion of analytic set with boundary. Let A be a pure p-dimensional analytic set on a complex manifold Ω, the closure $\bar{A}$ of which belongs to another pure p-dimensional analytic set $\tilde{A} \subset \Omega$. Assume that

1. the set $bA = \bar{A} \setminus A$ has locally finite $(2p-1)$-measure, i.e. $\mathcal{H}_{2p-1}(bA \cap K) < \infty$ for any compact set $K \subset \Omega$ (by A6.1, this property is independent of the Hermitian structure chosen on Ω);
2. there is a closed subset $E \subset \bar{A}$ such that $\mathcal{H}_{2p-1}(E) = 0$ and $A \setminus E$ is a C^1-submanifold with boundary $bA \setminus E$ on the manifold $\Omega \setminus E$.

Then we say that A is a p-dimensional *analytic set with boundary* bA. Without loss of generality we may assume that the special set E contains $\operatorname{sng} A$. The orientation of bA is assumed to be compatible with as the canonical orientation of $\operatorname{reg} A$.

T h e o r e m. *Let A be a p-dimensional analytic set with boundary on a complex manifold Ω. Then for any compactly supported $(2p-1)$-form ϕ of class $C^1(\Omega)$ the following equation (Stokes' formula) holds:*

$$\int_A d\phi = \int_{bA} \phi.$$

The integral over bA is understood as an improper integral, i.e. as the limit of the integrals over $bA \setminus U_\epsilon$, where U_ϵ are neighborhoods of the special set E contracting towards E as $\epsilon \to 0$. In particular, Stokes' formula shows that this improper integral exists, i.e. is independent of the neighborhoods U_ϵ chosen (in p.14.1 we have shown that the improper integral on the lefthand side exists).

■ Using partition of unity the statement clearly reduces to a local one, hence we may assume that Ω is a domain in $\mathbb{C}^n$, $0 \in A \subset \tilde{A}$, and we will prove Stokes' formula for forms ϕ supported in a sufficiently small neighborhood of 0 in $\mathbb{C}^n$. As in the previous Section, it suffices to consider forms ϕ represented by $(\alpha\, dz_p \wedge \beta\, d\bar{z}_p) \wedge dV_{p-1}$, where dV_{p-1} is the volume form in $\mathbb{C}_{1 \dots p}$ and α, β are smooth functions supported in a small neighborhood of 0. We may moreover assume that in some neighborhood $U \ni 0$ the projection

$\pi: \tilde{A} \cap U \to U' \subset \mathbf{C}_{1 \ldots p}$ is proper. Let E be the special set of the analytic set with boundary (A, bA), and let $\sigma \subset U'$ be the critical analytic set of the analytic cover $\pi|_{\tilde{A} \cap U}$. Then $E' = \pi(E \cap U) \cup \sigma$ is closed in U' and has $(2p-1)$-measure zero. Hence for any $\epsilon > 0$ there is a locally finite covering of E' by open balls $B(a_j, r_j)$, $a_j \in E'$, such that $\sum r_j^{2p-1} < \epsilon$. Put $U'_\epsilon = \cup B(a_j, r_j)$. Then $\partial U'_\epsilon$ is a locally finite union of parts of the spheres $\partial B(a_j, r_j)$, and

$$\mathcal{H}_{2p-1}(\partial U'_\epsilon) \leqslant c_p \sum r_j^{2p-1} < c_p \epsilon;$$

by changing r_j somewhat we can also prove that $(A \cap U) \setminus \pi^{-1}(U'_\epsilon)$ is a manifold with piecewise smooth boundary. Put $U_\epsilon = \pi^{-1}(U'_\epsilon) \cap U$. Since the map $\pi|_{\tilde{A} \cap U}$ is finitely-sheeted and locally biholomorphic above $U' \setminus \sigma$, the $\mathcal{H}_{2p}$-measure of the set $A \cap U_\epsilon$ tends to zero as $\epsilon \to 0$ ($\mathrm{vol}_{2p}(A \cap U_\epsilon) = \int_{U'_\epsilon} \lambda \, d\mathcal{H}_{2p}$ for a certain integrable function λ). Hence $\int_A d\phi$ is the limit of such integrals over $A \setminus U_\epsilon$. Since $A \setminus U_\epsilon$ has piecewise smooth boundary within $\operatorname{supp} \phi \subset U$, Stokes' classical formula (see A4.3) implies that

$$\int_{A \setminus U_\epsilon} d\phi = \int_{(bA) \setminus U_\epsilon} \phi - \int_{A \cap bU_\epsilon} \phi. \qquad (\star)$$

The projection $\pi: A \cap bU_\epsilon \to bU'_\epsilon$ is a local diffeomorphism and is k-sheeted, $k < \infty$. Hence

$$\int_{A \cap bU_\epsilon} \phi = \int_{A' \cap bU'_\epsilon} \phi_\star,$$

where $A' = \pi(A \cap U)$, $\phi_\star = (\alpha_\star dz_p + \beta_\star d\bar{z}_p) \wedge dV_{p-1}$, and $\alpha_\star(z')$, $\beta_\star(z')$ denote the sum of the values of α, respectively β, at all points of the set $\pi^{-1}(z') \cap A \cap U$. Since the number of these points is at most k, $\alpha_\star$ and $\beta_\star$ are bounded, piecewise continuous functions. Thus, the last integral can be written as $\int_{A' \cap bU'_\epsilon} \lambda \, d\mathcal{H}_{2p-1}$, where λ is a piecewise continuous function bounded by a constant depending on ϕ and k only (most important is that it is independent of ϵ). Hence

$$\left| \int_{A \cap bU_\epsilon} \phi \right| \leqslant C \mathcal{H}_{2p-1}(bU'_\epsilon) < C' \epsilon,$$

and Stokes' formula is obtained from $(\star)$ by limit transition as $\epsilon \to 0$. ■

C o r o l l a r y. *If A is a p-dimensional analytic set with boundary bA on a complex manifold* Ω, *then integration over bA defines on* Ω *a current* $\phi \to \int_{bA}\phi$ *of measure type and of dimension* $2p-1$. *The equality* $[bA] = -d[A]$ *holds (Stokes' formula for currents).*

■ The fact that $[bA]$ is a current of measure type follows from the local boundedness of the measure $\mathcal{H}_{2p-1}|_{bA}$. Stokes' formula in current notation is obtained from the definition of the operator d for currents. ■

We yet note another important situation in which analytic sets with boundary occur.

P r o p o s i t i o n 1. *Let A be a pure p-dimensional analytic subset of a complex manifold* Ω, *let* ρ *be a real-valued function of class* $C^{2p}(\Omega)$, *and let* $A_t = A \cap \{\rho<t\}$. *Then for almost all* $t \in \mathbb{R}$ *the set* A_t *is an analytic set with boundary; moreover, the boundary of* A_t *is the set* $A \cap \{\rho=t\}$.

In this case the basic fact is that for almost all t the set $\overline{A}_t \setminus A_t$ has locally finite $(2p-1)$-measure in Ω, and that outside a closed subset of $(2p-1)$-measure zero it is a C^1-manifold.

■ By induction with respect to p it is easy to prove that $A_t' = A \cap \{\rho=t\}$ belongs to the closure of A_t, for almost all t. Since $\rho|_{\text{reg}\,A}$ is a function of class C^{2p}, Sard's theorem implies that $d\rho|_{\text{reg}\,A}$ is zero free on $A_t' \cap \text{reg}\,A$, also for almost all t (see, e.g., [54], [85]). Let now K be an arbitrary compact set in Ω and χ_K its characteristic function. Then, by Fubini's theorem (A4.4) and Lelong's theorem (p.14.1),

$$\int\left[\int_{A_t'}\chi_K d\mathcal{H}_{2p-1}\right]dt = \int\chi_K\,|d\rho|_{\text{reg}\,A}|\,dV_A < \infty,$$

hence $\mathcal{H}_{2p-1}(A_t' \cap K) < \infty$, for almost all $t \in \mathbb{R}$. It remains to represent Ω as a countable union of compact sets, and to take into account that the countable union of sets of measure zero is a set of measure zero. ■

R e m a r k. If the level set $A \cap \{\rho=t\}$ has $2p$-measure zero (this is true for almost all t, e.g., in case ρ is a strictly plurisubharmonic function), then

$$\int_{A_t} d\phi = \lim_{\tau\to t}\int_{A_\tau} d\phi = \lim_{\tau\to t}\int_{bA_\tau}\phi,$$

where the limit is over regular values τ of the function $\rho|_{\text{reg}\,A}$. In the sequel the expression $\int_{bA_t}\phi$ in such a context always means this limit over regular (noncritical) values of ρ, and Stokes' formula $\int_{A_t} d\phi = \int_{bA_t}\phi$ is understood in this limiting sense for critical values; we will not mention this again in the sequel.

One of the often used versions of Stokes' formula is the following equation, which we will call *Green's formula* for analytic sets.

P r o p o s i t i o n 2. *Let A be a p-dimensional analytic set with boundary bA on a complex manifold Ω such that $\overline{A}$ is compact. Then for any function u and for any form ψ of bidegree $(p-1,p-1)$ and of class $C^2(\Omega)$ the following equation holds:*

$$\int_{bA}(ud^c\psi-\psi\wedge d^cu) = \int_A(udd^c\psi-\psi\wedge dd^cu).$$

In particular, if a form ϕ of bidegree $(p-1,p-1)$ is d-closed in Ω, then for any $u,v \in C^2(\Omega)$ the following equation holds:

$$\int_{bA}(ud^cv-vd^cu)\wedge\phi = \int_A(udd^cv-vdd^cu)\wedge\phi.$$

■ By Lemma 2, A4.2, at all regular points of A we have

$$du\wedge d^c\psi = d\psi\wedge d^cu, \quad du\wedge d^cv\wedge\phi = dv\wedge d^cu\wedge\phi$$

as forms on $\text{reg}\,A$. Hence, applying to the lefthand side Stokes' formula, we obtain the integrals on the righthand side. ■

As in the real case this formula is convenient for obtaining various integral representations (see p.15.2).

C o r o l l a r y. *Let A be a pure p-dimensional analytic set on a complex manifold Ω, let $u \in C^2(\Omega)$, and let ψ be a form of bidegree $(p-1,p-1)$ and of class $C^2(\Omega)$ such that* (supp ψ) $\cap$ *A is compact. Then*

$$\int_A udd^c\psi = \int_A \psi\wedge dd^cu.$$

■ By requirement there is a function ρ of class $C^\infty(\Omega)$ such that $A_0 = A \cap \{\rho<0\}$ is an analytic set with boundary, containing (supp ψ) $\cap$ A. Since $\psi = d^c\psi = 0$ on bA_0, our formula follows from Proposition 2. ■

14.4. Analytic sets as minimal surfaces. Wirtinger's inequality and Stokes' formula imply the following interesting minimality property of volumes of analytic sets in the class of "films" with given boundary.

P r o p o s i t i o n 1. *Let* (A,bA) *be a p-dimensional analytic set with boundary in* $\mathbb{C}^n$ *such that* $A \cup bA$ *is compact. Let M be a 2p-dimensional* (C^1-) *smooth oriented manifold in* $\mathbb{C}^n$ *such that* $\overline{M}$ *is compact,* $bA \subset \overline{M} \setminus M$, *and* $[bA] = -d[M]$ *in the sense of currents (i.e. Stokes' formula holds for the pair* (M,bA)*). Then* $\mathrm{vol}_{2p}A \leqslant \mathrm{vol}_{2p}M$.

■ By Wirtinger's theorem (p.13.2) and Corollary 13.3,

$$\mathrm{vol}_{2p}A = \frac{1}{p!}\int_A \omega^p = \frac{1}{4p!}\int_A d(d^c|z|^2 \wedge \omega^{p-1}) =$$

$$= \frac{1}{4p!}\int_{bA} d^c|z|^2 \wedge \omega^{p-1} = \frac{1}{p!}\int_M \omega^p \leqslant \mathrm{vol}_{2p}M. \quad ■$$

R e m a r k. From the proof it is clear that this minimality property holds on an arbitrary Hermitian manifold Ω with d-exact fundamental form ω (it suffices that ω^p be d-exact). The Harvey-Shiffman theorem (see [170], [164], and p.19.6) and the boundary uniqueness Theorem for analytic sets (p.19.2) imply that for rectifiable manifolds M the equation $\mathrm{vol}_{2p}A = \mathrm{vol}_{2p}M$ holds (under the conditions of the Proposition) if and only if $\overline{M} = \overline{A}$. The "films" of Proposition 1 can have singularities ($\overline{M} \setminus M$ contains bA but need not coincide with it). The condition $[bA] = -d[M]$ is fulfilled if $\overline{M} \setminus bA$ does not have "worse" singularities, e.g. if $\overline{M} \setminus bA = M \cup \sigma$, where σ is closed, $\mathcal{H}_{2p}(\sigma) = 0$, and there is a fundamental system of neighborhoods $U_\epsilon \supset \sigma$ such that $\mathcal{H}_{2p-1}((\partial U_\epsilon) \cap \overline{M}) \to 0$ as $\epsilon \to 0$ (cf. the proof of Stokes' theorem).

In relation to this minimality property of analytic sets we also note the following analog of the isoperimetric inequality.

P r o p o s i t i o n 2. *Let* Ω *be a Hermitian complex manifold with d-exact fundamental form* $\omega = d\gamma$, *and let A be a p-dimensional analytic set with boundary bA in* Ω *such that* $A \cup bA$ *is compact. Then*

$$\mathrm{vol}_{2p}\, A \leqslant \frac{1}{p}\left[\max_{bA} \| \gamma \| \right] \cdot \mathrm{vol}_{2p-1} bA,$$

where $||\gamma||(z) = \max\{|\gamma(v)|: v \in T_z\Omega, |v|=1\}$.

(Instead of $||\gamma||(z)$ we may in fact take $|\gamma(v(z))|$, where $v(z)$ is a unit vector tangent to $T_z(bA)$ and orthogonal to $T_z^c(bA)$, and instead of $\max_{bA}||\gamma||$ we may take the least upper bound of $|\gamma(v(z))|$ over the regular points $z \in bA$.)

■ By Stokes' formula, $\int_A \omega^p = \int_{bA} \gamma \wedge \omega^{p-1}$. At a regular point $a \in bA$ we choose local coordinates on Ω such that

$$\omega|_{T_a\Omega} = \sum_1^n dx_j \wedge dy_j,$$

$T_a(bA)$ is the plane $\{v: dx_1(v)=0\}$ in $T_a\Omega$, and T_aA is defined by the inequality $dx_1(v)<0$. Then $T_a^c(bA)$ is given by $dz_1(v) = 0$, hence

$$\omega^{p-1}|_{T_a(bA)} = (p-1)!\prod_2^n dx_j \wedge dy_j.$$

Since $\gamma = \sum_1^n (c_j dx_j + c_j' dy_j)$ in a neighborhood of a (c_j and c_j' are functions), we have

$$\frac{1}{(p-1)!}\gamma \wedge \omega^{p-1}|_{T_a(bA)} = c_1' dy_1 \wedge \prod_2^n dx_j \wedge dy_j = c_1' dV_{bA}.$$

If $v(a) = (\partial/\partial y_1)_a$, then $c_1'(a) = \gamma(v(a))$; in particular, $|c_1'(a)| \leqslant ||\gamma||(a)$. Thus

$$\mathrm{vol}_{2p} A = \frac{1}{p!}\int_A \omega^p \leqslant \frac{1}{p}\int_{bA} ||\gamma||(z) dV_{bA}. \quad ■$$

C o r o l l a r y. *If A is a p-dimensional analytic subset of the ball $|z|<R$ in $\mathbf{C}^n$ with boundary $bA \subset \{|z|=R\}$, then*

$$\mathrm{vol}_{2p} A \leqslant \frac{R}{2p}\mathrm{vol}_{2p} bA.$$

■ In this case we may take $\gamma = (d^c|z|^2)/4 = (|z|d^c|z|)/2$. Since $||d^c|z||| \equiv 1$ (this can be easily checked in suitable unitary coordinates), we have $||\gamma||(z) = |z|/2$. ■

14.5. Tangential and normal components of volume. Let ρ be a function of class C^1 on a Hermitian complex manifold. At each point where $d\rho \neq 0$ the fundamental form ω can be split into two mutually orthogonal terms: $\omega = \omega_\nu + \omega_\tau$ where $\omega_\nu = (d\rho \wedge d^c\rho) / |d\rho|^2$ and $\omega_\tau = \omega - \omega_\nu$. If we have chosen local orthonormal coordinates at a point a such that $d\rho = cdx$, then $\omega_\nu = dx_1 \wedge dy_1$ and $\omega_\tau = \sum_2^n dx_j \wedge dy_j$ (everything at a). Thus, $(\omega_\nu)_a$ vanishes on the complex tangent to the hypersurface $\rho = \rho(a)$, and $(\omega_\tau)_a$ vanishes on the complex normal $\mathbb{C}(\nabla \rho)_a$ to this hypersurface. The p-dimensional volume form $\omega^p / p!$ also naturally splits into two mutually orthogonal terms: $\omega^p = (\omega^p)_\nu + (\omega^p)_\tau$, where $(\omega^p)_\nu = \omega_\nu \wedge \omega^{p-1}$ and $(\omega^p)_\tau = \omega_\tau \wedge \omega^{p-1}$ (precisely because $\omega_\nu^p = 0$ for $p > 1$). Domination of some term in $\omega^p|_A$ reflects the position of the p-dimensional analytic set A with respect to the level sets of ρ.

Examples. (a) If ρ is a function in a domain $D \subset \mathbb{C}^n$ depending on z_1 only, then $(d\rho \wedge d^c\rho) / |d\rho|^2 = dx_1 \wedge dy_1$, hence $\omega_\nu = \omega(z_1)$, $\omega_\tau = \omega(z_2, \ldots, z_n)$.

(b) Let $\rho(z) = |z|^2$ in $\mathbb{C}^n$. Then $|d\rho| = 2|z|\,|d|z||$, hence $\omega_\nu = d|z| \wedge d^c|z|$. Since $d|z| \wedge d^c|z| = \omega - |z|^2\omega_0$, we have $\omega_\tau = |z|^2\omega_0$. Thus, the tangential component of ω with respect to the sphere $|z| = r$ is proportional to the "projective fundamental form" ω_0. Furthermore, $(\omega^p)_\nu = d|z| \wedge d^c|z| \wedge \omega^{p-1}$ and $(\omega^p)_\tau = |z|^2\omega_0 \wedge \omega^{p-1}$.

At first sight it may seem that the tangential and normal components are completely alike in the sense of contributing to the volume of an analytic set (especially in the case of $\mathbb{C}^2$, when a complex tangent plane to the level sets of ρ is one-dimensional, as is a complex normal). However, the following simple computation shows that in general this is not true.

Lemma 1. *Let ρ be a function of class C^1 on a complex manifold Ω, and let A be a pure p-dimensional analytic set in Ω. Assume that $A \cap \{\rho \leqslant 0\}$ is compact, and put $A_0 = A \cap \{\rho < 0\}$. Then*

$$\int_{A_0} d\rho \wedge \phi = \int_{A_0} |\rho|\, d\phi$$

for every form of ϕ of class C^1 in a neighborhood of $\overline{A}_0$.

■ Since in C^1 the function ρ can be approximated by functions $\tilde{\rho} > \rho$ of class C^∞, we may assume that for almost all $t < 0$ the set $A_t = A \cap \{\rho < t\}$ is an analytic set with boundary $A \cap \{\rho = t\}$ (or empty). Replacing ρ by $\rho - t$ and taking the limit as $t \to 0$ (this can be done because the $2p$-measure of A_0 is finite), we see that it suffices to investigate the case: A_0 an analytic set with boundary $A \cap \{\rho=0\}$. But then, by Stokes' formula,

$$\int_{A_0} d\rho \wedge \phi + \int_{A_0} \rho d\phi = \int_{A_0} d(\rho\phi) = 0. \quad \blacksquare$$

If A is an analytic subset of the set $\{\rho<0\}$ such that the level sets $A \cap \{\rho \leq t\}$ are compact, $t<0$, then the Lemma implies

$$\int_{A_t} d\rho \wedge \phi = \int_{A_t} |\rho - t| d\phi.$$

If the form $d\rho \wedge \phi$ is positive, the lefthand integral increases as t increases, hence has a limit as $t \to 0$ (finite or infinite). The righthand integral is equal to $\int_{A_t} |\rho| d\phi - |t| \int_{A_t} d\phi$. If $|t| \int_{A_t} d\phi \to 0$, in the limit we obtain the equation $\int_A d\rho \wedge \phi = \int_A |\rho| d\phi$ (the improper integrals converge simultaneously and are equal). In particular, under the same conditions as in Lemma 1 we have

L e m m a 2. *Let Ω be a Kähler manifold with fundamental form ω, let $\rho \in C^2(\Omega)$, and suppose $|t| \int_{A_t} dd^c\rho \wedge \omega^{p-1} \to 0$ as $t \to 0$. Then*

$$\int_{A_t} d\rho \wedge d^c\rho \wedge \omega^{p-1} = \int_{A_0} |\rho| dd^c\rho \wedge \omega^{p-1}$$

(as improper integrals).

Indeed, in this case the form $d\rho \wedge d^c\rho \wedge \omega^{p-1}$ is positive (see A4.5), $\phi = \rho d^c \wedge \omega^{p-1}$, and $d\phi = dd^c\rho \wedge \omega^{p-1}$ since ω is closed. Note that boundedness of the integrals is not stated here; they can be $+\infty$ (but then both).

Put $(\text{vol}\, A)_v = (\int_A (\omega^p)_v)/p!$ and $(\text{vol}\, A)_\tau = (\int_A (\omega^p)_\tau)/p!$.

Replacing ρ by $\rho - t$, Lemma 2 implies

$$\int_{A_t} d\rho \wedge d^c\rho \wedge \omega^{p-1} = \int_{A_t} |\rho - t| dd^c\rho \wedge \omega^{p-1}.$$

Since $|\rho - t| = |\rho| - |t|$ for $\rho < t < 0$, subtracting it from the inequality in Lemma 2 gives

$$\int_{A_{t0}} d\rho \wedge d^c\rho \wedge \omega^{p-1} = |t| \int_{A_t} dd^c\rho \wedge \omega^{p-1} + \int_{A_{t0}} |\rho| dd^c\rho \wedge \omega^{p-1},$$

where $A_{t0} = A \cap \{t<\rho<0\}$. If $d\rho \neq 0$ as $\rho = 0$, there are constants $c_1 > c_2 > 0$, depending on ρ only, such that

$$c_2(\mathrm{vol}\, A_{t0})_\nu \leqslant \int_{A_{t0}} d\rho \wedge d^c\rho \wedge \omega^{p-1} \leqslant c_1(\mathrm{vol}\, A_{t0})_\nu.$$

Since $dd^c\rho \leqslant c_3\omega$ for $\rho \leqslant 0$, with some constant c_3, we thus obtain

$$c_2(\mathrm{vol}\, A_{t0})_\nu \leqslant c' \int_{A_0} \min(|\rho|, |t|)\omega^p.$$

On the other hand, if ρ is strictly plurisubharmonic in a neighborhood of $\{\rho=0\}$, then

$$\int_{A_0} \min(|\rho|, |t|) dd^c\rho \wedge \omega^{p-1} \geqslant \int_{A_{t0}} |\rho| dd^c\rho \wedge \omega^{p-1} \geqslant c_4 \int_{A_{t0}} |\rho| \omega^p,$$

if $|t| < \epsilon = \epsilon(\rho)$, and then we have

$$(\mathrm{vol}\, A_{t0})_\nu \geqslant c'' \int_{A_{t0}} |\rho| \omega^p.$$

Since for $t < 0$ the volume of $A \cap \{\rho \leqslant t\}$ is finite, these inequalities imply

Corollary. *Let $\rho \in C^2(\Omega)$, let $\{\rho \leqslant 0\}$ be compact, and let $d\rho \neq 0$ as $\rho = 0$. If A is a pure p-dimensional analytic subset in $\{\rho<0\}$, then*

$$\mathrm{vol}\, A < \infty \ \textit{if and only if}\ (\mathrm{vol}\, A)_\tau < \infty.$$

If, moreover, ρ is strictly plurisubharmonic in a neighborhood of $\{\rho=0\}$, then

$$(\mathrm{vol}\, A)_\nu < \infty \ \textit{if and only if}\ \int_A |\rho| \omega^p < \infty.$$

14.6. Volumes of analytic subsets of a ball. The natural exhaustion function (with compact level sets) for the ball $|z| < R$ in $\mathbf{C}^n$ is the function $|z|^2$. As we proved in the previous Section, the normal component $(\omega^p)_\nu$ of the form ω^p for $\rho(z) = |z|^2$ is $d|z| \wedge d^c|z| \wedge \omega^{p-1}$, and the tangential component is $|z|^2 \omega_0 \wedge \omega^{p-1}$, where $\omega_0 = (dd^c \log |z|^2)/4$. It is easy to prove the following formula for the "normal part" of the volume, $(\mathrm{vol}\, A)_\nu = \int_A (\omega^p)_\nu / p!$.

P r o p o s i t i o n 1. *Let A be a pure p-dimensional analytic subset of the ball $|z|<R$ with $0 \notin A$. Then*

$$p!(\mathrm{vol}\,A)_v = 2\int_A \log\frac{R}{|z|}\omega^p =$$

$$= \int_A (R^2-|z|^2)\omega_0\wedge\omega^{p-1} = 2\int_0^R\left[\int_{A_t}\omega^p\right]\frac{dt}{t}.$$

■ Without loss of generality we may assume that A is an analytic set with boundary $bA \subset \{|z|=R\}$. Since

$$(\omega^p)_v = \frac{1}{2}d\log\frac{|z|}{R}\wedge d^c|z|^2\wedge\omega^{p-1} =$$

$$= \frac{1}{2}d\left[\log\frac{|z|}{R}d^c|z|^2\wedge\omega^{p-1}\right] - 2\log\frac{|z|}{R}\omega^p$$

and since $\log(|z|/R) = 0$ for $|z| = R$, the first equality follows from Stokes' formula. Similarly,

$$(\omega^p)_v = d|z|^2\wedge\frac{1}{4}d^c\log|z|^2\wedge\omega^{p-1} =$$

$$= d\left[\frac{1}{4}|z|^2 d^c\log|z|^2\wedge\omega^{p-1}\right] - |z|^2\omega_0\wedge\omega^{p-1}$$

and Stokes' formula gives

$$p!(\mathrm{vol}\,A)_v = R^2\int_{bA}\frac{1}{4}d^c\log|z|^2\wedge\omega^{p-1} - \int_A|z|^2\omega_0\wedge\omega^{p-1} =$$

$$= \int_A(R^2-|z|^2)\omega_0\wedge\omega^{p-1}.$$

The last equality follows from Fubini's theorem (A4.4). It implies

$$p!(\mathrm{vol}\,A)_v = \int_0^R\left[\int_{bA_t}d^c|z|\wedge\omega^{p-1}\right]dt.$$

Since $d^c|z| = (d^c|z|^2)/(2|z|)$, and since $|z| = t$ on bA_t, the remaining follows from Stokes' formula. ■

The formulas for the tangential components of volumes are less transparent.

Next to the Euclidean volume (and its components) it is convenient to consider also the projective volume of analytic sets in $\mathbb{C}^n$. The latter is, by definition, equal to $\int_A \omega_0^p / p!$, where $p = \dim A$. If A is a cone with vertex at 0, by Lemma 13.6, $\omega_0^p|_A = 0$. In general the volume of A is, obviously, equal to the volume of the image of A in $\mathbb{P}_{n-1}$, counted with multiplicities, under the canonical map $\Pi: \mathbb{C}_*^n \to \mathbb{P}_{n-1}$. Since ω_0^p is a positive form, the problem of convergence of the integral $\int_A \omega_0^p$ can be simply solved: for any compact exhaustion $A \setminus \{0\} = \cup K_j$, $K_j \subset K_{j+1}$, the limit $\lim_{j\to\infty} \int_{K_j} \omega_0^p$ exists (finite or equal to $+\infty$). Lemma 13.6 implies that $\int_A \omega_0^p < \infty$ if $\operatorname{vol} A < \infty$ and if A is at positive distance from 0. For analytic subsets of a ball these quantities are connected by the following simple relation.

Proposition 2. *Let A be a pure p-dimensional analytic subset of the ball $|z| < R$ with $0 \notin A$. Then*

$$\int_A \omega^p = R^{2p} \cdot \int_A \omega_0^p;$$

in particular, for analytic subsets of the unit ball not containing 0 the Euclidean and projective volumes coincide.

■ By Proposition 1, p.14.3, we may assume that A is an analytic set with boundary $bA \subset \{|z| = R\}$. Since

$$\omega_0^p = d\left[\frac{1}{4} d^c \log |z|^2 \wedge \omega_0^{p-1}\right],$$

$0 \notin A$, and $d|z|\,|_{bA} = 0$, Stokes' formula gives

$$\int_A \omega_0^p = \int_{bA} \frac{1}{4^p} \frac{d^c|z|^2}{|z|^2} \wedge \left[\frac{dd^c|z|^2}{|z|^2} - \frac{d|z|^2 \wedge d^c|z|^2}{|z|^4}\right]^{p-1} =$$

$$= \int_{bA} \frac{1}{4^p} \frac{d^c|z|^2}{|z|^2} \wedge \left[\frac{dd^c|z|^2}{|z|^2}\right]^{p-1} =$$

$$= \frac{1}{R^{2p}} \int_{bA} \frac{1}{4} d^c|z|^2 \wedge \omega^{p-1} = \frac{1}{R^{2p}} \int_A \omega^p. \quad ■$$

If $0 \in A$, in the formula obtained we must add to the projective volume the multiplicity of A at 0 (see p.15.1).

This Proposition implies a simple Crofton formula for analytic subsets of a ball (involving Euclidean volumes); it was in essence proved by H. Alexander [6] (in case $0 \notin A$).

P r o p o s i t i o n 3. *Let A be a pure p-dimensional analytic subset of the ball $|z| < R$ in $\mathbb{C}^n$. Then for any $k \geqslant n-p$ we have the equation*

$$\mathrm{vol}_{2p} A = (\pi R^2)^{n-k} \cdot \frac{(p+k-n)!}{p!} \int_{G(k,n)} \mathrm{vol}_{2(p+k-n)}(A \cap L)\, dL.$$

■ Assume that $0 \notin A$. By Proposition 2 and Proposition 1, p.13.6 (with $f \equiv 1$),

$$\mathrm{vol}_{2p} A = \frac{R^{2p}}{p!} \int_A \omega_0^p = \frac{R^{2p}}{p!} \pi^{n-k} \int_{G(k,n)} \left[\int_{A \cap L} \omega_0^{p+k-n} \right] dL =$$

$$= R^{2p} \pi^{n-k} \frac{1}{p!} \int_{G(k,n)} R^{-2(p+k-n)} \left[\int_{A \cap L} \omega^{p+k-n} \right] dL.$$

The case $0 \in A$ will be considered in p.15.1. ■

14.7. Volumes of algebraic sets. One of the quantities characterizing the complexity of a projective algebraic set is the degree of this set, i.e. the number of points of its intersection with a complex plane in general position and of complementary dimension (see p.11.3). Another real-valued function on algebraic sets is the metric characteristic, the projective volume of these sets, which also should reflect the complexity of an algebraic set. The following remarkable Theorem in complex algebraic geometry shows that in fact these quantities, being of different character, do coincide (see [141], [86]).

T h e o r e m. *Let $A \subset \mathbb{P}_n$ be a pure p-dimensional projective algebraic set, and let $\mathrm{vol}_{2p} A$ denote its volume with respect to the Fubini-Study metric in $\mathbb{P}_n$. Then*

$$\mathrm{vol}_{2p} A = \frac{\pi^p}{p!} \cdot \deg A.$$

Since the volume of a p-dimensional complex plane in $\mathbb{P}_n$ is $\pi^p / p!$ (p.13.4), this means that the degree of A is the quotient of the volume of A to the volume of a p-dimensional complex plane in $\mathbb{P}_n$, or, equivalently, to the volume of $\mathbb{P}_p$. If the Fubini-Study metric is normalized such that $\mathbb{P}_p \subset \mathbb{P}_n$ has volume 1 (regrettably, this cannot be done for all p simultaneously), we obtain that the degree of A and this normalized projective volume of A coincide.

■ By p.11.3 there is an $(n-p-1)$-dimensional complex plane $L \subset \mathbb{P}_n \setminus A$ such that the projection $\pi_L\colon A \to L^\perp$ is an analytic cover with number of sheets equal to $\deg A$. Without loss of generality we may assume that $L^\perp = \mathbb{P}_p \subset \mathbb{P}_n$. The homogeneous coordinates in $\mathbb{P}_n$ are divided into two groups: $z' = (z_0, \ldots, z_p)$, $z'' = (z_{p+1}, \ldots, z_n)$, and for each $t \in (0,1)$ we define in $\mathbb{P}_n$ a form

$$\omega_{1-t} = \Pi_* \left[\frac{1}{4} dd^c \log (|z'|^2 + |tz''|^2)\right];$$

it is the pre-image of ω_0 under the nonunitary automorphism $[z] \mapsto [z', z''/t]$ of $\mathbb{P}_n$. The form $\phi = \omega_0^p - (\omega_{1-t})^p$ can be written as

$$\phi = (\omega_0 - \omega_{1-t}) \wedge \sum_{k=0}^{p-1} \omega_{1-t}^k \omega_0^{p-1-k} =: (\omega_0 - \omega_{1-t}) \wedge \alpha.$$

Since $\log ((|z'|^2 + |tz''|^2)/|z|^2)$ is a function of class C^∞ on $\mathbb{P}_n$, the form

$$\psi = \frac{1}{4} d^c \left[\log \frac{|z'|^2 + |tz''|^2}{|z|^2}\right] \wedge \alpha$$

belongs to $\mathcal{D}^{2p-1}(\mathbb{P}_n)$. Since $d\alpha = 0$ in view of the fact that ω_{1-t} and ω_0 are d-closed, we have $d\psi = \phi$, hence Stokes' theorem (Proposition 3, p.14.2) implies $\int_A \phi = 0$. Thus, $\int_A \omega_{1-t}^p$ is independent of t. Writing this integral in affine coordinates (ζ', ζ'') in $\mathbb{C}^n$, where $\zeta_j = z_j / z_0$ and $\zeta' \in \mathbb{C}^p$, and letting t tend to 0, we thus obtain

$$\operatorname{vol}_{2p} A = \int_{A \cap \mathbb{C}^n} \frac{1}{p!} \left[\frac{1}{4} dd^c \log (1 + |\zeta'|^2)\right]^p.$$

Since the projection $(\zeta', \zeta'') \mapsto \zeta'$ (corresponding in $\mathbb{C}^n$ to π_L) is an analytic cover of $A \cap \mathbb{C}^n$ above $\mathbb{C}^p$ with number of sheets equal to $\deg A$, the righthand integral equals

$$\deg A \cdot \int_{\mathbb{C}^p} \frac{1}{p!} \left[\frac{1}{4} dd^c \log (1 + |\zeta'|^2)\right]^p = \deg A \cdot \int_{\mathbb{P}_p} \frac{1}{p!} \omega_0^p = \deg A \cdot \frac{\pi^p}{p!}. \quad ■$$

15. Lelong numbers and estimates from below

15.1. Lelong numbers. Let A be a pure p-dimensional analytic subset of the ball $|z|<R_0$ in $\mathbf{C}^n$ containing the coordinate origin. If $0<r<R<R_0$, then, by Stokes' theorem implies,

$$\frac{1}{R^{2p}}\int_{A_R}\omega^p - \frac{1}{r^{2p}}\int_{A_r}\omega^p =$$

$$= \frac{1}{4R^{2p}}\int_{bA_R} d^c|z|^2\wedge\omega^{p-1} - \frac{1}{4r^{2p}}\int_{bA_r} d^c|z|^2\wedge\omega^{p-1} =$$

$$= \int_{b(A_R\setminus \bar{A}_r)} \frac{1}{4}\frac{d^c|z|^2}{|z|^2}\wedge\omega_0^{p-1} = \int_{A_R\setminus A_r}\omega_0^p,$$

where $A_r = A\cap\{|z|<r\}$ (see Proposition 2, p.14.6). Since ω_0 is a positive form, we have thus proved

P r o p o s i t i o n 1. *The function* $r^{-2p}\mathrm{vol}_{2p}A$ *decreases monotone as* $r\to 0$, *hence the limit*

$$n(A,0) := \lim_{r\to 0}\frac{\mathrm{vol}_{2p}A_r}{c(p)r^{2p}}$$

exists. Here $c(p) = \pi^p/p!$ *is the volume of the unit ball in* $\mathbf{C}^n$.

This limit is called the Lelong number of the analytic set A at the point 0. The *Lelong number* $n(A,a)$ at a point a is defined similarly, where we take $A_r = A\cap\{|z-a|<r\}$. In the definition of $n(A,a)$ the denominator is the volume of the ball of radius r in $\mathbf{C}^p$, hence $n(A,a)$ is the $2p$-dimensional density of A at a (measured against the density of a $2p$-dimensional complex plane).

The existence of the limit in Proposition 1 also shows that the improper integral $\int_{A_R}\omega_0^p$ exists, and that the formula

$$\mathrm{vol}_{2p}A_r = n(A,0)\cdot c(p)r^{2p} + r^{2p}\left[\int_{A_r}\frac{\omega_0^p}{p!}\right] \qquad (\star)$$

holds. This formula naturally generalizes Proposition 2, p.14.6, to the case when $0\in A$. If A is a cone with vertex at 0, then $\omega_0^p|_A = 0$, hence $(1/r^{2p})\int_{A_r}\omega^p =$

$\int_{A_1} \omega^p$ is independent of r and $(\mathrm{vol}_{2p} A_1)/c(p) = n(A, 0)$.

The geometric meaning of the Lelong number is the following:

P r o p o s i t i o n 2. *The Lelong number of a pure p-dimensional analytic set $A \subset \mathbf{C}^n$ at a point $a \in A$ is equal to the multiplicity of A at this point: $n(A,a) = \mu_a(A)$. In particular, $n(A,a)$ is an integer.*

■ Suppose $a = 0$. We first investigate the case: A a cone. Since the form $\omega \wedge \omega_0^{p-1}$ is integrable over A_1, Stokes' theorem (taking into account that $|z| = 1$ on bA_1) and Fubini's theorem (applied to the canonical projection $\Pi: A \setminus \{0\} \to A_* \subset \mathbf{P}_{n-1}$) imply

$$\int_{A_1} \omega^p = \int_{A_1} \omega \wedge \omega_0^{p-1} = \int_{A_*} \left[\int_{A_1 \cap [z]} \omega \right] \omega_0^{p-1} = \pi \int_{A_*} \omega_0^{p-1},$$

where in the two last integrals ω_0 denotes the Fubini-Study metric in $\mathbf{P}_{n-1}$. By p.14.7, the last integral is equal to $\pi^{p-1} \cdot \deg A_*$. By the definition of degree (p.11.3), $\deg A_* = \mu_0(A)$, hence

$$n(A, 0) = \frac{\mathrm{vol}\, A_1}{c(p)} = \int_{A_1} \frac{\omega^p}{\pi^p} = \mu_0(A).$$

In the general case we consider the family of dilatations ϕ_r: $z \mapsto z/r$, $r > 0$. The image of A_r under ϕ_r is the analytic set A_r/r, and the pre-image of ω under this map is the form $\phi_r^*(\omega) = \omega(z/r) = \omega(z)/r^2$. Hence

$$\frac{1}{r^{2p}} \int_{A_r} \omega^p = \int_{A_r} (\phi_r^*(\omega))^p = \int_{A_r/r} \omega^p.$$

As $r \to 0$ the holomorphic chain $[A_r/r]$ tends to $C([A],0) \cap \{|z| < 1\}$, where $C([A],0)$ is the tangent cone to A at 0 in the sense of holomorphic chains (see p.11.6). Let $C([A],0) = \sum k_j S_j$ be the canonical representation of this chain. If the coordinates are chosen such that all $\mathbf{C}_I$, $\# I = n-p$, intersect with $C(A, 0)$ at 0 only, the definition of $C([A],0)$ and Wirtinger's theorem (p.14.1) clearly imply

$$n(A, 0) = \lim_{r \to 0} \frac{1}{\pi^p} \int_{A_r/r} \omega^p = \frac{1}{\pi^p} \sum k_j \int_{(S_j)_1} \omega^p.$$

By the case of cones sketched above and by p.11.6, the expression on the righthand side equals

$$\sum k_j \deg S_j = \deg C([A],0) = \mu_0(A). \quad \blacksquare$$

We now complete the proof of Proposition 3, p.14.6, for the case $0 \in A$.

$\blacksquare$ For convenience we assume $R = 1$. Formula $(\star)$ and Proposition 2 give

$$\operatorname{vol}_{2p} A = c(p)\cdot\mu_0(A) + \int_A \frac{\omega_0^p}{p!}$$

Crofton's formula $(\star\star)$ of p.13.6 with $f \equiv 1$ on $\operatorname{reg} A \setminus \{0\}$ gives

$$\int_A \frac{\omega_0^p}{p!} = \frac{\pi^{n-k}}{p!}(p+k-n)! \int_{G(k,n)} \left[\int_{A\cap L} \frac{\omega_0^{p+k-n}}{(p+k-n)!} \right] dL.$$

The inner integral on the righthand side is equal to $\operatorname{vol}(A\cap L) - c(p+k-n)\mu_0(A\cap L)$ for almost all L for which $\dim_0(A\cap L) = p+k-n$. By Proposition 3, p.11.4, $\mu_0(A\cap L) = \mu_0(A)$ for almost all L. Since $c(q) = \pi^q/q!$ and the measure dL is normalized, substituting this expression and the expression for $\int_A \omega_0^p/p!$ from $(\star)$ in the equation obtained, we find

$$\operatorname{vol}_{2p} A = \pi^{n-k}.\frac{(p+k-n)}{p!} \int_{G(k,n)} \operatorname{vol}_{2(p+k-n)}(A\cap L)\, dL. \quad \blacksquare$$

In view of Proposition 2 a new terminology (Lelong number) may seem unnecessary. For analytic sets this is true, in essence, but the new notion is more flexible than multiplicity, and can more easily be extended to more complicated objects (related also to analytic sets). E.g., for any current T of dimension $2p$ and of measure type in a domain $D \subset \mathbb{C}^n$, at any point $a \in D$ there is defined a function

$$n(T,a,r) = (\pi r^2)^{-p} \langle T, \chi_{a,r}\omega^p \rangle,$$

where $\chi_{a,r}$ is the characteristic function of the ball $\{|z-a|<r\} \subset D$. Its limit (if it exists), as $r \to 0$, is called the Lelong number of the current T at a, and is denoted by $n(T,a)$.

P r o p o s i t i o n 1′. *Let T be a positive, d-closed current in a domain D in $\mathbb{C}^n$. Then for each point $a \in D$ the function $n(T,a,r)$ decreases monotone as $r \to 0$, hence* *the Lelong number*

$$n(T,a) = \lim_{r\to 0} n(T,a,r)$$

exists.

In essence, the proof is the same as that for analytic sets, the difficulties involved in applying Stokes' theorem are overcome in the standard manner by smoothing the functions $\chi_{a,r}$ or the currents themselves. We will not dwell into this (see [164]).

15.2. Integral representations. Let ρ be a function of class C^2 on a complex manifold Ω, having compact level sets $\{\rho\leqslant t\}$ for all $t<\sup_\Omega\rho$. Such ρ are called exhaustion functions and are extensively used in various problems of complex analysis. In $\mathbf{C}^n$ (or in balls $|z|<R$ in $\mathbf{C}^n$) a natural exhaustion function is $|z|^2$. In various problems additional properties are required of exhaustion functions; most widespread are: plurisubharmonicity of $\log\rho$ (of course, it is now necessary that ρ be nonnegative), i.e. positivity of the form $\alpha = dd^c\log\rho$, and strict plurisubharmonicity of ρ itself, i.e. strict positivity of the form $\beta = dd^c\rho$. In applications we will almost always have $\Omega\subset\mathbf{C}^n$ and $\rho(z) = |z|^2$, and these additional conditions are clearly fulfilled.

Since $\alpha = \rho^{-1}\beta-\rho^{-2}d\rho\wedge d^c\rho$, and since the restrictions of $d\rho$ onto a surface $\rho = \text{const}$ vanishes, we have

On each surface $\rho = t$ we have (at regular points of the surface) $\alpha = t^{-1}\beta$, while on all of Ω the equality $d\rho\wedge\alpha = \rho^{-1}d\rho\wedge\beta$ holds.

(These relations were used already in p.14.6 and p.15.1.)

An exhaustion function is usually fixed once and for all, and for a set $E\subset\Omega$ the following notations are subsequently introduced:

$$E_r = E\cap\{\rho<r^2\},\quad E_{r,R} = E\cap\{r^2<\rho<R^2\}$$

(we want them to coincide in $\mathbf{C}^n$ with the standard notations $E_r = E\cap\{|z|<r\}$ in case $\rho = |z|^2$).

L e m m a. *Let A be a pure p-dimensional analytic subset of a complex manifold Ω with exhaustion function $\rho\geqslant 0$ of class $C^{2p}(\Omega)$, and let $R>r>0$ be such that $A_{r,R}$*

is an analytic set with boundary. Then for any function $u \in C^2(\Omega)$ *we have*

$$2\int_r^R \left[\int_{A_t} dd^c u \wedge \beta^{p-1}\right] \frac{dt}{t^{2p-1}} = \int_{bA_{r,R}} u \frac{d^c\rho}{\rho} \wedge \alpha^{p-1} - \int_{A_{r,R}} u\alpha^p. \qquad (\star)$$

■ By Stokes' theorem, the inner integral on the lefthand side is (at regular values t) equal to

$$\int_{bA_t} d^c u \wedge \beta^{p-1} = t^{2(p-1)} \int_{bA_t} d^c u \wedge \alpha^{p-1},$$

hence Fubini's theorem (A4.4) implies that the lefthand side of our equation equals

$$\int_{A_{r,R}} d\log \rho \wedge d^c u \wedge \alpha^{p-1} = \int_{A_{r,R}} du \wedge d^c \log \rho \wedge \alpha^{p-1}$$

(remember that $\rho = t^2$ on bA_t). The remaining clearly follows from Stokes' formula, since

$$du \wedge d^c \log \rho \wedge \alpha^{p-1} = d(u d^c \log \rho \wedge \alpha^{p-1}) - u\alpha^p. \quad ■$$

We apply this formula to the case Ω a ball in $\mathbf{C}^n$ and $\rho = |z|^2$. In this case we know the asymptotic behavior of α at 0, and we may pass to the limit as $r \to 0$. The second integral on the righthand side has limit equal to $\int_{A_R} u\alpha^p$, since α^p is positive and can be (improperly) integrated over $A_{0,R}$ (p.15.1), while u is a continuous function. The first integral on the righthand side in $(\star)$ can be represented as

$$\frac{1}{R^{2p}} \int_{bA_R} u d^c \rho \wedge \beta^{p-1} - u(0) \frac{1}{r^{2p}} \int_{A_r} \beta^p - \frac{1}{r^{2p}} \int_{bA_r} (u - u(0)) d^c \rho \wedge \beta^{p-1}.$$

Since $\rho = |z|^2$, we have $|d((u-u(0))d^c\rho \wedge \beta^{p-1})| \leqslant C|z|\beta^p$. Hence, by p.15.1, the last term tends to zero as $r \to 0$. Thus, the righthand side of formula $(\star)$ has a limit as $r \to 0$, equal to

$$\frac{1}{R^{2p}} \int_{bA_R} u d^c \rho \wedge \beta^{p-1} - u(0)(4\pi)^p \mu_0(A).$$

The inner integral on the lefthand side of $(\star)$ has order $O(t^{2p})$, $t \to 0$, hence its limit as $r \to 0$ also exists, and we obtain the following integral representation.

P r o p o s i t i o n . *Let A_R be a p-dimensional analytic subset of the ball $|z|<R$ with boundary $bA_R \subset \{|z|=R\}$. Then for any function u of class C^2 in a neighborhood of A_R, we have*

$$u(0)\cdot\pi^p\mu_0(A) =$$

$$= \frac{1}{2R^{2p-1}}\int_{bA_R} ud^c|z|\wedge\omega^{p-1} - \int_{A_R} u\omega_0^p - \frac{1}{2}\int_0^R\left[\int_{A_t} dd^c u\wedge\omega^{p-1}\right]\frac{dt}{t^{2p-1}}.$$

An interesting Corollary is obtained if u is a plurisubharmonic function.

C o r o l l a r y 1 (mean value Theorem). *If u is plurisubharmonic in a neighborhood of $\bar{A}_R$ and positive on A_R, then*

$$u(0) \leqslant \frac{1}{\mu_0(A)}\int_{bA_R} u\sigma.$$

Here σ is the Euclidean volume form on bA_R divided by the volume of the ball of radius R in $\mathbb{C}^p$.

■ The Euclidean volume form on A_R is equal to

$$\Sigma = \frac{\sigma\cdot 2\pi^p R^{2p-1}}{(p-1)!}.$$

By Lemma 13.2 (applied to the manifold reg $\tilde{A}$, where $\tilde{A} \supset \bar{A}_R$),

$$d^c|z|\wedge\omega^{p-1} = (p-1)!\Sigma\cdot|d|z||_{\tilde{A}}| \leqslant (p-1)!\Sigma,$$

since the norm of the restriction of $d|z|$ to $\tilde{A}$ does not exceed $|d|z|| = 1$ at each point of $\tilde{A}$. Thus,

$$u(0)\cdot\pi^p\mu_0(A_R) \leqslant \frac{1}{2R^{2p-1}}\int_{bA_R} ud^c|z|\wedge\omega^{p-1} \leqslant \pi^p\int_{bA_R} u\sigma,$$

and our inequality is proved under the additional restriction $u \in C^2(\bar{A}_R)$. Since any plurisubharmonic function is the limit of a monotone decreasing sequence of functions of class C^∞ which are also plurisubharmonic (see A1.2), this inequality also holds in the general situation. ■

Note that the condition that u be positive is essential, since in general the inequality obtained is not true, even with a negative constant: if A_R is not a cone with vertex at 0, then

$$\int_{bA_R} \sigma \geqslant \frac{1}{2\pi^p R^{2p-1}} \int_{bA_R} d^c|z| \wedge \omega^{p-1} = \frac{1}{\pi^p R^{2p}} \int_{A_R} \omega^p > \mu_0(A).$$

But on cones the situation is almost not different from that in planes.

C o r o l l a r y 2. *If A is a cone with vertex at 0, and if $u \in C^2(A_R)$, then*

$$u(0)\cdot\mu_0(A) = \int_{bA_R} u\cdot\sigma - \frac{1}{2\pi^p}\int_0^R \left[\int_{A_t} dd^c u \wedge \omega^{p-1}\right]\frac{dt}{t^{2p-1}}.$$

In particular, if u is pluriharmonic in a neighborhood of $\overline{A}_R$, then $u(0)\cdot\mu_0(A) = \int_{bA_R} u\sigma$, and if it is plurisubharmonic then $u(0)\cdot\mu_0(A) \leqslant \int_{bA_R} u\sigma$.

■ If A is a cone, we have $\omega_0^p|_A = 0$, $|d|z||_A| = 1$, hence $2\pi^p R^{2p-1}\cdot\sigma = d^c|z| \wedge \omega^{p-1}$. ■

15.3. Lower bounds for volumes. Formula (⋆) of p.15.1, taking into account the positivity of the form ω_0, implies the following important lower bound for volumes of analytic sets (see [76], [119], [22], [140]).

T h e o r e m. *Let A be a pure p-dimensional analytic subset of the ball $|z|<r$ in $\mathbf{C}^n$ containing the coordinate origin. Then*

$$\mathrm{vol}_{2p} A \geqslant \mu_0(A)\cdot c(p) r^{2p}, \qquad (\star)$$

in particular, the volume of A is at least the volume of $\{|z|<r\} \cap \mathbf{C}^p$, a p-dimensional linear section of the ball. Equality in (⋆) holds if and only if A is a cone; more precisely, if $tz \in A$ for all $z \in A$ and $t \in \mathbf{C}$, $|t|<1$.

■ We need only prove the last statement. If A is a cone, equality in (⋆) follows from p.15.1, since $\omega_0^p|_A = 0$ in this case. Conversely, equality in (⋆) implies that $\int_A \omega_0^p = 0$, and thus, by Lemma 13.6, $\sin^2(z,A) = 0$ for almost all $z \in \mathrm{reg}\, A$. Hence continuity implies that the rank of the canonical projection $\Pi: A \setminus \{0\} \to \mathbf{P}_{n-1}$ is strictly less than p at all regular points from $A \setminus \{0\}$. It cannot be smaller than $p-1$ since the rank of Π at all of $\mathbf{C}^n_*$ equals $n-1$. Hence $\mathrm{rank}\, \Pi|_{\mathrm{reg}\, A} \equiv p-1$; thus, all fibers of $\Pi|_{\mathrm{reg}\, A}$ are one-dimensional analytic subsets in $|z|<r$. But the fibers of Π in $0<|z|<r$ are punctured disks, hence the

uniqueness Theorem (p.2.2) implies that A contains with each regular point z^0 also the disk $\{|z|<r\} \cap \mathbf{C}\cdot z^0$. Since reg A is dense in A, the same holds for an arbitrary point $z^0 \in A$. ■

Since $\mu_0(A)$ is an integer, this implies

C o r o l l a r y 1. *If A is a pure p-dimensional analytic subset of the ball $|z|<r$ and if $\mathrm{vol}_{2p}A \leqslant (2-\epsilon)c(p)r^{2p}$ for some $\epsilon \in (0,1)$, then there are no singular points of A inside the ball $|z|<\epsilon r/8p$.*

■ If $a \in \mathrm{sng}\, A$ and $A' = A \cap \{|z-a|<r-|a|\}$, then, by the Theorem, $\mathrm{vol}_{2p}A' \geqslant 2c(p)(r-|a|)^{2p}$, since the multiplicity of an analytic set at a singular point is at least 2. By requirement, $2(r-|a|)^{2p} < (2-\epsilon)r^{2p}$, hence $r-|a| < (1-\epsilon/2)^{1/2p}\cdot r$, so that $|a| > r\cdot\epsilon/8p$. ■

This curious fact cannot be directly generalized: having volume of order $(2+\epsilon)c(p)r^{2p}$ is already compatible with the presence of any number of singularities in an arbitrary small neighborhood of 0 (obviously, we need a topological restriction on A in this case).

E x a m p l e. (a) Let A_ϵ be the analytic set in $\mathbf{C}^2$ defined by the equation $z_2^2 = \epsilon\prod_1^m(z_1-a_j)^3$, where $a_1,\ldots,a_m$ are arbitrary distinct points in the disk $|z_1|<r$ in the $\mathbf{C}_1$-plane. All points $(a_j,0)$ are singular for A_ϵ (of semicubic parabola type). On the other hand, the projection of A_ϵ onto $\mathbf{C}_1$ is two-sheeted, above $\mathbf{C}_2$ the number of sheets $\leqslant 3m$. Hence for ϵ sufficiently small the volume of the analytic set $A_\epsilon \cap \{|z|<r\}$ is arbitrarily close to $2\pi r^2$.

The lower bound for the volume of an analytic set and the isoperimetric inequality (p.14.4) imply an analogous lower bound for the volume of the boundary of the set.

C o r o l l a r y 2. *Let A be a pure p-dimensional analytic subset of the ball $\{|z|<r\}$ in $\mathbf{C}^n$ with boundary $bA \subset \{|z|=r\}$. Then if $0 \in A$,*

$$\mathcal{H}_{2p-1}(bA) \geqslant \mu_0(A)\cdot c'(p)r^{2p-1},$$

where $c'(p)$ is the volume of the unit sphere in $\mathbf{C}^p$.

Good lower bounds for volumes of analytic sets are available not only for sets in balls. However, the following problem has remained unsettled for sufficiently long time already.

For which convex domains $D \subset \mathbf{C}^n$ symmetric about the coordinate origin does the estimate

$$\mathrm{vol}_{2p}\, A \geqslant \min_{G(p,n)} \mathrm{vol}_{2p}\,(D \cap L)$$

holds for any pure p-dimensional analytic subset $A \subset D$ passing through 0?

It is known that this estimate holds for the unit cube in $\mathbf{C}^2$ ([60]), for tube domains in $\mathbf{C}^2$ with bounded base ([65]), etc. (see [8], [65]). Of course we need not require convexity and symmetry, but the problem has then not been considered yet. The fact that not every convex, symmetric domain satisfies this condition is shown by the following Example (which is, in essence, contained in [65]).*)

E x a m p l e. (b) Let $D = B_1 + iB_R$ be a "semitube" domain in $\mathbf{C}^3$, where B_1 is the unit ball in $\mathbf{R}^3_x$ and B_R is the ball $|y| < R$ in $\mathbf{R}^3_y$ $(z = x + iy)$. We estimate from below the volume of the sections of D by planes $L \in G(2,3)$. For symmetry reasons we may assume that L is given by equations $z_3 = a_1 z_1 + a_2 z_2$, where $|a_1|, |a_2| < 1$. The projection of $D \cap L$ to $\mathbf{C}_{12}$ is defined by the inequalities

$$x_1^2 + x_2^3 + (\mathrm{Re}(a_1 z_1 + a_2 z_2))^2 < 1, \quad y_1^2 + y_2^2 + (\mathrm{Im}(a_1 z_1 + a_2 z_2))^2 < R^2,$$

from which it can be seen that this projection contains the set

$$|x_1| < \frac{1}{4}, \quad |x_2| < \frac{1}{4}, \quad |\alpha_1 y_1 + \alpha_2 y_2| < \frac{1}{4},$$

$$|y_1| < \frac{R}{4}, \quad |y_2| < \frac{R}{4}$$

for certain constants $\alpha_1, \alpha_2 \in \mathbf{R}$, $|\alpha_1|, |\alpha_2| < 1$ (under the assumption that $R \geqslant 3$). Hence its volume is at least $R/4^4$. Thus, by Wirtinger's theorem, $\mathrm{vol}_4(D \cap L) \geqslant R/4^4$. On the other hand, consider the analytic set A defined by $z_3 = z_1^2 + z_2^2$, or, in real coordinates, by

$$x_3 = x_1^2 + x_2^2 - y_1^2 - y_2^2, \quad y_3 = 2x_1 y_1 + 2x_2 y_2.$$

Since $|x| < 1$ in D, this implies that $y_1^2 + y_2^2 < 2$ and (by the Schwarz inequality) $|y_3| < 2\sqrt{2}$ on $A \cap D$. Since the projection of this set to the coordinates planes

*)In this connection see also R. Zeinstra's paper 'On a question concerning zero sets of minimal area in domains of $\mathbf{C}^2$' Proc. KNAW A 87(3), Sept. 3 1984, 291-297 = Indag. Math. 46 (1984), 291-297.

is at most two-sheeted, these inequalities imply by Wirtinger's theorem that $\mathrm{vol}_4(A \cap D) < 20$. Thus, for $R > 20 \cdot 4^4$ the volume of $A \cap D$ is strictly less than that of any plane section $L \cap D$, $L \in G(2,3)$.

We yet note that there is also an unsettled problem for the unit sphere in $\mathbb{C}^n$. By Corollary 2, for any one-dimensional analytic subset $A \subset \{|z|<1\}$ with boundary and passing through 0, the length of its boundary is at least 2π.

Q u e s t i o n: is it true that the length of at last one of the connected components of bA is also at last 2π?

For the bidisk $\{|z_1|<1, |z_2|<1\}$ this is true (see [15]). For sets of dimension $p > 1$ the analogous statement is trivially solved by Corollary 2, since the boundary of such an irreducible analytic subset of the sphere must be connected (by the Harvey-Lawson theorem, see p.19.6 and [167], [164]).

15.4. Areas of projections. By Wirtinger's theorem, the volume of an analytic set in $\mathbb{C}^n$ is the sum of the volumes of its projections to the corresponding coordinate planes, counted with the multiplicities of the projections. A priori, volume may be accumulated precisely because of the multiplicities of the projections, while the volumes of the projection sets themselves could be uncontrollably small. However, this is not true, and in fact the sum of the volumes of the projections not counted with multiplicities also allows a good lower bound (see Alexander-Taylor-Ullman [9]). A basic step in the proof of such bounds is the following Lemma from the theory of functions of one complex variable (see [9]).

L e m m a. *Let f be a holomorphic function in the disk Δ: $|\zeta| < 1$ in $\mathbb{C}$, bounded and such that $f(0) = 0$. Then*

$$\mathcal{H}_2(f(\Delta)) \geqslant \frac{1}{2} \int_0^{2\pi} |f(e^{i\theta})|^2 d\theta.$$

(The integrand is the radial limit function of f on the unit circle.)

■ It suffices to investigate the case: $f \not\equiv 0$ and holomorphic in a neighborhood of $\bar{\Delta}$. Fix $t > 0$, $\epsilon > 0$, and put

$$E_t = \{e^{i\theta}: |f(e^{i\theta})| \geqslant t\}, \quad \Gamma_t = \{w \in \mathbb{C}: |w| = t\}.$$

The set $f(\bar{\Delta}) \cap \Gamma_t$ is covered by open arcs on Γ_t whose total length is at most

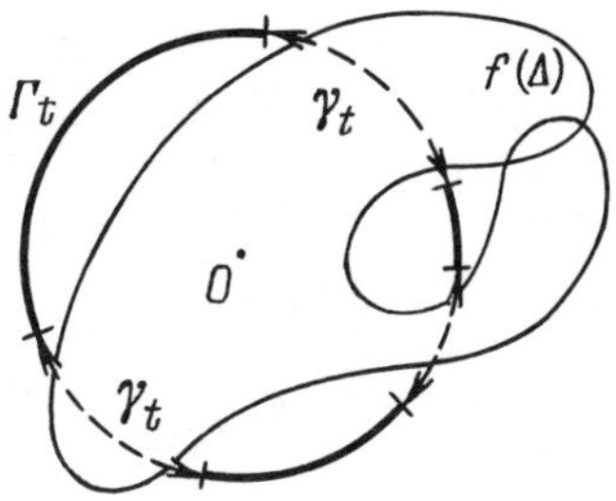

Figure 11.

$\mathcal{H}_1(f(\bar{\Delta})\cap\Gamma_t)+\epsilon$, the union of these arcs will be denoted by γ_t (Figure 11). Let h_t be the harmonic measure of γ_t in $|w|<t$, and let g_t be a holomorphic function in the disk such that $\operatorname{Re} g_t = h_t - 1$. Since $\operatorname{Re} g_t = 0$ on γ_t and $\operatorname{Re} g_t < 0$ in $|w|<t$, by the symmetry principle g_t can be analytically continued across every arc from γ_t, and the values of these continuations lie in the right halfplane. Hence there are a neighborhood U of $\{|w|<t\} \cup \gamma_t$ in $\mathbf{C}$ and a function $\tilde{g}_t$ holomorphic in it such that $\tilde{g}_t = g_t$ for $|w|<t$ and $\operatorname{Re}\tilde{g}_t > 0$ for $|w|>t$. Define h_t also in U, by putting $h_t = \operatorname{Re}\tilde{g}_t + 1$ in it, and put $\tilde{h}_t = h_t$ for $|w|<t$, $\tilde{h}_t = 1$ on $\{|w|>t\} \cup \gamma_t$. Since $\tilde{h}_t = \min(1, h_t)$ in U, $\tilde{h}_t$ is a superharmonic function in the domain $(\mathbf{C}\setminus\Gamma_t)\cup\gamma_t$. This domain contains $f(\bar{\Delta})$, hence $\tilde{h}_t \circ f$ is superharmonic in a neighborhood of $\bar{\Delta}$; moreover, $\tilde{h}_t\circ f \geqslant 0$, and $\tilde{h}_t \circ f|_{E_t} = 1$ since $|f| \geqslant t$ on E_t. The mean value Theorem implies

$$\tilde{h}_t\circ f(0) \geqslant \frac{1}{2\pi}\int_0^{2\pi} \tilde{h}_t\circ f(e^{i\theta})d\theta \geqslant \frac{1}{2\pi}\mathcal{H}_1(E_t).$$

On the other hand, $\tilde{h}_t\circ f(0) = h_t(0)$, and, by the mean value Theorem in $|w|<t$,

$$h_t(0) = \frac{1}{2\pi t}\mathcal{H}_1(\gamma_t).$$

Thus, $\mathcal{H}_1(E_t) \leqslant \mathcal{H}_1(\gamma_t)/t \leqslant (\mathcal{H}_1(f(\bar{\Delta}) \cap \Gamma_t + \epsilon)/t$. Since ϵ is arbitrary, for any $t>0$ we have the inequality

$$t\mathcal{H}_1(E_t) \leqslant \mathcal{H}_1(f(\bar{\Delta})\cap\Gamma_t). \qquad (\star)$$

Integrate it with respect to t (since f is bounded, the integral is over a finite interval). Denote by $\chi(\theta,t)$ the characteristic function of the set $\{\theta\colon |f(e^{i\theta})| \geq t\}$. By Fubini's theorem,

$$\int_0^\infty t\mathcal{H}_1(E_t)dt = \int_0^\infty \left[\int_0^{2\pi} \chi(\theta,t)d\theta\right] tdt = \int_0^{2\pi} \left[\int_0^\infty t\chi(\theta,t)dt\right] d\theta.$$

For fixed θ the function $\chi(\theta,t)$ is distinct from zero ($=1$) if and only if $t \leq |f(e^{i\theta})|$, hence the last integral is equal to

$$\int_0^{2\pi} \left[\int_0^{|f(e^{i\theta})|} tdt\right] d\theta = \frac{1}{2}\int_0^{2\pi} |f(e^{i\theta})|^2 d\theta.$$

Integrating the righthand side of ($\star$) in the same manner, we obtain by Fubini's theorem,

$$\int_0^\infty \mathcal{H}_1(f(\bar{\Delta})\cap\Gamma_t)\,dt = \mathcal{H}_2(f(\bar{\Delta})) = \mathcal{H}_2(f(\Delta)). \quad \blacksquare$$

T h e o r e m 1. *Let E be a closed, complete pluripolar subset of the ball $B\colon |z| < r$, and let A be a pure p-dimensional analytic subset in $B \setminus E$ passing through the coordinate origin $z = 0 \notin E$. Then*

$$\sum_{\#I=p}{}' \mathcal{H}_{2p}(\pi_I(A)) \geq c_p r^{2p},$$

where $c_p > 0$ are constants depending on p only (for $p = 1$ we may take $c_1 = \pi$).

(For the definition of complete pluripolar set see A1.2. Of course, the Theorem is meaningful also in case E is empty, this is its formulation in [9].)

■ We first consider the case $p = 1$. By slightly diminishing r (and then by limit transition as $r' \to r$) we may assume that there is, in a neighborhood of B, a subharmonic function ϕ such that $E = \{z \in B\colon \phi(z) = -\infty\}$. This function is uniformly bounded on $\bar{B}$, hence we may assume that $\phi < 0$ on B.

Without loss of generality we may assume that A is irreducible. The normalization $A^\star$ of A is a Riemann surface on which a proper holomorphic map $\nu\colon A^\star \to A$ is defined that is biholomorphic above reg A (see p.6.2). The universal covering surface of $A^\star$ is the unit disk Δ. Composition of the universal covering map $\pi\colon \Delta \to A^\star$ and ν defines a holomorphic map $f\colon \Delta \to \mathbf{C}^n$ for which $f(\Delta) = A$ and $f(0) = 0$ (see p.6.2). We show that almost everywhere on $\partial\Delta$ the function

$|f|^2 = |f_1|^2 + \cdots + |f_n|^2$ has radial limit values, equal to 1. By Fatou's theorem (see, e.g., [37], [98]), each f_j has radial limit values almost everywhere on $\partial\Delta$. Let $S \subset \partial\Delta$ be a set of full measure at the points of which all f_j have radial limits, and let $S' \subset S$ consist of the points $\alpha \in S$ for which $\lim_{t\to 1} f(t\alpha) \in E$. The function $\phi(f(\zeta))$ is subharmonic in Δ, negative, and its radial limits on S' are equal to $-\infty$. This implies that $\mathcal{H}_1(S') = 0$ (see [37]). Thus, $S \setminus S'$ has length 2π. Let $\alpha \in S \setminus S'$. Assume that $\lim_{t\to 0} f(t\alpha) = a$ with $|a| < 1$. Since $\bar{A} \subset A \cup E \cup bB$, we have $a \in A$, and $\{z = f(t\alpha): 0 \leqslant t \leqslant 1\}$ is a path on A joining 0 and a. It corresponds uniquely to a compact path γ^* on A^* with initial point $\pi(0)$. Since $\pi: \Delta \to A^*$ is a local biholomorphic cover, γ^* must uniquely correspond to a path $\tilde{\gamma}$ in Δ with initial point 0 (the existence of such a path covering γ^*, and thus γ, obviously follows from the local homeomorphy of π and the compactness of γ^*). However, the only path above γ^*, lying in Δ, and emanating from 0 is the radius $[0,1)$, and there is no compact covering path. This contradiction shows that $\lim_{t\to 1} |f(te^{i\theta})| = 1$ for almost all $\theta \in [0,2\pi)$.

Radial limit values will be denoted by the same symbols as the functions. Applying the Lemma to the components f_j of f we thus obtain

$$\sum_{j=1}^{n} \mathcal{H}_2(f_j(\Delta)) \geqslant \frac{1}{2}\int_0^{2\pi} |f(e^{i\theta})|^2 d\theta = \pi,$$

since $|f(e^{i\theta})|^2 = 1$ almost everywhere on $\partial\Delta$. Since $f(\Delta) = A$, we have $f_j(\Delta) = \pi_j(A)$ (where $\pi_j: z \mapsto z_j$), and the Theorem is proved for the case $p = 1$.

The general case is reduced to the case $p = 1$ by induction with respect to p, using Fubini's theorem. For any $p \geqslant 1$ and $\rho \leqslant r$ the sum of the areas of the sets $\pi_j(A_\rho)$, $j = 1, \ldots, n$, is at least $\pi\rho^2$, since $\pi_j(A_\rho) \supset \pi_j(A'_\rho)$, where A' is a one-dimensional analytic subset of A passing through 0 (e.g., the section of A by some $(n-p+1)$-dimensional plane). Take $\rho = r/2$ and $j = n$. For almost all $\zeta_n \in \pi_n(A_{r/2})$ the set $A_{\zeta_n} = A \cap \{z_n = \zeta_n\}$ has dimension $p-1$ and contains points from the ball $|z| < r/2$. Hence, by the induction hypothesis, the sum of the volumes of its projections to the $(p-1)$-dimensional coordinate planes is at least $c_{p-1}(r/2)^{2p-2}$. For the $(p-1)$-dimensional planes $\mathbf{C}_J$ containing the z_n-axis the volume $\pi_J(A_{\zeta_n})$ is zero, since this projection is contained in $\mathbf{C}_J \cap \{z_n = \zeta_n\}$. Thus,

$$\sum_{\substack{\#J = p-1 \\ n \notin J}}' \mathcal{H}_{2p-2}(\pi_J(A_{\zeta_n})) \geqslant c_{p-1}\left(\frac{r}{2}\right)^{2p-2}.$$

Integrating this over $\pi_n(A_{r/2})$ and applying Fubini's theorem gives

$$\sum_{\substack{\sharp J = p-1 \\ n \notin J}}' \mathcal{H}_{2p}(\pi_{J,n}(A)) \geqslant c_{p-1} \left(\frac{r}{2} \right)^{2p-2} \mathcal{H}_2(\pi_n(A_{r/2})).$$

The same procedure can be repeated with any value $j = k$ (not only with $j = n$). Summing all such inequalities over k, we find

$$\sum_{k=1}^{n} \sum_{\substack{\sharp J = p-1 \\ n \notin J}}' \mathcal{H}_{2p}(\pi_{J,k}(A)) \geqslant \frac{c_{p-1}\pi}{4^{p-1}} r^{2p}.$$

For a given ordered multi-index I with $\sharp I = p$ there are at most p pairs of the form (J,k) with $\sharp J = p-1$, J ordered, and $(J,k)' = 1$ (the prime denotes ordered sets). Thus,

$$\sum_{k=1}^{n} \sum_{\substack{\sharp J = p\,1 \\ n \notin J}}' \mathcal{H}_{2p}(\pi_{J,k}(A)) \leqslant p \sum_{\sharp I = p}' \mathcal{H}_{2p}(\pi_I(A)),$$

and the Theorem is proved. For c_p we may take $(\pi / p)\cdot(c_{p-1} / 4^{p-1})$. ■

The proof given above is taken from [9] with some modifications allowing for the presence of the set E. By the way, Theorem 1 and the theorem from [9] are equivalent: below we will prove (p.18.3) that if A has finite volume, then the closure of A in the ball B is analytic in B (Theorem 1 will be used in the very proof of this analytic continuation).

The following Theorem is obtained in a similar way (Alexander-Osserman [8]).

T h e o r e m 2. *Let A be a pure p-dimensional analytic subset of the ball $|z| < r$ in $\mathbb{C}^n$, and let d be the distance of A from* 0. *Then*

$$\sum_{1}^{n} \mathcal{H}_2(\pi_j(A)) \geqslant \pi(r^2 - d^2); \qquad (\star\star)$$

moreover, equality holds if and only if A is the intersection of the ball $|z| < r$ with a complex line.

■ If in the Lemma above the condition $f(0) = 0$ is not fulfilled (while the other conditions are the same), by replacing f with $f - f(0)$ we obtain

$$\mathcal{H}_2(f(\Delta)) \geqslant \frac{1}{2} \int_0^{2\pi} |f(e^{i\theta}) - f(0)|^2 d\theta =$$

$$= \frac{1}{2}\int_0^{2\pi} |f(e^{i\theta})|^2 d\theta + \pi |f(0)|^2 - \mathrm{Re}\left[\overline{f(0)}\int_0^{2\pi} f(e^{i\theta}) d\theta\right].$$

The mean value Theorem implies that the last integral equals $2\pi f(0)$, hence

$$\mathcal{H}_2(f(\Delta)) \geqslant \frac{1}{2}\int_0^{2\pi} |f(e^{i\theta})|^2 d\theta - \pi |f(0)|^2.$$

Repeating the preceding proof with this modification of the Lemma gives (**).

If A is the section of the ball by a complex line, equality clearly follows from Wirtinger's theorem. The proof of the converse statement (linearity of A from equality in (**)) is rather unwieldy (see [8]); we will not need it. ■

15.5. Sequences of analytic sets. The lower bound for volumes of analytic sets obtained in p.15.3 can be used to estimate from above the volume of limit sets. The beautiful idea of this application of the lower bound was advanced by Bishop [22] (see also the estimates of Hausdorff measures of boundary sets in [184]). The following notions are necessary in the proof of Bishop's theorem.

Let $\{E_j\}$ be a sequence of subsets of a metric space X. We say that it converges to a set $E \subset X$ (and write $E_j \to E$) if

1. E coincides with the limit set of the sequence $\{E_j\}$, i.e. consists of all points of the form $\lim_{j_\nu \to \infty} x_{j\nu}$, $x_{j_\nu} \in E_{j_\nu}$ (in particular, E is closed in X);
2. For any compact set $K \subset E$ and any $\epsilon > 0$ there is an index $j(\epsilon, K)$ such that K belongs to the ϵ-neighborhood of E_j in X for all $j > j(\epsilon, K)$.

(In the definition of convergence of p-chains, given in p.12.2, the supports of the chains converge in precisely this sense.)

T h e o r e m. *Let $\{A_j\}$ be a sequence of pure p-dimensional analytic subsets of a complex manifold Ω, converging to some set $A \subset \Omega$ and such that*

$$\mathcal{H}_{2p}(A_j \cap K) \leqslant M_K < \infty$$

for any compact set $K \subset \Omega$ (i.e. locally the volumes of A_j are uniformly bounded). Then A is also a pure p-dimensional analytic subset in Ω.

■ The statement is local, hence we may assume that Ω is a domain in $\mathbb{C}^n$ and that all $\mathcal{H}_{2p}(A_j) \leqslant M < \infty$. We first prove that $\mathcal{H}_{2p}(A \cap K)$ is also bounded, for any compact set $K \subset \Omega$.

Fix $\epsilon > 0$, $4\epsilon < \text{dist}\,(K, \partial\Omega)$, and take j such that $A \cap K$ lies in the ϵ-neighborhood of A_j. Cover $\mathbb{C}^n$ by equal cubes of diameter ϵ and with disjoint interiors, and extract all such cubes Q_ν which intersect A_j. Choose a point $a_\nu \in A_j \cap Q_\nu$ and put $B_\nu = B(a_\nu, 2\epsilon)$, for all ν. Then the balls B_ν cover A_j, their union contains the ϵ-neighborhood of A_j (in particular, $A \cap K$), and there is a constant, dependent on n only, such that each point $z \in \cup B_\nu$ belongs to at most C_n distinct B_ν's. Denote by N the number of balls B_ν. By Theorem 15.3, $\mathcal{H}_{2p}(A_j \cap B_\nu) \geqslant c(2p)\cdot(2\epsilon)^{2p}$. Since the multiplicity of the covering $\cup B_\nu$ does not exceed c_n, this implies

$$M \geqslant \mathcal{H}_{2p}(A_j) \geqslant \frac{1}{c_n} \sum_\nu \mathcal{H}_{2p}(A_j \cap B_\nu) \geqslant N \cdot c\epsilon^{-2p},$$

where c depends on n,p only. Thus, for the covering of $A \cap K$ by the balls B_ν of radii $r_\nu = 2\epsilon$ we obtain $\sum_\nu r_\nu^{2p} = N(2\epsilon)^{2p} \leqslant CM$, where the constant C depends on n,p only. Since ϵ is arbitrary, the definition of $\mathcal{H}_{2p}$ (A6.1) implies

$$\mathcal{H}_{2p}(A \cap K) \leqslant CM.$$

Let now a be an arbitrary point of A. We must prove that A is analytic and p-dimensional in a neighborhood of a. For this we first note that the set $\mathcal{Q} = A \cup (\cup A_j)$ is closed in Ω and has $(2p+1)$-measure zero. Hence we can choose coordinates in $\mathbb{C}^n$ such that the projection $\pi\colon \mathcal{Q} \cap U \to U' \subset \mathbb{C}^p$, for some neighborhood U of a in Ω, is proper. Since $A_j \cap U$ is a closed subset in $\mathcal{Q} \cap U$, the map $\pi\colon A_j \cap U \to U'$ is also proper, for each j. Since $a = \lim a_j$ for some sequence of points $a_j \in A_j$, we may assume that all $A_j \cap U$ are nonempty. But then, by Theorem 3.7, every $\pi|_{A_j \cap U}$ is an analytic cover. If k_j is the number of sheets of this cover, Wirtinger's theorem gives $k_j \mathcal{H}_{2p}(U') \leqslant \mathcal{H}_{2p}(A_j \cap U) \leqslant M$, i.e. all $k_j \leqslant k$ for some $k < \infty$. By passing to a subsequence we may assume that all $k_j = k$.

Let Φ_I^j, $|I| = k$, be canonical defining functions of the cover $\pi|_{A_j \cap U}$. Since they are uniformly bounded in U (see the definition in p.4.3), by again passing to a subsequence we may assume that on compact subsets in $U' \times \mathbb{C}^{n-p}_{z''}$ the Φ_I^j

converge uniformly to holomorphic functions Φ_I, $|I| = k$, respectively, which are polynomials of degree k in z''. If $a' \in U'$ is fixed, $\alpha_j^1(a'),...,\alpha_j^k(a')$ are the z''-coordinates of the points in the fiber $\pi^{-1}(a') \cap A_j \cap U$ (written with multiplicities), and $\alpha'(a'), \ldots, \alpha^k(a')$ is some limit system of these k-tuples as $j \to \infty$, then $\Phi_I(a',z'')$, $|I| = k$, clearly is the system of canonical defining functions for the set $\alpha^1(a'), \ldots, \alpha^k(\alpha')$ (see p.4.3). Thus, for each fixed $z' \in U'$ the $\Phi_I(z',z'')$, $|I| = k$, form the system of canonical defining functions for some tuple of k (not necessarily distinct) points in U''. By construction and Rouché's theorem (p.10.3), the analytic set $\{z \in U, \Phi_I(z',z'') = 0, |I| = k\}$ coincides with the limit set of the family $A_j \cap U$ in U, i.e. coincides with $A \cap U$. Since $\pi|_{A \cap U}$ is an analytic cover, the analytic set $A \cap U$ is pure p-dimensional. ■

C o r o l l a r y. *Let $\{A_j\}$ be a sequence of pure p-dimensional analytic subsets of a complex manifold Ω with locally uniformly bounded volumes:*

$$\mathcal{H}_{2p}(A_j \cap K) \leqslant M_K < \infty, \quad K \subset\subset \Omega.$$

Then we can extract a subsequence from $\{A_j\}$ converging in Ω to a pure p-dimensional analytic subset or to the empty set.

■ Representing Ω as a countable union of compact sets, we see that it suffices to investigate the case when Ω is a bounded domain in $\mathbf{C}^n$ and all $\mathcal{H}_{2p}(A_j) \leqslant M < \infty$.

Denote by μ_j the measure in Ω equal to $\mathcal{H}_{2p}|_{A_j}$ (i.e. $\mu_j(E) = \mathcal{H}_{2p}(E \cap A_j)$ for all Borel subsets $E \subset \overline{\Omega}$). This is a positive bounded measure on compact sets in $\overline{\Omega}$, i.e. an element of the space $C(\overline{\Omega})^*$ dual to $C(\overline{\Omega})$. Since in this Banach space balls are weak-$\star$ compact (see, e.g., [64]), we can extract a subsequence from $\{\mu_j\}$ weak-$\star$ converging to some measure μ (i.e. $\int f d\mu_j \to \int f d\mu$, $f \in C(\overline{\Omega})$). We will assume that $\mu_j \to \mu$ in this sense. Then μ is also a positive bounded measure (or the null measure). Let $A = (\operatorname{supp} \mu) \cap \Omega$. Since $(\operatorname{supp} \mu_j) \cap \Omega = A_j$, the set A belongs to the limit set in Ω of the family $\{A_j\}$. On the other hand, if $a = \lim_{j_\nu \to \infty} a_{j_\nu} \in \Omega$, where $a_{j_\nu} \in A_{j_\nu}$, Theorem 15.3 implies that for any neighborhood $U \ni a$ there is a constant $c > 0$ such that $\mu_{j_\nu}(U) \geqslant c$ from some ν_0 onwards. Hence $\mu(U) \geqslant c$, and $a \in A$. Thus, we have proved that A coincides with the limit set in Ω of the family $\{A_j\}$.

Furthermore, let K be an arbitrary compact set in Ω. For any point

$a \in A \cap K$ and any neighborhood $U \ni a$ we have $\mu(U) > 0$, hence $\mu_j(U) > 0$ for all $j > j_0$. Extract from the covering of $A \cap K$ by ϵ-neighborhoods of its points a finite subcovering, and find an index j_1 such that A_j intersects all these extracted neighborhoods if $j > j_1$. Then $A \cap K$ lies in the 2ϵ-neighborhood of A_j, and thus we find that $A_j \to A$ in Ω. Bishop's theorem implies that A is a pure p-dimensional analytic subset in Ω. ∎

16. Holomorphic chains

16.1. Sequences of holomorphic chains. By p.14.1, each holomorphic p-chain $T = \sum k_j A_j$ on a complex manifold Ω determines a current $\sum k_j [A_j]$ on it; in the sequel we will identify this current with the chain T. This current is of measure type, is d-closed and has bidegree (p,p) (cf. p.14.2). Thus, holomorphic p-chains can be thought of as belonging to the space $\mathcal{D}'_{2p}(\Omega)$ of currents of dimension $2p$, dual to $\mathcal{D}^{2p}(\Omega)$ and endowed with the topology of weak-$\star$ convergence (see A5.1). Convergence in the sense of currents is much weaker than the convergence introduced in p.12.2. It has the obvious advantage of being compatible with the arithmetical operations; however, it is easy to give examples of sequences of p-chains T_j converging to 0 in the sense of currents while the limit sets for the supports T_j coincide with, say, all of Ω (in the sense of p.12.2 such sequences do not have a limit). Another manner of getting rid of such inconveniencies is related to restrictions on the masses of currents, as in Bishop's theorem (p.15.5). To each holomorphic p-chain $T = \sum k_j A_j$ corresponds a measure $\sum k_j \mathcal{H}_{2p}|_{A_j}$. The mass of T on a compact set $K \subset \Omega$ is the nonnegative number $M_K T = \sum |k_j| \mathcal{H}_{2p}(K \cap A_j) < \infty$. It is easy to derive from 15.5 the following compactness property of families of holomorphic chains.

Proposition 1. *(1) Let $\{T_j\}$ be a sequence of holomorphic p-chains on a complex manifold Ω with locally uniformly bounded masses ($M_K T_j \leqslant M_K < \infty$, $K \subset\subset \Omega$) which converges to a certain current T in $\mathcal{D}'_{2p}(\Omega)$. Then T is a holomorphic p-chain also.*

(2) Every sequence of holomorphic p-chains in Ω with locally uniformly bounded masses contains a subsequence converging in the sense of currents to some holomorphic p-chain.

If, moreover, T_j are positive chains, then the limit chain T is also positive, and $|T_j| \to |T|$ in the sense of p.15.5.

■ The local uniform boundedness of the masses of the T_j and Theorem 15.3 clearly imply that the multiplicities $k_{j\nu}$ of the chains T_j and the volumes of their supports $|T_j|$ are also uniformly bounded on compact sets in Ω. By p.15.5, we can extract from $\{|T_j|\}$ a subsequence converging to an analytic subset $A \subset \Omega$. The support of the holomorphic limit chain T is the union of a certain family of irreducible components of A. The multiplicities of T at the points of A can be determined from the convergence $T_j \to T$: if $a \in \operatorname{reg} A$ and U_r is the r-neighborhood of a (everything may be considered in $\mathbf{C}^n$), then the multiplicity of T in a neighborhood of a is equal to $\lim_{r\to 0}\lim_{j\to\infty}\sum_\nu k_{j\nu}\, \mathcal{H}_{2p}(A_{j\nu} \cap U_r)/c(p)r^{2p}$. It easily follows from p.15.1 that this function is integer valued, and p.15.5 implies that it is locally constant on $\operatorname{reg} A$. The p-chain $\sum k_\nu A_\nu$ thus defined is, by construction, the limit of $\{T_j\}$ in the sense of currents.

The existence of a convergent subsequence clearly follows from what has been proved and from the local uniform boundedness of the multiplicities. The last statement (concerning positive chains) can be derived from Theorem 15.3, just as in p.15.5. ■

As an example we consider the tangent cone $C(T,0)$ to a holomorphic chain $T = \sum k_j A_j$ in a domain $D \ni 0$ in $\mathbf{C}^n$. In p.11.6 it was defined as the holomorphic chain corresponding to the formal sum $\sum k_j C([A_j],0)$, where $C([A_j],0)$ is the tangent cone at 0 in the sense of holomorphic chains to the analytic set A_j. From the point of view of currents the formal sum and its corresponding holomorphic chain are one and the same, since formal sums of the form $kA - kA$ correspond to the current equal to zero. Hence $C(T,0)$ corresponds to the current equal to $\sum k_j C([A_j],0)$. The construction of $C([A_j],0)$ in p.11.6 clearly implies that $C([A_j],0)$ is the limit of the currents $[A_j/r]$ as $r \to 0$, hence we have

P r o p o s i t i o n 2. *Let $T = \sum k_j[A_j]$ be a holomorphic p-chain in a neighborhood of 0 in $\mathbf{C}^n$ and put $(1/r)_* T = \sum k_j [A_j/r]$. Then, as $r \to 0$, the currents $(1/r)_* T$ converge to the chain-current $C(T,0) = \sum k_j C([A_j],0)$. If T is a positive current, the support of $C(T,0)$ coincides with $\cup C(A_j,0)$, while in the general case it is a subset of $\cup C(A_j,0)$.*

16.2. Intersection chains as currents. In view of the additivity of the definition of intersection chain, we will investigate intersections of analytic sets, which in the final end are defined by the intersection of a single analytic set and a plane of complementary dimension (see p.12.4). Thus, let first A be a pure p-dimensional analytic set in a neighborhood of 0 in $\mathbf{C}^n$ and let L be a plane of dimension $n-p$ such that $A \cap L = \{0\}$. In a neighborhood of 0 the chain $A \wedge L$ was defined in p.12.1 as $i_0(A,L)\cdot[0]$, where $i_0(A,L) = \mu_0(\pi_L|_A)$ is the multiplicity at 0 of the projection $\pi_L|_A$, i.e. the multiplicity of the analytic cover $\pi_L|_{A \cap U}$, where U is a sufficiently small neighborhood of 0. The value of this 0-chain (as a current) on an arbitrary continuous function ϕ supported in U is (according to the definition of the current [0]) equal to $i_0(A,L)\cdot\phi(0)$. Without loss of generality we may assume that $L = \mathbf{C}^{n-p}_{z''}$ in the sequel.

Let A_ϵ denote an ϵ-regularization of the current $[A]$ in a neighborhood of 0 (ϵ sufficiently small), i.e. the coefficientwise convolution of the generalized differential form $[A]$ with a function $\lambda_\epsilon \geqslant 0$ of class C^∞, supported in the ϵ-neighborhood of 0 and such that $\int\lambda_\epsilon dV = 1$ (see A5.3). Then A_ϵ is a C^∞ form of bidegree $(n-p,n-p)$, and we can define a current $[L]\wedge A_\epsilon$ of dimension 0 (of degree $2n$), acting on continuous functions ϕ supported in a neighborhood of 0 by the rule $<[L]\wedge A_\epsilon,\phi> = \int_L \phi A_\epsilon$. We show that $[L]\wedge A_\epsilon \to L\wedge A$ *in the sense of currents, i.e.* $\int_L \phi A_\epsilon \to i_0(A,L)\phi(0)$ *as* $\epsilon \to 0$ *for any such* ϕ.

■ As a generalized form the current $[A]$ can be written as $\mu dV(z'') + A'$, where A' contains certain dz_j or $d\bar{z}_j$ with $j \leqslant p$ and the measure μ is defined by $\int\psi d\mu = \int_A \psi dV(z')$ for any $\psi \in \mathfrak{D}(U)$ (here $z' \in \mathbf{C}_{1\ldots p}$ and

$$dV(z') = \prod_1^p dx_j \wedge dy_j;$$

for the representation of currents as generalized forms see A5.1 or, in more detail, [181]). The regularization A_ϵ was formed coefficientwise, i.e. it is equal to $\mu_\epsilon dV(z'') + (A')_\epsilon$, hence $\int_L \phi A_\epsilon = \int_L \phi(\mu)_\epsilon dV(z'')$, since $(A')_\epsilon|_L = 0$. Substituting in it the expression for $(\mu)_\epsilon$ (see A5.3), we find that

$$\int_L \phi A_\epsilon = \int_L \phi(0',z'') \left[\int_A \lambda_\epsilon(\zeta',\zeta''-z'')\, dV(\zeta') \right] dV(z'') =$$

$$= \int_A \left[\int_L \phi(0',z'')\lambda_\epsilon(\zeta',\zeta''-z'')\, dV(z'') \right] dV(\zeta') =$$

$$= \int_A \left[\int_L \phi(0', \zeta'' - z'') \lambda_\epsilon(\zeta', z'')\, dV(z'') \right] dV(\zeta').$$

For each fixed $\zeta' \in U'$ the fiber $\pi^{-1}(\zeta') \cap A \cap U$ counted with multiplicities consists of $k = i_0(A,L)$ points, whose z'' coordinates are equal to $\alpha^1(\zeta'), \ldots, \alpha^k(\zeta')$, respectively (the order is arbitrary). For any continuous function $\psi(\zeta', \zeta'')$ on A,

$$\int_A \psi(\zeta', \zeta'')\, dV(\zeta') = \int_{U'} \sum_j \psi(\zeta', \alpha^j(\zeta'))\, dV(\zeta'),$$

hence

$$\int_L \phi A_\epsilon = \int_{U'} \int_L \sum_j \phi(0', \alpha^j(\zeta') + z'') \lambda_\epsilon(\zeta', z'')\, dV(z'')\, dV(\zeta') = \int_U \phi_* \lambda_\epsilon dV,$$

where $\phi_*(\zeta', Z'')$ is a continuous function in a neighborhood of 0; moreover, $\phi_*(0) = k\phi(0)$. Since the λ_ϵ form a δ-sequence, we have

$$\lim_{\epsilon \to 0} \int_L \phi A_\epsilon = k\phi(0),$$

which was to be proved. ■

Now we investigate the case when $\dim L = k > n - p$ and the intersection of A and L is proper, i.e. $\dim A \cap L = p + k - n$. We may assume that $A \cap L$ is an irreducible analytic subset in a domain D. Then in D the current $A \wedge L$ is equal to $i(A,L) \cdot [A \cap L]$. Let, as above, A_ϵ be an ϵ-regularization of $[A]$. We show that $A \wedge L = \lim_{\epsilon \to 0} [L] \wedge A_\epsilon$ *in* D, *i.e.* $\int_L A_\epsilon \wedge \psi = i(A,L) \int_{A \cap L} \psi$ *for any* $\psi \in \mathcal{D}^{l,l}(D)$, $l = p + k - n$.

■ The statement being local we may assume that ψ is supported in an arbitrarily small neighborhood of $0 \in A \cap L$. By a rotation L is made to coincide with the coordinate plane of the variables $z_{n-k+1}, \ldots, z_n$. Since all integrations are over sets in L only, we may assume that ψ does not contain differentials of the variable $'z = (z_1, \ldots, z_{n-k})$. Let $\{\alpha_j\}$ be a basis of principal positive forms of bidegree (l,l) in L (see A4.5) and let L_j be the subspace in L corresponding to α_j (i.e. $\alpha_j = \pi_j^*(dV_{L_j})$, where dV_{L_j} is the Euclidean volume form on L_j and $\pi_j \colon L \to L_j$ is orthogonal projection). After a suitable unitary transformation of L we may assume that $A \cap L$ is transversal at 0 to all L_j. Decomposing ψ in the basis $\{\alpha_j\}$ we find that it suffices to prove our statement for forms $\phi\alpha_j$, where

$\phi \in \mathcal{D}(U)$ is a function supported in a small neighborhood of 0. Fix j. By an additional unitary transformation of L the plane L_j becomes the $z' = (z_{n-k+1}, \ldots, z_p)$-plane. Then we are in the following situation: $U = {}'U \times U' \times U''$ is a small neighborhood of 0, the projections $A \cap U \to {}'U \times U'$ and $A \cap L \cap U \to U'$ are analytic covers, and we have to prove that

$$\int_L A_\epsilon \wedge \phi \, dV(z') \to i(A,L) \int_{A \cap L} \phi \, dV(z') \quad \text{as } \epsilon \to 0.$$

As above we put $[A] = \mu \, dV(z'') + \cdots$, where $z'' = (z_{p+1}, \ldots, z_n)$ and $\int f d\mu = \int_A f(\zeta) \, dV('\zeta, \zeta')$ for $f \in \mathcal{D}(U)$. Then $A_\epsilon = (\mu)_\epsilon dV(z'') + \cdots$ and, similar to the above,

$$\int_L A_\epsilon \wedge \phi \, dV(z') = \int_L \phi(\mu)_\epsilon dV(z', z'') =$$

$$= \int_L \phi \left[\int_A \lambda_\epsilon(\zeta - z) \, dV('\zeta, \zeta') \right] dV(z', z'') =$$

$$= \int_{'U \times U'} \int_L \sum_1^k \phi('0, \zeta' - z', \alpha^j('\zeta, \zeta') - z'') \lambda_\epsilon('\zeta, z', z'') \, dV(z', z'') \, dV('\zeta, \zeta'),$$

where k is the number of sheets of the analytic cover $A \cap U \to {}'U \times U'$. By Fubini's theorem the last expression equals

$$\int_U \left[\int_{U'} \sum_1^k \phi('0, \zeta' - z', \alpha^j('\zeta, \zeta') - z'') \, dV(\zeta') \right] \lambda_\epsilon('\zeta, z', z'') \, dV('\zeta, z', z'').$$

And since λ_ϵ is a δ-sequence,

$$\lim_{\epsilon \to 0} \int_L A_\epsilon \wedge \phi \, dV(z') = \int_{U'} \sum_1^k \phi('0, \zeta', \alpha^j('0, \zeta')) \, dV(\zeta').$$

Let $\sigma' \subset U'$ be the critical analytic set of the cover $A \cap L \cap U \to U'$, and let k' be the number of sheets of this cover. If $z' \notin \sigma'$, the plane $\Lambda_{z'} = ('0, z', 0'') + \mathbb{C}^{n-p}_{z''}$ is transversal to $A \cap L \cap U$ at all points in common, and these points are regular on $A \cap L \cap U$. By Corollary 3, p.12.4, $i_a(A,L) = i_a(A, \Lambda_{z'})$ for each $a \in A \cap \Lambda_{z'} \cap U$. Since all these points are regular on $A \cap L \cap U$, we have $i_a(A, \Lambda_{z'}) = i(A,L)$ (p.12.3). Thus, at all points a of $A \cap \Lambda_{z'} \cap U$ the multiplicities of the projection $A \cap U \to {}'U \times U'$ (which are equal to $i_a(A, \Lambda_{z'})$) are one and the same, equal to $i(A,L)$. The number of these points is k', since $z' \notin \sigma'$, and the total multiplicity of the cover $A \cap U \to 'U \times U'$ is k. Thus we obtain $k = k' \cdot i(A,L)$, and $\sum_1^k \phi('0,z',\alpha^j('0,z')) = i(A,L)\sum_1^{k'} \phi('0,z',\beta^i('0,z'))$, where $\beta^1('0,z'), \ldots, \beta^{k'}('0,z')$ are the z''-coordinates of the points in the fiber of $A \cap L \cap U \to U'$ above z' (counted with the multiplicities of this cover, if $z' \in \sigma'$). Substituting this in the expression obtained above we find

$$\lim_{\epsilon \to 0} \int_L A_\epsilon \wedge \phi \, dV(z') = i(A,L) \int_{U'} \sum_1^{k'} \phi('0,z',\beta^i('0,z'))\, dV(z') =$$

$$= i(A,L) \int_{A \cap L} \phi \, dV(z'). \quad \blacksquare$$

Finally we investigate the general case of proper intersection of several analytic sets. In view of the associativity and the commutativity of intersection of chains and of the exterior product of differential forms of even degrees, it suffices to investigate the intersection of two sets. Thus, let in a domain $D \subset \mathbf{C}^n$ an intersection chain $A_1 \wedge A_2$ be defined; by Proposition 2, p.12.4, it is the projection in $\mathbf{C}^n$ of the chain $(A_1 \times A_2) \cap \Delta$, defined in $D \times D \subset \mathbf{C}^{2n}$. If $A_{j\epsilon_j}$ are regularizations of the currents $[A_j]$, then $A_{1\epsilon_1}(z^1) \wedge A_{2\epsilon_2}(z^2)$ is a regularization of $[A_1 \times A_2]$ (with kernel $\lambda_{\epsilon_1}(z^1) \cdot \lambda_{\epsilon_2}(z^2)$). According to what has already been proved, $[\Delta] \wedge A_{1\epsilon_1} \wedge A_{2\epsilon_2} \to [A_1 \times A_2] \wedge \Delta$ in $D \times D$ in the sense of currents if $\epsilon_j > 0$ simultaneously tend to 0. This convergence is uniform in $\epsilon = (\epsilon_1, \epsilon_2)$. Indeed, let a form ϕ be supported and continuous in a small neighborhood of some point of $[A_1 \times A_2] \wedge \Delta$, and let $\omega(\delta)$ be the common modulus of continuity of the coefficients of ϕ. The proof of the preceding statement implies that the volume of the current $[\Delta] \wedge A_{1\epsilon_1} \wedge A_{2\epsilon_2} - (A_1 \times A_2) \cap \Delta$ on ϕ does not exceed $C\omega(|\epsilon|)$ in absolute value, where $|\epsilon|^2 = \epsilon_1^2 + \epsilon_2^2$ and the constant C only depends on geometric properties of $A_1 \times A_2$ in a neighborhood of the support of ϕ.

We find the limit of $[\Delta] \wedge A_{1\epsilon_1} \wedge A_{2\epsilon_2}$ as $\epsilon_1 \to 0$. Since $A_{2\epsilon_2}$ is a smooth form and since $A_{1\epsilon_1} \to [A_1 \times D]$ in $D \times D$ in the sense of currents, according to what has already been proved,

$$\lim_{\epsilon_1 \to 0} [\Delta] \wedge A_{1\epsilon_1} \wedge A_{2\epsilon_2} = \left[\lim_{\epsilon_1 \to 0} [\Delta] \wedge A_{1\epsilon_1}\right] \wedge A_{2\epsilon_2} = ([A_1 \times D] \wedge \Delta) \wedge A_{2\epsilon_2}.$$

In view of the uniform convergence with respect to ϵ, the value of this limit current on ϕ deviates from the value of $([A_1 \times A_2] \wedge \Delta$ on ϕ by at most $C\omega(\epsilon_2)$, which tends to zero as $\epsilon_2 \to 0$.

Since the projection $\Delta \to \mathbb{C}^n$ is biholomorphic and since, clearly, the image of $([A_1 \times D] \cap \Delta) \wedge A_{2\epsilon_2}$ is $[A_1] \wedge A_{2\epsilon_2}$, we have that $[A_1] \wedge A_{2\epsilon_2} \to A_1 \wedge A_2$ in D in the sense of currents. Thus, if we agree to put $A_{j\epsilon_j} = [A_j]$ as $\epsilon_j = 0$, we obtain $A_{1\epsilon_1} \wedge A_{2\epsilon_2} \to A_1 \wedge A_2$ as $\epsilon = (\epsilon_1, \epsilon_2) \to 0$ in arbitrary manner, $\epsilon_1 \geqslant 0$, $\epsilon_2 \geqslant 0$.

In the light of what has been proved it becomes understandable why we have used the symbol $\wedge$ of exterior differentiation in order to denote intersections of holomorphic chains. Now it becomes clear that if we regard chains as generalized differential forms (currents), then their intersection is this exterior product of generalized forms, which is naturally defined as the limit of the exterior products of corresponding regularizations. Therefore, everywhere in the sequel the sign $\wedge$ will be understood as this exterior product of (generalized) forms, if it is correctly defined (i.e. if the stated limit of the regularizations exists).

Taking into account the properties of associativity and commutativity, the above proved may be summarized as follows.

T h e o r e m. *Let $T_1, \ldots, T_k$ be holomorphic chains of pure dimensions in a domain $D \subset \mathbb{C}^n$, properly intersecting in D, and let $T_{j\epsilon_j}$ be regularizations of them. Then*

$$T_{1\epsilon_1} \wedge \cdots \wedge T_{l\epsilon_l} \wedge T_{l+1} \wedge \cdots \wedge T_k \to T_1 \wedge \cdots \wedge T_k, \quad l = 1, \ldots, k,$$

in D in the sense of currents as $\epsilon = (\epsilon_1, \ldots, \epsilon_k) \to 0$ in arbitrary manner; this convergence is moreover uniform on any equicontinuous family of uniformly compactly supported forms.

The currents at the lefthand side are defined not in all of D, but as $\epsilon \to 0$ their domain of definition, which grows, fills D: the chains $T_{l+1} \wedge \cdots \wedge T_k$ are defined in a neighborhood of the set $\cap_1^k |T_j|$ (see the beginning of p.12.3), hence for any subdomain $G \subset\subset D$, for sufficiently small ϵ the restriction of $T_{1\epsilon_1} \wedge \cdots \wedge T_{l\epsilon_l} \wedge T_{l+1} \wedge \cdots \wedge T_k$ to G is correctly defined.

16.3. Formulas of Poincaré-Lelong. The divisor of a meromorphic function f on a complex manifold Ω is locally defined as follows. In a neighborhood of an arbitrary point $a \in \Omega$ the function f has a representation $f = h/g$ with h, g holomorphic functions. If $h = h_1^{n_1} \cdots h_k^{n_k}$ and $g = g_1^{p_1} \cdots g_l^{p_l}$ are factorizations into factors irreducible at a, then in a neighborhood of a the divisor of f is the holomorphic $(n-1)$-chain $\sum_i n_i[Z_{h_i}] - \sum_j p_j[Z_{g_j}]$. These local holomorphic chains are combined to a unique $(n-1)$-chain in Ω, called also the *divisor* D_f *of the meromorphic function* f (see p.1.5). From such a local definition the globally defined function f itself cannot be visualized. In order to express D_f directly in terms of f we first prove the following statement.

L e m m a. *Let f be a meromorphic function on a complex manifold Ω, and let A be a pure p-dimensional analytic subset in Ω such that* $\dim A \cap |D_f| < p$. *Then* $\log |f|$ *is locally absolutely integrable on A, i.e.*

$$\int_K |\log|f|| \, d\mathcal{H}_{2p} < \infty$$

for every compact set $K \subset A$.

■ The statement being local we may assume that Ω is a domain in $\mathbb{C}^n$ and f a holomorphic function in Ω. We first investigate the case $A = \Omega$.

Let $a \in \Omega$ be an arbitrary point and $U \subset\subset \Omega$ a neighborhood of a. By choosing suitable coordinates and by shrinking U, we may assume that f has, in $U = U' \times U_n$, a representation by the Weierstrass preparation theorem: $f = h \cdot P$, where h is zero free in U and P is a polynomial in z_n with leading coefficient 1. For each fixed $z_0' \in U'$ this can be written as

$$f(z_0', z_n) = h(z_0', z_n) \prod_1^k (z_n - z_{nj}(z_0')),$$

where all $z_{nj}(z_0')$, $j = 1, \ldots, k$, are uniformly bounded. Since $\log|\zeta - a|$ is absolutely integrable in disks in $\mathbb{C}_\zeta$, the function $\log|f(z_0', z_n)|$ is absolutely integrable in U_n and its integral is uniformly bounded with respect to $z_0' \in U'$. By Fubini's theorem, $\log|f|$ is hence integrable in U.

In the general case we may also assume that f is holomorphic and that in suitable neighborhoods of $a \in \Omega \subset \mathbb{C}^n$ the set A can be properly projected to all p-dimensional coordinate planes. Since, by Wirtinger's theorem,

$$dV_A = \frac{1}{p!} \omega^p |_A = \sum_{\#I = p}' dV(z_I)|_A,$$

it suffices to elucidate the convergence of the integral

$$\int_{A \cap U} \log|f|\,dV(z'),\quad z' = (z_1, \ldots, z_p),$$

under the assumption that the projection $\pi\colon A \cap U \to U' \subset \mathbf{C}^p$ is a k-sheeted analytic cover in a neighborhood $U \ni a$. For each fixed $z' \in U'$ we denote by $\alpha^1(z'), \ldots, \alpha^k(z')$ the z''-coordinates of the points in the fiber of $\pi|_{A \cap U}$ above z' counted with multiplicities and in arbitrary order. Then

$$\int_{A \cap U} \log|f|\,dV(z') = \int_{U'} \log\left|\prod_1^k f(z', \alpha^j(z'))\right| dV(z').$$

The function $g(z') = \prod_1^k f(z', \alpha^j(z'))$, being holomorphic and uniformly bounded outside the critical analytic set $\sigma \subset U'$, can be holomorphically continued onto all of U'. According to what was proved above,

$$\int_{U'} |\log|g||\,dV(z') < \infty. \ \blacksquare$$

Thus, the function $\log|f|$ defines on Ω a current of degree 0, and hence it can be differentiated in the sense of currents.

Theorem 1 (Poincaré-Lelong formula). *Let f be a meromorphic function on a complex manifold Ω. Then*

$$D_f = \frac{i}{\pi}\partial\bar{\partial}\log|f|$$

on Ω in the sense of currents.

■ The statement being local we may assume that $\Omega = U$ is a neighborhood of 0 in $\mathbf{C}^n$ and that f is holomorphic in U. We first investigate the case $n = 1$, when $f(\zeta) = e^{h(\zeta)}\prod_1^k(\zeta - a_j)$, where $a_j \in U$ and $h \in \mathcal{O}(U)$. Since $\partial\bar{\partial}\mathrm{Re}\, h = 0$ it suffices to consider the case $f(\zeta) = \zeta - a$. The function $1/(\zeta - a)$ is integrable in U, hence

$$\partial\bar{\partial}\log|\zeta - a| = -\frac{1}{2}d\partial\log|\zeta - a|^2 = -\frac{1}{2}d\left(\frac{d\zeta}{\zeta - a}\right),$$

i.e.

$$\left\langle \frac{i}{\pi}\partial\bar{\partial}\log|\zeta - a|, \phi \right\rangle = \frac{i}{2\pi}\int_U d\phi \wedge \frac{d\zeta}{\zeta - a},\quad \phi \in \mathfrak{D}(U).$$

The righthand integral is the limit of the integrals over annuli $U_\epsilon = \{\zeta \in U\colon |\zeta - a| > \epsilon\}$ as $\epsilon \to 0$, while the integral over an annulus is equal to

$$\frac{i}{2\pi}\int_{U_\epsilon} d\phi \wedge \frac{d\zeta}{\zeta - a} = \frac{-1}{2\pi i}\int_{U_\epsilon} d\left(\frac{\phi\, d\zeta}{\zeta - a}\right) = \frac{-1}{2\pi i}\int_{\partial U_\epsilon} \frac{\phi\, d\zeta}{\zeta - a} =$$

$$= \frac{1}{2\pi i}\int_{|\zeta - a| = \epsilon} \frac{\phi(\zeta) - \phi(a)}{\zeta - a}\, d\zeta + \phi(a).$$

Since ϕ is differentiable, the function $(\phi(\zeta) - \phi(a))/(\zeta - a)$ is uniformly bounded in U, hence the first integral on the right tends to zero as $\epsilon \to 0$. Thus

$$< \frac{i}{\pi}\partial\bar\partial \log|\zeta - a|, \phi > = \phi(a).$$

Thus, if $f = e^h \cdot \prod_1^k (\zeta - a_j)$, then $(i/\pi)\partial\bar\partial\log|f|$ is the sum of the δ-functions at the points a_j, written with multiplicities, i.e. the divisor of f.

In $\mathbf{C}^n$ we first prove that the currents under investigation coincide on forms $\psi = \phi(z)\, dV(z')$, where $z = (z', z_n)$ and $\phi \in \mathfrak{D}(U)$, assuming thereby that f satisfies in $U = U' \times U_n$ the conditions of the Weierstrass preparation theorem. By definition,

$$< \frac{i}{\pi}\partial\bar\partial \log|f|, \psi > = \frac{i}{\pi}\int_U \log|f|\, \partial\bar\partial\psi =$$

$$= \frac{i}{\pi}\int_U \log|f(z)| \cdot ((\partial\bar\partial)_n \phi(z)) \wedge dV(z'),$$

where the subscript n indicates that differentiation is with respect to z_n, $\bar z_n$ only. For each fixed $z' \in U'$, the function f has in U_n identical number of zeros $z_{nj}(z')$, written with multiplicities, hence by the case $n = 1$ and Fubini's theorem.

$$< \frac{i}{\pi}\partial\bar\partial \log|f|, \psi > = \int_{U'} \sum_j \phi(z', z_{nj}(z'))\, dV(z').$$

The definition of multiplicity of the divisor of f implies that the righthand integral is just equal to $< D_f, \psi >$.

The general case is obtained from the above by using expansion with respect to a positive basis. Let $\{\alpha_j\}$ be a basis of positive forms of bidegree $(n-1, n-1)$ in $\mathbf{C}^n$ (see A4.5) and let L_j be the complex hyperplanes corresponding to the forms

α_j (i.e. $\alpha_j = \pi_j^*(dV_{L_j})$, where π_j: $\mathbf{C}^n \to L_j$ is orthogonal projection). By p.3.4 we may assume that for each j the projection π_j: $Z_f \cap U_j \to U_j' \subset L_j$ is proper in some neighborhood $U_j \ni 0$. Since α_j is the volume form on L_j, we have proved above that

$$<\frac{i}{\pi}\partial\bar{\partial}\log|f|, \phi\alpha_j> = <D_f, \phi\alpha_j>$$

for any function $\phi \in \mathcal{D}(U)$, with U a sufficiently small neighborhood of 0. By A4.5, any form $\psi \in \mathcal{D}^{n-1,n-1}(U)$ has a representation $\psi = \sum\phi_j\alpha_j$, where $\phi_j \in \mathcal{D}(U)$, hence the Poincaré-Lelong formula has been proved for all forms of bidegree $(n-1, n-1)$. This, however, is sufficient, since both D_f and $(i/\pi)\partial\bar{\partial}\log|f|$ are of bidimension $(n-1, n-1)$. ■

According to the Lemma, under natural restrictions on a meromorphic function f on a pure p-dimensional analytic set A the current $[A]\cdot\log|f|$ is defined, acting according to the formula

$$<[A]\cdot\log|f|, \phi> = \int_A (\log|f|)\cdot\phi.$$

Under the same conditions the intersection chain $A \wedge D_f$ is defined. The natural question of the relation between these currents leads to the following generalized Poincaré-Lelong formula (cf. [42]).

T h e o r e m 2. *Let f be a meromorphic function on a complex manifold Ω, and let A be a pure p-dimensional analytic subset of Ω such that* $\dim A \cap |D_f| < p$. *Then*

$$\frac{i}{\pi}\partial\bar{\partial}([A]\cdot\log|f|) = A \wedge D_f$$

on Ω in the sense of currents.

■ It suffices to investigate the case when Ω is a domain in $\mathbf{C}^n$ and f is a holomorphic function. Let $u_\epsilon = (\log|f|)_\epsilon$ be an ϵ-regularization of $\log|f|$. Then $\partial\bar{\partial}([A]u_\epsilon) = [A]\wedge\partial\bar{\partial}u_\epsilon$, since Green's formula (p.14.3) implies $\int_A u_\epsilon\partial\bar{\partial}\phi = \int_A(\partial\bar{\partial}u_\epsilon)\wedge\phi$. Since regularization commutes with the differentiation operators (A5.3), $(i/\pi)\partial\bar{\partial}u_\epsilon$ is an ϵ-regularization of the current $(i/\pi)\partial\bar{\partial}\log|f| = D_f$. By Theorem 16.2, $[A]\wedge D := \lim_{\epsilon\to 0}[A]\wedge(D_f)_\epsilon = A \wedge D_f$ is the current of integration over the chain $A \wedge D_f$. ■

By additivity we obtain

C o r o l l a r y. *Let f be a meromorphic function and T a holomorphic p-chain on a complex manifold Ω such that* $\dim |T| \cap |D_f| < p$. *Then* $(i/\pi)\partial\bar{\partial}(T\cdot\log|f|) = T\wedge D_f$ *on Ω in the sense of currents.*

In the case of several functions there are also Poincaré-Lelong type formulas. We restrict ourselves to the formulation of the corresponding statement for holomorphic functions; regrettably, lack of space does not allow us to give the rather tedious proof here (for an outline of the proof see [42]).

T h e o r e m 3. *Let A be a pure p-dimensional analytic subset of an n-dimensional complex manifold Ω, let $f_1,\ldots,f_k$ be holomorphic functions on Ω, and put* $|f|^2 = |f_1|^2 + \cdots + |f_k|^2$. *Suppose that* $\dim \cap_1^k Z_{f_j} = n-k$ *and* $\dim A \cap (\cap_1^k Z_{f_j}) = p-k$. *Then the coefficients of the form* $\log|f|\cdot(\partial\bar{\partial}\log|f|)^{k-1}$ *are locally absolutely integrable in Ω and on A, and the following equalities between currents in Ω holds:*

$$Z_f := D_{f_1}\wedge\cdots\wedge D_{f_k} = \frac{1}{k!}\frac{i}{\pi}\partial\bar{\partial}\left[\log|f|\cdot\left[\frac{i}{\pi}\partial\bar{\partial}\log|f|\right]^{k-1}\right],$$

$$A\wedge Z_f = \frac{1}{k!}\frac{i}{\pi}\partial\bar{\partial}\left[[A]\wedge\log|f|\cdot\left[\frac{i}{\pi}\partial\bar{\partial}\log|f|\right]^{k-1}\right].$$

(The first formula can be written in short as $Z_f = (1/k!)\{(i/\pi)\partial\bar{\partial}\log|f|\}^k$.)

16.4. Jensen formulas. Let A be a pure p-dimensional analytic subset of the ball $|z|<R$, let f be a holomorphic function in this ball, and suppose $\dim A \cap Z_f = p-1$. Then for $|z|<R$ the holomorphic $(p-1)$-chain $A\wedge D_f$ is defined, where D_f is the divisor of f. The Poincaré-Lelong equality and p.16.2 imply

$$A\wedge D_f = \lim_{\epsilon\to 0}[A]\wedge\frac{i}{\pi}\partial\bar{\partial}(\log|f|)_\epsilon,$$

since regularizaton commutes with differentiation operators. To the function $(\log|f|)_\epsilon$, which is plurisubharmonic and infinitely differentiable in the ball $|z|<R-\epsilon$, we may apply the integral representation of Proposition 15.2: if A_r,

$r<R-\epsilon$, is an analytic set with boundary, then

$$(\log|f|)_\epsilon(0)\pi^p\mu_0(A) = \frac{1}{2r^{2p-1}}\int_{bA_r}(\log|f|)_\epsilon d^c|z|\wedge\omega^{p-1} -$$

$$-\int_{A_r}(\log|f|)_\epsilon\omega_0^p - \frac{1}{2}\int_0^r\left[\int_{A}dd^c(\log|f|)_\epsilon\wedge\omega^{p-1}\right]\frac{dt}{t^{2p-1}}.$$

In this formula we may pass to the limit as $\epsilon\to 0$. If $f(0)\neq 0$, then $(\log|f|)_\epsilon(0)\to\log|f(0)|$ and $\int_A(\log|f|)_\epsilon\omega_0^p\to\int_{A_r}\log|f|\omega_0^p$ (by p.15.1 this integral converges if $f(0)\neq 0$). Further, as already has been proved, for $\epsilon\to 0$ the expression

$$\frac{1}{2}\int_{A_t}dd^c(\log|f|)_\epsilon\wedge\omega^{p-1} = \int_A(i\partial\bar{\partial}\log|f|)_\epsilon\wedge\chi_t\omega^{p-1}$$

tends to $<[A\wedge D_f], \pi\chi_t\omega^{p-1}>$, where χ_t is the characteristic function of the ball $|z|<t$. Up to the factor $\pi(p-1)!$ this is the volume of $(A\cap D_f)\cap\{|z|<t\}$, counted with the multiplicities of the intersection chain $A\wedge D_f$. Dividing this volume by the volume of the $(2p-2)$-dimensional ball of radius t, we obtain the normalized volume of the chain $A\wedge D_f$ in the ball $|z|<t$,

$$n(A\wedge D,t) = \frac{1}{(\pi t^2)^{p-1}}<A\wedge D_f,\chi_t\omega^{p-1}>,$$

which is also called the nonintegrated counting function of the chain $A\wedge D_f$. With these notations the limit of the double integral in the original formula is equal to $\pi^p\int_0^r n(A\wedge D_f,t)\,dt/t$. (If $f(0)\neq 0$ this limit exists since the forms $(\partial\bar{\partial}\log|f|)_\epsilon$ are positive because $\log|f|$ is plurisubharmonic and since the inner integrals are uniformly bounded by $2\pi<A\wedge D_f,\chi_r\omega^{p-1}>+o_\epsilon(1)$, where $o_\epsilon(1)\to 0$ as $\epsilon\to 0$.) Again by the plurisubharmonicity of $\log|f|$, the sequence $(\log|f|)_\epsilon$ converges to $\log|f|$ and is monotone decreasing, hence the existence of all limits indicated above also implies that the first integral in the initial formula tends to $\int_{bA_r}\log|f|d^c|z|\wedge\omega^{p-1}$ (this implies, in particular, that the form $\log|f|d^c|z|\wedge\omega^{p-1}$ is integrable on bA_r).

For all inferences made above it suffices that f be holomorphic in a neighborhood of A only. We introduce yet the notation

$$\sigma_p = \frac{1}{2\pi^p}d^c\log|z|\wedge\omega_0^{p-1},$$

and obtain as result the following "generalized Jensen formula".

T h e o r e m. *Let A be a pure p-dimensional analytic subset of the ball $|z|<R$, let f be a function holomorphic in a neighborhood of A, let* $\dim A\cap Z_f = p-1$, *and* $f(0)\neq 0$ *if* $0\in A$. *Then*

$$\int_0^r n(A\wedge D_f,t)\frac{dt}{t} = \int_{bA_r}\log|f|\cdot\sigma_p - \frac{1}{\pi^p}\int_{A_r}\log|f|\cdot\omega_0^p - \mu_0(A)\cdot\log|f(0)|.$$

The form σ_p, positive on bA_r, is in general smaller than the Euclidean volume form on bA_r, and coincides with it only if A is a cone (see p.15.2). For cones the second integral at the righthand side is omitted, since $\omega_0^p|_A = 0$ (see p.13.6), and we obtain

C o r o l l a r y 1. *If A is a cone with vertex at 0, and f is a function holomorphic in a neighborhood of A with $f(0)\neq 0$, then for $r<R$,*

$$\int_0^r n(A\wedge D_f,t)\frac{dt}{t} = \int_{bA_r}\log|f|\cdot\sigma - \log|f(0)|\cdot\mu_0(A).$$

In particular, for $p = n$ we obtain the classical Jensen formula for several complex variables:

C o r o l l a r y 2. *If f is a function meromorphic in the ball $|z|<R$ with* $0\notin|D_f|$, *then for $r<R$,*

$$\int_0^r n(D_f,t)\frac{dt}{t} = \int_{|z|=r}\log|f|\cdot\sigma - \log|f(0)|.$$

■ A function meromorphic in the ball can be represented as $f = h/g$, where h,g are holomorphic and $\dim|D_h|\cap|D_g|<n-1$ (see, e.g., [35]). Substituting h and g in the formula of Corollary 1 with $A = \{|z|<R\}$ and subtracting the equalities obtained we arrive at the same formula for meromorphic functions, since $n(D_f,t) = n(D_h,t) - n(D_g,t)$ by definition. ■

It is clear that the formulas from the Theorem and Corollary 1 are also valid for meromorphic functions if the condition $f(0)\neq 0$ is replaced by $0\notin|D_f|$.

For an arbitrary p-chain T in a neighborhood of a ball $|z| \leqslant r$ the quantity $n(T,r)$ is defined by

$$n(T,r) = \frac{1}{(\pi r^2)^p} < T, \chi_r \omega^p >.$$

Its logarithmic mean

$$N(T,r) = \int_{r_0}^{r} n(T,t)\frac{dt}{t},$$

where $r_0 \geqslant 0$ is a fixed constant (if $0 \notin |T|$ the value $r_0 = 0$ is commonly taken), is called the (integrated) *counting function* of the chain T. Thus, the left-hand side of Jensen's formula is the counting function (of the chain $A \wedge D_f$ in the Theorem and in Corollary 1, and of the divisor D_f in Corollary 2).

Jensen's formulas are commonly used to estimate the growth of the zero set of a holomorphic function f as $r \to R$ in terms of the growth of f itself (see p.17.1), and vice versa. Similarly we can estimate the density of the zero of f at a point, i.e. the Lelong number for D_f (see [61]):

P r o p o s i t i o n. *Let a function f be meromorphic in a neighborhood of $0 \in \mathbb{C}^n$. Then*

$$n(D_f, 0) = \lim_{r \to 0} \frac{1}{\log r} \int_{|z|=r} \log|f| \cdot \sigma,$$

where $\sigma = (1/2\pi^n) d^c \log|z| \wedge \omega_0^{n-1}$ is the normalized spherical volume form (i.e. $\int_{|z|=r}\sigma = 1$ for all $r > 0$).

■ Since both expressions are invariant under dilation and since f is the quotient of two holomorphic functions, we may assume that f is holomorphic in a neighborhood of the closed unit ball $\overline{B}_1$. Let $u = \log|f|$ and let $\epsilon > 0$ be small such that an ϵ-regularization u_ϵ is defined in a neighborhood of $\overline{B}_1$. We apply Green's formula (Proposition 2, p.14.3) to the domain $B_1 \setminus \overline{B}_r$, with u_ϵ and $\psi = \log|z| \cdot \omega_0^{n-1}$. Since $\log|z| = 0$ on $S_1 = \partial B_1$, and $\omega_0^n = 0$ for $z \neq 0$, this formula leads to

$$\int_{S_1} u_\epsilon d^c\psi - \int_{S_r} u_\epsilon d^c\psi + \log r \cdot \int_{S_r} d^c u_\epsilon \wedge \omega_0^{n-1} = -\int_{B_1 \setminus B_r} dd^c u_\epsilon \wedge \psi.$$

Since $\omega_0^{n-1}|_{S_r} = (\omega / r^2)^{n-1}|_{S_r}$, by Stokes' formula the third integral on the lefthand side equals

$$\frac{1}{r^{2n-2}} \int_{B_r} dd^c u_\epsilon \wedge \omega^{n-1}.$$

By Theorem 16.2 we may pass to the limit as $\epsilon \to 0$ in the equations obtained. Since by the Poincaré-Lelong formula $dd^c u = 2\pi\{(i/\pi)\partial\bar{\partial}u\} = 2\pi D_f$, as the result we obtain

$$\int_{S_1} ud^c\psi - \int_{S_r} ud^c\psi + 2\pi\frac{\log r}{r^{2n-2}} \int_{(D_f)_r} \omega^{n-1} = -2\pi \int_{(D_f)_{r,1}} \log|z|\cdot\omega_0^{n-1}$$

(the integrals over the chain D_f are defined by additivity, i.e. are taken with the multiplicities of D_f). Divide this relation by $\log r$ and pass to the limit as $r \to 0$. Since ω_0^{n-1} is a positive form, integrable over $(Z_f)_1$ (p.15.1), while the function $\chi_{r,1}\cdot(\log|z|)/\log r \leqslant 1$ tends to zero on $Z_f \setminus \{0\}$ as $r \to 0$, Lebesgue's dominated convergence theorem implies that the righthand side tends to zero. The first term on the lefthand side also gives zero, and as a result we obtain

$$-\lim_{r \to 0} \frac{1}{\log r} \int_{S_r} ud^c\psi + 2\pi^n \cdot n(D_f, 0) = 0,$$

which was exactly what we had to prove. ■

An analogous formula for the Lelong number holds for currents of the form $T = (i/\pi)\partial\bar{\partial}u$, where u is an arbitrary function that is plurisubharmonic in a neighborhood of 0 (see [61]). The proof is essentially the same, only formula (*) in p.15.1 (implying integrability of ω_0^{n-1} over Z_f) must be replaced by an analogous formula for T, ensuring integrability of the coefficients of ω_0^{n-1} with respect to the coefficient-measures of T.

C o r o l l a r y. *If f is meromorphic in a ball $|z| < R$, then for arbitrary $0 < r_0 < r < R$ we have the estimate*

$$\int_{|z|=r} \log|f|\cdot\sigma = \int_{r_0}^{r} n(D_f, t)\frac{dt}{t} + n(D_f, 0)\cdot\log r_0 + o(\log r_0) =$$

$$= \int_{r_0}^{r} (n(D_f, t) - n(D_f, 0))\frac{dt}{t} - n(D_f, 0)\cdot\log r + o(\log r_0).$$

■ Replace f by $f_a(z) = f(z+a)$ where $a \notin |D_f|$ is small so that f_a is meromorphic in a neighborhood of $|z| \leqslant r$. Apply the formula of Corollary 2 to f_a, and subtract from it the same formula with r_0 instead of r. As the result we obtain

$$\int_{r_0}^{r} n(D_f,t)\frac{dt}{t} = \int_{|z|=r} \log|f|\cdot\sigma - \int_{|z|=r_0} \log|f|\cdot\sigma.$$

According to the Proposition, the last integral equals $n(D_f,0)\cdot\log r_0 + o(\log r_0)$, hence we obtain the second equality. The first one obviously follows from

$$n(D_f,0)\cdot\log r_0 = n(D_f,0)\cdot\log r - \int_{r_0}^{r} n(D_f,0)\frac{dt}{t}. \quad ■$$

17. Growth estimates of analytic sets

17.1. Blaschke's condition. Let a function f be holomorphic in the ball $|z| < R$. Proposition 1, p.14.6, implies that in case $f(0) \neq 0$, the counting function $N(D_f,r)$ coincides, up to a constant, with the normal component of D_f in the ball $|z| < r$, i.e. with $<D_f,\chi_r d|z| \wedge d^c|z| \wedge \omega^{n-2}>$. Equality was proved for analytic sets (i.e. for currents $[A]$), but it can clearly be extended by additivity to arbitrary p-chains, in particular to divisors. If $f(0) = 0$, then upon replacing f by $f_a(z) = f(z+a)$, $f(a) \neq 0$, subtracting the equations in Proposition 1, p.14.6 with $R = r$ and with $R = r_0$, and then letting a tend to zero we obtain that, up to a constant,

$$N(D_f,r) = \int_{r_0}^{r} n(D_f,t)\frac{dt}{t}$$

is the normal component of the volume of D_f in the annular fiber $r_0 < |z| < r$. Combining the relation in Proposition 1, p.14.6, with Jensen's formula (p.16.4), we obtain the following statement (see [177], [178]).

P r o p o s i t i o n 1. *Let f be a holomorphic function in the ball $|z| < R$ such that $\overline{\lim}_{r\to R}\int_{|z|=r}\log|f|\cdot\sigma < \infty$ (e.g., f bounded). Then the following equivalent* *inequalities hold:*

$$N(D_f, R) < \infty,$$

$$\int_{D_f} d|z| \wedge d^c|z| \wedge \omega^{n-2} < \infty,$$

$$\int_{D_f} (R - |z|)\omega^{n-1} < \infty.$$

(We use the natural notation $\int_{\sum k_j A_j} := \sum k_j \int_{A_j}$ for integrals over holomorphic chains.)

Each of these equivalent conditions is called the *Blaschke condition*. For $n = 1$ and $R = 1$ the last inequality actually is the classical Blaschke condition for $\{a_j\}$ to be the sequence of zeros of a bounded holomorphic function in the unit disk: $\sum(1 - |a_j|) < \infty$.

Blaschke's condition can be naturally generalized to a wider class of domains (see [173], [81]); we have

P r o p o s i t i o n 2. *Let* Ω: $\rho < 0$ *be a bounded domain in* $\mathbf{C}^n$, *where* $\rho \in C^2(\Omega)$ *and* $d\rho \neq 0$ *on* $\partial\Omega$, *let* $\Omega_\delta = \Omega \cap \{|\rho| > \delta\}$, *and let* $\Gamma_\delta = \partial\Omega_\delta$. *Let* f *be a holomorphic function in* Ω *such that* $\overline{\lim}_{\delta \to 0} \int_{T_\delta} \log^+|f| \cdot \sigma_\delta < \infty$, *where* $\log^+|f| = \max(0, \log|f|)$ *and* σ_δ *is the Euclidean volume form on* Γ_δ. *Then the divisor of* f *satisfies the Blaschke condition:* $\int_{D_f} |\rho| \omega^{n-1} < \infty$.

(The functions $f \in \mathcal{O}(\Omega)$ for which $\log^+|f|$ satisfies the integral estimate stated are called functions of Nevanlinna class in Ω.)

■ Suppose $\int_{\Gamma_\delta} \log^+|f| \cdot \sigma_\delta \leqslant M$ for all $\delta < \delta_0$. Fix $\delta \in (0, \delta_0)$ and denote by u_ϵ an ϵ-regularization of $\log|f|$ with $\epsilon \in (0, \delta/2)$. Green's formula (A5.2) implies that

$$\int_{\Omega_\delta} g_\delta dd^c u_\epsilon \wedge \omega^{n-1} = -\int_{\Gamma_\delta} u_\epsilon d^c g_\delta \wedge \omega^{n-1} - c_n \cdot u_\epsilon(z^0),$$

where g_δ is the Green function of the domain Ω_δ with pole at a point z^0 for which $f(z^0) \neq 0$. Letting $\epsilon \to 0$ and taking into account that $(1/2\pi) dd^c u_\epsilon$ then weakly converges to the current D_f (in a neighborhood of $\overline{\Omega}_\delta$), we obtain

$$\int_{D_f \wedge \Omega_\delta} g_\delta \omega^{n-1} = -\int_{\Gamma_\delta} \log|f| \cdot d^c g_\delta \wedge \omega^{n-1} - c_n \log|f(z^0)| \leqslant$$

$$\leqslant -\int_{\Gamma_\delta} \log^+ |f| \cdot d^c g_\delta \wedge \omega^{n-1} + \text{const},$$

since $-d^c g_\delta \wedge \omega^{n-1}$ is a positive form on Γ_δ. As $\delta \to 0$, the function g_δ and its first partial derivatives tend to the Green function g of Ω and its corresponding partial derivatives, respectively (recall that $\partial\Omega \in C^2$). For the same reason the derivative of g_δ along the normal to Γ_δ is uniformly bounded on Γ_δ and bounded away from 0 on Γ_δ, moreover, both uniformly in δ. Therefore $-d^c g_\delta \wedge \omega^{n-1} = \phi_\delta \sigma_\delta$, where $0 < c_1 \leqslant \phi_\delta \leqslant c_2 < \infty$. Taking the limes superior as $\delta \to 0$ we thus obtain the estimate $\int_{D_f} g\omega^{n-1} < \infty$. Finally, it is well known that g behaves like the distance to $\partial\Omega$, i.e. $g(z) \geqslant c|\rho(z)|$ for all $z \in \Omega$, with some constant $c > 0$. ■

In exactly the same manner we can prove the following, more general theorem (cf. L. Lempert [81]).

Let f be meromorphic in Ω, let D_f^+ be its divisor of zeros and D_f^- its divisor of poles in Ω. Then if

$$\overline{\lim_{\delta \to 0}} \int_{\Gamma_\delta} \log^+ |f| \cdot \sigma_\delta < \infty,$$

the conditions $\int_{D_f^+} |\rho|\omega^{n-1} < \infty$ and $\int_{-D_f^-} |\rho|\omega^{n-1} < \infty$ are equivalent.

In distinction to the one-dimensional case, the Blaschke condition for a complex hyperplane (or a divisor) is not sufficient in order that this hyperplane be the zero set (divisor) of a bounded holomorphic function (see [118], [10]). By investigating explicit formulas for solutions of the Poincaré-Lelong equation, G.M. Khenkin [172], [173] and H. Skoda [130] have shown that any pure $(n-1)$-dimensional analytic subset of a strictly pseudoconvex, simply coconnected (in particular, of every strictly convex) domain Ω in $\mathbf{C}^n$, satisfying the Blaschke condition, is the zero set of a holomorphic function in Ω of Nevanlinna class (see also [118]).

Metrical properties of divisors of various classes of holomorphic functions of several complex variables have been intensively studied in the last years; many beautiful results concerning this theorem can be found in W. Rudin's books [117], [118].

17.2. Metrical conditions of algebraicity. If A is a pure p-dimensional analytic set in $\mathbf{C}^n$, then its limit points at ∞ form a $(p-1)$-dimensional algebraic set in

$\mathbf{P}_n \setminus \mathbf{C}^n$, hence we can choose coordinates in $\mathbf{C}^n$ such that all projections $\pi_I: A \to \mathbf{C}_I$, $\# I = p$, are analytic covers, in particular, finitely-sheeted (for a detailed proof see below). By Wirtinger's theorem this implies that vol $A_r \leqslant Cr^{2p}$ for a certain constant $C < \infty$. Using the expression

$$n(A,r) = \frac{1}{(\pi r^2)^p}\int_{A_r}\omega^p = \frac{1}{\pi^p}\int_{A_r}\omega_0^p + n(A,0),$$

for the nonintegrated counting function, we thus obtain that

For an affine algebraic set the function $n(A,r)$ is uniformly bounded with respect to $r \in [0,\infty)$.

For an arbitrary analytic set the counting function $N(A,r)$ is related with $n(A,r)$ by the two-sided estimate

$$\frac{1}{\log(r/r_0)}N(A,r) \leqslant n(A,r) \leqslant \frac{1}{\log k}N(A,kr), \quad k > 1,$$

which clearly follows from the positivity and the monotonicity of $n(A,r)$ (e.g., $N(a,kr) \geqslant \int_r^{kr} n(A,t)\,dt/t \geqslant n(A,r)\log k$)). If $n(A,r) < C$, this implies that $N(A,r) = O(\log r)$. Conversely, if $N(A,r) \leqslant C\log(1+r)$, then $n(A,r) \leqslant N(A,r^2)/\log r \leqslant C'$. Thus, boundedness of $n(A,r)$ is equivalent to boundedness of $N(A,r)/\log(1+r)$, whence

If A is an affine algebraic set, then $N(A,r) \leqslant C\log(1+r)$.

Finally, if A is an affine algebraic set in $\mathbf{C}^n$, then its closure in $\mathbf{P}_n$ is a projective algebraic set whose volume with respect to the Fubini-Study metric in $\mathbf{P}_n$ is bounded by the degree of A, up to a constant (see p.14.7). In the coordinates of $\mathbf{C}^n$ the Fubini-Study metric $\tilde{\omega}_0$ in $\mathbf{P}_n$ has the form $(dd^c\log(1+|z|^2))/4$ (see p.13.1), while its k-th power can be expressed in terms of ω and ω_0 by the formula

$$\omega_0^k = (1+|z|^2)^{-k-1}(\omega^k + |z|^{2k+2}\omega_0^k),$$

which can be verified immediately. Using this equation, Stokes' formula, and also the relation $\omega = r^2\omega_0$ on bA_r, we find that

$$\int_{A_r}\tilde{\omega}_0^p = \int_{bA_r}\frac{1}{4}d^c\log(1+|z|^2)\wedge\tilde{\omega}_0^{p-1} =$$

$$= \frac{1}{4(1+r^2)^{p+1}}\int_{bA_r}d^c|z|^2\wedge(\omega^{p-1} + |z|^{2p}\omega_0^{p-1}) =$$

$$= \frac{1}{(1+r^2)^p} \int_{A_r} \omega^p = \left[\frac{\pi r^2}{1+r^2}\right]^p n(A,r).$$

Since $\int_A \tilde{\omega}_0^p = \pi^p \deg A$ (p.14.7), we hence find that

For an affine algebraic set A we have

$$\deg A = \lim_{r\to\infty} n(A,r) = \lim_{r\to\infty} \frac{N(A,r)}{\log r}.$$

Moreover, we see that the volume of A in $\mathbb{P}_n$ is bounded if and only if $n(A,r)$ is a bounded function or, equivalently, if and only if $N(A,r)/\log(1+r)$ is a bounded function. It turns out that this boundedness property is characteristic for affine algebraic sets; the following Theorem holds (Stoll [199], Bishop [22]).

T h e o r e m. *Let A be a pure p-dimensional analytic subset in* $\mathbb{C}^n$ *whose volume in* $\mathbb{P}_n$ *is bounded (equivalently, for which* $n(A,r) < C$*). Then A is an algebraic set (of degree at most C).*

■ Bishop's theorem on removable singularities of analytic sets (p.18.3) implies that the closure $\bar{A}$ of A in $\mathbb{P}_n$ is an analytic subset of $\mathbb{P}_n$; by Chow's theorem $\bar{A}$ is projective algebraic, hence A is an affine algebraic set. ■

In its subject-matter, Bishop's theorem belongs to the next Chapter; its proof (not using results of this paragraph, of course) is given in p.18.3. Proofs based on other considerations can be found in, e.g., [42], [186].

Using this Theorem it is easy to obtain the following simple criterion for being algebraic (see [139]).

C o r o l l a r y. *A pure p-dimensional analytic subset A in* $\mathbb{C}^n$ *is algebraic if and only if there is a linear transformation* $l\colon \mathbb{C}^n \to \mathbb{C}^n$ *such that the projections of* $l(A)$ *on all coordinate planes* $\mathbb{C}_I$, $\sharp I = k$, *are proper.*

■ If A is algebraic then $\bar{A}\setminus A$ in $\mathbb{P}_n$ is an algebraic subset of dimension $p-1$ on the plane $H_0 = \mathbb{P}^n \setminus \mathbb{C}^n$. If the orthogonal projection of A on a p-dimensional plane $L \ni 0$ in $\mathbb{C}^n$ is not proper, then $\overline{L^\perp} \cap (\bar{A}\setminus A)$ is nonempty. But $\overline{L^\perp} \cap H_0$ is a complex plane in $H_0 \approx \mathbb{P}_{n-1}$ of dimension $n-p-1$, while $\dim \bar{A}\setminus A = p-1$. Hence such L form a closed nowhere dense set Σ in the

Grassmannian $G(p,n)$. Let $\mathbb{U}(n)$ be the set of unitary transformations of $\mathbb{C}^n$, endowed with the natural topology, and let $I \subset \{1, \ldots, n\}$, $\# I = p$, be fixed. Then the transformations $l \in \mathbb{U}(n)$ such that $l^{-1}(\mathbb{C}_I) \in \Sigma$ clearly also form a closed nowhere dense subset Σ_I in $\mathbb{U}(n)$. But then $\mathbb{U}(n) \setminus \cup_{\# I = k} \Sigma_I$ is nonempty, i.e. there is an $l \in \mathbb{U}(n)$ such that $l^{-1}(\mathbb{C}_I) \notin \Sigma$ for no I. This implies that all projections $\pi_I \colon l(A) \to \mathbb{C}_I$, $\# I = p$, are proper.

Conversely, let l be a linear transformation of $\mathbb{C}^n$ such that all projections $\pi_I \colon l(A) \to \mathbb{C}_I$, $\# I = p$, are proper. Then $\pi_I |_{l(A)}$ is an analytic cover with number of sheets $k_I < \infty$. Let $U_R = \{z\colon |z_j| < R,\ j = 1, \ldots, n\}$. Wirtinger's theorem implies that

$$\mathrm{vol}_{2p}(l(A) \cap U_R) = \sum_{\# I = p}{}' \int_{l(A) \cap U_R} dV_1 \leqslant \sum{}' k_I \mathrm{vol}\, \pi_I(U_R) =$$

$$= (\pi R^2)^p \sum{}' k_I = c R^{2p}.$$

Since U_R contains the ball $|z| < R$, this implies that $n(l(A), R) \leqslant C$ for some constant C and all $R > 0$. By the theorem of Stoll-Bishop, $l(A)$ (and thus, A) is an algebraic set. ■

17.3. Growth estimates of hyperplane sections. Let A be a pure p-dimensional analytic subset in $\mathbb{C}^n$ and L a complex plane of dimension q such that $\dim A \cap L = p + q - n$. If A is algebraic and of degree d, by Corollary 12.5 the set $A \cap L$ has degree $\leqslant d$, hence (p.14.7), the projective volume of $A \cap L$ does not exceed an absolute constant times the projective volume of A. In the transcendental case both A and $A \cap L$ have, in general, infinite projective volume, and it is natural to compare their counting functions. As a quantitative characteristic of the growth of A at infinity we use in the transcendental case the standard notion of order of the analytic subset A in $\mathbb{C}^n$. By definition it equals $\overline{\lim}_{r \to \infty} (\log N(A,r) / \log r) \in [0, \infty]$. In a certain sense, analytic sets of finite order are similar to entire functions of finite order, and are to the same extent more accessible for analysis in comparison to the immense complexity of analytic sets of infinite order of growth.

As distinct from the algebraic case, in general the counting function of $A \cap L$ for a concrete fixed L cannot be estimated in terms of the counting function of A itself. Cornalba and Shiffman [66] have constructed the example of an analytic subset A in $\mathbf{C}^n$ of zero order and such that the counting function of $A \cap \mathbf{C}^q$ has prescribed fast growth (see also [186]). Converse estimates for concrete L also do not hold, even not in the algebraic case ($A \cap L$ can be empty, for example). On the other hand there are Crofton's formulas, which show that the projective volume of A_r coincides with the mean projective volume of its sections by q-dimensional planes $L \in G(q,n)$. Our nearest aim is to give a two-sided estimate of $N(A,r)$ in terms of the mean values of $N(A \wedge L,r)$ over certain subsets in $G(q,n)$. It is obvious that the 'finer' these subsets, the stronger will be the inference on smallness of the set of exceptional planes L, for which $N(A \wedge L,r)$ is sharply distinct from 'mean' behavior. We restrict ourselves to the case of hyperplanes, $q = n-1$; in higher dimensions this problem has been investigated incompletely (see p.17.5).

The counting functions of an analytic set A and its hyperplane section $A \wedge L^w$, where $L^w = \{z \in \mathbf{C}^n\colon (z,w)=0\}$, are related by a simple equation (here $N(C,r) = \int_{r_0}^r n(C,t)\,dt\,/\,t$).

L e m m a. *If A is a pure p-dimensional analytic subset of the ball $|z| < R$ and if $p \geqslant 1$, then for $r < R$,*

$$N(A \cap L^w,r) - N(A,r) = \int_{bA_{r_0,r}} \log\frac{|(z,w)|}{|z|\cdot|w|}\cdot\sigma_p - \frac{1}{\pi^p}\int_{A_{r_0,r}} \frac{\log|(z,w)|\cdot\omega_0^p}{|z|\cdot|w|}.$$

■ Jensen's formula (p.16.4) clearly implies that

$$N(A \wedge L^w,r) = \int_{bA_{r_0,r}} \log\left\|\left(z,\frac{w}{|w|}\right)\right\|\cdot\sigma_p - \frac{1}{\pi^p}\int_{A_{r_0,r}} \log\left\|\left(z,\frac{w}{|w|}\right)\right\|\cdot\omega_0^p.$$

On the other hand,

$$n(A,t) = \frac{1}{4(\pi r^2)^p}\int_{bA_t} d^c|z|^2 \wedge \omega^{p-1} = \frac{1}{2\pi^p}\int_{bA_t} d^c\log|z| \wedge \omega_0^{p-1}$$

and thus, by Fubini's theorem and Stokes' formula,

$$N(A,r) = \frac{1}{2\pi r^p}\int_{A_{r_0,r}} d\log|z| \wedge d^c\log|z| \wedge \omega_0^{p-1} =$$

$$= \int_{bA_{r_0,r}} \log|z|\cdot\sigma_p - \frac{1}{\pi^p}\int_{A_{r_0,r}} \log|z|\cdot\omega_0^p.$$

Subtracting the formulas obtained leads to the relation stated. ■

The integrands in this equation have the same sign (negative), and the mean estimates given below show the possibility of mutual interference of the terms at the righthand side of this formula. Moreover, we cannot control individually each of the integrals mentioned, hence we cannot obtain good estimates for individual L (which is confirmed by the examples mentioned above). Thus, let's turn to averaging.

The manifold $G(n-1,n)$ is isomorphic to $\mathbb{P}_{n-1}$: each point $[w]\in\mathbb{P}_{n-1}$ corresponds uniquely to the hyperplane L^w: $(z,w)=0$ in $\mathbb{C}^n$. Let μ be a probability measure in $\mathbb{P}_{n-1}$ (i.e. $\mu\geqslant 0$ and $\int_{\mathbb{P}_{n-1}}d\mu=1$). The potential of μ is the function in $\mathbb{C}^n$ defined by

$$u_\mu(z)=\int_{\mathbb{P}_{n-1}}\log\frac{|z|\cdot|w|}{|(z,w)|}\cdot d\mu(w).$$

In order to simplify notations we write points in $\mathbb{P}_{n-1}$ in homogeneous coordinates (w instead of $[w]$, etc.). If μ is supported on a complex hyperplane $\{(w,a)=0\}$ in $\mathbb{P}_{n-1}$ or, more generally, if $\mu(\{(w,a)=0\})>0$, then clearly $u_\mu(a)=+\infty$. We are interested in measures μ that are 'not too singular', for which such points do not exist. It turns out that already this weak condition is sufficient in order to estimate $N(A,r)$ in terms of the mean of $N(A\wedge L^w,r)$ with respect to the measure μ (cf. [89]).

P r o p o s i t i o n. *Let A be a pure p-dimensional subset of the ball $|z|<R$ in $\mathbb{C}^n$, let $p>1$, and let μ be a probability measure in $\mathbb{P}_{n-1}$ such that $u_\mu\leqslant c<\infty$. Then*

$$\int_{\mathbb{P}_{n-1}}N(A\wedge L^w,r)\,d\mu(w)\leqslant\left(\frac{c}{\log k}+1\right)N(A,kr)$$

for any $r<R$, $k>1$ such that $kr<R$.

■ First of all we note that if $\dim A\cap L^w=p$, then L^w contains an irreducible component of A, hence such w form an at most countable set. Since $u_\mu\leqslant c$, $\mu(\{w\})=0$ for any point $w\in\mathbb{P}_{n-1}$, hence for any countable set of points. Thus, for μ-almost all w the set $A\cap L^w$ has dimension $p-1$, and only such w will be considered in the sequel. By the Lemma,

$$N(A\wedge L^w,r)-N(A,r)\leqslant$$

$$\leqslant -\int_{bA_{r_0}}\log\frac{|(z,w)|}{|z|\cdot|w|}\cdot\sigma_p-\frac{1}{\pi^p}\int_{A_{r_0},r}\log\frac{|(z,w)|}{|z|\cdot|w|}\cdot\omega_0^p,$$

where $\sigma_p=(1/2\pi^p)d^c\log|z|\wedge\omega_0^{p-1}$. Integrating with respect to μ gives (using Fubini's theorem) a righthand side not exceeding

$$c\left[\int_{bA_{r_0,r}}\sigma_p+\frac{1}{\pi^p}\int_{A_{r_0}}\omega_0^p\right]=c\int_{bA_r}\sigma_p=\frac{c}{4(\pi r^2)^p}\int_{bA_r}d^c|z|^2\wedge\omega^{p-1}=c\,n(A,r).$$

Thus

$$\int_{\mathbb{P}_{n-1}}N(A\wedge L^w,r)\,d\mu(w)\leqslant N(A,r)+c\,n(A,r)\leqslant\left[1+\frac{c}{\log k}\right]N(A,kr).\ \blacksquare$$

Similarly we obtain for the nonintegrated counting function the estimate

$$\int_{\mathbb{P}_{n-1}}n(A\wedge L^w,r^{1-\theta})\,d\mu(w)\leqslant$$

$$\leqslant\left[\frac{c+\log r/r_0}{\theta\log r}\right]n(A,r)=\left[\frac{1}{\theta}+o_r(1)\right]n(A,r)$$

for any θ, $0<\theta<1$, such that $r^{1-\theta}\geqslant r_0$ $(r>1)$.

Growth estimates of plane sections in terms of the growth of the set itself can be regarded as analogs of Nevanlinna's first main theorem from the value distribution theory of holomorphic maps. The Proposition proved is so to speak an averaged version of this theorem for the case of analytic sets (as noted above, a direct generalization for individual sections does not apply here). The converse estimates, of the type of Nevanlinna's second main theorem, are more difficult, but also richer. For analytic sets, as also noted above, it is in general impossible to estimate $N(A,r)$ in terms of a finite number of $N(A\wedge L^\omega,r')$, therefore the converse estimates will also be average estimates, with respect to the same measures as considered above.

17.4. Converse estimates. We first prove the following important lemma of Gruman [45].

L e m m a. *Let A be a pure p-dimensional analytic subset of the ball $|z|<R$, and let v be a plurisubharmonic function in this ball of class C^2 such that $\log|z|-c\leqslant v(z)\leqslant\log|z|$ on a set $A\setminus A_{r_1}$ for some $r_1\geqslant 0$. Then*

$$r^2\int_{A_r} dd^c v\wedge\omega^{p-1}\geqslant 4c\int_{A_{r_1,r'}}\omega^p,\quad r'=re^{-2c},$$

for any $r, r_1<r<R$.

■ Put $A_r' = \{z\in A\colon v(z)<\log r-c\}$. Since $r>r_1$ we have $v(z)\geqslant\log r-c$ on $A\setminus A_r$, hence $A_r'\subset A_r$. Since v is plurisubharmonic, the form $dd^c v$ is positive, hence for $u=v-\log r+c$ we have

$$r^2\int_{A_r} dd^c v\wedge\omega^{p-1}\geqslant$$

$$\geqslant r^2\int_{A_r'} d^c v\wedge\omega^{p-1} = r^2\int_{A_r'} dd^2 u\wedge\omega^{p-1} = r^2\int_{bA_r'} d^c u\wedge\omega^{p-1}\geqslant$$

$$\geqslant\int_{bA_r'}|z|^2 d^c u\wedge\omega^{p-1} = \int_{A_r'}|z|^2 dd^c u\wedge\omega^{p-1}+\int_{A_r'} d|z|^2\wedge d^c u\wedge\omega^{p-1}.$$

Since u is plurisubharmonic, this and Stokes' formula imply

$$r^2\int_{A_r} dd^c v\wedge\omega^{p-1}\geqslant$$

$$\geqslant\int_{A_r'} d|z|^2\wedge d^c u\wedge\omega^{p-1} = \int_{A_r'} du\wedge d^c|z|^2\wedge\omega^{p-1} = -4\int_{A_r'} u\omega^p,$$

since $u=0$ on bA_r'. Now note that $A_{r_1,r'}\subset A_r'$, that $u\leqslant 0$ on A_r' and that $u=v-\log r+c\leqslant\log|z|-\log r+c\leqslant -c$ on $A_{r_1,r'}$. Hence

$$-\int_{A_r'} u\omega^p\geqslant -\int_{A_{r_1,r'}} u\omega^p\geqslant c\int_{A_{r_1,r'}}\omega^p.\ ■$$

The following estimate was obtained by Molzon, Shiffman and Sibony [89].

P r o p o s i t i o n. *Let A be a pure p-dimensional analytic subset of the ball $|z|<R$ in $\mathbb{C}^n$, let $p\geqslant 1$, and let μ be a probability measure in $\mathbb{P}_{n-1}$ such that $u_\mu\leqslant c<\infty$. Then*

$$n(A,e^{-2c}r)\leqslant\frac{e^{4pc}}{2c}\int_{\mathbb{P}_{n-1}} n(A\wedge L^w,r)\,d\mu(w).$$

■ Fix a point in $\mathbb{P}_{n-1}$ with homogeneous coordinates w, $|w| = 1$, and fix the plane L^w in $\mathbb{C}^n$ given by the equation $(z,w) = 0$. By Theorem 2, p.16.3,

$$A \wedge L^w = \frac{1}{2\pi} dd^c([A]\cdot \log|(z,w)|).$$

For each $\theta < 1$ we choose a function $\lambda_\theta \in \mathfrak{D}(|z| < r)$ for which $0 \leqslant \lambda_\theta \leqslant 1$ and $\lambda_\theta = 1$ for $|z| \leqslant \theta r$. Then

$$/\lim_{\theta \to 1} < A \wedge L^w, \lambda_\theta \omega^{p-1} > = (\pi r^2)^{p-1} n(A \wedge L^w, r)$$

and

$$< A \wedge L^w, \lambda_\theta \omega^{p-1} > = \frac{1}{2\pi} \int_A \log |(z,w)| dd^c \lambda_\theta \wedge \omega^{p-1}.$$

This and Fubini's theorem imply

$$\int < A \wedge L^w, \lambda_\theta \omega^{p-1} > d\mu(w) = \frac{1}{2\pi} \lim_{\epsilon \to 0} \int_A v_\epsilon dd^c \lambda_\theta \wedge \omega^{p-1},$$

where $v_\epsilon(z) = (1/2) \int \log (\epsilon^2 + |(z,w)|^2) \, d\mu(w)$. By Green's theorem (p.14.3),

$$\int_A v_\epsilon dd^c \lambda_\theta \wedge \omega^{p-1} = \int_A (dd^c v_\epsilon) \wedge \lambda_\theta \omega^{p-1} \geqslant \int_{A_{\theta r}} dd^c v_\epsilon \wedge \omega^{p-1},$$

since v_ϵ is clearly plurisubharmonic. Since

$$v_\epsilon - \log|z| = \frac{1}{2} \int \log \frac{\epsilon^2 + |(z,w)|^2}{|z|^2} d\mu(w) \geqslant -u_\mu(z) \geqslant -c$$

for all z and since $v_\epsilon \leqslant (1/2) \log (\epsilon^2 + |z|^2) \leqslant \log|z| + \epsilon$ as $|z|^2 \geqslant \epsilon/2$, the function $v_\epsilon - \epsilon$ satisfies the conditions of the Lemma (with $c + \epsilon$ instead of c and with $r_1 = \sqrt{\epsilon/2}$). Hence

$$\int_A v_\epsilon dd^c \lambda_\theta \wedge \omega^{p-1} \geqslant \frac{4(c+\epsilon)}{(\theta r)^2} \int_{A_{r_1, r''}} \omega^p,$$

where $r'' = \theta r e^{-2(c+\epsilon)}$. Limit transition as $\epsilon \to 0$ gives

$$\int < A \wedge L^w, \lambda_\theta \omega^{p-1} > d\mu(w) \geqslant \frac{2c}{\pi(\theta r)^2} \int_{A_{\theta r'}} \omega^p,$$

where $r' = re^{-2c}$. Since, as $\theta \to 1$, λ_θ tends to the characteristic function χ_r of the ball $|z|<r$ and does not exceed it, $<A\wedge L^w,\lambda_\theta\omega^{p-1}>$ tends to $(\pi r^2)^{p-1}n(A\wedge L^w,r)$ and does not exceed this limit function. Since the limit function is integrable with respect to μ (according to the preceding Section), Lebesgue's theorem implies that we may pass to the limit as $\theta\to 1$ in the last inequality obtained. As the result we obtain

$$(\pi r^2)^{p-1}\int n(A\wedge L^w,r)d\mu(w)\geqslant\frac{2c}{\pi r^2}(\pi(r')^p n(A,r'),$$

which is the estimate which had to be proved. ■

Logarithmically averaging the inequality proved leads to the same relation for the counting functions. Combining the estimate with what was proved in the preceding Section leads us to the following theorem of Molzon, Shiffman and Sibony [89].

Theorem. *Let A be a pure p-dimensional analytic subset of the ball $|z|<R$ in $\mathbb{C}^n$, and let μ be a probability measure in $\mathbb{P}_{n-1}$ with bounded potential: $u_\mu\leqslant c<\infty$. Then*

$$2ce^{-4pc}N(A,e^{-2c}r)\leqslant\int_{\mathbb{P}_{n-1}}N(A\wedge L^w,r)\,d\mu(w)\leqslant\left[\frac{c}{\log k}+1\right]N(A,kr)$$

for all $r>r_0$, $k>1$ such that $kr<R$.

We note that in the case of measures absolutely continuous with respect to the Fubini-Study volume this Theorem was proved by Gruman [45] (Gruman's lemma is a basic stage in the general case also).

17.5. Corollaries and generalizations. By definition, a compact set $K\subset\mathbb{P}_{n-1}$ has positive (projective) logarithmic capacity if there is a probability measure μ supported on it and with uniformly bounded potential: $||u_\mu||=\sup u_\mu<\infty$. (The logarithmic capacity of K is the supremum of $1/||u_\mu||$ over all such measures μ, cf. [89].) E.g., any nonpluripolar compact set in $\mathbb{P}_{n-1}$ has positive logarithmic capacity; however, projective logarithmic capacity distinguishes between much "finer" properties. (E.g., any compact set of positive length on an $\mathbb{R}$-analytic arc not contained in any hyperplane also has positive logarithmic capacity.) More details can be found in [89], from which we also take the following

P r o p o s i t i o n 1. *Let A be a pure p-dimensional analytic subset in $\mathbb{C}^n$, let $p \geqslant 1$, and let $E \subset \mathbb{P}_{n-1}$ be a compact set of positive logarithmic capacity. If $A \cap \{(z,w)=0\}$ is, for each $[w] \in E$, an algebraic set, then A is also an algebraic set.*

■ Let L^w: $(z,w) = 0$. If $\dim A \cap L^w = p$, then L^w contains an irreducible component of A, hence the set of such $[w]$ is at most countable; we denote it by E'. Let ν be a probability measure on E with bounded potential. Since the ν-measure of any point is zero, $\nu(E') = 0$. For a $[w] \in E \setminus E'$ the set $A \cap L^w$ is algebraic, hence $n(A \wedge L^w, r)$ is uniformly bounded with respect to $r > 0$. Put $E_j = \{[w] \in E \setminus E'\colon\ n(A \wedge L^w, r) \leqslant j \text{ for all } r\}$. Then $E \setminus E' = \cup E_j$, hence $\nu(E_j) > 0$ for at least one j. Fix such a j and put $\mu = (\chi / \nu(E_j))\nu$, where χ is the characteristic function of E_j. Then the potential $u_\mu(z) \leqslant u_\nu(z) / \nu(E_j)$ is bounded, hence, by Proposition 17.4,

$$n(A,r) \leqslant \frac{e^{4pc}}{2c} \int n(A \wedge L^w, e^{2c} r)\, d\mu(w) \leqslant j \frac{e^{4pc}}{2c} < \infty$$

for all $r > 0$. The Bishop-Stoll theorem (p.17.7) now implies that A is an algebraic set. ■

Theorem 17.4 also readily implies the following Proposition, which we state without proof (cf. [89]).

P r o p o s i t i o n 2. *Let A be a pure p-dimensional subset in $\mathbb{C}^n$, let $p \geqslant 1$, and let μ be a probability measure in $\mathbb{P}_{n-1}$ with bounded potential u_μ. Then* $\operatorname{ord} A \wedge L^w = \operatorname{ord} A$ *for μ-almost all $w \in \mathbb{P}_{n-1}$.*

(Here $\operatorname{ord} C$ denotes the order of the analytic set C, i.e. $\overline{\lim}_{r\to\infty} (\log N(C,r)) / \log r$.)

For generalizations to plane sections of codimensions $k > 1$ the logarithmic potential and logarithmic capacity are not very suited (if only for the simple fact that $A \cap L$ may have codimension $> p-k$ for planes L forming in $G(n-k,n)$ an analytic set of positive dimension). Here we give one result of Gruman [45], which is true for plane sections of arbitrary codimensions.

T h e o r e m. *Let A be a pure p-dimensional analytic subset of the ball $|z| < R$ in $\mathbb{C}^n$, let $p \geqslant 1$, and let E be a Borel set in $G(q,n)$ of positive Fubini-Study volume dL, $p+q \geqslant n$. Then there are positive constants c_1, c_2, depending on the volume of E, on* *n, and on q only, such that*

$$c_1 N(A, c_2 r) \leqslant \int_E N(A \cap L, r)\, dL$$

for all r, $r_0 < r < R$.

The proof may proceed by, e.g., induction over $k = n - q$, using Theorem 17.4 as initial step, and using Fubini's theorem on the incidence manifold $\{(L, \Lambda)\colon L \subset \Lambda\}$ in $G(q,n) \times G(q+1,n)$ in order to reduce q. Because the purely technical details are rather awkward, we do not give a proof (cf. [45]).

The consequences from this Theorem are similar to those from Theorem 17.4 and are proved along the same plan. In particular, in order that a pure p-dimensional analytic subset A in $\mathbf{C}^n$ be algebraic it suffices that its sections $A \cap L$ be algebraic for all L from some Borel set $E \subset G(q,n)$ of positive volume, $p + q \geqslant n$. (Besides, for $p = n - 1$ and $q = 1$ there is the stronger theorem of Stoll [198], in which E is only required to be nonpluripolar.)

Chapter 4

ANALYTIC CONTINUATION AND BOUNDARY PROPERTIES

18. Removable singularities of analytic sets

18.1. Singularities of small codimensions. The general problem concerning removing singularities of analytic sets is as follows.

Let E be a closed subset of a complex manifold Ω and A a pure p-dimensional analytic subset in $\Omega \setminus E$ $(p>0)$. Under what conditions on E and A is the closure $\bar{A}$ of A in Ω an analytic subset in Ω?

Maybe the best known result in this direction is the Remmert-Stein theorem (p.4.4), in which analyticity of $\bar{A}$ is obtained under the condition that E be an analytic subset in Ω of dimension $<p$. This result also follows from Shiffman's theorem (p.4.4) on removability of the singularities of A on E under the condition that the $(2p-1)$-measure of E vanishes; we already repeatedly used this Theorem in the previous Chapters. We derived Shiffman's theorem from Theorem 4.3 (on the analyticity of generalized covers), which can be strengthened as follows.

P r o p o s i t i o n. *Let E be a closed subset of a domain $U = U' \times U''$ in $\mathbb{C}^n$, $U' \subset \mathbb{C}^p$, such that the projection π: $E \to U'$ is proper and such that $\pi(E)$ is a locally removable subset in U'. Let A be a pure p-dimensional analytic subset in $U \setminus E$ such that $A \cap \pi^{-1}(z')$ is zero-dimensional or empty for all $z' \in \pi(E)$. Then $\bar{A} \cap U$ is an analytic subset in U.*

■ Let $a \in \bar{A} \cap E$, $a' = \pi(a)$, and $\phi(z'')$ a real-valued continuous function with compact support in U'', equal to 1 on $E \cap \pi^{-1}(a')$. Since the conditions imply that $A \cap \pi^{-1}(a')$ is at most countable, the range of ϕ on it is at most countable, hence $A \cap \pi^{-1}(a')$ does not intersect some level set $\{\phi(z'')=c\}$, $0<c<1$. But then $\overline{A} \cap \pi^{-1}(a')$ also does not intersect this level set, thus there is a domain $V' \ni a'$ in U' such that $\overline{A} \cap \{z\colon z' \in V', \phi(z'')=c\}$ is empty. Put $V'' = \{z'' \in U''\colon \phi(z'')<c\}$ and $V = V' \times V''$. Then $a \in V$ and $\pi\colon \overline{A} \cap V \to V'$ is a proper map. Since $\pi(E)$ is locally removable, this is a generalized analytic cover (see p.4.1). By Theorem 4.3, $\overline{A} \cap V$ is an analytic subset in V. Since a was arbitrary from $\overline{A} \cap E$, $\overline{A} \cap U$ is an analytic subset in U. ■

Under addition restrictions on E the conditions on A can be removed completely. E.g.,

Corollary. *Let E be a closed subset of a domain $U' \times U_n$ in $\mathbf{C}^n$, $U' \subset \mathbf{C}^{n-1}$, such that the projection $\pi\colon E \to U'$ is proper, $\pi(E)$ is a locally removable subset in U', and such that $(z' \times U_n) \setminus E$ is connected and everywhere dense in $z' \times U_n$ for all $z' \in \pi(E)$. Then for any pure $(n-1)$-dimensional analytic subset A in $U \setminus E$ its closure $\overline{A} \cap U$ is an analytic subset in U.*

■ Let Σ' consist of the points $z' \in U'$ for which $A \cap \pi^{-1}(z')$ is one-dimensional. Since U_n is a domain in $\mathbf{C}$, for such z' the set $A \cap \pi^{-1}(z')$ coincides with $(z' \times U_n) \setminus E$, hence $\Sigma' \times U_n \subset \overline{A}$. It is obvious that Σ' is closed in U'. If $a' \in \Sigma'$, $(a', a_n) \in A$, and $V = V' \times V_n$ is a neighborhood of (a', a_n) in U disjoint from E, then $\Sigma' \cap V' = \bigcap_{c \in V_n} \pi(A \cap \{z_n = c\})$ is an analytic subset in V', by p.5.6. Thus, Σ' is an analytic subset in U' of dimension $\leqslant n-2$. According to the Proposition (with $U' \setminus \Sigma'$ instead of U'), the set

$$\overline{A} \cap ((U' \setminus \Sigma') \times U_n)$$

is analytic. If $\dim \Sigma' < n-2$, then

$$\dim \Sigma' \times U_n < n-1,$$

hence $\overline{A} \cap U$ is analytic by the Remmert-Stein theorem. If $\dim \Sigma' = n-2$, A can be represented as $A_1 \cup A_2$, where A_2 consists of the irreducible components of A belonging to $\Sigma' \times U_n$, and

$$\dim (\Sigma' \times U_n) \cap A_1 < n-1.$$

The closure of A_2 in U coincides with the set of points at which $\Sigma' \times U_n$ has dimension $n-1$. By p.5.2 this set is analytic. For A_1 we can repeat the above. Since

$$\dim (\Sigma' \times U_n) \cap A_1 < n-1,$$

the analytic set $\Sigma_1' \subset \Sigma'$, constructed for A_1 similarly to the construction of Σ' for A, has dimension $<n-2$, hence $\overline{A}_1 \cap U$ is analytic in U by the Remmert-Stein theorem. ■

18.2. Infectiousness of continuation. A set of Hausdorff dimension $2p-1$ can be the boundary of a pure p-dimensional analytic set, so to expect the removability of singularities of such large dimension is, in general, not justified. (It is also not clear whether there is for analytic sets some analog of Theorem A1.5 on removable singularities of continuous functions.) The boundary behavior of pure p-dimensional analytic sets in a neighborhood of a smooth $2p-1$-dimensional manifold will be investigated in §19, while here we give the following result on the "infectiousness" of continuation across such manifolds.

Theorem 1. *Let M be a connected $(2p-1)$-dimensional C^1-submanifold of a complex manifold Ω, and let A be a pure p-dimensional analytic subset in $\Omega \setminus M$. Suppose that the set $\overline{A}$ is analytic in a neighborhood of some point $z^0 \in M$. Then $\overline{A}$ is an analytic subset in Ω.*

■ Let M' be the set of points of M in a neighborhood of which $\overline{A}$ is analytic. Then M' is open in M and, by requirement, nonempty. Let b be a limit point of M', and let z be a local coordinate in a neighborhood of b, $z' = (z_1, \ldots, z_p)$. Since $\overline{A} \subset A \cup M$ has $(2p+1)$-measure zero, after a linear change of coordinates z we can choose a neighborhood $U \ni b$ such that

$$\pi \colon \overline{A} \cap U \to U' \subset \mathbf{C}^p$$

is a proper map and $\pi \colon M \cap U \to \Gamma$ is a diffeomorphism onto a C^1-hypersurface Γ partitioning U into two domains $U_{\pm}'$ (here $\pi \colon \zeta \mapsto (z_1(\zeta), \ldots, z_p(\zeta))$, $\zeta \in U$). Further, without loss of generality we may assume that $U = U' \times U''$ is a bounded domain in $\mathbf{C}^n$. Since M' is open in M, $\Gamma' = \pi(M' \cap U)$ is open and nonempty on Γ, hence $V' = U_+' \cup \Gamma' \cup U_-'$ is a domain in U'. Since $\overline{A} \setminus (M \setminus M')$ is an analytic set, the map $\pi|_{\overline{A} \cap U}$ is an analytic cover above V'. Let k denote number of sheets of this cover, let Φ_I, $|I| = k$, denote the canonical defining functions, and let $\phi_{IJ}(z')$ be the coefficients of Φ_I regarded as a polynomial in z'' (see p.4.3). Then the ϕ_{IJ} are bounded holomorphic functions in V'.

For each $a' \in \Gamma$ the set $\pi^{-1}(a') \cap A \cap U$ is zero-dimensional or empty (if S is an irreducible component of this set of positive dimension and if $a \in M \cap U$, $\pi(a) = a'$, then $S \cup \{a\}$ is analytic by the Remmert-Stein theorem; but $S \cup \{a\}$ is compact, hence zero-dimensional; a contradiction). Hence for each point $a^j \in \pi^{-1}(a') \cap A \cap U$ there is a neighborhood $U^j \subset U$ such that $\pi: A \cap U^j \to \pi(U^j)$ is an analytic cover. Since $\pi(U^j)$ is a neighborhood of a', this implies that $\pi^{-1}(V') \cap A \cap U$ is everywhere dense in $A \cap U$ and that $k(a')$, the number of points in the fiber $\pi^{-1}(a') \cap A \cap U$ counted with multiplicities, does not exceed k. If $k(a') = k$, then a neighborhood of $a = \pi^{-1}(a') \cap M \cap U$ is disjoint with A, hence $a' \in \Gamma'$. If $k(a') < k$, then in a sufficiently small neighborhood $W \ni a$ the projection

$$\pi: A \cap W \setminus \pi^{-1}(\Gamma) \to W' \setminus \Gamma$$

has $k - k(a')$ sheets, and all these sheets tend to a as $z' \to a'$, $z' \notin \Gamma$ (the point a has, so to speak, multiplicity $k - k(a')$). Thus, for any limit approach $z' \to a'$, $z' \in V'$, the k-tuples of points from $\pi^{-1}(z') \cap A \cap U$, written with multiplicities, have a limit, and this limit is a k-tuple of points from the fiber $\pi^{-1}(a') \cap \bar{A} \cap U$, written with multiplicities (see above). Hence, by Lemma 2, p.4.2, the functions $\phi_{IJ}(z')$ have a continuous extension onto all of U'. By Theorem A1.5 these extensions are holomorphic in U', hence $\bar{A} \cap U$ is defined as the zero set of the holomorphic functions $\sum_{|J| \leqslant k} \phi_{IJ}(z') (z'')^J$, $|I| = k$, i.e. is analytic in U.

Thus, we have proved that M' is closed in M, hence $M' = M$, i.e. $\bar{A} \subset A \cup M$ is analytic in a neighborhood of each of its points. ■

Singularities of yet larger dimensions may also be removed under additional restrictions; one such restriction is, e.g., the condition of pluripolarity (see A1.2). The following Lemma was proved by Bishop [22] (see also [115]).

L e m m a. *Let E be a closed complete pluripolar subset of a bounded domain $U = U' \times U''$ in $\mathbb{C}^n$, $U' \subset \mathbb{C}^p$, and let A be a pure p-dimensional analytic subset in $U \setminus E$ without limit points on $U' \times \partial U''$. Suppose that U' contains a nonempty* *subdomain V' such that $\bar{A} \cap (V' \times U'')$ is an analytic set. Then $\bar{A} \cap U$ is an analytic subset in U.*

■ Let $E = \{z: \phi(z) = -\infty\}$, where the function ϕ is plurisubharmonic in U (see A1.2). Our statement being local we may assume that ϕ is plurisubharmonic in a neighborhood of $\bar{U}$, hence bounded from above on U.

We first prove that $A \cap (V' \times U'')$ is nonempty. Let W' be the maximal subdomain in U' containing V' and such that $\bar{A} \cap (W' \times U'')$ is an analytic set. Then

$$\pi: \bar{A} \cap (W' \times U'') \to W'$$

is an analytic cover. Hence it suffices to prove that $\bar{A} \cap (W' \times U'')$ is nonempty. Assume it is not. Since A is nonempty ($\dim A = p > 0$) and W' is maximal, there is a ball $B' = B(a', r') \subset U$ with center $a' = (a_1, \ldots, a_p) \in W'$ such that $A \cap (B' \times U'')$ is nonempty. Hence for some complex line $L' \ni a'$ the set $A_1 = A \cap (L' \times U'')$ is nonempty too; we may assume $L' = \mathbb{C}_1$. The boundary of the analytic set A_1 consists of two parts: $E_1 = \bar{A}_1 \cap E$ and $E_2 = (\bar{A}_1 \setminus A_1) \setminus E_1$. The function ϕ, which is plurisubharmonic and bounded on $\bar{A}_1$, is equal to $-\infty$ on E_1. Hence by the maximum principle (p.6.3, Corollary), the function $1/|z_1 - a_1|$ cannot exceed $1/r'$ on all of A_1. But at all points of A_1 it is strictly larger than $1/r'$ because $A_1 \subset B' \times U''$. This contradiction shows that $A \cap (V' \times U'')$ is nonempty.

Put $W = W' \times U''$. The projection $\pi: \bar{A} \cap W \to W'$ is a k-sheeted cover, $1 \leq k < \infty$. For each $z' \in W'$ we write the points of the fiber $\pi^{-1}(z') \cap A$ with multiplicities: $\alpha^1(z'), \ldots, \alpha^k(z')$, and we define on W' the function $\psi(z') = \sum_1^k \phi(\alpha^j(z'))$. If σ is the critical analytic set of the cover $\pi: \bar{A} \cap W \to W'$, then in $W' \setminus \sigma$ the functions $\alpha^j(z')$ can be chosen locally holomorphic; hence ψ is plurisubharmonic or $\equiv -\infty$ on $W' \setminus \sigma$. Since A does not intersect E and $A \cap W$ is everywhere dense in $\bar{A} \cap W$, the set $E \cap \bar{A} \cap W$ has $2p$-measure zero (e.g., by the mean value Theorem for $\phi|_{\bar{A}}$ in local coordinates). Consequently, $\pi(E \cap \bar{A} \cap W)$ also has $2p$-measure zero, and thus $\psi \not\equiv -\infty$ on $W' \setminus \sigma$. Since ψ is uniformly bounded from above and, clearly, upper semicontinuous on W', we have (A1.2) that ψ is plurisubharmonic in all of W'. We show that all limit values of ψ on the boundary of W' in U' are equal to $-\infty$.

Let $a' \in U' \cap \partial W'$; in view of the maximality of W' there is a point $a \in \bar{A} \cap E$ such that $\pi(a) = a'$. Since $A \cap (U' \times \partial U'')$ is empty, the whole boundary of the analytic set $\pi^{-1}(a') \cap A$ belongs to E, hence, by the maximum principle (p.6.3), $\pi^{-1}(a') \cap A$ is zero-dimensional. If $a^j \in \pi^{-1}(a') \cap A$, then there is a neighborhood $U^j \ni a^j$ in U such that $\pi: A \cap U^j \to \pi(U^j)$ is an analytic cover. Since W' is a domain and a' is a limit point of W', this implies that the number of points in the fiber $\pi^{-1}(a') \cap A$ is at most $k < \infty$. Hence there is a neighborhood $U^o \ni a$ such that $\pi^{-1}(a') \cap A \cap U^o$ is empty and $\pi: \overline{A} \cap U^o \to \pi(U^o)$ is a proper map. The sets U^o, $E \cap U^o$, and $W' \cap \pi(U^o)$ satisfy the conditions of the Lemma and thus, as was shown in the first part of the proof, $A \cap U^o \cap W$ is nonempty. Since $\pi^{-1}(a') \cap \overline{A} \cap U^o \subset E$, as $z' \to a'$, $z' \in W$, all points in the (nonempty) fibers $\pi^{-1}(z') \cap A \cap U^o$ tend to E. Since $\phi|_E = -\infty$, this implies $\psi(z') \to -\infty$.

Now we can define an upper semicontinuous function Ψ on U', by putting $\Psi = \psi$ on W' and $\Psi = -\infty$ on $U' \setminus W'$. By A1.2, Ψ is a plurisubharmonic function in U', in particular, $E' = U' \setminus W'$ is pluripolar in U'.

Let Φ_I, $|I| = k$, be canonical defining functions of the analytic cover $\pi: \overline{A} \cap W \to W'$,

$$\Phi_I(z',z'') = \sum_{|J| \leqslant k} \phi_{IJ}(z')(z'')^J.$$

Then the ϕ_{IJ} are holomorphic and bounded in W'. By Corollary A1.4 they have a holomorphic continuation to U'. The functions $\tilde{\Phi}_I$, $|I| = k$, thus continued define in U' an analytic subset $\tilde{A}$ coinciding with A above W'. Since $A|_W$ is dense in A and $\tilde{A}|_W$ is dense in $\tilde{A}$, we have $\tilde{A} = \overline{A} \cap U$. ■

C o r o l l a r y. *Let M be a connected p-dimensional complex submanifold of a complex manifold Ω, and let A be a pure p-dimensional analytic subset in $\Omega \setminus M$. Suppose that M contains a point a in a neighborhood of which $\overline{A}$ is analytic (possibly, empty). Then $\overline{A}$ is an analytic subset of Ω.*

■ Let M' be the set of points on M in a neighborhood of which $\overline{A}$ is analytic. Then M' is open in M and, by requirement, nonempty. We show that an arbitrary limit point b of M' belongs to it. This statement being local we may assume that $b = 0$ in $\mathbb{C}^n$, $\Omega = U' \times U''$, and $M = U'$ is a domain in $\mathbb{C}^p$. After a suitable linear change of coordinates of the form $(z',z'') \mapsto (z', l(z',z''))$ and after shrinking U we may assume that $\pi: \overline{A} \cap U \to U'$ is a proper map. Since $U' \cap M'$ is nonempty, the conditions of the Lemma are satisfied (with $E = M$), hence

$\overline{A} \cap U$ is an analytic subset in U. Thus, M' is closed in M, hence $M' = M$ since M is connected. ■

This readily implies the following theorem of Thullen on the infectiousness of continuation across an analytic set (see [146]).

Theorem 2. *Let S be a p-dimensional analytic subset of a complex manifold Ω, and let A be a pure p-dimensional analytic subset of $\Omega \setminus S$. Suppose that Ω contains an open subset U intersecting every irreducible component of S and such that $\overline{A} \cap U$ is analytic in U. Then $\overline{A}$ is an analytic subset in Ω.*

■ By the Corollary $\overline{A} \setminus \mathbb{S}(S)$ is an analytic subset in $\Omega \setminus \mathbb{S}(S)$, of pure dimension p since A is everywhere dense in it. Since $\mathbb{S}(S)$ is an analytic subset in Ω of dimension $<p$ (see p.5.2), the remaining follows from the Remmert-Stein theorem. ■

In relation to this we yet note the following Proposition, which can be regarded as an analog for analytic sets of Sokhotskiĭ's theorem on essential singularities (see [146], [114]).

Corollary. *Let S be an irreducible p-dimensional analytic subset of a complex manifold Ω, and let A be a pure p-dimensional analytic subset in $\Omega \setminus S$. Then either $\overline{A}$ is an analytic subset in Ω or $\overline{A} \supset S$.*

■ If $S \not\subset \overline{A}$, there is an open set U in Ω intersecting S and such that $\overline{A}$ is analytic in U. ■

18.3. Removing pluripolar singularities. Bishop's theorems. Lemma 18.2 readily implies the following two Theorems on continuation across pluripolar singularities. In essence, these Theorems were proved by Bishop [22].

Theorem 1. *Let E be a closed locally complete pluripolar subset of a complex manifold Ω, and let A be a pure p-dimensional analytic subset in $\Omega \setminus E$. If $\overline{A} \cap E$ has $2p$-measure zero, then $\overline{A}$ is an analytic subset in Ω.*

■ The statement being local we may assume that E is a complete pluripolar subset of a bounded domain $U = U' \times U''$ in $\mathbb{C}^n$, and that A is an analytic subset in $U \setminus E$ such that $\overline{A} \cap (U' \times \partial U'')$ is empty (for an isolated point of $\overline{A}$ such a neighborhood exists since $\overline{A} = A \cup (\overline{A} \cap E)$ has $2p$-measure zero). Since $\overline{A} \cap E$

has $2p$-measure zero, $\pi(\bar{A} \cap E)$ is a nowhere dense subset of U', hence the conditions of Lemma 18.2 are fulfilled. According to this Lemma, $\bar{A} \cap U$ is an analytic subset in U. ■

T h e o r e m 2. *Let E be a closed locally complete pluripolar subset of a complex manifold Ω, and let A be a pure p-dimensional analytic subset in $\Omega \setminus E$ such that $\mathcal{H}_{2p}(A \cap K) < \infty$ for any compact set $K \subset \Omega$. Then $\bar{A}$ is an analytic subset in Ω*[*)].

■ The statement being local we may assume that Ω is a domain in $\mathbb{C}^n$, E is a complete pluripolar subset of Ω, and $\mathcal{H}_{2p}(A) < \infty$. Then, denoting by U_ϵ the ϵ-neighborhood of E, we have that $\mathcal{H}_{2p}(A \cap U_\epsilon) \to 0$ as $\epsilon \to 0$. We show that $\mathcal{H}_{2p}(\bar{A} \cap E) = 0$.

For this we cover $\mathbb{C}^n$ by equal cubes of side length ϵ (with disjoint interiors), select the cubes Q_j among them which intersect $\bar{A} \cap E$, and denote their number by N_ϵ. Denote by $3Q_j$ the cube obtained from Q_j by triple inflation from its center. Then $3Q_j$ contains some ball $B_j = B(a_j, \epsilon)$ with center $a_j \in A$. By Theorem 1, p.15.5 (and Wirtinger's theorem), $\mathcal{H}_{2p}(A \cap B_j) \geqslant c_p \epsilon^{2p}$. Since each point $z \in \mathbb{C}^n$ is contained in at most 3^{2n} distinct B_j and since all B_j belong to the $c\epsilon$-neighborhood of $\bar{A} \cap E$, where $c = 3\sqrt{2n}$, we have $N_\epsilon c_p \epsilon^{2p} \leqslant 3^{2n} \mathcal{H}_{2p}(A \cap U_{c\epsilon})$. Hence $N_\epsilon \leqslant C\epsilon^{-2p} \mathcal{H}_{2p}(A \cap U_{c\epsilon}) = o(\epsilon^{-2p})$ as $\epsilon \to 0$. Thus, $\bar{A} \cap E$ is covered by the balls $\tilde{B}_j$ (circumscribed around the corresponding cubes Q_j) with radii $r_j = \epsilon\sqrt{2n}$, with $\sum r_j^{2p} = N_\epsilon \cdot \epsilon^{2p} (2n)^p \to 0$ as $\epsilon \to 0$. According to the definition of Hausdorff measure this implies $\mathcal{H}_{2p}(\bar{A} \cap E) = 0$.

Hence, by Theorem 1, $\bar{A}$ is analytic. ■

Along the scheme outlined above Bishop [22] proved the following important Theorem on removing singularities of analytic sets of locally finite volume (we already used it in p.17.2).

T h e o r e m 2′. *Let S be an analytic subset of a complex manifold Ω, and let A be a pure p-dimensional analytic subset in $\Omega \setminus S$ of locally finite $2p$-measure in Ω (i.e. $\mathcal{H}_{2p}(A \cap K) < \infty$ for all $K \subset\subset \Omega$). Then $\bar{A}$ is an analytic subset in Ω.*

Any analytic subset is a locally complete pluripolar set, hence Theorem 2′ is a

---

[*)] For positive closed currents of locally finite mass an analogous theorem has recently been obtained by El Mir [202].

particular instance of Theorem 2. However, in Bishop's proof [22] it is necessary practically in full generality in order to prove Theorem 2, including a necessary estimate from below for the volume of an analytic subset in $B_j \setminus E$ passing through the center of B_j (we have used the rather involved theorem of Alexander-Taylor-Ullman, p.15.5, giving an estimate from below for areas of projections, instead). Therefore Theorems 1, 2 are naturally called Bishop's theorems.

18.4. Continuation across $\mathbb{R}^n$. Obstructions of sufficiently large dimensions for analytic sets cannot, in general, be removed. In order that they can additional conditions on the singularity and on the set itself are necessary. As an example we give the following result of Alexander [5].

Theorem 1. *Let D be a domain in $\mathbb{C}^n$, and A a pure one-dimensional analytic subset in $D \setminus \mathbb{R}^n$ that is symmetric about $\mathbb{R}^n$ (i.e. for each $z \in A$ the complex conjugate point $\bar{z} = (\bar{z}_1, \ldots, \bar{z}_n)$ also belongs to A). Then $\bar{A} \cap D$ is an analytic subset in D.*

■ The statement being local we may assume that $0 \in \bar{A} \cap D$, and it suffices to prove that $\bar{A}$ is analytic in a neighborhood of 0. Let $g(z) = z_1^2 + \cdots + z_n^2$, and let Z be the zero set of g. If $\dim A \cap Z = 1$, then Z contains an irreducible component A' of A. Since $Z \cap \mathbb{R}^n = \{0\}$, the set $A' \cup \{0\}$ is an analytic subset in D. The union of the remaining irreducible components of A satisfies the conditions of the theorem, hence in the sequel we may assume that $Z \cap A$ is at most countable. Since $Z \cap \bar{A} \cap D = (Z \cap A) \cup \{0\}$, the set $Z \cap \bar{A} \cap D$ is also at most countable, hence there are neighborhoods $U \ni 0$ in $\mathbb{C}_z^n$ and $V \ni 0$ in $\mathbb{C}_w$ such that U is symmetric about $\mathbb{R}^n$ and such that $g\colon \bar{A} \cap U \to V$ is a proper map. Without loss of generality we may assume that V is the disk $|w| < 1$.

The image of $\mathbb{R}^n$ under the map $z \mapsto g(z)$ is the positive half-line $\mathbb{R}_+$, and the map $g|_{A \cap U}$ is a k-sheeted analytic cover, $1 \leqslant k < \infty$, above $V \setminus \mathbb{R}_+$ (since $g \not\equiv 0$ on $A \cap U$, also $\operatorname{Im} g \not\equiv 0$ on $A \cap U$). By Proposition 4, p.3.5, U can be shrinked such that for each $a \in \mathbb{R}_+$ the set $g^{-1}(a) \cap A \cap U$ is at most countable. For each $a^j \in A \cap U$ such that $g(a^j) = a$, there is a neighborhood $U_j \ni a^j$ in U such that $g\colon A \cap U_j \to g(U_j)$ is an analytic cover above the domain $g(U_j) \subset V$. This clearly implies that the number of such points a^j (counted with multiplicities) is at most k, and that $A \cap U \setminus g^{-1}(\mathbb{R}_+)$ is everywhere dense in $A \cap U$.

Let $\alpha^1(w), \ldots, \alpha^k(w)$ be the points in the fiber $g^{-1}(w) \cap A \cap U$, $w \in V \setminus \mathbb{R}_+$, counted with multiplicities, and let

$$\sigma_{js}(w) = \sum_{\nu=1}^{k} (\alpha_j^\nu(w))^s,$$

$s = 1, \ldots, k$, by the elementary symmetric functions of the j-th coordinates of these points, $j = 1, \ldots, n$. Then the $\sigma_{js}(w)$ are clearly holomorphic and bounded in $V \setminus \mathbb{R}_+$. Since $g(\overline{w}) = \overline{g(w)}$ and A is symmetric about $\mathbb{R}^n$, we have $\{\alpha_j^\nu(\overline{w})\} = \{\overline{\alpha_j^\nu(w)}\}$, hence $\sigma_{js}(\overline{w}) = \overline{\sigma_{js}(w)}$. If $\{\alpha_+^\nu\}$ and $\{\alpha_-^\nu\}$ are certain limit values of k-tuples $\{\alpha^\nu(w)\}$ as $w \to a \in \mathbb{R}_+$ from, respectively, the upper and lower sides of $\mathbb{R}$, the above (concerning a fiber $g^{-1}(a) \cap A \cap U$) implies that

$$\{\alpha_+^\nu\} \setminus \mathbb{R}^n = \{\alpha_-^\nu\} \setminus \mathbb{R}^n.$$

Hence the difference between any two limit values of σ_{js} at a point $a \in \mathbb{R}_+$ from the upper and lower sides of $\mathbb{R}$ is a real number, so that on $V \cap \mathbb{R}_+$ all limit values of the pure imaginary functions $\sigma_{js}(w) - \sigma_{js}(\overline{w})$ are zero, i.e. these functions have continuous extensions onto V (vanishing on $\mathbb{R}_+$). This and Cauchy's integral formula in the domains $V \setminus \{\operatorname{Re} w \geqslant -\epsilon, \ |\operatorname{Im} w| \leqslant \epsilon\}$ clearly imply that as $\epsilon \to 0$ we obtain

$$\sigma_{js}(w) = \frac{1}{2\pi i} \int_{|\zeta|=r} \frac{\sigma_{js}(\zeta)}{\zeta - w} d\zeta, \quad w \in V \setminus \mathbb{R}_+,$$

hence the σ_{js} have holomorphic continuations onto all of V.

In the domain $\mathbb{C}_z^n \times (V \setminus \mathbb{R}_+)$ there are defined the holomorphic functions

$$p_j(z,w) = \prod_{\nu=1}^{k} (z_j - \alpha_j^\nu(w)) = z_j^k + c_{j1}(w) z_j^{k-1} + \cdots + c_{jk}(w),$$

$1 \leqslant j \leqslant n$. Since the c_{ji} can be written as polynomials in the elementary symmetric functions σ_{js} with integer coefficients (see [67]), the c_{ji} also have holomorphic continuations onto V. The p_j thus continued define in $\mathbb{C}_z^n \times V$ a set $\tilde{A}$, which is one-dimensional (and an analytic cover above V). By construction $\tilde{A}$ contains all points $(z, g(z))$, $z \in A \cap U \setminus g^{-1}(\mathbb{R}_+)$. Since $A \cap U \setminus g^{-1}(\mathbb{R}_+)$ is everywhere dense in $A \cap U$, the projection $\pi(\tilde{A})$ of $\tilde{A} \cap \{w = g(z)\}$ to $\mathbb{C}^n$ (i.e. the analytic subset $\{p_j(z, g(z)) = 0, \ 1 \leqslant j \leqslant n\}$ in U) is one-dimensional and contains $A \cap U$. Let S be an irreducible component of $\pi(\tilde{A})$ whose intersections with A is one-dimensional. The boundary of $S \cap \bar{A} \cap U$ on S belongs to $S \cap \mathbb{R}^n$, i.e. consists of isolated points and analytic arcs. If $\zeta \in (\operatorname{reg} S) \cap \mathbb{R}^n$ and if the arc $S \cap \mathbb{R}^n$ is smooth in a neighborhood of ζ, then in a neighborhood of $\zeta, S \setminus \mathbb{R}^n$ is the union of two manifolds with boundary on $\mathbb{R}^n$. If $\zeta \in \bar{A} \cap U$, the symmetry of A about $\mathbb{R}^n$ implies that A intersects both these manifolds (the uniqueness Theorem for analytic sets implies that S is also symmetric about $\mathbb{R}^n$ in a neighborhood of ζ). Since A is closed in $D \setminus \mathbb{R}^n$, the uniqueness Theorem implies that $\bar{A}$ contains a neighborhood of ζ in S. Hence $\bar{A} \cap S$ is everywhere dense in S, hence $\bar{A} \supset S$.

Thus we have proved that $\bar{A} \cap U$ is the union of the irreducible components of the analytic set $\pi(\tilde{A}) \cap U$ in U whose intersection with A are one-dimensional. By p.5.4, $\bar{A} \cap U$ is an analytic subset in U. ■

C o r o l l a r y. *Let D be a domain in $\mathbb{C}^n$, and A a pure one-dimensional analytic subset in $D \setminus \mathbb{R}^n$. Then A has an analytic continuation across $D \cap \mathbb{R}^n$, i.e. in some neighborhood $U \supset D \cap \mathbb{R}^n$ there is a pure one-dimensional analytic subset $\tilde{A}$ containing $A \cap U$.*

■ Put $D^- = \{\bar{z}\colon z \in D\}$, $A^- = \{\bar{z}\colon z \in A\}$, and $U = D \cap D^-$. The set A^- is analytic in D^- (if A is defined in a neighborhood of $a \in D$ by holomorphic functions $\{f_j\}$, then A^- is defined in a neighborhood of $\bar{a}$ by the functions $\{\overline{f_j(\bar{z})}\}$, which are also holomorphic). Hence $(A \cup A^-) \cap U$ satisfies the condition of Theorem 1, i.e. has analytic closure in U, which is the required analytic continuation of A. ■

The following interesting result of Shiffman [192] on continuation across a torus touches upon the themes here presented.

T h e o r e m 2. *Let A be a pure one-dimensional analytic subset of the polydisk $|z_j| \leq 1$, $j = 1, \ldots, n$, in $\mathbb{C}^n$ such that $\bar{A} \subset U \cap \Gamma$, where $\Gamma\colon |z_j| = 1$, $j = 1, \ldots, n$, is the distinguished boundary of U. Then in $\mathbb{C}^n$ there is a one-dimensional algebraic set $\tilde{A}$ such that $\tilde{A} \cap U = A$. (The set $\tilde{A} \setminus \bar{A}$ is obtained from A by symmetry about Γ, i.e. consists of the points $z^* = (1/\bar{z}_1, \ldots, 1/\bar{z}_n)$ such that $z = (z_1, \ldots, z_n) \in A$.)*

■ If in a neighborhood of one of its points a the set A is defined by holomorphic functions $\{f_j\}$, then the set $A^\star$ symmetric (about Γ) to it is defined in a neighborhood of $a^\star$ by the functions $\{f_j(1/\bar{z}_1, \ldots, 1/\bar{z}_n)\}$, which are holomorphic too. Hence the set $A^\star = \{z^\star : z \in A\}$ is analytic in $\mathbb{C}^n \setminus \Gamma$. Since $\bar{A} \subset U \cup \Gamma$, the set $A \cap \{z_1 \ldots z_n = 0\}$ is finite. Hence the set of limit points of $A^\star$ at infinity in $(\mathbb{P}_1)^n = (\mathbb{C} \cup \{\infty\})^n$ is also finite. By the Remmert-Stein theorem, the closure of $A^\star$ in the complex manifold $(\mathbb{P}_1)^n \supset \mathbb{C}^n$ is analytic in $(\mathbb{P}_1)^n \setminus \Gamma$.

By a coordinatewise fractional linear transformation $l(z) = (l_1(z_1), \ldots, l_n(z_n))$ of $(\mathbb{P}_1)^n$ the torus Γ can be mapped into $\mathbb{R}^n$ leaving an arbitrary pre-given point on Γ finite. In accordance with the properties of fractional linear transformations of the plane, a pair $z, z^\star$ is mapped to a pair that is symmetric about $\mathbb{R}^n$. The closure of $A \cup A^\star$ in $(\mathbb{P}_1)^n$ is mapped under such a transformation into a set that is analytic outside $\mathbb{R}^n$ (in $\mathbb{C}^n$) and symmetric about $\mathbb{R}^n$. This and Theorem 1 imply that $\overline{A \cup A^\star}$ is an analytic subset in $(\mathbb{P}_1)^n$. As is well known, the manifold $(\mathbb{P}_1)^n$ is algebraic, i.e. there is an embedding $(\mathbb{P}_1)^n \to \mathbb{P}_N$ that is polynomial in homogeneous coordinates (in $\mathbb{P}_N$ and in all copies of $\mathbb{P}_1$); see, e.g., [188]. According to Chow's theorem the image of $\overline{A \cup A^\star}$ under this embedding is defined by polynomial identities in $\mathbb{P}_N$. But then $\overline{A \cup A^\star}$ is also defined in $(\mathbb{P}_1)^n$ by polynomial identities; in $\mathbb{C}^n$ these identities define an affine algebraic set which clearly coincides with $\bar{A} \cup A^\star$. ■

C o r o l l a r y. *Let A be an analytic subset of the polydisk U such that $\bar{A} \subset U \cup \Gamma$ and such that $A \cap \{z_1 \ldots z_n = 0\}$ is empty. Then A is an at most countable set.*

■ Assume that $\dim A = p > 0$. Then there is a plane $L \ni 0$ such that $A \cap L$ is a one-dimensional analytic set. Without loss of generality we may assume that L can be bijectively projected onto a plane $\mathbb{C}^m \subset \mathbb{C}^n$; moreover, the projection of U is contained in the polydisk $U_1 = U \cap \mathbb{C}^m$, and the projection of Γ is contained in the distinguished boundary Γ_1 of U_1. Let A_1 denote the projection to $\mathbb{C}^m$ of a one-dimensional irreducible component of $A \cap L$. Since $\bar{A}_1 \subset U_1 \cup \Gamma_1$, Theorem 2 implies that $\bar{A}_1 \cup A_1^\star$ is a one-dimensional analytic subset in $\mathbb{C}^m$. But $\bar{A}_1 \cap \{z_1 \ldots z_m = 0\}$ is empty, hence $\bar{A}_1 \cup A_1^\star$ is compact. This contradiction shows that $p \leqslant 0$. ■

Analytic sets of pure dimensions $p > 1$ can be continued both across tori

(Shiffman [193]) and across $\mathbb{R}^n$ (Becker [12]). More details can be found in the next Section.

18.5. Obstructions of small CR-dimensions. When the set E has a rich complex structure, an analytic subset in $\Omega \setminus E$ can approach E, uncontrollably "oscillating" around it (as, e.g., the graph of $w = e^{1/z}$ along the $z = 0$ axis in $\mathbb{C}^2$). Hence additional conditions are necessary in order to remove such singularities too (conditions of the type of those in Bishop's theorems, p.18.3). On the other hand, if E has poor complex structure, say E a smooth manifold not containing a maximal complex manifold of necessary dimension (the candidate for the boundary of an analytic set), one may hope for an analytic subset A in $\Omega \setminus E$ "not to take notice of" E, i.e. for $\bar{A}$ to be analytic. This is supported by the results of Shiffman [193] and Becker [12] concerning continuation of analytic sets of dimensions $p > 1$ across the distinguished boundary of the polydisk and across $\mathbb{R}^n$. Funahasi [160] has shown that for pure p-dimensional analytic sets in $\mathbb{C}^n$, $p > 1$, singularities of the form $\mathbb{R}^m \times \mathbb{C}^{p-2}$ are also removable. The following Theorem (see [183]) generalizes these results and partially substantiates the hypothesis made above.

Theorem. *Let M be a C^1-submanifold of a complex manifold Ω, and let A be a pure p-dimensional analytic subset in $\Omega \setminus M$, $p > 1$. Suppose that the complex tangent plane $T_\zeta^c M$ has (complex) dimension less than $p-1$ at all points $\zeta \in M \cap \bar{A}$, with the possible exception of the points of a set of $(2p-1)$-measure zero. Then $\bar{A}$ is an analytic subset in Ω.*

■ The set of points $\zeta \in M$ at which $\dim T_\zeta^c M \geqslant p-1$ is clearly closed in M, while, by requirement, its intersection with $\bar{A}$ has $(2p-1)$-measure zero. Hence, by Shiffman's theorem (p.4.4) it suffices to prove analyticity of $\bar{A}$ in a neighborhood of every point $\zeta^\circ \in \bar{A} \cap M$ at which $\dim T_\zeta^c M < p-1$.

Step 1. Choice of coordinates. The statement being local we may assume that $\zeta^\circ = 0$ is the coordinate origin in $\mathbb{C}^n$ and that $T_0^c M$ is the $z' = (z_1, \ldots, z_q)$-plane, $q < p-1$. By a linear change of variables leaving $\mathbb{C}_{1 \ldots q}$ fixed we can secure that $A \cap \mathbb{C}_{(q+1) \cdots n}$ has dimension $\leqslant p-q$. After this linear change of variables $z'' = (z_{q+1}, \ldots, z_n)$ the totally real plane $T_0 M \cap \mathbb{C}_{(q+1) \cdots n}$ is taken to the plane of real variables $'x = (x_{q+1}, \ldots, x_m)$, $m \leqslant n$ ($x_j = \operatorname{Re} z_j$). In these coordinates $T_0 M$ is the $(z', 'x)$-plane, hence in some neighborhood

$U = U' \times U'' \ni 0$, $U' \subset \mathbb{C}^q$, the manifold $M \cap U$ is given by the system of equations

$$'y = '\phi(z','x), \quad z_j = \psi_j(z','x), \quad j > m,$$

where ϕ_i and ψ_j are functions of class C^1 of order $o(|z'| + |'x|)$. Leaving from the last $n-m$ equations only the imaginary parts gives a manifold $\tilde{M}$ in U containing $M \cap U$ and defined by the system $y'' = \phi(z',x'')$, where $\phi \in C^1$, $|\phi| = o(|z'| + |x''|)$. Since $\dim T_\zeta^c \tilde{M} \leqslant q < p-1$ at all points $\zeta \in \tilde{M}$ near 0, we find that $\tilde{M} \cap A$ is nowhere dense in $A \cap U$. Hence we may assume that $M \cap U = \tilde{M}$, that $|\phi(z',x'')| \leqslant (|z'| + |x''|)/20$ and that $|y'' - \phi(z',\alpha'')| \leqslant |x'' - \alpha''|$ for all $z \in M \cap U$ and $\alpha'' \in U'' \cap \mathbb{R}^{n-q}$.

Step 2. Transversality of the approach of A to M. Fix an arbitrary $r > 0$ such that $B_r = B(0,r) \subset U$. Let S be an irreducible component of $A \cap U$ intersecting the ball $B_{r/4}$, and let $a \in S$, $|a| < r/4$. The function

$$\rho(a',z'') = |y'' - \phi(a',\alpha'')|^2 - |x'' - \alpha''|^2, \text{ where } \alpha'' = \operatorname{Re} a'',$$

differs from $\operatorname{Re} \sum_{q+1}^n z_j^2$ by a linear term, hence is pluriharmonic. The boundary of $S \cap B_r \cap \{z' = a'\}$, considered as a set in the plane $z' = a'$, partially lies on $M \cap B_r$ and partially lies on the sphere $|z| = r$. Since $a \notin M$, we have $\rho(a',a'') > 0$. Since $\rho \leqslant 0$ on $M \cap U \cap \{z' = a'\}$, by the maximum principle (p.6.3) we can find a point $b \in S \cap \partial B_r$ such that $b' = a'$ and $|y''(b) - \phi(a',\alpha'')| > |x''(b) - \alpha''|$. Hence $|y''(b)| > |x''(b)| - |\alpha''| - |\phi(a',\alpha'')| > |x''(b)| - 2|a| > |x''(b)| - r/2$. Since $|b| = r$ we thus have $|y''(b)| \geqslant r/8$. Because $|y''| < r/10$ on $M \cap B_r$, the compact set $\bar{B}_r \cap \{|y''| > r/8\}$ does not intersect with M, hence only finitely many irreducible components of $A \cap U$ intersect with it (see p.5.4). Since $0 \in \bar{A}$, 0 is a limit point for some such component, and we may further assume that $A \cap U$ is irreducible.

Since $0 = \lim a_j$, $a_j \in A$, the above implies that there are $b_j \in A \cap \partial B_r$ such that $y''(b_j) \geqslant r/8$ and $b_j' = a_j'$. The limit points of the sequence $\{b_j\}$ cannot belong to $M \cap A$, hence belong to A. Therefore the set $A \cap U \cap \{z' = 0\}$ is nonempty, and contains a point c for which $|c| = r$ and $|y''(c)| \geqslant r/8$. Denote by c_j such a point with $r = 1/j$. Let $v = (0',v'')$ be a limit vector for the sequence $\{c_j/|c_j|\}$. By construction, $v \notin T_0 M$, i.e. $y''(v) > 0$. Thus, the tangent cone to $A \cap \{z'=0\}$ at its limit point 0 (see p.8.1) contains a vector not belonging to T_0M.

Step 3. Construction of a proper map. Choose constants $a_{q+1}, \ldots, a_n$, $1/2 < a_j < 1$, such that $g(z) := a_{q+1}z_{q+1}^2 + \cdots + a_n z_n^2 \not\equiv 0$ on $A \cap U \cap \{z'=0\}$; if $x''(v) = 0$ no other conditions need be imposed on a_j, while if $x''(v) \neq 0$ we also require $\operatorname{Im} g(v) \neq 0$ (since $y''(v) \neq 0$ and $A \cap U$ is irreducible, such a_j clearly exist). Put $\theta = \arcsin \operatorname{Im}(g(v)/|g(v)|)$.

Since

$$\operatorname{Re} g \geqslant \frac{1}{2}|x''|^2 - |y''|^2 \geqslant \frac{1}{4}|x''|^2$$

on $M \cap U \cap \{z'=0\}$, 0 is a limit point of $M \cap U \cap \{z'=g(z)=0\}$. Since

$$\dim A \cap U \cap \{z'=g(z)=0\} \leqslant p-q-1,$$

we can choose linear functions $l = (l_{q+1}, \ldots, l_{p-1})$ such that the set $\bar{A} \cap U \cap \{z'=g(z)=l(z)=0\}$ is at most countable. Let $f: z \mapsto (z', l(z), g(z))$ be a map $\mathbf{C}_z^n \to \mathbf{C}_w^p$. According to the above said, the neighborhood U can be shrinked so that $f: \bar{A} \cap U \to V$ is a proper map into some domain $V \ni 0$ in $\mathbf{C}^p$. Since $|y''| = o(|x''|)$ as $z \to 0$ on $M \cap \{z'=0\}$, we may assume that the image of $M \cap \{z'=0\}$ under this map lies in the "wedge" $w' = 0$, $|\arg w_p| \leqslant \theta/2$ (here $w' = (w_1, \ldots, w_q)$). According to the above we may also assume that $A \cap U$ is irreducible.

Step 4. Analyticity of $\bar{A}$ in a neighborhood of 0. Put $M' = f(M \cap U)$. Since $c_j \in A$ and $\arg g(c_j) > \theta/2$ for all sufficiently large j, we have that $f: A \cap U \setminus f^{-1}(M') \to V \setminus M'$ is a nonempty analytic cover. If Γ denotes the graph of f in $\mathbf{C}_z^n \times \mathbf{C}_w^p$, and

$$\mathcal{Q} = \{(z, f(z)) \colon z \in A \cap U \setminus f^{-1}(M')\},$$

the projection $\pi_2: \mathcal{Q} \to V \setminus M'$ is also a k-sheeted cover, $1 \leqslant k < \infty$. Let $\Phi_I(z,w) = \sum_{|J| \leqslant k} \phi_{IJ}(w) z^J$, $|I| = k$, be canonical defining functions for this cover; the $\phi_{IJ}(w)$ are holomorphic in $V \setminus M'$. Since $M' \cap \{w'=w_p=0\} = \{0\}$, for all sufficiently small c', c_p the sets $M' \cap \{w'=c, w_p=c_p\}$ are compact. Since $M' \cap \{w'=0, \operatorname{Re} w_p < 0\}$ is empty, by the continuity principle (A1.3) every function holomorphic in $V \setminus M'$ has a holomorphic continuation to the domain

$(V \setminus M') \cup W$, where W is a neighborhood of 0. Since the only condition on V is that it be small, we may assume that $V = (V \setminus M') \cup W$. Continuing in this way the coefficients of the functions Φ_I we obtain an analytic subset $\tilde{\mathscr{Q}}$ in $\Gamma \cap \pi_2^{-1}(V)$, defined by these extended $\Phi_I(z,w)$, $|I| = k$. Since the projection π_1 of $\Gamma \cap \pi_2^{-1}(V)$ into $\mathbf{C}_z^n$ is single-sheeted and its image is $f^{-1}(V)$, the set $\pi_1(\tilde{\mathscr{Q}})$ is a pure p-dimensional analytic subset in $f^{-1}(V)$. Since $U \subset f^{-1}(V)$, $\tilde{\mathscr{Q}} \supset \mathscr{Q}$ and $\pi_1(\mathscr{Q}) = A \cap U \setminus f^{-1}(M')$, in view of the irreducibility of $A \cap U$, the set $\bar{A} \cap U$ is contained in an irreducible component $\tilde{A}$ of $\pi_1(\tilde{\mathscr{Q}}) \cap U$. Since $\dim \tilde{A} > 1$, by A2.4 the set $\tilde{A} \setminus M$ is connected. According to the uniqueness Theorem, $\tilde{A} \setminus M \subset A$. Since $\tilde{A} \cap M$ is nowhere dense in $\tilde{A}$ (A2.4), we find $\tilde{A} \subset \bar{A}$. Thus, $\bar{A} \cap U = \tilde{A}$. ■

C o r o l l a r y. *Let D be a domain in $\mathbf{C}^n$, and A a pure p-dimensional analytic subset in $D \setminus \mathbb{R}^n$, with, moreover, $p > 1$. Then $\bar{A} \cap D$ is an analytic subset in D.*

18.6. "Hartogs' lemma" for analytic sets. Problems in which it is required to establish analyticity of a given set under conditions related to "partial" analyticity of it can also be regarded as problems on removing singularities of analytic sets. As an example we give the following geometric analog of Hartogs' lemma from A1.1 (see Rothstein [114]).

L e m m a. *Let A be a closed subset of a domain $U = U' \times U''$ in $\mathbf{C}^n$, $U' \subset \mathbf{C}^m$, and let V'' be a subdomain of U''. Suppose that:*

1) $A \cap (U' \times V'')$ is a pure p-dimensional analytic subset in $U' \times V''$;

2) $A_c = A \cap \{z' = c\}$ is a pure q-dimensional, or empty, analytic subset in $c \times U''$, where, moreover, $q \geqslant 1$ is independent of $c \in U'$; and

3) each irreducible component of A_c intersects with $c \times V''$. Then A is an analytic subset in U.

Of course, this is a very weak analog of Hartogs' lemma, since the condition that A be closed is analogous to the function in Hartogs' lemma being continuous (and under this additional requirement Hartogs' lemma easily follows from Cauchy's integral formula in a polydisk). We have required that A be closed only in order to simplify the proof. Indeed, the Lemma holds also without this restriction (the other requirements being, in general, necessary); however, the proof of this general version is too involved to give here (it was proved under additional assumptions by Rothstein [114]). For our purposes this weak version in the following more general form suffices.

P r o p o s i t i o n. *Let $\tilde{D}$, D be domains in $\mathbf{C}^n$ with common projection D' in $\mathbf{C}^m \subset \mathbf{C}^n$, and suppose, moreover, that $D \cap \{z'=c\} \subset \tilde{D}$ for all $c \in D'$. Let A be a closed subset in $\tilde{D}$ such that:*

1) $A \cap D$ is a pure p-dimensional analytic subset in D;

2) $A_c = A \cap \{z'=c\}$ is a pure q-dimensional, or empty, analytic subset in $\tilde{D} \cap \{z'=c\}$, where $q \geqslant 1$ is independent of c; and

3) each irreducible component of A_c intersects with D.

Then A is an analytic subset in $\tilde{D}$.

■ We rename the coordinates in $\mathbf{C}^n = \mathbf{C}^m \times \mathbf{C}^{n-m}$: the points in $\mathbf{C}^m$ will be denoted by z, those in $\mathbf{C}^{n-m}$ by w. Let G be the set of points in $\tilde{D}$ in a neighborhood of which A is analytic (possibly, empty); we must show that $G = \tilde{D}$.

Let (c,b) be a limit point for $G \cap A_c$ belonging to reg A_c. Without loss of generality we may assume that in a neighborhood of (c,b) on the plane $z = c$ the set A_c is defined by a system $w'' = f(w')$, where $w' = (w_1, \ldots, w_q)$ and $w = (w',w'')$. Let W' be a small ball with center b' in $\mathbf{C}^q_{w'}$ (f holomorphic in $\overline{W'}$), and let $a' \in W'$ be such that $(c,a',f(a')) \in G$. Then there is an $\epsilon > 0$ such that all points (z,w) with $|z-c| < \epsilon$, $|w'-a'| < \epsilon$ and $|w''-f(w')| < \epsilon$ belong to G, and such that A_c is disjoint from the compact set $\{z=c,\ w' \in W',\ |w''-f(w')| = \epsilon\}$. Since A is closed in $\tilde{D}$, there is a δ, $0 < \delta < \epsilon$, such that $A_z \cap \{w' \in W',\ |w''-f(w')| = \epsilon\}$ is also empty for all z for which $|z-c| < \delta$. Let

$$V_0:\ |z-c| < \delta,$$

$$U_0 = \{(z,w):\ |z-c| < \delta,\ w' \in W',\ |w''-f(w')| < \epsilon\},$$

and $\tilde{\pi}$: $(z,w) \mapsto (z,w')$. By construction, $\tilde{\pi}$: $A \cap U_0 \to V_0 \times W'$ is a proper map. Since A_z is pure q-dimensional or empty and W' is a domain in $\mathbf{C}^q_{w'}$, the uniqueness Theorem implies that the analytic set $\tilde{\pi}(A_z \cap U_0)$ is either empty or coincides with $z \times W'$. Since $U \cap \{|w'-a'| < \epsilon\} \subset G$, we have that $\tilde{\pi}(A \cap U_0) \cap \{|w'-a'| < \epsilon\}$ is a pure p-dimensional analytic set. According to the above said, it has the form $S\times\{|w'-a'|<\epsilon\}$, where S is a pure $(p-q)$-dimensional analytic subset in V_0. After a suitable linear change of z-coordinates we may assume that in some domain $V = V'\times V'' \subset V_0$ containing c the projection $S\cap V\to V'\subset \mathbf{C}_{1\cdots(p-q)} \subset \mathbf{C}_z^m$ is proper. Put

$$z' = (z_1,\ldots,z_{p-q}),\quad z = (z',z''),$$

$$\pi\colon (z,w)\mapsto(z',w') \text{ and } U = U_0\cap\{z\in V\};$$

then the map $\pi\colon A\cap U\to V'\times W'$ in a domain of p-dimensional space is proper. For each $c'\in V'$ the set $A\cap U\cap\{z'=c'\}$ is a finite union of sets $A_z\cap U$ (where z runs through the finite set $S\cap\{z'=c'\}$). Hence all restrictions $\pi\colon A\cap U\cap\{z'=c'\}\to c'\times W'$ are analytic covers. Since above $|w'-a'|<\epsilon$ the map $\pi|_{A\cap U}$ is an analytic cover also, the number of sheets of these covers is independent of $c'\in V'$; we denote it by k. Let $\Phi_I(c';w',w'')$, $|I| = k$, be canonical defining functions of the cover $\pi|_{A\cap U\cap\{z'=c'\}}$,

$$\Phi_I = \sum_{|J|\leqslant k} \phi_{IJ}(c',w')(w'')^J.$$

Then the $\phi_{IJ}(c',w')$ are holomorphic in W' for each fixed $c'\in V'$. Since above $|w'-a'|<\epsilon$ the map $\pi|_{A\cap U}$ is also a k-sheeted analytic cover, the $\Phi_I(z';w',w'')$, $|I| = k$, are its canonical defining functions, hence the $\phi_{IJ}(z',w')$ are holomorphic in the domain $\{z'\in V',\ |w'-a'|<\epsilon\}$. By Hartogs' lemma (A1.1) the $\phi_{IJ}(z',w')$ are holomorphic in $V'\times W'$, hence $A\cap U$ is an analytic subset in U. Since each component of A_c intersects with $D\subset G$, we have thereby proved that G contains $\cup_{c\in D'}\operatorname{reg} A_c$, hence $E = \tilde{D}\setminus G$ belongs to $\cup_{c\in D'}\operatorname{sng} A_c$.

Let $(c,b)\in E$. Then, as above, after a suitable coordinate change there are neighborhoods $U\ni(c,b)$, $V = V'\times V''\ni c$ and $W'\ni b'$ such that the map $\pi\colon A\cap U\to V'\times W'$ is proper (and $\tilde{\pi}(A\cap U) = S\times W'$, where S is a pure $(p-q)$-dimensional analytic subset in V). The set $\sigma = \pi(E\cap U)$ is closed in $V'\times W'$, and for each $c'\in V'$ the set $\sigma\cap\{z'=c'\}$ is contained in a finite union of analytic sets $\pi(\operatorname{sng} A_c\cap U)$ (where c runs through the finite set $S\cap\{z'=c'\}$). Hence (Theorem A1.4 and Hartogs' lemma A1.1), σ is a removable subset in $V'\times W'$, so that $\pi\colon A\cap U\to V'\times W'$ is a generalized analytic cover (see p.4.1). By Theorem 4.3, $A\cap U$ is an analytic subset in U. We have thus proved that indeed E is empty, i.e. A is analytic at all points of $\tilde{D}$. ■

19. Boundaries of analytic sets

19.1. Regularity near the boundary. In classical function theory the following boundary regularity property is well known: if f is holomorphic in the disk Δ, continuous on its closure $\bar{\Delta}$ and of class C^k, $k > 0$, on the boundary, then $f \in C^k(\bar{\Delta})$. The following statement may be regarded as an analog of this result for analytic set.

Theorem. *Let M be a connected $(2p-1)$-dimensional C^k-submanifold of a complex manifold Ω, $p \geqslant 1$, and let A be a pure p-dimensional analytic subset in $\Omega \setminus M$ such that $M \subset \bar{A}$. Then A is either an analytic subset in Ω or there is a closed subset $E \subset \bar{A}$ with $\mathcal{H}_{2p-1}(E) = 0$ such that $(A \setminus E, M \setminus E)$ is a C^k-manifold with boundary in $\Omega \setminus E$.*

■ The statement being local we may assume that Ω is a neighborhood of 0 in $\mathbb{C}^n$ and $0 \in M$. The set $\bar{A} = A \cup M$ has $(2p+1)$-measure zero, hence, after a suitable unitary transformation the projection $\pi: \bar{A} \cap U \to U' \subset \mathbb{C}^p$ is proper in some neighborhood $U \ni 0$ in Ω (see A6.4). The planes L intersecting T_0M at 0 only form in $G(n-p,n)$ an everywhere dense open set (a system of p linear equations on T_0M with complex coefficients has, in general, the unique solution $z = 0$). Hence, by A6.4, we may assume that $T_0M \cap \mathbb{C}^{n-p}_{z''} = \{0\}$, and we find that by suitably shrinking U the map $\pi: M \cap U \to \Gamma$ is a diffeomorphism onto some connected C^k-hypersurfaces in U'. Shrinking U yet further we may assume that Γ splits U' into two domains, $U'_{\pm}$. Put

$$A_{\pm} = (A \cap U) \cap \pi^{-1}(U'_{\pm}).$$

Then $\pi: A_{\pm} \to U'_{\pm}$ are analytic covers with numbers of sheets $k_{\pm}$, respectively; we will assume that $k_+ \geqslant k_-$.

For $a' \in \Gamma$ the set $\pi^{-1}(a') \cap A \cap U$ is zero-dimensional (if S is an irreducible component of this set of positive dimension and if $a \in M \cap U$, $\pi(a) = a'$, the Remmert-Stein theorem, p.4.4, implies that $S \cup \{a\}$ is analytic; however, it is compact and hence cannot have positive dimension, a contradiction). Therefore there is for each point $a^j \in \pi^{-1}(a') \cap A \cap U$ a neighborhood $U_j \subset U$ such that

$$\pi: A \cap U_j \to \pi(U_j)$$

is an analytic cover. Since $\pi(U_j)$ intersects with both U'_+ and U'_-, this implies that $A_+ \cup A_-$ is everywhere dense in $A \cap U$ and that the number of points in each fiber $\pi^{-1}(a') \cap A \cap U$ counted with multiplicities is at most k_+.

Now assume $n = p+1$. By p.2.8, $A_\pm$ are the zero sets of Weierstrass polynomials $F_\pm(z',z_n)$ in $U'_\pm \times \mathbf{C}_n$, respectively, and for almost each fixed $z' \in U'_\pm$ the polynomial $F_\pm$ has only simple roots in z_n. Let $\Delta_\pm$ be the discriminants of $F_\pm$ as polynomials in z_n, and let $E_\pm \subset \Gamma$ be the sets of points $a' \in \Gamma$ at which the limits of $\Delta_\pm$ as $z' \to a'$, $z \in U'_\pm$, respectively, exist and are equal to zero. Since $\Delta_+ \not\equiv 0$ in U'_+, the boundary uniqueness Theorem (A1.1) implies $\mathcal{H}_{2p-1}(E_+) = 0$; moreover, clearly the $E_\pm$ are closed on Γ.

For $a' \in \Gamma \setminus E_+$ there is a sequence $a^j \to a'$, $a'_j \in U'_+$, such that the numbers $\Delta_+(a^j)$ have a limit distinct from zero. By passing to subsequences we may assume that the polynomials $F_+(a^j,z_n)$ also have a limit as $j \to \infty$. This limit polynomial will be denoted by $F_+(a',z_n)$; its discriminant is $\Delta_+(a')$ and is thus distinct from zero. Because the leading coefficients of the polynomials $F_+(a^j,z_n)$ of degree k_+ are 1, the degree of $F_+(a',z_n)$ is k_+ also. It is clear that the zeros of $F_+(a',z_n)$ are the z_n-coordinates of certain points in the fiber $\pi^{-1}(a') \cap A \cap U$. Since there are at most k_+ such points, as was proved above, we find

$$\pi^{-1}(a') \cap \bar{A} \cap U = \{a^1, \ldots, a^{k_+}\} = \{(a',z_n) : F_+(a',z_n)=0\}.$$

Let $V_j \subset U$ be pairwise disjoint neighborhoods of the points a^j such that $V_j \cap M$ is empty if $a^j \notin M$, and such that $\pi \colon \bar{A} \cap V_j \to V'$ are proper maps into a domain $V' \subset U'$, $V' \ni a'$ (such V_j, V' obviously exist). Since $M \subset \bar{A}$, there is one point among the a^j, say a^1, belonging to M; it is a limit point for A, hence above $V'_+ = V' \cap U'_+$ the cover $\pi|_{A \cap V_1}$ is at least single-sheeted. Since there are k_+ domains V_j and since all $A \cap V_j$, $j > 1$, are nonempty, this implies that $\pi \colon A \cap V_j \to V'_j$, $j > 1$, are single-sheeted analytic covers, while above V'_+ the cover $\pi|_{A \cap V_1}$ is single-sheeted and above V'_- it is either single-sheeted (if $k_- = k_+$) or has empty domain of definition (if $k_- < k_+$).

If $k_- < k_+$, then $\bar{A} \cap V_1$ is the graph of a continuous function above $V'_+ \cup (\Gamma \cap V')$ which is holomorphic above V'_+. Since $M \in C^k$, above $\Gamma \cap V'$ this function is of class C^k. By the property of boundary regularity for holomorphic functions (derived in $\mathbf{C}^n$ from the Bochner-Martinelli integral representation just as it is derived in $\mathbb{C}^1$ from the Cauchy integral formula), $\bar{A} \cap V_1$ is a C^k-submanifold with boundary in V_1.

If $k_- = k_+$, then $\bar{A} \cap V_1$ is the graph of a continuous function above V' which is holomorphic above $V' \setminus \Gamma$. By Theorem A1.5 this function is holomorphic in V', hence $\bar{A} \cap V_1$ is a complex submanifold in V_1.

Thus, for $k_- = k_+$ the set $\bar{A}$ is analytic in Ω by Theorem 1, p.18.2.

So, for $p = n-1$ in a sufficiently small neighborhood U of an arbitrary point $a \in M$ the set $\bar{A} \cap U$ is either analytic or is a C^k-submanifold with boundary outside a closed subset of $(2p-1)$-measure zero. The same conclusion is true for any $p \geqslant 1$, since the general case clearly reduces to that of codimension 1, using almost single-sheeted projections in $\mathbb{C}^{p+1}$ (see p.3.6).

Now define E to be the set of points $z \in \bar{A}$ in no neighborhood of which $\bar{A}$ is a C^k-manifold (with or without boundary). As proved above, $\mathcal{H}_{2p-1}(E) = 0$, and (by definition) E is closed.

The set $M \setminus E$ can be represented as $M' \cup M''$, where M'' consists of the removable singularities of A, while in a neighborhood of each point of M' the set $\bar{A}$ is a C^k-manifold with boundary. If M'' is nonempty, $\bar{A}$ is analytic by Theorem 1, p.18.2. In the opposite case $M = M' \cup E$. ■

C o r o l l a r y. *Let M be a connected $(2p-1)$-dimensional C^k-submanifold of a complex manifold Ω, $p \geqslant 1$, $k \geqslant 1$, and let A be a pure p-dimensional analytic subset in $\Omega \setminus M$ such that $\bar{A}$ is not analytic in Ω. Then $\bar{A}$ has locally finite 2p-measure in Ω and M can be oriented such that Stokes' theorem holds for the pair (A,M): $d[A] = -[M]$.*

■ By the Theorem, there is a closed set $E \subset \bar{A}$ of $(2p-1)$-measure zero such that $(A \setminus E, M \setminus E)$ is a C^k-submanifold with boundary in $\Omega \setminus E$. Choose coordinates in a neighborhood of a point $a \in M$ such that $a = 0$ and the projections $\pi_I: \bar{A} \cap U_I \to U'_I \subset \mathbb{C}_I$, $\# I = p$, are proper in suitable neighborhoods $U_I \ni 0$ (this can be done because $\mathcal{H}_{2p+1}(\bar{A}) = 0$, see p.3.4). We may, moreover, assume that $\pi_I: M \cap U_I \to \Gamma_I \subset U_I'$ are diffeomorphisms onto corresponding hypersurfaces in U_I', partitioning U_I' into two domains, $U_{I\pm}'$. Since the number of sheets of $\pi_I|_{A \cap U_I}$ above $U_{I\pm}'$ is finite and since $\pi_I^{-1}(\Gamma_I) \cap \bar{A} \cap U_I$ has $2p$-measure zero (see the proof of the Theorem), by Wirtinger's theorem the $2p$-measure of

$\bar{A} \cap (\cap_{\#I=p} U_I)$ is finite.

Choose the orientation on $M \setminus E$ to be that of the boundary of $A \setminus E$. Analysis of the local situation (proof of the Theorem) shows that orientation can be extended onto all of M. Stokes' formula is also a local property, and in our situation its proof is the same as in 14.3 (expansion with respect to a positive basis of forms, locally finitely-sheeted covers, etc.). ■

19.2. Boundary uniqueness theorems. Theorem 19.1 readily implies the following two uniqueness Theorems.

P r o p o s i t i o n 1. *Let M be a connected* $(2p-1)$*-dimensional* C^1*-submanifold of a complex manifold* Ω, $p \geqslant 1$, *and let* A_1, A_2 *be two irreducible p-dimensional analytic subsets in* $\Omega \setminus M$ *whose closures in* Ω *contain M. Then either* $A_1 = A_2$ *or* $A_1 \cup M \cup A_2$ *is an analytic subset of* Ω.

■ Put $A = A_1 \cup A_2$. If $A \cup M$ is not analytic in Ω, by Theorem 19.1 there is a closed subset $E \subset A \cup M$ with $\mathcal{H}_{2p-1}(E) = 0$ such that $\bar{A} \setminus E$ are submanifolds with boundary in $\Omega \setminus E$. But then neither $A_1 \cup M$ nor $A_2 \cup M$ are analytic in Ω, hence, by the same Theorem 19.1, $\bar{A}_j \setminus E$, $j = 1,2$, is a submanifold with boundary in $\Omega \setminus E$. Since $\bar{A} = \bar{A}_1 \cup \bar{A}_2$, this implies $A_1 = A_2$ in a neighborhood of $M \setminus E$. Since A_1, A_2 are irreducible, $A_1 = A_2$. ■

P r o p o s i t i o n 2. *Let* M_1, M_2 *be* C^1*-submanifold of dimension* $2p-1$ *on a complex manifold* Ω, *and let* A_1, A_2 *be irreducible p-dimensional analytic subsets in* $\Omega \setminus (M_1 \cup M_2)$ *such that* $\bar{A}_1 = A_1 \cup M_1$ *and* $\bar{A}_2 = A_2 \cup M_2$. *Suppose that* $\bar{A}_1$ *and* $\bar{A}_2$ *are not analytic, that* $\mathcal{H}_{2p-1}(M_1 \cap M_2) > 0$, *and that the orientations of* M_1 *and* M_2 *compatible with, respectively,* A_1 *and* A_2 *are the same at the points of tangency of* M_1 *and* M_2. *Then* $A_1 = A_2$.

(By p.19.1, the canonical orientation of A_j induces almost everywhere on M_j an orientation, and this orientation can be continuously extended onto all of M_j.)

■ In view of the irreducibility of A_j it suffices to prove that $A_1 = A_2$ in a neighborhood of some point $a \in M_1 \cap M_2$. Hence we will assume that Ω is a domain in $\mathbb{C}^n$. Since $\mathcal{H}_{2p-1}(M_1 \cap M_2) > 0$, there is a point

$$a \in (M_1 \cap M_2) \setminus (E_1 \cup E_2)$$

in any neighborhood of which $M_1 \cap M_2$ has positive $(2p-1)$-measure. This, in particular, implies $T_a M_1 = T_a M_2$. Choose coordinates such that $a = 0$ and such that

$$T_a M_1 \cap \mathbf{C}^{n-p}_{z''} = \{0\}$$

(see the proof of Theorem 19.1). Then in a sufficiently small neighborhood $U \ni 0$ the sets A_j can be represented as graphs of functions above $U_j' \subset U' \subset \mathbf{C}^p$, where

$$\partial U_j' \cap U' = \Gamma_j = \pi(M_j \cap U)$$

are smooth hypersurfaces in U'. Since at a the orientations of $T_a M_1$ and $T_a M_2$ are the same, the orientations of Γ_1 and Γ_2 (being the boundaries of U_1' and U_2', respectively) at $a' = \pi(a)$ are also the same. Hence for U' sufficiently small, $V' = U_1' \cap U_2'$ is a nonempty domain in U' whose boundary in U' belongs to $\Gamma_1 \cup \Gamma_2$ and contains $\Gamma_1 \cap \Gamma_2$. Since both Γ_1 and Γ_2 are tangent to $T_{a'}\Gamma_1$, the boundary of V' inside U' is bijectively projected onto $T_{a'}\Gamma_1$, the projection satisfying a Lipschitz condition in both directions (U' is sufficiently small). Since $\mathcal{H}_{2p-1}(\Gamma_1 \cap \Gamma_2) > 0$, the boundary uniqueness Theorem (A1.1) implies that the holomorphic vector functions whose graphs are $A_j \cap U$ coincide in V', hence $A_1 = A_2$ in $U \cap \pi^{-1}(V')$. ■

19.3. Plateau's problem for analytic sets. Let A be a pure p-dimensional analytic set with boundary bA on a complex manifold Ω. According to the definition (p.14.3), the $(2p-1)$-measure of bA is locally finite and there is a closed set $E \subset \bar{A}$ with $\mathcal{H}_{2p-1}(E) = 0$ such that $\bar{A} \setminus E$ is a C^1-submanifold with boundary $(bA) \setminus E$ on the complex manifold $\Omega \setminus E$. By Proposition 1, A2.4, $M^\circ = (bA) \setminus E$ is a maximal complex manifold, i.e. the complex tangent plane $T_\zeta M^\circ \subset T_\zeta \Omega$ has, at each point $\zeta \in M^\circ$, maximum possible dimension $p-1$. (In this paragraph we will mainly consider analytic sets of dimensions $p > 1$, since for $p = 1$ the complex structure on bA degenerates and the boundary theory is slightly different.) Since $[bA] = -d[A]$ in the sense of currents in Ω (Stokes' formula, p.14.3), $[bA]$ is a d-closed current. In the sequel it is helpful to introduce a class of "manifolds with singularities", having the properties listed here.

D e f i n i t i o n. Let M be a closed subset of a complex manifold Ω of locally finite $(2p-1)$-measure, $p>1$, and let Σ be a nowhere dense closed subset in M with $\mathcal{H}_{2p-1}(\Sigma)=0$ such that $M^\circ = M\setminus\Sigma$ is an orientable maximal complex submanifold in $\Omega\setminus E$. If the current $[M]$ of integration over M° (which is defined in view of the local boundedness of $\mathcal{H}_{2p-1}|_M$) is closed, M is called a *maximal complex cycle* (MC-cycle for short) in Ω.

Apart from the example $M = bA$ given above, any orientable maximal complex submanifold in Ω (the case $M^\circ = M$) is a maximal complex cycle. The condition that $[M]$ be d-closed imposes a certain restriction on the behavior of M in a neighborhood of the singular set Σ. This condition is satisfied if, e.g., M is a piecewise smooth submanifold in Ω, if the $(2p-1)$-measure of the δ-neighborhood of Σ on M has order $o(\delta)$ as $\delta\to 0$, etc.

A function f defined on a $(2p-1)$-dimensional MC-cycle M in Ω and locally integrable with respect to $\mathcal{H}_{2p-1}|_M$ will be called a CR-function on M if f satisfies the tangential Cauchy-Riemann equations on M:

$$\int_M f\bar\partial\phi = 0, \quad \phi\in\mathcal{D}^{p,p-2}(\Omega)$$

(see A2.4 and A5.2; an integral over M is, by definition, the integral over M°).

An MC-cycle M will be called *irreducible* if it cannot be represented as $M_1\cup M_2$ where M_1, M_2 are also MC-cycles, distinct from M.

MC-cycles are the natural "candidates" for boundaries of analytic sets. If the topological boundary of an analytic set of dimension $p>1$ is a smooth $(2p-1)$-dimensional manifold, then by Theorem 19.1 and A2.4, this manifold is maximal complex. It is natural to state the converse problem: Under what additional conditions will a given maximal complex manifold, or, more general, a given MC-cycle, be the boundary of an analytic set of corresponding dimension? The problem of existence and construction of such "analytic films" is an analog of the classical Plateau problem, especially since analytic sets are minimal surfaces (see p.14.4).

For compact MC-cycles on Stein manifolds this problem was completely solved by Harvey and Lawson ([164], [167]): there are no additional restrictions necessary; a solution with compact support always exists and is unique (see also [168]). We prove the theorem of Harvey and Lawson (with certain generalizations) in the next two Sections, while we now investigate the simplest model situation in which the Plateau problem for analytic sets is solvable.

P r o p o s i t i o n. *Let M be a maximal complex manifold of dimension $2p-1>1$ in $U = U' \times \mathbf{C}^{n-p}$, where U' is a convex domain in $\mathbf{C}^p$. Let the projection $\pi\colon M \to \Gamma \subset U'$ be a diffeomorphism onto a hypersurface Γ of class C^1, partitioning U' into two domains, U'_+ and U'_-. Suppose that for any holomorphic function on U'_- there is a holomorphic continuation onto all of U'. Then M is the boundary in U of a complex manifold-graph above U'_+ (the boundary is to be understood in the topological sense and in the sense of currents, under a corresponding orientation).*

We immediately note that the condition on U'_- is satisfied when, e.g., $U'_+ \cup \Gamma$ is compact (this follows from removability of compact singularities for holomorphic functions in $\mathbf{C}^p$, $p>1$, see A1.3).

■ The manifold M is the graph above Γ of a vector function $(f_{p+1}, \ldots, f_n)$. Since

$$\int_\Gamma f_j \psi = \int_M z_j \cdot (\pi^* \psi), \quad \psi \in \mathfrak{D}^{2p-1}(U'),$$

and $\pi^* \bar\partial \phi = \bar\partial(\pi^* \phi)$, $\phi \in \mathfrak{D}^{p,p-2}(U')$, the f_j satisfy on Γ the tangential Cauchy-Riemann equations. By the jump Theorem (A5.4) there are functions $h_j^{\pm}$, continuous in $U'_\pm \cup \Gamma$ and holomorphic in $U'_\pm$, respectively, such that $f_j = h_j^+ - h_j^-$ on Γ. By requirement the h_j^- have holomorphic continuations onto U'; subtracting these from h_j^+ we obtain new functions F_j, continuous in $U'_+ \cup \Gamma$ and holomorphic in U'_+, with the values f_j on Γ. The graph of the vector function $(F_{p+1}, \ldots, F_n)$ in $U' \times \mathbf{C}^{n-p}$ is, obviously, the required complex manifold with boundary M. ■

We will not use this Proposition in the sequel; we have stated it only to demonstrate by what methods the given problem concerning analytic films may be solved.

19.4. Preparation lemmas. We will solve Plateau's problem for analytic sets along the lines of Harvey-Lawson ([167], [164]), using suitable projections and the construction of analytic covers with given boundary values. In order to do this we need the following geometric lemmas on the structure of projections of MC-cycles.

A map $f\colon M \to \Gamma$ is called *almost single-sheeted* if $f^{-1}(f(z)) = \{z\}$ for the points z in a certain everywhere dense subset on M.

L e m m a 1. *Let M be a C^1-manifold of dimension $2p-1$ in $\mathbb{C}^n$ whose closure has finite $(2p-1)$-measure. Then the set of linear maps $l\colon \mathbb{C}^n \to \mathbb{C}^p$ for which the restriction $l|_{\overline{M}}$ is almost single-sheeted is everywhere dense in the space of linear maps $\mathbb{C}^n \to \mathbb{C}^p$.*

■ Fix a countable everywhere dense subset $\{a_j\} \subset M$; by induction with respect to $m = n, \ldots, p$ we will prove that almost every linear map $l\colon \mathbb{C}^n \to \mathbb{C}^m$ has the following properties: for certain neighborhoods $M_j \ni a_j$ on M (depending on l), all restrictions $l|_{M_j}$ are diffeomorphisms, the $l(M_j)$ large open in $l(\overline{M})$, and $l^{-1}(l(M_j)) \cap \overline{M} = M_j$. For $m = n$ the statement is trivial.

$m \Rightarrow m-1 \geqslant p$. Let $\tilde{l}\colon \mathbb{C}^n \to \mathbb{C}^m$ have the properties indicated with certain M_j, and let $\tilde{a}_j = \tilde{l}(a_j)$, $\tilde{M}_j = \tilde{l}(M_j)$. For each j a map

$$\Pi_j\colon \mathbb{C}^m \setminus \{\tilde{a}_j\} \to \mathbb{P}_{m-1}$$

is defined, under which $\zeta \to [\zeta - \tilde{a}_j]$. Since $\tilde{M}_j$ is regular at $\tilde{a}_j$ and open in $\tilde{l}(\overline{M})$, while $\tilde{l}(\overline{M})$ has finite $(2p-1)$-measure, the closure of $\Pi_j(\tilde{M}_j \setminus \{\tilde{a}_j\})$ in $\mathbb{P}_{m-1}$ has $2p$-measure zero (see A6.1, property 4). Since $m-1 \geqslant p$, the union of the sets over all j has $2(m-1)$-measure zero in $\mathbb{P}_{m-1}$, hence its complement is everywhere dense. If $[c]$ belongs to this complement and if $\lambda\colon \mathbb{C}^m \to \mathbb{C}^{m-1}$ is a map with fibers parallel to c, then

$$\operatorname{rank}_{\tilde{a}_j} \lambda|_{\tilde{M}_j} = 2p-1 \quad \text{and} \quad \lambda^{-1}(\lambda(\tilde{a}_j)) \cap \tilde{l}(\overline{M}) = \{\tilde{a}_j\}$$

for all j. This readily implies that $\lambda \circ l$ has the necessary properties; in particular, it is almost single-sheeted on M. Clearly, the maps $\lambda \circ l$ form an everywhere dense subset in the space of linear maps $\mathbb{C}^n \to \mathbb{C}^{m-1}$. ■

L e m m a 2. *Let $M = M^0 \cup \Sigma$ be a closed subset of the half-space $U\colon x_1 > 0$ in $\mathbb{C}^n$, where Σ is closed in M, $\mathcal{H}_{2p-1}(\Sigma) = 0$, and M^0 is a C^1-manifold of dimension $2p-1 \geqslant 1$ that is everywhere dense in M. Suppose that the projection $\pi|_M$ into the half-space $U' = U \cap \mathbb{C}^p_{z'}$ is proper and almost single-sheeted. Then there is a closed subset $\Sigma' \subset \Gamma = \pi(M)$, nowhere dense in Γ and such that $\Gamma \setminus \Sigma'$ is a C^1-manifold, nowhere dense in Γ, and such that $\Gamma \setminus \Sigma'$ is a C^1-manifold, $\pi(\Sigma) \subset \Sigma'$, and*

$$\pi\colon M \setminus \pi^{-1}(\Sigma') \to \Gamma \setminus \Sigma'$$

is a diffeomorphism. Moreover, the coordinate origin (in $\{x_1=0\}$ *in* $\mathbb{C}^p$*) can be chosen such that the image of* Σ' *under the projection*

$$\rho\colon z' \mapsto \frac{z'}{|z'|}$$

from $\mathbb{C}^p_*$ *onto the unit sphere* S^{2p-1} *is closed and nowhere dense in* $S^{2p-1} \cap U'$*, while the rank of* $\rho|_{\Gamma\setminus\Sigma'}$ *is identically equal to* $2p-1$.

■ Let Σ_1' be the set of points on Γ in no neighborhood of which Γ is a C^1-manifold. Choose the coordinate origin in $\{x_1=0\}$ in $\mathbb{C}^p$ such that the rank of $\rho|_{\Gamma\setminus\Sigma_1'}$ is equal to $2p-1$ on an everywhere dense subset (this can clearly be done). Denote by $\Sigma_0 \subset M$ the union of Σ and the set of critical points of the map $\rho\circ\pi$; put $\Sigma' = \Sigma_1' \cup \pi(\Sigma_0)$. The almost single-sheetedness of $\pi|_M$ and Sard's theorem readily imply that Σ' is the required set. ■

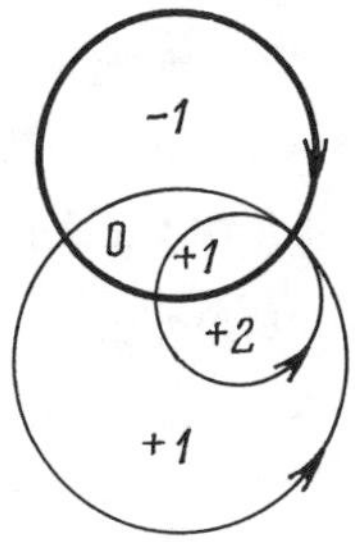

Figure 12.

Viewed as currents, MC-cycles can be the boundaries of holomorphic chains whose multiplicities need not be of one sign (see the schematic Figure 12, where the chain is the sum of the components of the complement to the three circle-chains; the multiplicities of these components are given, ensuring Stokes' formula to hold). Hence, in general an analytic film with given MC-boundary cannot be defined by Weierstrass polynomials (as analytic covers), but is defined by quotients of such polynomials. We analyze the situation using the example of divisors.

L e m m a 3. *Let* T *be a holomorphic* p*-chain in a domain* $U'\times\mathbb{C}_w \subset \mathbb{C}^{p+1}$ *such that the projection* $\pi\colon |T| \to U' \subset \mathbb{C}^p$ *of its support is an* m*-sheeted analytic cover with critical set* $\sigma \subset U'$. *Let* $w_1(z'), \ldots, w_m(z')$ *be the* w*-coordinates of the points in the fiber*

$$\pi^{-1}(z') \cap |T|, \quad z' \in U' \setminus \sigma,$$

and let $k_j(z') \in \mathbb{Z}$ *be the multiplicity of* T *at* $(z', w_j(z'))$. *Then*

$$s_k(z') = \sum_1^m k_j(z') \cdot w_j(z')^k,$$

$k = 0, 1, \ldots$, *has a continuation* S_k *which is holomorphic in* U'; *moreover,* S_0 *is an integer. The chain* T *is the divisor of a function* $R(z', w)$ *that is meromorphic in* $U' \times \mathbb{C}$ *and rational in* w; *moreover, in a neighborhood of* $w = \infty$ *this function has the representation*

$$R(z', w) = w^{S_0} \exp\left[-\sum_1^\infty \frac{S_k(z')}{k w^k} \right].$$

■ Let $T = \sum m_j A_j = T^+ - T^-$, where $T^+ = \sum_{m_j > 0} m_j A_j$ and T^- are positive chains. Construct the defining Weierstrass polynomials, $P(z', w)$ and $Q(z', w)$, respectively, for the covers π: $T^+ \to U'$ and π: $T^- \to U'$ (counted with the multiplicities of $T^{\pm}$). Then T is the divisor of the function $R = P / Q$.

In small domains $V' \subset U' \setminus \sigma$ the functions $w_j(z')$ can be chosen holomorphic in z'. The multiplicity of T on a sheet $w = w_j(z')$, $z' \in V'$, is constant, i.e. $k_j(z') \equiv k_j \in \mathbb{Z}$, hence

$$R(z', w) = \prod_1^m (w - w_j(z'))^{k_j}$$

and $S_0 = \sum_1^m k_j = \deg P - \deg Q$ is a constant. If $|w| > \max_j |w_j(z')|$, we can define

$$\log \frac{R(z', w)}{w^{S_0}} = \sum_{j=1}^m k_j \log \left[1 - \frac{w_j(z')}{w} \right] = -\sum_{k=1}^\infty \frac{S_k(z')}{k w^k}.$$

This equality holds on σ also, by continuity; it clearly implies the required representation of R in a neighborhood of $w = \infty$ in $U' \times \mathbb{C}_w$. ■

The essence of this Lemma is as follows. When the chain T itself is absent and we only have its boundary M, we will not succeed in directly constructing the polynomials P and Q for future T^+ and T^-. On the other hand, the functions S_k can easily be defined by the boundary values on the projection of M (see below); Lemma 3 indicates a way to pass from the S_k to functions defining the divisor required. We will develop this program in the next Section.

19.5. Boundaries of analytic covers. The basic step in the proof of the Harvey-Lawson theorem is the following

L e m m a. *Let M be a maximal complex cycle of dimension $2p-1>1$ in the half-space U: $x_1>0$ in $\mathbf{C}^n$ with compact closure in $\mathbf{C}^n$, let $\lambda(z'')$ be a linear function on $\mathbf{C}^{n-p}_{z''}$, and let M_λ be the image of M under the map $z \mapsto (z,w)$, $w = \lambda(z'')$. If the projection $\pi|_M$ into the half-space $U' = U \cap \mathbf{C}^p_{z'}$ is almost single-sheeted, then there are a closed subset Σ', nowhere dense in $\pi(M)$, and a pure p-dimensional analytic subset A_λ in $((U'\setminus\Sigma')\times\mathbf{C}_w)\setminus M_\lambda$ such that $M_\lambda \subset \overline{A}_\lambda$, $(\overline{A}_\lambda\setminus A_\lambda)\cap((U'\setminus\Sigma')\times\mathbf{C}_w)\subset M_\lambda$, and such that the projection of $\overline{A}_\lambda$ into $\mathbf{C}^p_{z'}$ is a compact set.*

■ Since $\pi|_M$ is proper, for each $k\in\mathbf{Z}_+$ there are defined currents $T_k = -\pi_*([M]w^k)$ in U' ($w=\lambda(z'')$), acting by the formula

$$<T_k,\psi> = -\int_M w^k\cdot(\pi^*\psi), \quad \psi\in\mathcal{D}^{2p-1}(U').$$

All these currents are $\bar\partial$-closed since $\bar\partial\phi = \bar\partial\phi^{p,p-1}$ and $w^k(\pi^*\bar\partial\phi) = \bar\partial(w^k(\pi^*\phi)^{p,p-1}) = d(w^k(\pi^*\phi)^{p,p-1})$ on M, while the current $[M]$ is closed in U.

By A5.4 there are generalized functions (distributions) S_k in U', of bounded supports in $\mathbf{C}^p$, such that $\bar\partial S_k = T_k$. Since the T_k are currents of measure type, the S_k are ordinary integrable functions, holomorphic outside the supports of all T_k, i.e. outside $\Gamma = \pi(M)$. All S_k vanish on the unbounded component of $U'\setminus\Gamma$.

We first investigate S_0. If Σ_0' is the exceptional set of Lemma 2, p.19.4, the current $\pi_*[M]$ coincides within $U'\setminus\Sigma_0'$ with $[\Gamma]$, under suitable orientations of the connected components of $\Gamma\setminus\Sigma_0'$. Since $\bar\partial S_0 = -[\Gamma]$ in $U'\setminus\Sigma_0'$, the jump Theorem (A5.4) implies that S_0 has limit values on $\Gamma\setminus\Sigma_0'$ from both sides of this manifold, and the difference between these values is 1. Since $S_0 = 0$ on the unbounded component of $U'\setminus\Gamma$, and since $\rho(\Sigma_0')$ is nowhere dense on the unit sphere, this implies that S_0 is integer valued and constant on the components of $U'\setminus\Gamma$.

We define above $U'\setminus\Gamma$ a function $R(z',w)$, at first for sufficiently large w. For convenience we will assume that M lies in the unit ball of $\mathbf{C}^n$ and that $||\lambda|| = 1$.

Since the currents $[M]w^k$, and hence the T_k, have locally bounded masses, by A5.4 the functions S_k are uniformly bounded on each compact set in $U' \setminus \Gamma$. Hence the function

$$R(z',w) = w^{S_0} \exp\left[-\sum_1^\infty \frac{1}{k} S_k(z') w^{-k}\right]$$

is defined and holomorphic in $(U \setminus \Gamma') \times \{|w|>1\}$, and is zero free in it.

Let U_0, $U_1, \ldots$, be the connected components of $U' \setminus \Gamma$, with U_0 unbounded, and let R_j be the restriction of R onto U_j. Since all S_k have a continuous extension onto $\overline{U}_j \cap (U' \setminus \Sigma_0')$ (see A5.4), the R_j also have continuous extensions, onto

$$(\overline{U}_j \cap (U' \setminus \Sigma_0')) \times \{|w|>1\}$$

using the formula above. If γ is a connected component of $\Gamma \setminus \Sigma_0'$ belonging simultaneously to the boundaries of U_j and U_i, then $\pi^{-1}(\gamma) \cap M_\lambda$ is defined in $\mathbb{C}^{p+1}$ by an equation $w = \phi(z')$ with $\phi \in C^1(\gamma)$. Hence the current T_k coincides, in a neighborhood of γ, with $-[\gamma]\phi^k$ (with on γ the orientation corresponding to $\pi_*[M]$). Since $\bar\partial S_k = T_k$, the jump Theorem (A5.4) implies that the difference of the limit values of S_k on γ from the side of U_j and from the side of U_i is equal to $\pm\phi^k$ (the sign depending on whether or not the orientations of γ as part of $\pi_*[M]$ and as part of the boundary of U_j are the same). Hence, for $z' \in \gamma$ and $|w|>1$,

$$\frac{R_j(z',w)}{R_i(z',w)} = w^{\pm 1} \exp\left[\mp \sum_1^\infty \frac{1}{k}\left[\frac{\phi(z')}{w}\right]^k\right] = (w-\phi(z'))^{\pm 1}, \qquad (\star)$$

i.e. when passing across γ the function R is multiplied by a (fractional-linear) factor $(w-\phi(z'))^{\pm 1}$. Since $R_0 \equiv 1$, formula $(\star)$ and E. Levi's boundary theorem (see A1.1) imply that R_j has a meromorphic continuation onto $U_j \times \mathbb{C}$ (which is fractional-linear in w) if U_j and U_0 have a boundary part in common that belongs to $\Gamma \setminus \Sigma_0'$. Since R_j has a continuous extension onto

$$(\overline{U}_j \cap U' \setminus \Sigma_0') \times \{|w|>1\},$$

which remains holomorphic in w, and since the condition of being a rational function (with denominator of given degree) can be analytically expressed in terms of the coefficients of the Taylor or Laurent series (see the proof of Levi's theorem in A1.1), for such j the function $R_j(z',w)$, $z' \in \partial U_j \cap (U' \setminus \Sigma_0')$, has a rational continuation onto $\mathbf{C}_w$ with denominator of degree $\leqslant 1$. If U_i has with such an U_j a boundary part γ in common that belongs to $\Gamma\setminus\Sigma_0'$, formula (*) implies that R_i has a meromorphic continuation onto $U_i\times\mathbf{C}_w$ and $R_i = P_i/Q_i$, where P_i, Q_i are Weierstrass polynomials of degree $\leqslant 2$. Since any component of $U'\setminus\Gamma$ can be joined with U_0 by a segment intersecting Γ in finitely many points all of which belong to $\Gamma\setminus\Sigma_0'$ (Lemma 2, p.19.4), proceeding in the above manner by induction we obtain that each R_j has a meromorphic continuation onto $U_j\cup\mathbf{C}$ and can be represented as the quotient of two Weierstrass polynomials, P_i/Q_i. We may assume that this quotient is irreducible, i.e. that Q_i has lowest possible degree.

If $a'\in\partial U_j\cap(\Gamma\setminus\Sigma_0')$ and $z_\nu'\in U_j$, $z_\nu'\to a'$ as $\nu\to\infty$, then we can choose a subsequence from $R_j(z'_\nu,w)$ converging to a function that is also rational in w. Since R_j is continuous on $(\overline{U}_j\cap U'\setminus\Sigma_0')\times\{|w|>1\}$, the uniqueness Theorem implies that the limit function thus obtained is independent of the choice of the sequence $z_\nu'\to a'$; we thus denote it by $R_j(a',w)$. The number of its poles can be less than $\deg Q_j$, but there are few such a': if $R_j(a',w)$ has $\leqslant m$ poles in w (counted with multiplicities) for a' from a set of positive $(2p-1)$-measure on $\partial U_j\cap(\Gamma\setminus\Sigma_0')$, Levi's boundary Theorem (A1.1) implies that this is true for all $z'\in U_j$. If $R_j(a',w)$ has maximum possible number of poles (equal to $\deg Q_j$), Hurwitz' theorem (in a neighborhood of these poles) implies that $R_j(z',w)$ also has $\deg Q_j$ poles for all z' in a neighborhood of a' in $\overline{U}_j$. Since the poles of $R_j(z',w)$ tend to poles of $R_j(a',w)$ as $z'\to a$ (Hurwitz' theorem), $Q_j(z',w)$ also has a limit (by Viète's formulas). This limit polynomial in w, of $\deg Q_j$, will be denoted by $Q_j(a',w)$. Thus, by adding to Σ_0' a necessarily closed subset of $\Gamma\setminus\Sigma_0'$ of $(2p-1)$-measure zero, we obtain a set Σ', closed and nowhere dense on Γ, such that all Q_j (hence all P_j) have a continuous extension onto $\overline{U}_j\times(U'\setminus\Sigma')$; moreover the zeros of $Q_j(a',w)$ and of $P_j(a',w)$ for $a'\in\Gamma\setminus\Sigma'$ do not intersect.

We define in $(U'\setminus\Gamma)\times\mathbf{C}_w$ a holomorphic p-chain T, by putting $T|_{U_j\times\mathbf{C}} = D_{R_j}$ (the divisor of the meromorphic function R_j). Since $R_0\equiv 1$, the projection of the support of T in $\mathbf{C}^p$ has compact closure $(\subset\overline{U'}\setminus U_0)$. We show that T has an extension to a holomorphic p-chain in $((U'\setminus\Sigma)\times M_\lambda$.

Let γ be a connected component of $\Gamma\setminus\Sigma_0'$, part of the common boundaries of U_i and U_j, and suppose that, for the sake of being specific, in formula (*) the + sign holds, i.e. $R_j(z',w) = (w-\phi(z'))R_i(z',w)$ for $z'\in\gamma$. Fix an $a'\in\gamma\setminus\Sigma'$. If some point $(a',b)\notin M_\lambda$ is a limit point for the support of T, the definition of Σ'

implies that b is a zero or a pole of $R_i(a',w)$. Hence there is a polydisk $V = V' \times W \ni (a',b)$ disjoint from M such that $|T|$ does not have limit points on $V' \times \partial W$ and such that the disk W does not contain zeros and poles of R_i other than $w = b$. Suppose, for the sake of being specific, that $R_i(a',b) = 0$. Then $Q_i(z',w)$ is zero free in W for any $z' \in \overline{U}_i \cap V'$, while $P_i(z',w)$ necessarily has a zero in it (when V' is sufficiently small). Formula (*) implies that $Q_j(z',w)$ is also zero free in W, while $P_j(z',w)$ with $z' \in \overline{U}_j \cap V'$ necessarily has a zero in W; moreover, for $\zeta \in \gamma \cap V'$, the zeros of $P_i(\zeta,w)$ and $P_j(\zeta,w)$ in W coincide (counted with multiplicities). For $z' \in \overline{U}_j \cap V'$ the polynomial P_i has the zeros $\alpha^1(z'), \ldots, \alpha^{m_i}(z')$ (written with multiplicities) in W. The function $p_i(z',w) = \prod_1^{m_i}(w - \alpha^\nu(z'))$ is holomorphic in $(U_i \cap V) \times \mathbf{C}$ and continuous in $(\overline{U}_i \cap V) \times \mathbf{C}$. We define $p_j(z',w)$ analogously. By what has been proved above, $p_i(\zeta,w) = p_j(\zeta,w)$ for all $\zeta \in \gamma \cap V'$. Thus, the function $p(z',w)$ equal to p_i in $(\overline{U}_i \cap V') \times \mathbf{C}$ and to p_j in $(U_j \cap V') \times \mathbf{C}$ is continuous in $V' \times \mathbf{C}$ and is holomorphic outside the smooth hypersurface $\gamma \times \mathbf{C}$. By Theorem A1.5, $p(z',w)$ is holomorphic in $V' \times \mathbf{C}$. By construction, the divisor of p coincides with the chain T in $(U_i \cap V') \times W$ and $(U_j \cap V') \times W$; hence T has an extension to a p-chain in V. It is clear from the construction that these local p-chains coincide in pairwise intersections, and T is thus extended to a p-chain in $((U' \times \Sigma') \times \mathbf{C}) \setminus M_\lambda$.

The support of this chain is a pure p-dimensional analytic subset A_λ in $((U' \setminus \Sigma') \times \mathbf{C}) \setminus M_\lambda$, hence within $(U' \setminus \Sigma') \times \mathbf{C}$ the topological boundary of A_λ belongs to M_λ. But formula (*) implies that all points of $M_\lambda \setminus (\Sigma' \times \mathbf{C})$ are limit point for A_λ, hence $M_\lambda \subset \overline{A}_\lambda$. ■

19.6. The Harvey-Lawson theorem. We first give the original formulation of the Harvey-Lawson theorem, concerning complex analytic films with given compact boundaries (see [167], [164]).

Theorem 1. *Let M be a compact maximal complex cycle of dimension $2p-1 > 1$ on a Stein manifold Ω. Then there is a unique holomorphic p-chain T in $\Omega \setminus M$, of finite mass and of compact support in Ω, whose boundary in the sense of currents in Ω equals M (i.e. $dT = -[M]$); if the MC-cycle M is irreducible, T is irreducible too.*

(The formulation of the Harvey-Lawson theorem includes also the case $p = 1$

and the regularity of $|T|$ up to the boundary, which follows from Theorem 19.2, see [167], [164].)

We will prove the following statement, partly generalizing Theorem 1 in the spirit of Stolzenberg's theorem on polynomial hulls of smooth curves (cf. [134]).

Theorem 2. *Let Ω be a Stein manifold, K a compact set, convex with respect to $\mathcal{O}(\Omega)$, and M a maximal complex cycle of dimension $2p-1>1$ in $\Omega \setminus K$ such that $K \cup M$ is compact. Then there is a pure p-dimensional analytic subset A in $\Omega \setminus (K \cup M)$ such that $A \cup K \cup M$ is compact and such that M belongs to $\bar{A}$ (the closure of A). Moreover, there is a closed subset $E \subset \bar{A} \setminus K$ (possibly empty) such that $\mathcal{H}_{2p-1}(E) = 0$ and $\bar{A} \setminus E$ is a submanifold with boundary (belonging to M) in $\Omega \setminus K$, of the same smoothness class as M.*

The last statement clearly follows from Theorem 19.1, and we will not prove it. We clarify some terminology. A compact set $K \subset \Omega$ is called convex with respect to $\mathcal{O}(\Omega)$ if for each point $a \in \Omega \setminus K$ there is a function $f \in \mathcal{O}(\Omega)$ for which $|f(a)| > \max_K |f|$. If $\Omega = \mathbf{C}^n$, this property is called *polynomial convexity*. A Stein manifold is a complex manifold that is biholomorphically equivalent to a complex submanifold of $\mathbf{C}^n$ (for some n). It is known that every submanifold of $\mathbf{C}^n$ is the set of common zeros of a family of entire functions (cf., e.g., [35]). Hence, if Ω is a complex submanifold in $\mathbf{C}^n$ and A is a pure p-dimensional analytic set in $\mathbf{C}^n$ with compact closure and with boundary belonging to Ω, then the maximum principle (for $f|_A$, f entire) implies that all of A also belongs to Ω. Thus, Plateau's problem on Stein manifolds (with compact boundaries) reduces to the similar problem in $\mathbf{C}^n$. Here we can ask: why at all consider it on such abstract manifolds, if the problem is properly contained in $\mathbf{C}^n$ itself? To this we can reply the following: any domain of holomorphy in $\mathbf{C}^n$ is a Stein manifold (see [35]), hence the considerations given lead to the following supplementary information concerning the solution of Plateau's problem for analytic sets:

The set A in Theorem 2 is contained in the envelope of holomorphy of $K \cup M$, i.e. in the intersection of all domains of holomorphy $D \subset \Omega$ containing $K \cup M$.

We yet give two simplifying remarks.

1) If $K_j \supset K_{j+1}$ is a sequence of polynomially convex compact sets in $\mathbf{C}^n$ converging to K as $j \to \infty$ (i.e. $K = \cap K_j$), and if A_j are the analytic sets from

Theorem 2 constructed for $K_j \cup M$, the maximum principle (p.6.3) and the boundary uniqueness Theorem (p.19.2) clearly imply that $A_{j+1} = A_j$ in $\mathbf{C}^n \setminus (K_j \cup M)$, and hence $A = \cup A_j$ is the solution of our problem for $K \cup M$. The definition of polynomial convexity easily implies that K is the intersection of some, inclusion decreasing, sequence of compact polynomial polyhedra K_j, i.e. sets of the form $\{|f_{j1}| \leqslant 1, \ldots, |f_{jk}| < 1\}$, where $f_{j\nu}$ are polynomials. Hence it suffices in Theorem 2 to investigate the case when K is a compact polynomial polyhedron $\{|f_1| \leqslant 1, \ldots, |f_k| \leqslant 1\}$ in $\mathbf{C}^n$ and when the $(2p-1)$-measure of M is finite.

2) Let G be the graph of the vector function $(f_1, \ldots, f_k)$ in $\mathbf{C}_z^n \times \mathbf{C}_w^k$, and let π be projection of G into $\mathbf{C}_z^n$. Then

$$\tilde{K} = \pi^{-1}(K) \cap G = G \cap \{|w_j| \leqslant 1, \ 1 \leqslant j \leqslant k\}.$$

Since the polydisk is convex, $\tilde{K}$ coincides with conv$(\tilde{K}) \cap G$, where conv denotes the convex hull (in $\mathbf{C}^{n+k}$). The pair conv $\tilde{K}$, $\tilde{M}$, with $\tilde{M} = \pi^{-1}(M) \cap G$, satisfies the conditions of Theorem 2. If $\tilde{A}$ is the analytic set in $\mathbf{C}^{n+k}$ corresponding by Theorem 2 to this pair, the maximum principle implies that every irreducible component of $\tilde{A}$ has an open part on the boundary belonging to $\tilde{M}$. Since the polynomial equations $w_j = f_j(z)$, $j = 1, \ldots, k$, hold on $\tilde{M}$, the boundary and ordinary uniqueness Theorems for holomorphic functions (see A1.1) and p.19.1 imply that these equations hold at all points of A, hence $\tilde{A} \subset G$. Thus, the projection of $\tilde{A}$ in $\mathbf{C}^n$ gives the required A for the pair (K,M). In other words, we have reduced Theorem 2 to the situation of convex compact sets K and MC-cycles with closures of finite $(2p-1)$-measure in $\mathbf{C}^n$. We investigate this situation.

■ Let $l = (l_1, \ldots, l_p)$: $\mathbf{C}^n \to \mathbf{C}^p$ be a linear map, almost single-sheeted on M, and let λ be an arbitrary linear function independent of $l_1, \ldots, l_p$. Then $l_1 - \max_K \operatorname{Re} l_1, \ l_2, \ldots, l_p$ can be taken as new first coordinates $w' = (w_1, \ldots, w_p)$ in $\mathbf{C}^n$, and we may assume that λ depends on w'' only. Now $M \cap \{\operatorname{Re} w_1 > 0\}$ and λ satisfy the conditions of Lemma 19.5. Let Σ' be the exceptional set from Lemma 19.5 for this data, let

$$\Sigma_{l,\lambda} = \{z \in \mathbf{C}^n \colon w'(z) \in \Sigma'\} \text{ and let } A_{l,\lambda} = \{z \in \mathbf{C}^n \colon w'(z) \in A_\lambda\}.$$

Let $\lambda_{p+1}, \ldots, \lambda_n$ be linear functions forming together with $l_1, \ldots, l_p$ a basis in $\mathbf{C}^n$. Since $A_{l,\lambda}$ is locally defined by meromorphic functions in the variables l,λ, the analytic set

$$A_l = \bigcap_{j=p+1}^{n} A_{l,\lambda_j}$$

has pure dimension p. It is an analytic subset of the domain $U_l \setminus M$, where

$$U_l = \{z \in \mathbf{C}^n \setminus \Sigma_l\colon \operatorname{Re} w_1 > 0\}$$

and $\Sigma_l = \cup_{j=p+1}^{n} \Sigma_{l,\lambda_j}$. By construction, the closure of A_l contains $M \cap \{\operatorname{Re} w_1 > 0\}$. Since $l|_{M \setminus \Sigma_l}$ is a bijective map (see Lemma 19.5), the topological boundary of A_l inside U_l belongs to M. Put

$$U_l^\epsilon = U_1 \cap \{\operatorname{Re} w_1 > \epsilon\}, \quad \epsilon > 0.$$

Let l' be another linear map $\mathbf{C}^n \to \mathbf{C}^p$, almost single-sheeted on M, and let $A_{l'}$ be its corresponding analytic subset in $U_{l'} \setminus M$. If l' is sufficiently close (in norm) to l, we have $U_{l'} \cup \Sigma_{l'} \supset M \cap U_l^\epsilon$. By the maximum principle (for $\operatorname{Re} l_1$), the boundary of each irreducible component of $A_l \cap U_l^\epsilon$ contains a relatively open subset in M. Since $M \subset \overline{A}_l$ and the orientations induced on the parts of M^0 from A_l and $A_{l'}$ are the same if l,l' are sufficiently close, the boundary uniqueness Theorem (p.19.2) implies $A_l \cap U_l^\epsilon \subset A_{l'}$. Similarly, $A_{l'} \cap U_{l'}^\epsilon \subset A_l$, hence $A_l = A_{l'}$ in $U_l^\epsilon \cap U_{l'}^\epsilon$ if l,l' are sufficiently close. Since any l,l' can be "joined" by a finite chain l^j of arbitrarily close l^j, l^{j+1}, this implies that $A_l = A_{l'}$ in $U_l^\epsilon \cap U_{l'}^\epsilon$. Thus, the sets A_l can be glued to one pure p-dimensional analytic subset A in the union of the U_l^ϵ over all l for which $l|_M$ is almost single-sheeted. Since $U_{cl}^\epsilon = U_l^{\epsilon/c}$, $c > 0$, we have $\cup U_l^\epsilon = \cup U_l$. The maps that are almost single-sheeted on M form an everywhere dense set in the space of all linear maps $\mathbf{C}^n \to \mathbf{C}^p$ (Lemma 1, p.19.4). Hence for each point $z^0 \in \mathbf{C}^n \setminus (K \cup M)$ there is such an l with the additional properties: $\operatorname{Re} l_1(z^0) > \max_K \operatorname{Re} l_1$, and $l_1(z^0) \notin l_1(M)$, i.e. $z^0 \in U_l$. Thus,

$$\cup U_l \supset \mathbf{C}^n \setminus (K \cup M),$$

and also A, is an analytic subset in $\mathbf{C}^n \setminus (K \cup M)$. By construction $\overline{A}$ contains $\overline{M}$. Since the intersection $\cap \Sigma_l$ over all l indicated belongs to M, we have $\overline{A} \setminus A \subset K \cup M$, hence M is the topological boundary of A in $\mathbf{C}^n \setminus K$. Finally, since $A_l = A_{l'}$ in $U_l^\epsilon \cap U_{l'}^\epsilon$ for any arbitrary close l,l', all sets $l(\overline{A})$ are compact; hence $A \cup K \cup M$ is compact also. ■

R e m a r k. Theorem 2 can be strengthened to the construction of a holomorphic p-chain T in $\mathbb{C}^n \setminus (K \cup M)$, with locally finite mass in $\mathbb{C}^n \setminus K$ and with compact $|T|$, such that $dT = -[M]$. In order to do this we must strengthen Lemma 19.5 correspondingly, and repeat the final steps from the proof of the Harvey-Lawson theorem in [164]. Essentially, we have proved this in case M is a connected maximal complex manifold (without singularities); see Proposition 2, p.19.2.

19.7. On singularities of analytic films. What conditions on the manifold M would ensure that the solution of Plateau's problem for analytic sets is smooth, i.e. that the set $A = |T|$ in the Harvey-Lawson theorem is singularity free (K empty)? In general the answer is very complicated (see [206]). We give two descriptive geometric results concerning this theme. They are due to M.P. Gambaryan.

P r o p o s i t i o n 1. *Let M be a compact $(2p-1)$-dimensional maximal complex manifold on the boundary of a strictly pseudoconvex domain D in $\mathbb{C}^n$, and let A be a pure p-dimensional analytic subset in $\mathbb{C}^n \setminus M$ such that $M = \overline{A} \setminus A$ and $\overline{A}$ is compact. Then, in some neighborhood of M the pair (A, M) is a C^1-manifold with boundary, and* sng A *consists of finitely many points.*

■ If D is defined by $\rho < 0$, where ρ is strictly plurisubharmonic in a neighborhood of $\overline{D}$, then $\overline{D}$ is the intersection of the domains of holomorphy $\{\rho < \epsilon\}$, containing M, and hence $A \subset \overline{D}$ (see the beginning of p.19.6). Since D is strictly plurisubharmonic, ∂D does not contain any analytic set of positive dimension, hence $\rho|_A \not\equiv 0$. By the maximum principle (p.6.3), $A \subset D$.

Fix an arbitrary point $a \in M$ and choose local holomorphic coordinates in a neighborhood of a, such that in U the domain D is strictly pseudoconvex (see A1.2) and such that the projection $\pi\colon M \cap U \to \Gamma$ is a diffeomorphism onto a hypersurface Γ partitioning the domain $U' = \pi(U)$ into two subdomains $U'_{\pm}$, with $U'_{+} \cap \pi(\overline{A} \cap D)$ nonempty (this can clearly be done). Since $\mathcal{H}_{2p+1}(\overline{A}) = 0$, we may assume that $\pi\colon \overline{A} \cap U \to U'$ is a proper map. Then the restrictions $\pi|_{A \cap U}$ are analytic covers above $U'_{\pm}$. If U is sufficiently small and the numbers of sheets above $U'_{\pm}$ are positive, p.19.1 and p.19.2 imply the existence of a p-dimensional analytic subset $A_1 \subset (A \cup M) \cap U$ such that $A_1 \cap M$ is nonempty. But $A \cup M \subset \overline{D}$, and this contradicts the maximum principle for $\rho|_{A_1}$ (p.6.3).

Thus $U'_- \cap \pi(A \cap U)$ is empty and (p.19.2) the number of sheets of $\pi|_{A \cap U}$ above U'_+ is 1 (U sufficiently small). Hence $A \cap U$ is a manifold with boundary $M \cap U$. Since a is arbitrary in M, all singularities of A lie at a positive distance from M, hence $\operatorname{sng} A$ is a compact analytic set in $\mathbf{C}^n$, i.e. $\# \operatorname{sng} A < \infty$. ■

P r o p o s i t i o n 2. *Let M be a compact maximal complex manifold of dimension $2p-1$ in $\mathbf{C}^n$, lying on the boundary of a strictly pseudoconvex domain, and let A be a pure p-dimensional analytic subset in $\mathbf{C}^n \setminus M$ such that $M = \bar{A} \setminus A$ and $\bar{A}$ is compact. Suppose that $2p > n$ and that there is an $(n-p)$-dimensional subspace $L \subset \mathbf{C}^n$ such that $L \cap T_\zeta M = \{0\}$ for all $\zeta \in M$. Then A is a complex submanifold in $\mathbf{C}^n \setminus M$ and $\bar{A}$ is a compact C^1-manifold with boundary M.*

■ We may assume that $L = \mathbf{C}^{n-p}_{z''}$. Let π be projection into $\mathbf{C}^p_{z'}$. Suppose that the set of critical points $\operatorname{br} \pi|_A$ has a limit point ζ on M. Since in a neighborhood of ζ the set A is regular (Proposition 1), $\operatorname{br} \pi|_A$ consists, in this neighborhood, of the points z for which $L \cap T_z A \neq \{0\}$. Let $z^j \in \operatorname{br} \pi|_A$, $z^j \to \zeta$ as $j \to \infty$, and let $v^j \in L \cap T_{z^j} A$, $|v^j| = 1$, tend to some vector v. Since (A,M) is a C^1-manifold with boundary, in a neighborhood of ζ, the vector v belongs to the $\mathbf{C}$-linear hull of the real space $T_\zeta M$ in $\mathbf{C}^n$. Since M is a maximal complex manifold, this hull equals $(T^{\mathbf{C}}_\zeta M) \oplus \mathbf{C}e$ for some vector e (see A2.4); hence $cv \in T_\zeta M$ for some number $c \in \mathbf{C}_*$. But then $cv \in L \cap T_\zeta M$, a contradiction. Thus we have proved that $\operatorname{br} \pi|_A$ is a compact, hence finite, set.

Since $2p > n > 1$, we have $p \geq 2$, and $\operatorname{br} \pi|_A$ is zero-dimensional or empty. By Proposition 4.5, in a neighborhood of each point $a \in \operatorname{br} \pi|_A$ the set A is the union of finitely many complex manifolds S_j; moreover, $\operatorname{br} \pi|_A = \cup_{i \neq j} (S_i \cap S_j)$ in a neighborhood of a. But these manifolds are p-dimensional, hence $\dim_a S_i \cap S_j \geq 2p - n > 0$, a contradiction with what we proved above. Thus, $\operatorname{br} \pi|_A$, and we hence $\operatorname{sng} A$, is empty. ■

It is clear from the proof that for $p \geq 2$, without the condition $2p > n$, the film can have singularities of selfintersection type only. In particular, it is a manifold immersed in $\mathbf{C}^n \setminus M$.

20. Analytic continuation

20.1. On continuation of analytic sets. Something concerning terminology. Let A,

$\tilde{A}$ be pure p-dimensional analytic sets on a complex manifold Ω. We will say that $\tilde{A}$ is an analytic continuation of A if: 1) $A \subset \tilde{A}$; and 2) each irreducible component of $\tilde{A}$ contains a nonempty open subset belonging to A (i.e. its intersection with A has dimension p). If 1) holds, 2) may be obtained by suitably shrinking $\tilde{A}$. More precisely, the following simple statement, obviously following from the definition and Theorem 5.4, holds.

Lemma 1. *Let $A \subset A_0$ be pure p-dimensional analytic sets on a complex manifold, and let $\tilde{A}$ be the union of the irreducible components of A_0 that have p-dimensional intersection with A. Then $\tilde{A}$ is an analytic continuation of A.*

We will say that an analytic set $A \subset \Omega$ has an *analytic continuation* in(to) an open set $U \subset \Omega$ if there is an analytic continuation of $A \cap U$ in U which is also an analytic subset in U (i.e. is closed in U). The uniqueness Theorem for analytic sets clearly implies that an analytic continuation of A in U, if existing, is uniquely defined.

It is well known that the problem of uniqueness of analytic continuation of analytic functions is rather troublesome. However, analytic sets are geometric analogs of multivalued analytic functions, hence for them the similar problems do not arise. More precisely, the following Lemma concerning locality of continuations of analytic sets holds.

Lemma 2. *Let E be a closed subset of a complex manifold Ω, and let A be a pure p-dimensional analytic subset in $\Omega \setminus E$. Suppose that for each point $\zeta \in \bar{A} \cap E$ there is a neighborhood $U_\zeta \ni \zeta$ in Ω, in which A has an analytic continuation. Then there is a neighborhood U of $\bar{A}$ in Ω, in which A has an analytic continuation.*

■ Fix on Ω some distance function $d(z,w)$ compatible with its smooth structure (e.g. induced by properly embedding Ω in some $\mathbb{R}^N$).

Denote by $r(z)$, $z \in \bar{A}$, the least upper bound of the numbers r such that A has an analytic continuation into the ball $B(z,r)$; by requirement, $r(z) > 0$ for all $z \in \bar{A}$. Obviously, $r(z)$ is continuous on $\bar{A}$ (see Lemma 1). Denote by A_z the analytic continuation of A into $B(z,r/2)$.

Let $\tilde{A}_z$ be the limit set for a family A_w as $w \to z$, $w \in \bar{A}$; more precisely, it is the set of points in $B(z,r(z)/2)$ of the form $\lim_{j\to\infty} z_j$, where $z_j \in A_{w_j}$, $w_j \in \bar{A}$, and $w_j \to z$ as $j \to \infty$. For each $w \in \bar{A}$ for which $d(z,w) < r(z)/6$, the set A_w belongs to a pure p-dimensional analytic set $\mathcal{A}_z$, the analytic continuation of A into $B(z, 2r(z)/3)$. Hence p.15.5 implies that $\tilde{A}_z$ is a pure p-dimensional analytic subset of $B(z,r(z)/2)$. On the same grounds the definition of $\tilde{A}_z$ implies the existence of a continuous function $\epsilon(z) > 0$ on $\bar{A}$ such that $\tilde{A}_w \cap B(z,r(z)/3) \subset \tilde{A}_z$ for all $w \in B(z, \epsilon(z))$. Without loss of generality we assume that $\epsilon(z) < r(z)/3$. Put $\delta(z) = \inf_{w \in \bar{A}} (\epsilon(w) + d(z,w))/4$, and put $B_z = B(z, \delta(z))$. If $d(z,w) \geqslant \epsilon(z)$, then

$$\delta(z) + \delta(w) \leqslant \frac{1}{4}\epsilon(z) + \frac{1}{4}(\epsilon(z) + d(w,z)) < d(z,w)$$

hence the balls B_z and B_w are disjoint.

Since $\delta(z) > 0$, the set $U = \cup_{z \in \bar{A}} B_z$ is a neighborhood of $\bar{A}$ in Ω. Put $\tilde{A} = \cup_{z \in A} (\tilde{A}_z \cap B_z)$. Then, by construction, $\tilde{A} \cap B_z = \tilde{A}_z \cap B_z$ (if $d(z,w) < \epsilon(z)$, then $\tilde{A}_w \cap B_z \subset \tilde{A}_z$, while if $d(z,w) \geqslant \epsilon(z)$ the balls B_z, B_w are disjoint). Thus, $\tilde{A}$ is a pure p-dimensional analytic subset in U; by construction it is the analytic continuation of A in U. ■

20.2. Compact singularities. For holomorphic functions of two or more variables compact singularities are removable (for a precise formulation see A1.3). An analogous property holds for analytic sets also, where the role of the number of variables is played by the dimension of the set.

T h e o r e m. *Let D be a domain on a Stein manifold (e.g. in $\mathbf{C}^n$), K a compact set in D, and A a pure p-dimensional analytic subset in $D \setminus K$; let, moreover, $p \geqslant 2$ and suppose that the closure in D of every irreducible component of A is noncompact. Then A has an analytic continuation into D.*

It is easy to give examples showing that all conditions on A are necessary; in particular, for $p < 2$ the Theorem is not true (the simplest example is given by the graph in $\mathbf{C}^2_{z,w}$ of a function holomorphic for $|z| > 1$, continuous for $|z| \geqslant 1$, and without analytic continuation to each point on the circle $|z| = 1$).

■ It turns out that the shortest proof is obtained by invoking the Harvey-Lawson theorem (p.19.6). Let ϕ be a real valued function of class C^∞ in D, with compact support belonging to D, and equal to 1 in a neighborhood of K. By Proposition 1, p.14.3, there is a number c, $0 < c < 1$, such that $A \cap \{\phi < c\}$ is an analytic subset with boundary $M = A \cap \{\phi=c\}$ in $D \setminus K$ (with corresponding orientation). By Stokes' theorem (p.14.3) and Proposition 1, A2.4, M is a $(2p-1)$-dimensional maximal complex cycle, compact because ϕ has compact support and $c<1$. By the Harvey-Lawson theorem there is a pure p-dimensional analytic subset A_1 in $\Omega \setminus M$ such that $M = bA_1$ and $A \cup M$ is compact. Put $A_0 = A \cup (A_1 \cap D)$; we show that A_0 is an analytic subset in D.

Let Σ be the set of singularities of the MC-cycle M. Then $\mathcal{H}_{2p-1}(\Sigma) = 0$. Since $M \subset A$ and $M = bA_1$, the boundary uniqueness Theorem (p.19.2) implies that there is a neighborhood $U \supset M \setminus \Sigma$ such that $A_1 \cap U \subset A$. Since $\mathcal{H}_{2p-1}(\Sigma) = 0$ this implies that every irreducible component of $(A_1 \cap D) \setminus K$ having a limit point on M intersects A in a nonempty open subset, hence belongs to A (by the uniqueness Theorem, see p.5.3). The remaining irreducible components of $(A_1 \cap D) \setminus K$ have boundaries on K, hence form an analytic subset A_2 in $D \setminus K$. Thus, $A_0 \setminus K = A \cup A_2$ is an analytic subset in $D \setminus K$.

On the other hand, if S is an irreducible component of $A \setminus M$ with compact closure in D, the limit points of S on M contain a nonempty open subset in M (otherwise $bS \subset K$, contradicting the condition that the closures of the irreducible components of A in D are noncompact). By the boundary uniqueness Theorem (p.19.2), $S \subset A_1$. The remaining components of $A \setminus M$ belong to the set $\{\phi<c\}$, which does not intersect K, hence form an analytic subset A_3 in $D \setminus M$. Thus, in $D \setminus M$ the set $A_0 = A \cup (A_1 \cap D)$ coincides with $A_1 \cup A_3$, hence is analytic also. Since $D \setminus M$ and $D \setminus K$ cover D, the set A_0 is analytic in D and, clearly, of pure dimension p and containing A. By Lemma 1, p.20.1, A_0 contains the analytic subset $\tilde{A}$ in D, which is the analytic continuation of A in D. ■

20.3. Continuation across pseudoconcave surfaces. One of the central theorems concerning continuation of holomorphic functions of several variables is the theorem concerning continuation across a strictly pseudoconcave hypersurface (see A1.3). This remarkable fact also applies for analytic sets of dimensions exceeding 1. We state it in a somewhat more general form.

Let $G \subset D$ be domains in $\mathbf{C}^n$ where $\partial G \cap D = \Gamma$ is a *smooth* (C^1) hypersurface, oriented as the boundary of G. We call Γ *q-pseudoconcave* if for each point $\zeta \in \Gamma$ there are a neighborhood $U \ni \zeta$ in D and a real valued function $u \in C^2(U)$ such that $\{u<0\} \cap U = G \cap U$, $du(\zeta) \neq 0$, and the restriction of the Levi form of u (see A1.2) onto the complex tangent plane $T_\zeta^c\Gamma = (T_\zeta\Gamma) \cap i(T_\zeta\Gamma)$ has at least q negative eigenvalues (for $q = n-1 = \dim_{\mathbf{C}}(T_\zeta^c\Gamma)$ we obtain a *strictly pseudoconcave* hypersurface). It is not difficult to prove that this property does not depend on the choice of u, i.e. is determined by the oriented hypersurface Γ itself.

Theorem. *Let $G \subset D$ be domains in $\mathbf{C}^n$ with $\Gamma = (\partial G) \cap D$ an $(n-p+1)$-pseudoconcave hypersurface, $p \geqslant 2$. Then for each pure p-dimensional analytic set A in G there is a neighborhood $U \supset \Gamma$ in D such that A has an analytic continuation into $G \cup U$. In particular, if Γ is strictly pseudoconcave, every analytic subset in G of pure dimension $p \geqslant 2$ has an analytic continuation into some neighborhood of $G \cup \Gamma$.*

■ By Lemma 2, p.20.1, it suffices to prove the local statement, i.e. that A has an analytic continuation into a neighborhood of any point $\zeta \in \Gamma$; for convenience we will regard ζ to be the coordinate origin in $\mathbf{C}^n$. The proof of this statement is largely analogous to that of Theorem 18.5.

We first investigate the leading particular case $p = 2$ (when Γ is strictly pseudoconcave). Let G be defined, in a neighborhood of 0, by a function u (as above). We choose coordinates in $\mathbf{C}^n$ such that $u(z) = x_1 + o(|z|)$ in a neighborhood of 0. Then $T_0\Gamma$ is defined by $x_1 = 0$, and $T_0^c\Gamma$ by $z_1 = 0$. Since on the plane $T_0^c\Gamma$ the Levi form of $-u$ is positive definite, for sufficiently large $c > 0$ the Levi form of $-u + cu^2$ is positive definite on all of $\mathbf{C}^n$, i.e. this function is strictly plurisubharmonic in a neighborhood of 0. Replacing u by $u - cu^2$ in a neighborhood of 0 does not change the zero set of u, hence we may assume that $-u$ is strictly plurisubharmonic in a neighborhood of 0. By A1.2, there is a biholomorphic change of coordinates, with linear part that is the identity, such that $-u$ becomes strictly convex in some ball $|z| < r_0$.

We yet want to transform the coordinates such that projection of $\overline{A}$ to $\mathbf{C}_{12}$ is proper in a neighborhood of 0. For this we first need to study the limit behaviour of A on the plane $L = \{z_1 = 0\}$ (in a neighborhood of 0).

If the set $A \cap L \cap B(0,r_0)$ is two-dimensional, the plane L contains an irreducible component of $A \cap B(0,r_0)$. Since 0 is isolated in the set $(D \setminus G) \cap L$ (in view of the strict convexity of $-u$), the Remmert-Stein theorem (p.4.4) implies that the closure of this component is analytic in a neighborhood of 0. We can remove this component from $A \cap B(0,r_0)$ and prove that the union of the remaining components has an analytic continuation. Hence we may assume in the sequel that $\dim A \cap L \cap B(0,r_0) \leqslant 1$. Choose δ, $0<\delta<r_0/4$, small such that $(B(0,r_0)\setminus G) \cap \{|z_1|<\delta\}$ is contained in the ball $|z|<r_0/4$. Put $V = B(0,r_0) \cap \{|z_1|<\delta\}$, and let S be an irreducible component of $A \cap V$ intersecting the ball $|z|<\delta/2$, i.e. containing a point a with $|a|<\delta/2$. The analytic set $S \cap \{z_1=a_1\}$ is then nonempty, hence of dimension $\geqslant 1$ at each of its points. Since the function $-u(a_1,z_2,\ldots,z_n)$ is strictly convex on $V \cap \{z_1=a_1\}$, negative on $V\setminus \overline{G}$, and since $-u(a)>0$ (because $A \subset G$, and $u<0$ in G), the maximum principle implies that S has nonempty intersection with the sphere $|z| = r_0/2$. Since, by construction, the compact set $K = \{|z_1| \leqslant \delta/2,\ |z|=r_0/2\}$ belongs to $V \cap G$, only finitely many irreducible components of the set $A \cap V$ intersect with it. Thus, we have proved that only finitely many components of $A \cap V$ can intersect the ball $|z|<\delta/2$. It suffices for us to prove that each of these components has an analytic continuation into a neighborhood of zero. Hence in the sequel we may assume that $A \cap V$ is irreducible and that 0 is a limit point of A (otherwise there is nothing to prove).

As proved above, for a sufficiently small $a \in A$ the set $A \cap \{z_1=a_1\}$ has positive dimension at each of its points, and contains a component joining a with K. Since, by agreement, $\dim A \cap V \cap L<2$, the limit set as $a \to 0$ of the family $A \cap \{z_1=a_1\}$, $a \in A$ is, within V, equal to

$$\overline{A} \cap L \cap L = (A \cap V \cap L) \cup \{0\}.$$

By Proposition 4, p.3.5, $\dim A \cap V \cap L \geqslant 1$ at all its points, hence $A \cap V \cap L$ is a pure one-dimensional analytic subset in $V\setminus\{0\}$. By the Remmert-Stein theorem, $\overline{A} \cap V \cap L$ is a pure one-dimensional analytic subset in $V \cap L$; moreover $0 \in \overline{A} \cap V \cap L$.

Now we perform a unitary transformation of the coordinates $z_2,\ldots,z_n$ after which $\overline{A} \cap V \cap \mathbb{C}_{3\ldots n}$ is zero-dimensional. Then we find a convex neighborhood $U_0 = U' \times U'' \ni 0$ in V such that the projection $\pi\colon \overline{A} \cap U_0 \to U' \subset \mathbb{C}_{12}$ is proper. Since $U_0 \setminus \overline{G}$ is (within U_0) strictly convex and lies in the half-space $x_1>0$, its projection into $\mathbb{C}_{12}$ has the same properties. Therefore there exist r_1, $r_2>0$ such that the polydisk $|z_1|<r_1$, $|z_2|<r_2$ belongs to U', while the set

$$\Sigma = \pi(U_0 \setminus \overline{G}) \cap \{|z_1|<r_1\}$$

lies in the smaller polydisk $|z_1|<r_1$, $|z_2|<r_2/2$. Since U' can be chosen arbitrarily (but small), we may further assume that $U' = \{|z_1|<r_1, |z_2|<r_2\}$. Repeating the reasoning involving the maximum principle above, we are led to the conclusion that $A \cap U_0$ may also be regarded as being irreducible.

Since $\bar{A} \cap U_0 \cap \{z_1=0\}$ is a pure one-dimensional analytic subset in U_0 intersecting $\mathbf{C}_{3\cdots n}$ at 0 only, $\pi(A \cap U_0)\setminus\bar{\Sigma}$ is nonempty; hence

$$\pi\colon A \cap U_0 \setminus \pi^{-1}(\bar{\Sigma}) \to U'\setminus\bar{\Sigma}$$

is an analytic cover with number of sheets $k \geqslant 1$. Let $\Phi_I(z',z'') = \sum \phi_{IJ}(z')(z'')^J$, $|I| = k$, be canonical defining functions of this cover ($z'=(z_1,z_2)$, $z=(z',z'')$). The ϕ_{IJ} are then holomorphic in $U'\setminus\bar{\Sigma}$. Since $\bar{\Sigma}$ is convex and belongs to the set $\{x_1 \geqslant 0,\ |z_1|\leqslant r_1,\ |z_2|\leqslant r_2/2\}$, by Lemma A1.3 (with $D = U'$ and $G = U'\setminus\bar{\Sigma}$) all ϕ_{IJ} have holomorphic continuations onto all of U'. The functions Φ_I, $|I| = k$, thus continued define in U_0 a pure two-dimensional analytic subset $\tilde{A}$, containing $A \cap U_0\setminus\pi^{-1}(\bar{\Sigma})$. Thus, the analytic subsets $A \cap U_0$ and $\tilde{A} \cap G$ of $U_0 \cap G$ have two-dimensional intersection. Since, by agreement, $A \cap U_0$ is irreducible, the uniqueness Theorem (p.5.3) implies that $A \cap U_0 \subset \tilde{A}$, hence $\tilde{A}$ is the required analytic continuation of A into a neighborhood of 0.

The case of an arbitrary $p \geqslant 2$ can be reduced to that of $p = 2$ by Hartogs' lemma for analytic sets (p.18.6). We may also repeat the proof given above, with minor complications: first choose coordinates in $\mathbf{C}^n$ such that $u(z) = x_1 + o(|z|)$ and such that the Levi form of $-u$ is positive definite on $\mathbf{C}_{p\cdots n} \subset T_0^c\Gamma$; replacing u by $u - cu^2$, $c>0$, we may assume that $-u$ is strictly plurisubharmonic on $\mathbf{C}_{1p\cdots n}$. A biholomorphic change makes $-u(z_1,'0,z_p,\ldots,z_n)$ strictly convex in a neighborhood of 0 in $\mathbf{C}_{1p\cdots n}$; since $u \in C^2$, all functions $-u(z_1,'c,z_p,\ldots,z_n)$ with sufficiently small $'c \in \mathbf{C}_{2\cdots p-1}$ are such. As in the two-dimensional case it turns out that $\bar{A} \cap \{z_1= \cdots =z_{p-1}=0\}$ is, in a neighborhood of 0, analytic and of positive dimension. After a suitable unitary change of coordinates $z_p, \cdots, z_n$, we thus obtain that 0 is isolated in $\bar{A} \cap \mathbf{C}_{(p+1)\cdots n}$, hence the projection $\pi\colon \bar{A} \cap U_0 \to U' \subset \mathbf{C}_{1\cdots p}$ is proper in a neighborhood $U_0 = U'\times U'' \ni 0$. The remainder (including the nuance involving the maximum principle) is as above. ■

R e m a r k s. 1) For $p = 1$ the effect of "forced" analytic continuation described in the Theorem is missing. Here is a simple example. Let U_1 be a domain in $\mathbb{C}$ containing the exterior of the unit disk and with nowhere analytic boundary of class C^∞. Let f be a function holomorphic in U_1, continuous on $\overline{U}_1$, and such that $|f(z)|^2 = 1+|z|^2$ on ∂U_1 (the existence of such a function readily follows from the solvability in U_1 of the corresponding Dirichlet problem). Let

$$A = \{(z,w)\in\mathbb{C}^2 \colon |z|^2 + |w|^2 > 1,\ w=f(z)\}.$$

Then A is a pure one-dimensional analytic subset of the exterior of the unit ball in $\mathbb{C}^2$. All points $(z,f(z))$, $z \in \partial U_1$, belong to the boundary of A; however, at no one of them there is an analytic continuation of A, since the curve $\{(z,f(z))\colon z\in\partial U_1\}$ is nowhere analytic.

2) It is obvious from the proof of the Theorem that for each point $\zeta \in \Gamma$ there is a unique neighborhood (depending on Γ only) into which all analytic subsets of G of corresponding dimension have an analytic continuation. However, to globally continue A we must "glue together" the continuations into these standard neighborhoods, and it may thereby turn out necessary to shrink the neighborhood of Γ, generally speaking in dependence on A (see Lemma 2, p.20.2).

20.4. Continuation across an edge. The following result is an analog of Kneser's theorem on the "embedded edge" (see [32], [185]). It generalizes Theorem 18.5 (with additional restrictions on the smoothness, however) and, in turn, follows from analytic continuation across a pseudoconcave hypersurface.

T h e o r e m. *Let D be a domain in $\mathbb{C}^n$, and let $\rho_1, \ldots, \rho_k$ be real valued functions of class $C^2(D)$ such that $\partial\rho_1\wedge\cdots\wedge\partial\rho_k \neq 0$ at every point of the manifolds*

$$M = \{z\in D\colon \rho_j(z)=0,\ 1\leqslant j\leqslant k\},$$

$$K = \{z\in D\colon \rho_j(z)\geqslant 0,\ 1\leqslant j\leqslant k\},$$

and

$$K_{\epsilon,\delta} = \{z\in D\colon \rho_j(z)\geqslant\epsilon\sum_1^k|\rho_i(z)|^{2-\delta},\ 1\leqslant j\leqslant k\}.$$

Let A be a pure p-dimensional analytic subset of $D \setminus K_{\epsilon,\delta}$ *for some* $\epsilon > 0$, $\delta > 0$; *moreover, let* $p \geqslant n-k+2$. *Then the set* $A \setminus K$ *has an analytic continuation into a neighborhood of M.*

■ The condition $\partial\rho_1 \wedge \cdots \wedge \partial\rho_k \neq 0$ on M means (Figure 13) that at each point $\zeta \in M$ the complex tangent planes to the hypersurfaces $\rho_j = 0$ intersect transversally (remember that the complex tangent plane to $\rho_j = 0$ at ζ consists of the vectors v for which $\sum_1^n (\partial\rho_j / \partial z_i)(\zeta)v_i = 0$). Hence $\dim T_\zeta^0 M = n-k < p-1$ for all $\zeta \in M$. Such singularities are removable for p-dimensional analytic subsets in $D \setminus M$ (Theorem 18.5), but A is defined in $D \setminus K_{\epsilon,\delta}$ only.

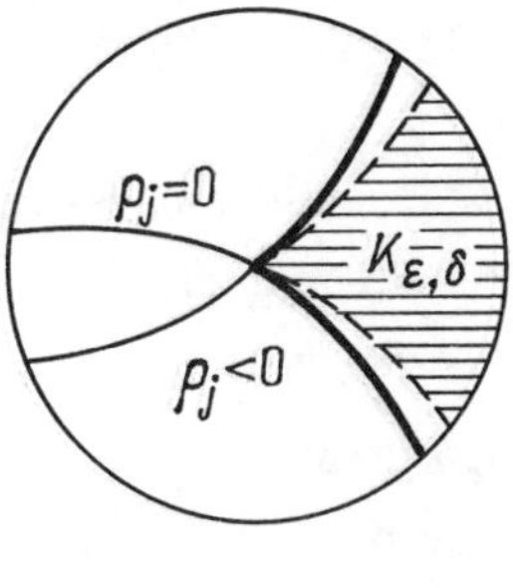

Figure 13.

As before, it suffices to prove that $A \setminus K$ has a continuation into a neighborhood of an arbitrary point of $\overline{A} \cap M$. Let $0 \in \overline{A \setminus K} \cap M$ and choose coordinates such that $\rho_j = x_j + o(|z|)$ in a neighborhood of 0 (this can be done in view of the transversality). By Taylor's formula, $\rho_j = x_j + q_j(z,\overline{z}) + o(|z|^2)$ where the q_j are homogeneous polynomials in $z,\overline{z}$ of degree 2. Put $\tilde{\rho}_j = \rho_j - C_j \sum_1^k \rho_i^2$, then

$$\tilde{\rho}_j(z) = x_j + q_j(z,\overline{z}) - C_j \sum_1^k x_i^2 + o(|z|^2);$$

hence for sufficiently large C_j the Levi form of $\tilde{\rho}_j$ at 0 has at least k negative eigenvalues. Fix such a C_j for $j = 1, \ldots, k$. Then Γ_j: $\tilde{\rho}_j = 0$ is a hypersurface, $(k-1)$-pseudoconcave in a neighborhood $U \ni 0$ (being the boundary of $\tilde{\rho}_j < 0$). Shrinking $U \subset D$ we may assume that the inequalities $C_j \sum_1^k \rho_i^2 \leqslant \epsilon \sum_1^k |\rho_i|^{2-\delta}$, $j = 1, \ldots, k$, are fulfilled in it. Then for each j the set $U \cap K_{\epsilon,\delta}$ belongs to $U \cap \{\tilde{\rho}_j \geqslant 0\}$, hence $A_j = A \cap U \cap \{\tilde{\rho}_j < 0\}$ is a pure p-dimensional analytic subset of $G_j = U \cap \{\tilde{\rho}_j < 0\}$.

By Theorem 20.3 there is a neighborhood $V \ni 0$ in U such that A_j has an analytic continuation into V; denote this continuation by $\tilde{A}_j$, and put $\tilde{A} = \cup_1^k \tilde{A}_j$. Then $\tilde{A}$ is a pure p-dimensional analytic subset in V, and $\tilde{A}$ contains $(A \setminus K) \cap V$ since $\cup_1^k A_j \supset (A \setminus K) \cap U$. Further (by the definition of analytic continuation), every irreducible component of $\tilde{A}$, being a component of one of $\tilde{A}_j$, has p-dimensional intersection with $A \cap V$, hence $\tilde{A}$ is the required analytic continuation of $A \setminus K$ into a neighborhood of 0. ■

20.5. The symmetry principle. We already used symmetry for continuing analytic sets across $\mathbb{R}^n$ (p.18.4). This method works in more general situations as well. We confine ourselves to the following simple Proposition (see [184]).

P r o p o s i t i o n. *Let M be a totally real $\mathbb{R}$-analytic submanifold of a complex manifold Ω, and let A be a pure one-dimensional analytic subset in $\Omega \setminus M$. Then A has an analytic continuation into a neighborhood of M in Ω.*

Recall that for pure p-dimensional analytic subsets in $\Omega \setminus M$ with $p > 1$ the singularities on M are removable by Theorem 18.5 (hence the Proposition is formulated for $p = 1$ only).

■ By p.20.1, it suffices to prove the corresponding local statement, therefore we will assume in the sequel that Ω is a neighborhood of 0 in $\mathbb{C}^n$ and $0 \in \bar{A} \cap M$. The coordinates in $\mathbb{C}^n$ are chosen such that $T_0 M \subset \mathbb{R}^n$ (this is possible because M is totally real). Then, in a neighborhood of 0, M belongs to a manifold $\tilde{M}$ defined by a system of equations $y = \phi(x)$ with ϕ a real analytic vector function $(x + iy = z)$ and $\phi(0) = d\phi(0) = 0$. We want to "straighten" $\tilde{M}$ by a biholomorphic transformation, with linear part that is the identity at 0.

For this we consider in a neighborhood of 0 in $\mathbb{C}_z^n \times \mathbb{C}_w^n$ the system of n holomorphic equations

$$w - z + 2i\phi\left[\frac{1}{2}(z + w)\right] = 0, \qquad (\star)$$

in which the components ϕ are regarded as power series in which $(z + w)/2$ is substituted for x. By the implicit function Theorem (A2.2) this system can be solved for w; more precisely, there is a vector function ψ holomorphic in a neighborhood of 0, $\psi(0) = d\psi(0) = 0$, such that $(\star)$ is equivalent to the system $w = z + \psi(z)$ in a neighborhood of 0. On the real plane $w = \bar{z}$ the system $(\star)$ takes the form $y = \phi(x)$, and the system equivalent to it takes the form $\bar{z} = z + \psi(z)$. Thus, in a neighborhood of 0 the manifold $\tilde{M}$ is defined by $\bar{z} = z + \psi(z)$.

The map

$$z \mapsto \zeta = z + \frac{1}{2}\psi(z)$$

is biholomorphic in a neighborhood of 0 in $\mathbb{C}^n$. Let $\eta = \operatorname{Im} \zeta$. Then on $\tilde{M}$ we have

$$\eta = \frac{1}{2}\operatorname{Im}(z + (z + \psi(z))) = \frac{1}{2}\operatorname{Im}(z + \bar{z}) = 0$$

and hence the manifold $\tilde{M}$ turns (in a neighborhood of 0) into the real subspace $\mathbb{R}^n \subset \mathbb{C}^n$ under the biholomorphism constructed.

Let A_0 be the image of A (in a neighborhood of 0) under this biholomorphism, and put $A^- = \{\bar{\zeta}: \zeta \in A_0\}$. By Alexander's theorem (p.18.4), the closure of $A_0 \cup A^-$ is analytic in a neighborhood of 0. Let $\tilde{A}$ be the image of this closure under the inverse map $\zeta \mapsto z$. Then $\tilde{A} \cap U$ is an analytic subset of some neighborhood $U \ni 0$, and $\tilde{A} \supset (A \setminus \tilde{M}) \cap U$. Since $A \cap \tilde{M}$ is nowhere dense on A ($\tilde{M}$ is totally real and reg A is a complex manifold), $\tilde{A} \supset (A \setminus M) \cap U$, hence $\tilde{A}$ is the required continuation of A into a neighborhood of 0. ■

In a more general situation the following Theorem holds.

T h e o r e m. *Let M be an $\mathbb{R}$-analytic submanifold of a complex manifold Ω, and let A be a pure p-dimensional analytic subset in $\Omega \setminus M$. Then if* $\dim_{\mathbb{C}} T_\zeta^c M \leqslant p - 1$ *at every point $\zeta \in \bar{A} \cap M$, the set A has an analytic continuation into some neighborhood of $\bar{A}$ in Ω.*

The proof follows from the Proposition and Hartogs' lemma for analytic sets (p.18.6). The idea of the proof is clear (reduction to the preceding by taking plane sections), while the details are rather involved. Therefore we leave this Theorem without proof.

Appendix

ELEMENTS OF MULTI-DIMENSIONAL COMPLEX ANALYSIS

A1. Removable singularities of holomorphic functions

A1.1. Holomorphic functions in $\mathbb{C}^n$. Linear functions in $\mathbb{C}^n$ are functions $l(z) = a_1 z_1 + \cdots + a_n z_n$, or, for short, $l(z) = <z,a>$. They form the $\mathbb{C}$-linear space $(\mathbb{C}^n)^*$, dual to $\mathbb{C}^n$; in it the coordinate functions z_k form a canonical basis, dual to the standard basis in $\mathbb{C}^n$. Any $\mathbb{R}$-linear function l: $\mathbb{C}^n \to \mathbb{C}$ has a unique representation $l = l_1 + \overline{l}_2$, where l_1, l_2 are complex linear functions. The complex space of all complex valued $\mathbb{R}$-linear functions in $\mathbb{C}^n \approx \mathbb{R}^{2n}$ will be denoted by $(\mathbb{C}^n)^*_{\mathbb{R}}$. There are two standard bases (over the field $\mathbb{C}$) in it: the real valued coordinate functions $x_1, y_1, \ldots, x_n, y_n$, and the complex valued functions z_1, $\overline{z}_1, \ldots, z_n, \overline{z}_n$.

The differentials of complex valued functions of class C^1 in a domain $D \subset \mathbb{C}^n$ are the $\mathbb{R}$-linear functions of increment. The $\mathbb{C}$-linear part of the differential of such a function f is denoted by ∂f; the $\mathbb{C}$-antilinear part (of the form $\overline{l}$ with l complex linear) is denoted by $\overline{\partial} f$. Hence $df = \partial f + \overline{\partial} f$, where

$$\partial f = \sum \frac{\partial f}{\partial z_k} dz_k, \quad \overline{\partial} f = \sum \frac{\partial f}{\partial \overline{z}_k} d\overline{z}_k$$

with the formal derivatives

$$\frac{\partial f}{\partial z_k} := \frac{1}{2}\left[\frac{\partial f}{\partial x_k} - i\frac{\partial f}{\partial y_k}\right], \quad \frac{\partial f}{\partial \overline{z}_k} := \frac{1}{2}\left[\frac{\partial f}{\partial x_k} + i\frac{\partial f}{\partial y_k}\right].$$

A function f is called *holomorphic* (complex analytic, or, simply, analytic) in an open set $G \subset \mathbb{C}^n$ if it is continuously differentiable ($f \in C^1(G)$) and complex differentiable at every point $z \in G$, i.e. $\partial f / \partial \overline{z}_k = 0$, $k = 1, \ldots, n$, or, for short,

$\bar{\partial} f \equiv 0$ in G (the Cauchy-Riemann conditions). The set of all functions holomorphic in G forms, with the pointwise operations, the algebra $\mathcal{O}(G)$. If E is an arbitrary set in $\mathbb{C}^n$, $\mathcal{O}(E)$ denotes the set of functions holomorphic in a neighborhood of E (each in its own neighborhood). For holomorphic functions of several variables the following *Cauchy integral formula* is valid. It can be easily proved by induction with respect to n (see, e.g., [32], [185]).

Let f be a function holomorphic in a neighborhood of the closed polydisk $\overline{U} \subset \mathbb{C}^n$ with distinguished boundary Γ. Then

$$f(z) = \frac{1}{(2\pi i)^n} \int_{\Gamma} \frac{f(\zeta) d\zeta}{(\zeta - z)}, \quad z \in U,$$

where $d\zeta = d\zeta_1 \cdots d\zeta_n$ and $(\zeta - z) = (\zeta_1 - z_1) \cdots (\zeta_n - z_n)$.

The righthand side of this formula makes sense for, say, an arbitrary continuous function on Γ; such a Cauchy-type integral also represents a holomorphic function in U. We will use the following more general statement.

Lemma 1. *Let $F(z,t)$ be a function that is continuously differentiable in a neighborhood of a set $G \times \Gamma$, where G is an open set in $\mathbb{C}^n$ and Γ is a compact set in $\mathbb{R}^m$. Suppose that for each fixed $t_0 \in \Gamma$ the function $F(z,t_0)$ is holomorphic in G. Then for an arbitrary measure μ on Γ the integral*

$$f(z) = \int_{\Gamma} F(z,t)\, d\mu(t)$$

is a function that is holomorphic in G.

■ Starting from the definition of partial derivatives, write down the corresponding increment relations for $\Delta f / \Delta x_k$. The conditions on F readily imply that as $\Delta x_k \to 0$ these relations tend to $\int_\Gamma (\partial F(z,t) / \partial x_k)\, d\mu(t)$, respectively. Similarly for the derivatives with respect to y_k. The integrals obtained are continuous in z, hence $f \in C^1(G)$. Since $\partial F / \partial \bar{z}_k \equiv 0$ by requirement, all $\partial f / \partial \bar{z}_k \equiv 0$ in G, i.e. f is holomorphic in G. ■

This Lemma readily implies the following useful result concerning analytic continuation.

Lemma 2. *Let f be a function defined in a neighborhood of a set $D' \times \overline{D}_n$, where D' is a domain in $\mathbb{C}^{n-1}$ and $\overline{D}_n$ is a closed, bounded domain in the z_n-plane.*

Suppose that f is holomorphic in a neighborhood of $D' \times \partial D_n$, and that for each fixed $z' \in D'$ it is holomorphic with respect to z_n in D_n. Then f is holomorphic in $D' \times D_n$.

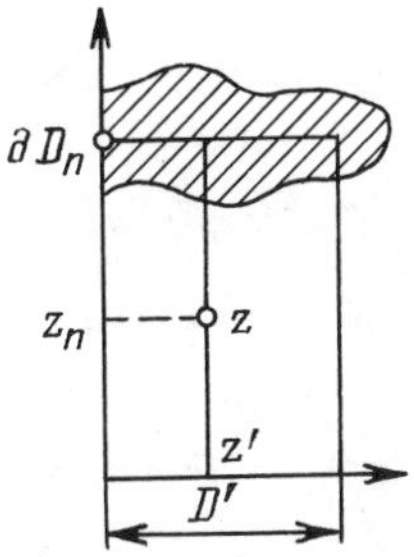

Figure 14.

■ By Cauchy's formula for one variable,

$$f(z', z_n) = \frac{1}{2\pi i} \int_{\partial D_n} \frac{f(z', \zeta_n)\, d\zeta_n}{\zeta_n - z_n}, \quad z' \in D', \quad z_n \in D_n$$

(since f is holomorphic in a neighborhood of $D' \times \partial D_n$, we may assume that ∂D_n is smooth, see Figure 14). The integrand is holomorphic in a neighborhood of $D' \times D_n \times \Gamma$, where $\Gamma = \partial D_n$, since $\zeta_n \neq z_n$ and $f(z', \zeta_n)$ is holomorphic (and independent of z_n) in it. By Lemma 1, the integral represents a function holomorphic in $D' \times D_n$. ■

This Lemma allows us to reduce many Theorems concerning analytic continuation to the corresponding statements for functions of one variable.

Differentiating under the integral sign in Cauchy's formula we obtain, as in the one-dimensional case, that every function f holomorphic in a domain $D \subset \mathbf{C}^n$ is infinitely differentiable in D, and its partial derivatives can, in any polydisk $U \subset\subset D$ be expressed by the formulas

$$\frac{\partial^{|k|} f}{\partial z^k}(z) = \frac{k!}{(2\pi i)^n} \int_{\Gamma} \frac{f(\zeta)\, d\zeta}{(\zeta - z)^{k+1}}, \quad z \in U,$$

where Γ is the distinguished boundary of U (here $k+1 = (k_1+1, \ldots, k_n+1)$). These formulas imply in the standard manner.

W e i e r s t r a s s' t h e o r e m. *If a sequence of functions f_j, holomorphic in a domain $D \subset \mathbf{C}^n$, converges uniformly on compact sets in D to a function f, then f is holomorphic in D, and*

$$\frac{\partial^{|k|} f_j}{\partial z^k} \to \frac{\partial^{|k|} f}{\partial z^k}$$

uniformly on compact sets in D, for any multi-index k.

The restriction of a function $f \in \mathcal{O}(G)$ to an arbitrary complex line $z = a+\lambda v$ ($a,v \in \mathbf{C}^n$ fixed, $v \neq 0$, $\lambda \in \mathbf{C}$ a parameter) clearly is a holomorphic function $f(a+\lambda v)$ of one complex variable λ in the open set $\{\lambda\colon a+\lambda v \in G\} \subset \mathbf{C}$. This simple remark allows us to reduce many theorems concerning holomorphic functions in $\mathbf{C}^n$ to the corresponding theorems for one complex variables. We demonstrate this by the example of the uniqueness Theorems.

U n i q u e n e s s t h e o r e m. *Let f be a function holomorphic in a domain $D \subset \mathbf{C}^n$, and suppose that at some point $z^0 \in D$ all its partial derivatives $\partial^{|k|} f / \partial z^k$ vanish, $|k| \geqslant 0$. Then $f \equiv 0$ in D.*

■ Let E be the set of points in D at which all partial derivatives of f vanish; by requirement E is nonempty. Since the derivatives of f are continuous, E is closed in D. On the other hand, let $a \in E$ and let $B = B(a,r) \subset D$. For any unit vector $v \in \mathbf{C}^n$ the function $f(a+\lambda v)$ is holomorphic in a disk $|\lambda| < r$ in the complex plane, and all its partial derivatives with respect to λ vanish for $\lambda = 0$, since they can be linearly expressed in terms of the partial derivatives of f at $a \in E$. By the uniqueness Theorem for functions of one variable, $f(a+\lambda v) \equiv 0$ for $|\lambda| < r$. Since v is arbitrary, this gives $f \equiv 0$ in B. Thus, $B \subset E$, hence E is open. Since D is connected and E is nonempty, this implies $E = D$, i.e. $f \equiv 0$ in D. ■

C o r o l l a r y. *If a function f is holomorphic in a domain D in $\mathbf{C}^n$, and if its zero set in D has positive volume (in $\mathbf{C}^n \approx \mathbb{R}^{2n}$), then $f \equiv 0$ in D.*

■ The proof proceeds by induction with respect to n. For $n = 1$ the statement clearly follows from the classical uniqueness Theorem. Let now $n > 1$. By requirement there is a point $c \in Z_f$ such that $\mathcal{H}_2(Z_f \cap V) > 0$ for any neighborhood V of c. Let $U = U' \times U_n$ be a neighborhood of c in D. By Fubini's theorem there is a set $E' \subset U'$ of positive volume in $\mathbf{C}^{n-1}$ such that $\mathcal{H}_{2n}(Z_f \cap (\{a'\} \times U_n)) > 0$ for all $a' \in E'$. By the uniqueness Theorem for functions of one variable, $f(a',z_n) \equiv 0$ in U_n if $a' \in E'$, hence $E' \times U_n \subset Z_f$. The function $f(z',a_n)$, $a_n \in U_n$, is holomorphic in U' and vanishes on the set E' of positive volume in U'. By the induction hypothesis $f(z',a_n) \equiv 0$ in U', hence $U \subset Z_f$. By the ordinary uniqueness Theorem we have $f \equiv 0$ in D. ■

This, of course, is a very weak statement; indeed, Z_f has locally finite Hausdorff $(2n-2)$-measure in D if $f \not\equiv 0$ (see p.14.1).

B o u n d a r y u n i q u e n e s s t h e o r e m. *Let D be a domain with boundary of class C^1 in $\mathbb{C}^n$, let f be a function holomorphic and bounded in D, and suppose that its nontangential limit values (exist and) vanish on some set $E \subset \partial D$ of positive $(2n-1)$-measure. Then $f \equiv 0$.*

■ Since $\mathcal{H}_{2n-1}(E) > 0$ there is a point $c \in \overline{E}$ such that $\mathcal{H}_{2n-1}(E \cap V) > 0$ for any neighborhood $V \ni c$. Since $\partial D \in C^1$, after a suitable linear change of coordinates we may assume that in a neighborhood of c the surface ∂D is the graph $y_n = \phi(z',x_n)$ of a C^1-function. Then there is a neighborhood $U = U' \times U_n \ni c$ such that for every $a' \in U'$ the set $D \cap (\{a'\} \times U_n)$ is a plane domain with piecewise smooth boundary. Since $\mathcal{H}_{2n-1}(E \cap U) > 0$ and $E \cap U$ is bijectively projected into the (z',x_n)-plane, Fubini's theorem implies that there is a set $E' \subset U'$ of positive $(2n-2)$-measure such that

$$\mathcal{H}_1(E \cap (\{a'\} \times U_n)) > 0 \quad \text{for all} \quad a' \in E'.$$

By the Luzin-Privalov boundary uniqueness Theorem for functions f of one variable (see, e.g., [98], [37]), for every a' the function $f(a',z_n)$ vanishes in the whole domain $\{z_n \in U_n\colon (a',z_n) \in D\}$. Thus, the zero set of f in $D \cap U$ has positive volume, and hence $f \equiv 0$ in D. ■

The following well-known properties of holomorphic functions can also readily be proved using plane sections (cf. [32], [185]).

M a x i m u m p r i n c i p l e. *Let a function f be holomorphic in a domain $D \subset \mathbb{C}^n$, and suppose that the modulus of f attains its maximum at a point $a \in D$: $|f(a)| = \max_D |f|$. Then $f \equiv f(a)$ in D.*

S c h w a r z ' l e m m a. *Let a function f be holomorphic in a ball B_r: $|z| < r$,* *and vanish, together with all its partial derivatives up to order* $k-1$ *inclusive* $(k \geqslant 1)$, *at* $z=0$. *If* $|f(z)| \leqslant M$ *in* B_r, *then*

$$|f(z)| \leqslant M\left[\frac{|z|}{r}\right]^k, \quad z \in B_r.$$

Expanding the kernel of the Cauchy integral formula in a multiple geometric series and integrating termwise we obtain an expansion of f in a multiple Taylor series:

$$f(z) = \sum_{|k|=0}^{\infty} c_k(z-a)^k, \quad c_k = \frac{1}{(2\pi i)^n}\int_\Gamma \frac{f(\zeta)\,d\zeta}{(\zeta-a)^{k+1}}.$$

This series converges to f, uniformly on compact sets in $U = U(a,r)$. The formula for the coefficients obviously implies Cauchy's inequalities:

$$|c_k| \leqslant \frac{\max_\Gamma |f|}{r_1^{k_1} \cdots r_n^{k_n}}.$$

Grouping in the Taylor expansion the monomials of identical degrees into homogeneous polynomials,

$$(f)_m(z-a) := \sum_{|k|=m} c_k(z-a)^k,$$

we obtain the expansion

$$f(z) = \sum_{m=0}^{\infty} (f)_m(z-a)$$

in a series of homogeneous polynomials in $z-a$; the polynomials are called homogeneous because $(f)_m(\lambda z) = \lambda^m (f)_m(z)$ for all $\lambda \in \mathbf{C}$, $z \in \mathbf{C}^n$. In general, the domain of convergence of the series after this grouping is larger than before (see below).

If f is holomorphic in the ball $B(a,r)$, and if $v \in \mathbf{C}^n$ is a vector with $|v| = 1$, then

$$f(a+\lambda v) = \sum_0^{\infty}(f)_m(\lambda v) = \sum_0^{\infty}((f)_m(v))\lambda^m, \quad |\lambda| < r.$$

Uniqueness of power series expansion for holomorphic functions of one variable therefore implies that

The restriction onto an arbitrary complex line $z = a+\lambda v$ of the series of homogeneous polynomials in $z-a$ is the Taylor series for the function $f(a+\lambda v)$ of one variable λ.

By Cauchy's inequalities for such functions, $|(f)_m(v)| \leqslant (\sup_{B(a,r)} |f|)/r^m$. Substituting $v = (z-a)/|z-a|$ in it gives Cauchy's inequalities for series of homogeneous polynomials:

$$|(f)_m(z-a)| \leqslant \left[\sup_{B(a,r)} |f|\right] \left[\frac{|z-a|}{r}\right]^m .$$

These inequalities imply, in particular, that if $f \in \mathcal{O}(B(a,r))$, then its series in homogeneous polynomials in $z-a$ converges to f, uniformly on compact subsets of $B(a,r)$.

As usual, the Cauchy inequalities imply

L i o u v i l l e ' s t h e o r e m. *If f is a function holomorphic in all of $\mathbb{C}^n$, and if $|f(z)| < C(1+|z|)^N$ for all $z \in \mathbb{C}^n$, with some constants C, N, then f is a polynomial in z of degree at most N, i.e. $f(z) = \sum_{m \leqslant N} (f)_m(z)$.*

If the terms of the power series are grouped according to the degree of one variable, e.g. z_n, we obtain the so-called Hartogs series $\sum_0^\infty c_k(z')z_n^k$, whose coefficients are holomorphic functions of $z' = (z_1, \ldots, z_{n-1})$. Using Hartogs series we can prove the following well-known Lemma, which can be regarded as a statement concerning analytic continuation.

H a r t o g s ' l e m m a. *Let a function $f(z',z_n)$ be defined in a polydisk $U = U' \times U_n$, be holomorphic in a smaller polydisk $U' \times V_n$, $V_n \subset\subset U_n$, and let, for each fixed $z' \in U'$, it be holomorphic with respect to z_n in the disk U_n. Then f is holomorphic in U.*

For a *proof* see, e.g., [32], [185].

Using Hartogs' lemma and the uniqueness Theorem it is easy to prove the following boundary version of E. Levi's theorem concerning rationality with respect to one variable. We have used this version in p.19.5 in the proof of the Harvey-Lawson theorem (see [164]).

L e v i ' s b o u n d a r y t h e o r e m. *Let a function f be holomorphic in a domain $U = U' \times U_n$ with piecewise smooth boundary in $\mathbb{C}^n$, and let it be continuous in $\overline{U}$. Suppose that $\partial U'$ contains a set E' of positive $(2n-3)$-measure such that for each fixed $\zeta' \in E'$ the function $f(\zeta', z_n)$ is rational with respect to z_n (i.e. can be continued to a rational function in $\mathbb{C}_{z_n}$). Then for each fixed $z' \in U'$ the function $f(z', z_n)$ is also rational with respect to z_n.*

■ Denote by E_m' the set of points $\zeta' \in E'$ for which $f(\zeta', z_n)$ is representable as the quotient of two polynomials in z_n of degrees at most m. Since $E' = \cup_1^{\infty} E_m'$, at least one E_m' has positive $(2n-3)$-measure; hence we may assume that $E' = E_m'$ in the sequel. Without loss of generality we may also assume that U_n is the disk $|z_n| < r$.

Expand f in a Hartogs series $\sum_0^{\infty} c_k(z')z_n^k$ in U. Cauchy's formula for the coefficients of the Taylor series clearly implies that all $c_k(z')$ can be continuously extended to functions in $\overline{U'}$. For each fixed $\zeta' \in E'$ the series $\sum_0^{\infty} c_k(\zeta')z_n^k$ represents a rational function with numerator and denominator of degrees at most m. By methods of linear algebra it can be relatively elementary proved that this condition is equivalent to satisfaction of the infinite system of algebraic equations in the coefficients: $\det C^K(\zeta') = 0$, where $K = (k_1, \ldots, k_{m+1})$ is an arbitrary multi-index with $k_j > m$, and $C^K(\zeta')$ is the matrix with columns $(c_{k_j}(\zeta'), c_{k_j-1}(\zeta'), \ldots, c_{k_j-m}(\zeta'))$ (see, e.g., [164]). Each function $\det C^K(\zeta')$ is holomorphic in U', continuous in $\overline{U'}$, and, by requirement, vanishes on E'. By the boundary uniqueness Theorem, $\det C^K(z') \equiv 0$ in U', hence $f(z', z_n)$ is rational with respect to z_n for each $z' \in U'$. ■

A1.2. Plurisubharmonic functions. A real valued function u defined on a domain $D \subset \mathbb{C}^n$, and with values in $[-\infty, +\infty)$, that is not identically equal to $-\infty$ is called *plurisubharmonic in D* if

1) u is upper semicontinuous in D (i.e. all sets $\{z \in D: u(z) < t\}$, $t \in \mathbb{R}$, are open);
2) the restriction of u to each complex line $z = a + \lambda v$ is a function that is either subharmonic or $\equiv -\infty$ on the components of the open set $\{\lambda: a + \lambda v \in D\} \subset \mathbb{C}$.

Typical examples are: functions of the form $\log |f|$, $|f|^{\alpha}$, $\alpha > 0$, where $f \in \mathcal{O}(D)$; their sums, maxima, etc.

For twice continuously differentiable functions u plurisubharmonicity is equivalent to the *Levi form* $\sum(\partial^2 u / \partial z_j \partial \bar{z}_k)\,(z) v_j \bar{v}_k$ being positive definite at all points of the domain. Clearly, the Laplacian of such a function u is nonnegative, hence u is, as a function of the real variables $(x_1, \ldots, y_n)$, subharmonic.

An arbitrary plurisubharmonic function u in a domain $D \subset \mathbf{C}^n$ can on any compact set $K \subset D$ be approximated by plurisubharmonic functions of class C^∞ in a neighborhood of K. Indeed, the means

$$u_\delta(z) := \int u(z+\delta\zeta)\lambda(|\zeta|)\,dV(\zeta) = \delta^{-2n}\int u(\eta)\lambda\left[\frac{|\eta-z|}{\delta}\right]dV(\eta)$$

with a weight $\lambda(|z|) \in C^\infty(\mathbf{C}^n)$ such that $\lambda \geqslant 0$, $\lambda(|\zeta|) = 0$ for $|\zeta| > 1$, and $\int \lambda\, dV = 1$, are defined and infinitely differentiable in the open set

$$D_\delta = \{z \in D\colon \operatorname{dist}(z, \partial D) > \delta\}.$$

Since $\lambda \geqslant 0$, the u_δ are plurisubharmonic in D_δ. As $\delta \to 0$ these functions tend to u in $L^1_{\mathrm{Loc}}(D)$, which implies that, in particular, any (not necessarily twice continuously differentiable) plurisubharmonic function is also subharmonic. Hence the mean value Theorem holds for them:

$$u(z) \leqslant \frac{1}{c(n) r^{2n}} \int_{B(z,r)} u(\zeta)\, dV(\zeta), \quad B(z,r) \subset\subset D,$$

which, in particular, implies the maximum principle for such functions. (By the way, both also follow from the analogous properties in one-dimensional plane sections of D.) The stronger theorem of Green (see A5.2) implies that the sequence u_δ tends to u pointwise and monotone decreasing.

The mean value Theorem is, as is well known, a characteristic property of subharmonic functions (in the class of all upper semicontinuous functions), see, e.g., [32]. Hence the following statement holds.

If E is a closed subset of domain D, $E \neq D$, and a function u is semicontinuous in D, (pluri)subharmonic on $D \setminus E$, and equal to $-\infty$ on E, then u is (pluri)subharmonic in all of D.

For plurisubharmonic functions this property also follows from the one-dimensional case.

The property of the Levi form being positive definite does not change under a biholomorphic (holomorphic in both directions) change of variables, as is easily verified. Hence the approximability of general plurisubharmonic functions by smooth ones implies that the property of being plurisubharmonic is invariant under biholomorphic transformations.

A set $E \subset \mathbb{R}^n$ is called *polar* if there is a subharmonic function in some neighborhood of it that is identically $-\infty$ on E. The mean value Theorem implies that such a set has volume zero in $\mathbb{R}^N$ (indeed, its Hausdorff dimension is at most $N-2$). A set $E \subset \mathbb{C}^n$ is called *pluripolar* (or $\mathbb{C}^n$-polar) if there is a plurisubharmonic function in a neighborhood of it that is identically $-\infty$ on E; of course, a pluripolar set in $\mathbb{C}^n$ is polar (as a set in $\mathbb{R}^{2n}$). A pluripolar set $E \subset D$ is called *complete* in the domain D if there is in D a plurisubharmonic function ϕ such that $E = \{z \in D \colon \phi(z) = -\infty\}$; a pluripolar set $E \subset D$ is called *locally complete* in D if it is complete in a neighborhood of each of its limit points in D.

As an example we can invoke the set of common zeros of an arbitrary finite set of holomorphic functions $f_1, \ldots, f_k \not\equiv 0$ in a domain D: the function $\phi = \log \Sigma_1^k |f_j|$ is plurisubharmonic in D and equal to $-\infty$ precisely on $\cap_1^k Z_{f_j}$. (Of course, this implies that every analytic set of positive codimension is locally complete pluripolar in some neighborhood of it.)

Pluripolar sets are removable singularities for a wide class of holomorphic and plurisubharmonic functions (see A1.4, A1.5). Here we note the following property

Let u be a locally bounded function in a domain D, plurisubharmonic in $D \setminus E$ with E a closed pluripolar subset in D. Then the function

$$u^*(z) = \overline{\lim_{\substack{\zeta \to z \\ \zeta \notin E}}}\ u(\zeta)$$

(equal to u in $D \setminus E$), is plurisubharmonic in D.

■ Let ϕ be plurisubharmonic in D with $\phi|_E = -\infty$. Put $u_\epsilon = u + \epsilon\phi$, $\epsilon > 0$. Then u_ϵ is plurisubharmonic in $D \setminus E$, equal to $-\infty$ on E, and hence plurisubharmonic in D (see above). As $\epsilon \to 0$ the functions u_ϵ tend to u. Hence the upper regularization u^* of u, being upper semicontinuous in D, satisfies the same criterion for being plurisubharmonic: the mean value Theorem on plane sections. ■

A function $u \in C^2(D)$ is called *strictly plurisubharmonic* in D if its Levi form is strictly positive definite at all points of D, i.e. if there is a function $a(z)$, positive and continuous in D, such that

$$\sum \frac{\partial^2 u}{\partial z_j \partial \bar{z}_k}(z) v_j \bar{v}_k \geqslant a(z) \cdot |v|^2$$

for all $v \in \mathbf{C}^n$. An example is $|z|^2$. The function $\log |z|$ is plurisubharmonic in $\mathbf{C}^n_\star$, but not strictly plurisubharmonic in it: at every point $z \neq 0$ its Levi form vanishes on the vectors that are complex proportional to z.

Let u be a function defined and strictly plurisubharmonic in a neighborhood of the coordinate origin 0 in $\mathbf{C}^n$, $u(0) = 0$ and $du(0) \neq 0$. By Taylor's formula,

$$u(z) = 2\text{Re}\left[\sum u_j'(0) z_j + \sum u_{jk}''(0) z_j z_k\right] + \sum u_{j\bar{k}}''(0) z_j \bar{z}_k + o(|z|^2),$$

where $u_j' = \partial u / \partial z_j$, $u_{jk}'' = \partial^2 u / \partial z_j \partial z_k$ and $u_{j\bar{k}}' = \partial^2 u / \partial z_j \partial \bar{z}_k$. Since $du(0) \neq 0$, after a suitable linear change of coordinates we may assume that $u_1'(0) = 1$, $u_j'(0) = 0$ for $j > 1$. Now we make another change,

$$z_1^\star = z_1 + \sum u_{jk}''(0) z_j z_k, \quad z_j^\star = z_j, \quad j > 1,$$

which is biholomorphic and one-to-one in a neighborhood of 0. In the new coordinates (omitting the asterix) the expansion of u is:

$$u(z) = 2\text{Re}\, z_1 + \sum u_{j\bar{k}}''(0) z_j \bar{z}_k + o(|z|^2),$$

i.e. the quadratic term in the Taylor expansion coincides with the Levi form of u at 0. Since u is strictly plurisubharmonic, this form is strictly positive definite, hence in the new coordinates u is a strictly convex function in a small neighborhood of 0. In particular, the real hypersurface $u = 0$ is strictly convex (from the side of $u < 0$). For this reason the level sets of a strictly plurisubharmonic function are called *strictly pseudoconvex hypersurfaces*; the domains bounded by these surfaces are called strictly pseudoconvex domains. The class of strictly pseudoconvex domains plays an important role in the modern theory of functions of several complex variables.

The fundamentals of the theory of plurisubharmonic functions are presented in detail in V.S. Vladimirov's book [32], see also [78], [123].

A1.3. Holomorphic continuation along sections. We say that a function f defined in a domain G has a *holomorphic continuation* in (to) a domain $D \supset G$ if there is a function $\tilde{f} \in \mathcal{O}(D)$ for which $\tilde{f} = f$ in G. Thus, in distinction from an analytic continuation, a holomorphic continuation is, required to be one-to-one. By the uniqueness Theorem, if a holomorphic continuation exists, it is unique. As distinct from the one-dimensional case, there are in $\mathbb{C}^n$ (with $n > 1$) domains from which every holomorphic function has a holomorphic continuation into a larger domain. We start with a well-known situation.

Proposition 1 (Continuation across a pseudocave hypersurface). *Let u be a function defined and strictly plurisubharmonic in a neighborhood $U \ni 0$ in $\mathbb{C}^n$, let $u(0) = 0$ and $du(0) \neq 0$, and put $U_+ = \{z \in U\colon u(z) > 0\}$. Then there is a neighborhood $V \ni 0$ such that every function f holomorphic in U_+ has a holomorphic continuation into $U_+ \cup V$.*

■ By A1.2 we may assume that in a neighborhood of 0,

$$u(z) = 2x_1 + L(z,\bar{z}) + o(|z|^2),$$

where $x_1 = \operatorname{Re} z_1$ and L is the Levi form of u at 0. Since $L(z,\bar{z}) \geqslant a|z|^2$ for some constant $a > 0$, the hypersurface $\Gamma\colon u = 0$ lies, in a neighborhood of 0, in the half-space $x_1 \leqslant 0$ and intersects the plane $x_1 = 0$ at $z = 0$ only. Let $r > 0$ be small such that the ball $|z| \leqslant r$ lies in the intersection of this neighborhood of 0 and U. Then the circle $|z_n| = r$, $z' := (z_1, \ldots, z_n) = 0$ is disjoint from Γ, hence there is an $r' > 0$ such that the set $|z'| \leqslant r'$, $|z_n| = r$ belongs to U_+. Put $V' = B(0', r')$ in $\mathbb{C}^{n-1}$, and put $\gamma_n\colon |z_n| = r$, a circle in $\mathbb{C}_{z_n}$. The function f is holomorphic in a neighborhood of $V' \times \gamma$, hence for each fixed $z' \in V'$ it has a Laurent expansion in z_n:

$$f(z) = \sum_{-\infty}^{\infty} c_k(z') z_n^k, \quad c_k(z') = \frac{1}{2\pi i} \int_{\gamma_n} f(z', \zeta_n) \zeta_n^{-k-1} d\zeta_n,$$

with coefficients that are holomorphic in V', by Lemma 1, A1.1. Since f is holomorphic in U_+, for $z' \in V' \cap \{x_1 > 0\}$ the coefficients with negative indices are zero. By the uniqueness Theorem, $c_k(z') \equiv 0$ in V' for $k < 0$. By the maximum principle (for each fixed $z' \in V'$), the series $\sum_0^\infty c_k(z') z_n^k$ converges uniformly on compact sets in $V = V' \times V_n$, where $V_n\colon |z_n| < r$ in $\mathbb{C}_{z_n}$. By Weierstrass' theorem its sum, F, is holomorphic in V. Since $F = f$ in a neighborhood of $V' \times \gamma_n$ belonging to U_+, and since by construction $V \cap U_+$ is connected, the uniqueness Theorem implies that $F = f$ in $V \cap U_+$. Setting $\tilde{f} = f$ in U_+ and $\tilde{f} = F$ in V we obtain a function $\tilde{f}$ that is holomorphic in $U_+ \cup V$. ■

Continuation along a family of sections is a particular instance of the so-called continuity principle (see [32], [185]). Instead of series we may use some other analytic tool like, e.g., in the following important technical Lemma.

L e m m a. *Let D be a domain in $\mathbb{C}^n$, D' its projection in $\mathbb{C}^{n-1}_{z'}$, $z' = (z_1, \ldots, z_{n-1})$, and let $D_{\zeta'} = D \cap \{z'=\zeta'\}$. Let G be a subdomain of D such that for every $z' \in D'$ the set $D_{z'} \setminus G_{z'}$ is compact, and such that there is an $a' \in D'$ with $D_{a'} = G_{a'}$. Then every function f holomorphic in G has a holomorphic continuation into D.*

■ The conditions imply that there is a domain $\Omega \subset D$ with $\mathbb{R}$-analytic boundary containing $D \setminus G$, and such that for every $z' \in D'$ the set $\overline{\Omega}_{z'}$ is compact (Figure 15). Then $(\partial\Omega)_{z'}$, $z' \in D'$, belongs to G, consists of finitely many piecewise smooth curves (we will denote their union by $\Gamma_{z'}$) and, possibly, finitely many isolated points. By suitably enlarging Ω in the direction of G we may assume that $\Omega \cap G$ is connected.

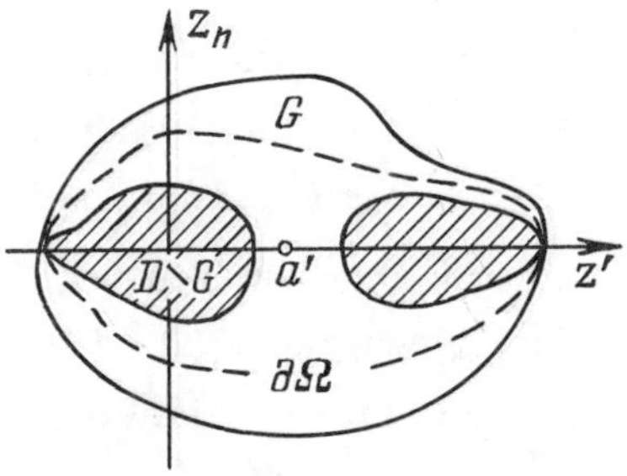

Figure 15.

We denote by $\gamma_{z'}$ the projection of $\Gamma_{z'}$ in the z_n-plane, and define in Ω the function

$$F(z',z_n) = \frac{1}{2\pi i}\int_{\gamma_{z'}} \frac{f(z',\zeta_n)\,d\zeta_n}{\zeta_n - z_n}, \quad z_n \in \Omega_{z'}.$$

If z' is sufficiently close to the fixed point $\zeta' \in \Omega'$, the integration contour $\gamma_{z'}$ can be replaced by $\gamma_{\zeta'}$, by Cauchy's theorem (here we use the fact that $(D \setminus G)_{\zeta'}$ is a compact set, being at positive distance from $\Gamma_{\zeta'}$). Hence, by Lemma 1, A1.1, F is holomorphic in Ω.

By requirement, the projection of $D \setminus G$ in $\mathbb{C}^{n-1}$ does not contain a neighborhood of a'. Since $D \setminus G \subset \Omega$, this implies that there is in D' an open subset U' for which $D \cap \{z' \in U'\} \subset G$ and $\Omega \cap \{z' \in U'\}$ is nonempty. For $z' \in U'$ the function $f(z', z_n)$ is holomorphic in z_n from $D_{z'} \supset \Omega_{z'}$, hence (by Cauchy's formula) coincides in $\Omega_{z'}$ with $F(z', z_n)$. Thus, $F = f$ on the open subset $\Omega \cap \{z' \in U'\} \subset G$. Since $\Omega \cap G$ is connected, the uniqueness Theorem implies $F = f$ in $\Omega \cap G$, hence the function equal to f in G and to F in Ω is the required holomorphic continuation of f into D. ■

We give two well-known results, which readily follow from this Lemma.

P r o p o s i t i o n 2 (Removing compact singularities). *Let D be a domain in $\mathbb{C}^n$, $n > 1$, and let $K \subset D$ be a compact set such that $D \setminus K$ is connected. Then every function holomorphic in $D \setminus K$ has a holomorphic continuation into D.*

■ The domains D and $G = D \setminus K$ satisfy the conditions of the Lemma since the projection of K in $\mathbb{C}^{n-1}_{z'}$ is a compact set $K' \subset D'$ and $G_{z'} = D_{z'}$ for all $z' \in D' \setminus K'$. ■

P r o p o s i t i o n 3 (Removing singularities of large codimensions). *Let D be a domain in $\mathbb{C}^n$, and E a closed subset in D of Hausdorff measure $\mathcal{H}_{2n-2}(E) = 0$. Then every function f holomorphic in $D \setminus E$ has a holomorphic continuation into D.*

■ We first prove the local statement: Let $0 \in E$; we must show that f has a holomorphic situation into a neighborhood of 0. The requirements and A6.4 imply that after a suitable unitary transformation of $\mathbb{C}^n$ the projection $\pi\colon E \cap U \to U' \subset \mathbb{C}^{n-1}_{z'}$ is a proper map in some polydisk $U \ni 0$. In particular, all fibers $\pi^{-1}(z') \cap E \cap U$ are compact, and $E' = \pi(E \cap U)$ is a closed set in U' (see p.3.1). Hausdorff measures do not increase under projection, hence E' is nowhere dense in U'. Thus, the domains U and $G = U \setminus E$ satisfy the conditions of the Lemma, since $G_{z'} = U_{z'}$ for $z' \in U' \setminus E'$ (G' is connected by A6.1). Therefore f has a holomorphic continuation from $U \setminus E$ into U. If now $U_1, U_2 \subset D$ are two such convex domains, and if $\tilde{f}_1, \tilde{f}_2$ are the corresponding holomorphic continuations, then $f = \tilde{f}_1 = \tilde{f}_2$ on $(U_1 \cap U_2) \setminus E$. Since E is nowhere dense in D, and since $(U_1 \cap U_2) \setminus E$ is connected, $\tilde{f}_1 = \tilde{f}_2$ in $U_1 \cap U_2$. Hence the local continuations of f thus constructed give in totality one function $\tilde{f} \in \mathcal{O}(D)$. ■

A1.4. Removable singularities of bounded functions.

Theorem. *Let D be a domain in $\mathbf{C}^n$, and let E be a closed subset of D of Hausdorff measure $\mathcal{H}_{2n-1}(E) = 0$. Then every function f holomorphic and uniformly bounded in $D \setminus E$ has a holomorphic continuation into D.*

■ We first investigate the case $n = 1$, $D \subset \mathbf{C}$. Since E is nowhere dense in D, by A1.3 it suffices to prove the local statement: f has a holomorphic continuation into a neighborhood of an arbitrary point of E, which we will take as the coordinate origin. Under projection $z \mapsto |z|$ of the disk $\{|z| < R\} \subset D$ onto the radius $[0, R)$, the Hausdorff length clearly does not increase. Since $\mathcal{H}_1(E) = 0$, the image of $E \cap \{|z| < R\}$ under this map is a nowhere dense subset of the interval $[0, R)$. Hence there is an $r \in (0, R)$ such that the circle γ: $|z| = r$ does not intersect E, consequently $E_r = E \cap \{|z| < r\}$ is compact. Since $\mathcal{H}_1(E_r) = 0$, for every $\epsilon > 0$ there is a finite covering of E_r by disks, disjoint from γ, with total sum of the radii $< \epsilon$. The union of these disks is denoted by V_ϵ, and we put $U_\epsilon = \{|z| < r\} \setminus \overline{V}_\epsilon$. Since f is holomorphic on the closure of U_ϵ, we have

$$f(z) = \frac{1}{2\pi i} \int_\gamma \frac{f(\zeta)\, d\zeta}{\zeta - z} - \frac{1}{2\pi i} \int_{\partial V_\epsilon} \frac{f(\zeta)\, d\zeta}{\zeta - z}, \quad z \in U_\epsilon.$$

If $z \in U_\epsilon$ is at distance $\delta > \epsilon$ from E and if $M = \sup_{D \setminus E} |f|$, the modulus of the second integral is at most $M\epsilon/\delta$. Since ϵ can be chosen arbitrarily small, the function f is represented on $\{|z| < r\} \setminus E$ by the first integral only. But this integral is a function holomorphic in all of the disk $|z| < r$, hence determines the required holomorphic continuation of f into a neighborhood of 0.

In the case $n > 1$ we again may restrict ourselves to the local statement. Let $0 \in E$. Since $\mathcal{H}_{2n-1}(E) = 0$, by A6.4 there are a unitary transformation l: $\mathbf{C}^n \to \mathbf{C}^n$ and an arbitrarily small polydisk $U = U' \times U_n \ni 0$ such that the set $l(E)$ is disjoint from the compact set $\overline{U'} \times \partial U_n \subset D$. Without loss of generality we may assume l to be the identity transformation $z \mapsto z$. Put $E_{\zeta'} = E \cap U \cap \{z' = \zeta'\}$. Since $\mathcal{H}_{2n-1}(E) = 0$, there is an everywhere dense set $Q' \subset U'$ such that $\mathcal{H}_1(E_{\zeta'}) = 0$ for all $\zeta' \in Q'$. We define on U the function

$$\tilde{f}(z', z_n) = \frac{1}{2\pi i} \int_{\partial U_n} \frac{f(z', \zeta_n)\, d\zeta_n}{\zeta_n - z_n}, \quad z' \in U', \quad z_n \in U_n.$$

Since f is holomorphic in a neighborhood of $U' \times \partial U_n$, by Lemma 1, A1.1, the function $\tilde{f}$ is holomorphic in U. From the proof of the case $n = 1$ we see that $\tilde{f} = f$ on the everywhere dense subset $(U \setminus E) \cap \{z' \in Q'\}$ of $U \setminus E$, hence, by continuity, $\tilde{f} = f$ on $U \setminus E$. ■

L e m m a. *Let D be a domain in $\mathbb{R}^N$ and let E be a polar set that is closed in D. Then every function u harmonic and bounded in $D \setminus E$ continues to a harmonic function in D.*

■ Let U be a neighborhood of E in D, and let ϕ be a function subharmonic in U such that $\phi|_E \equiv -\infty$. Then the functions $u + \epsilon\phi$, $\epsilon > 0$, taken to be $-\infty$ on E, are subharmonic in U. Denote by $(u)_\delta$ a mean with C^∞-kernel, as in A1.2; then $(u)_\delta$ and $(u \pm \epsilon\phi)_\delta$ are C^∞-functions in

$$U_\delta = \{x \in U: \text{dist}\,(x, \partial U) > \delta\};$$

moreover, $(u + \epsilon\phi)_\delta$ is a subharmonic and $(u - \epsilon\phi)_\delta$ is a superharmonic function in U_δ. Since $u \pm \epsilon\phi$ tend to u in $L^1_{\text{loc}}(U)$, we have that $(u \pm \epsilon\phi)_\delta \to (u)_\delta$, as $\epsilon \to 0$, uniformly on compact sets in U_δ. This clearly implies that $(u)_\delta$ is a harmonic function in U_δ. Since $(u)_\delta \to u$ in $L^1_{\text{loc}}(U)$, as $\delta \to 0$, the mean value Theorem implies that the $(u)_\delta$ converge uniformly on compact subsets in U, hence the limit function $\tilde{u}$ is harmonic in U. Thus, the function equal to u in $D \setminus E$ and to $\tilde{u}$ in U is the required harmonic continuation of u into D. ■

C o r o l l a r y. *Let D be a domain in $\mathbb{C}^n$, and let E be a polar set (as a set in $\mathbb{R}^{2n}$) that is closed in D. Then every function f holomorphic and bounded in $D \setminus E$ has a holomorphic continuation into D.*

■ The real and the imaginar part of f are functions that are harmonic and bounded in $D \setminus E$. By the Lemma they have harmonic continuations onto all of D. Since harmonic functions are infinitely differentiable, this implies that f continues to a function $\tilde{f} \in C^\infty(D)$. Since $\bar{\partial}\tilde{f} \equiv 0$ in $D \setminus E$, and since E is nowhere dense in D, by continuity $\bar{\partial}\tilde{f} \equiv 0$ in D, i.e. $\tilde{f}$ is the required holomorphic continuation of f. ■

This Corollary follows, of course, also from the Theorem, since $\mathcal{H}_{2n-1}(E) = 0$ for any polar set E in $\mathbb{R}^{2n}$. However, we have given another proof, in the spirit of the well-known proof of Radó's theorem (see below).

A1.5. Removable singularities of continuous functions.

Theorem. *Let D be a domain in $\mathbb{C}^n$, and let E be a closed subset in D such that $\mathcal{H}_{2n-1}(E \cap K) < \infty$ for any compact set $K \subset D$. Then every function f continuous in D and holomorphic in $D \setminus E$ is holomorphic in all of D.*

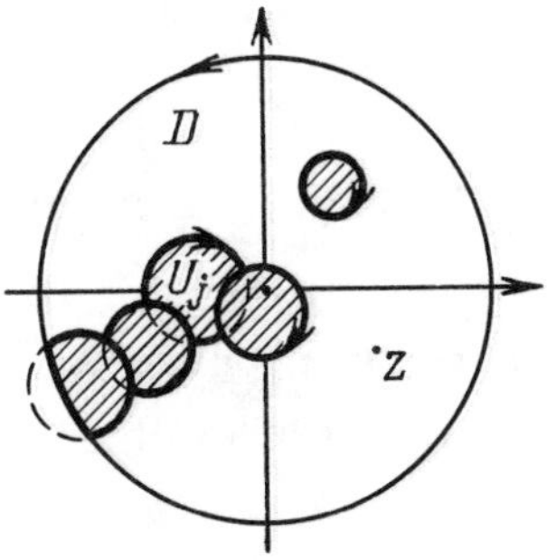

Figure 16.

■ As before, the general case reduces to the local case and the one-dimensional case. We will assume that D is the disk $|z| < r$ in $\mathbb{C}$, f is continuous in $\overline{D}$ and holomorphic in $D \setminus E$, where $E \subset \overline{D}$ is a compact set of finite Hausdorff length, $\mathcal{H}_1(E) < L$. Fix an arbitrary $\delta > 0$. By the definition of $\mathcal{H}_1$ there is a finite covering of E by disks of radii $< \delta$ and with total sum of the radii $< L$. Denote these disks by V_j (Figure 16). We inductively define open sets U_j, putting

$$U_1 = V_1 \cap D, \ldots, U_j = (V_j \cap D) \setminus \bigcup_{k<j} \overline{V}_k.$$

The boundary of each U_j consists of finitely many arcs of circles, and $\overline{U(\delta)} := \cup \overline{U}_j = \cup \overline{V}_j \cap \overline{D}$ contains E. The function f is holomorphic in $D \setminus \overline{U(\delta)}$, and it can be given there by Cauchy's formula. Since on common arcs of adjoining sets $\overline{U}_j$ opposite orientations are induced, while on arcs common to ∂D and ∂V_j the orientations are the same, by adding zero terms to Cauchy's formula we find that

$$f(z) = \frac{1}{2\pi i}\int_{\partial D} \frac{f(\zeta)\,d\zeta}{\zeta - z} - \frac{1}{2\pi i}\sum \int_{\partial U_j} \frac{f(\zeta)\,d\zeta}{\zeta - z}, \quad z \in D \setminus \overline{U(\delta)}.$$

Let a_j be the center of V_j. Then $|f(\zeta) - f(a_j)| \leqslant \omega(\delta)$ for all $\zeta \in \overline{V}_j \cap \overline{D} \supset \overline{U}_j$, where $\omega(\delta)$ is the modulus of continuity of f in $\overline{D}$, hence

$$\left| \frac{1}{2\pi i}\sum \int_{\partial U_j} \frac{f(\zeta)\,d\zeta}{\zeta - z} \right| \leqslant \frac{1}{2\pi}\sum \left| \int_{\partial U_j} \frac{f(\zeta) - f(a_j)}{\zeta - z}\,d\zeta \right| \leqslant \frac{\omega(\delta)}{2\pi\rho(z)} \sum |\partial U_j|,$$

where $|\partial U_j|$ is the length of ∂U_j and $\rho(z)$ is the distance from z to $\overline{U(\delta)}$. Since $\cup\partial U_j \subset \partial D \cup (\cup\partial V_j)$, while each arc occurring in $\cup\partial U_j$ does locally belong to at most two ∂U_j, we have

$$\sum|\partial U_j| \leqslant 2\sum|\partial V_j| \leqslant 4\pi L,$$

since the sum of the radii of the V_j is at most L. By limit transition as $\delta \to 0$ we find

$$f(z) = \frac{1}{2\pi i}\int_{\partial D}\frac{f(\zeta)\,d\zeta}{\zeta - z} \quad \text{for all } z \in D\setminus E.$$

Since both f and this integral are continuous in D, and since E is nowhere dense, equality holds throughout D, hence f is holomorphic in D. ■

Radó's theorem. *Let D be a domain in $\mathbf{C}^n$, and let f be a function continuous in D and holomorphic everywhere outside its zero set. Then f is holomorphic in D.*

■ We may assume $f \not\equiv 0$. Then $\log|f|$ is subharmonic in $D\setminus Z_f$, equal to $-\infty$ on Z_f: $f = 0$, and upper semicontinuous in D. By A1.2 it is (pluri)subharmonic in D, hence Z_f is closed in D and is a polar set. By A1.4, f is holomorphic in D. ■

Radó's theorem can also be easily reduced to the one-dimensional case (using Hartogs' lemma from A1.1). A singular proof in the one-dimensional case can be given; in any case, readers not familiar with the theory of subharmonic functions in space may use this reduction to the case $n = 1$. Radó's theorem, along with Theorem A1.4, is used in an essential way in the construction of the fundamentals of the theory of analytic sets.

A2. Holomorphic maps. Manifolds in $\mathbf{C}^n$

A2.1. Holomorphic maps. Linear maps $l: \mathbf{C}^n \to \mathbf{C}^m$ have a representation $l(z) = A\cdot z$, where z is a column vector and $A = (a_j^k)$ is an $m \times n$ matrix with columns a^k. Thus, $l = (l_1, \ldots, l_m)$, where $l_k(z) = \langle z, a^k\rangle$. The image of $\mathbf{C}^n$ under a $\mathbf{C}$-linear map is a complex linear subspace in $\mathbf{C}^m$, of dimension equal to the rank of A. The kernel (zero set) of l is a $\mathbf{C}$-linear subspace of $\mathbf{C}^n$ of dimension

$n -$rank A. A map $z \mapsto A \cdot z$ is called *nonsingular* (*nondegenerate*) if rank $A =$ min (n,m) is maximal. Nonsingular (complex) linear maps l: $\mathbb{C}^n \to \mathbb{C}^n$ are called linear transformations of $\mathbb{C}^n$; the condition of being nonsingular is in this case equivalent to det $A \neq 0$.

A transformation $z \mapsto A \cdot z$ is called *unitary* if $A \cdot \overline{A}^t = E$, where $\overline{A}^t = (\overline{a_k^j})$ is the matrix Hermitian conjugate to A and E is the identity matrix. This condition means that the columns of A (= the images of the vectors of a canonical basis) have length 1 and are pairwise complex orthogonal. This readily implies that a transformation $z \mapsto A \cdot z$ is unitary if and only if it is norm preserving: $|A \cdot z| = |z|$ for all $z \in \mathbb{C}^n$. For $n = 1$ the unitary transformations are the rotations $z \mapsto e^{i\alpha} \cdot z$. The unitary transformations of $\mathbb{C}^n$ (so to say, the complex rotations) form the group $\mathbb{U}(n)$ under composition of maps. This group is sufficiently rich, e.g. any p-dimensional complex subspace in $\mathbb{C}^n$ can clearly be mapped to the $(z_1, \ldots, z_p)$-plane by a unitary transformation.

The coordinate planes in $\mathbb{C}^n$ are conveniently regarded as spaces on their own, using only the coordinates that act in them. So, the coordinate plane of the variable $z_I := (z_{i_1}, \ldots, z_{i_p})$, $i_i < \cdots < i_p$, which is usually denoted by $\mathbb{C}_I = \mathbb{C}_{i_1 \cdots i_p}$, is regarded as the space $\mathbb{C}^p$ with coordinates $z_{i_1}, \ldots, z_{i_p}$, and its points are written as p-dimensional vectors. The projection onto this plane is denoted by π_I and has, by agreement, the form $z \mapsto z_I$. (Formally, the spaces $\mathbb{C}_I$ thus defined are, of course, not subspaces of $\mathbb{C}^n$, and the π_I are not projections in this space; however, such agreement is convenient, especially when $\mathbb{C}^n$ is regarded as a direct product $\mathbb{C}^p \times \mathbb{C}^{n-p}$ with two groups of variables, $z' = (z_1, \ldots, z_p) \in \mathbb{C}^p$ and $z'' = (z_{p+1}, \ldots, z_n) \in \mathbb{C}^{n-p}$.)

Let G be an open set in $\mathbb{C}^n$. A map f: $G \to \mathbb{C}^m$ has a representation $f(z) = (f_1(z), \ldots, f_m(z))$, where the f_k are complex valued functions (the components of f). The map f is called *holomorphic* if all f_k are holomorphic in G. This is clearly equivalent to the condition: the map f is continuously differentiable in G and at every point $a \in G$ the differential of f is a $\mathbb{C}$-linear map $(df)_a$: $\mathbb{C}^n \to \mathbb{C}^m$. The differential of a holomorphic map at a point $a \in G$ has the form $z \mapsto \partial f / \partial z(a) \cdot z$, where $\partial f / \partial z := (\partial f_k / \partial z_j)$ is the Jacobi $m \times n$ matrix. Its rank (i.e. the rank of the tangential $\mathbb{C}$-linear map $(df)_a$) is called the *rank* of the holomorphic map at a, and is denoted by rank$_a f$. It is the maximum number of $\mathbb{C}$-linear independent differentials among the $(df_1)_a, \ldots, (df_m)_a$. A map f is called *nonsingular* (*nondegenerate*) at a point a if $\operatorname{rank}_a f = \min(n,m)$ is maximal. For $m = n$ it is the condition $\det \partial f / \partial z(a) \neq 0$.

By forgetting the complex structure, a map $f: G \to \mathbb{C}^m$ can be regarded as a smooth map $\mathbb{R}^{2n} \supset G \to \mathbb{R}^{2m}$. How do the notions of being nonsingular in the complex and in the real version relate? In the case of equal dimensions the answer is given by the following

P r o p o s i t i o n. *Let D be a domain in $\mathbb{C}^n$, let f: $D \to \mathbb{C}^n$ be a holomorphic map, and let $f_k = u_k + iv_k$, $z_j = x_j + iy_j$, be the decompositions in real and imaginary parts. Then*

$$\frac{\partial(u_1, v_1, \ldots, u_n, v_n)}{\partial(x_1, y_1, \ldots, x_n, y_n)} = \left|\det \frac{\partial f}{\partial z}\right|^2.$$

■ The Jacobian of f as a real map is the determinant of the matrix consisting of blocks

$$\begin{pmatrix} \dfrac{\partial u_k}{\partial x_j} & \dfrac{\partial u_k}{\partial y_j} \\ \dfrac{\partial v_k}{\partial x_j} & \dfrac{\partial v_k}{\partial y_j} \end{pmatrix}.$$

Elementary transformations (with complex coefficients) of rows do not change the determinant, but using them the Jacobian can be transformed as

$$\frac{1}{(2i)^n}\det\left(\begin{pmatrix} \dfrac{\partial f_k}{\partial x_j} & \dfrac{\partial f_k}{\partial y_j} \\ \dfrac{\overline{\partial f_k}}{\partial x_j} & \dfrac{\overline{\partial f_k}}{\partial y_j} \end{pmatrix}\right) = \frac{1}{2^n}\det\left(\begin{pmatrix} \dfrac{\partial f_k}{\partial z_j} & -\dfrac{\partial f_k}{\partial z_j} \\ \dfrac{\overline{\partial f_k}}{\partial z_j} & \dfrac{\overline{\partial f_k}}{\partial z_j} \end{pmatrix}\right) =$$

$$= \det\left(\begin{pmatrix} \dfrac{\partial f_k}{\partial z_j} & 0 \\ 0 & \dfrac{\overline{\partial f_k}}{\partial z_j} \end{pmatrix}\right)$$

(the first equality follows from the Cauchy-Riemann conditions: $\partial f_k / \partial z_j = \partial f_k / \partial x_j = -i\partial f_k / \partial y_j$). Permuting in the last matrix the rows, and subsequently the columns, we are led, without changing the determinant, to the matrix

$$\begin{bmatrix} \dfrac{\partial f}{\partial z} & 0 \\ 0 & \overline{\dfrac{\partial f}{\partial z}} \end{bmatrix},$$

whose determinant is clearly equal to $(\det \partial f / \partial z) \cdot (\det \overline{\partial f / \partial z}) = |\det \partial f / \partial z|^2$. ■

A2.2. The implicit function theorem and the rank theorem.

Theorem 1. *Let $f: D \to \mathbb{C}^m$ be a holomorphic map defined in a neighborhood D of a point $a \in \mathbb{C}^n$, let $p := n - m > 0$ and $z = (z', z'')$, where $z' \in \mathbb{C}^p$, $z'' \in \mathbb{C}^m$. Suppose that*

$$\det \frac{\partial f}{\partial z''}(a) := \det \left[\frac{\partial f_k}{\partial z_{p+j}}(a) \right]_{k,j=1}^{m} \neq 0.$$

Then there are a neighborhood $U = U' \times U'' \ni a$ and a holomorphic map $g: U' \to U''$ such that the level set $\{z \in U: f(z) = f(a)\}$ coincides with the graph $\{(z', z''): z' \in U', z'' = g(z')\}$ of the map g.

The condition of the Theorem means that the restrictions of the linear functions $(df_k)_a(z)$ onto the coordinate plane $\mathbb{C}_{z''}$ are linearly independent.

■ Since $\det \partial f / \partial z''(a) \neq 0$, by regarding z' as parameter and f as a smooth map

$$z'' \mapsto (u_1(z', z''), v_1(z', z''), \ldots, u_m(z', z''), v_m(z', z'')),$$

we find that

$$\det \frac{\partial (u, v)}{\partial (x'', y'')}(a) = \left| \det \frac{\partial f}{\partial z''}(a) \right|^2 \neq 0.$$

Hence we may use the implicit function Theorem for smooth maps. According to this Theorem, there are a neighborhood $U = U' \times U'' \ni a$ and a smooth map $g: U' \to U''$ such that the set $\{z \in U: f(z) = f(a)\}$ coincides with the graph of g above U'. Writing g with respect to complex coordinates, $g = (g_1, \ldots, g_m)$, $z'' = g(z')$, and by differentiation of the identity $f(z', g(z')) \equiv f(a)$ in U', we obtain

$$\left[\frac{\partial f}{\partial z'} \bigg|_{z''=g(z)} \right] \cdot dz' + \left[\frac{\partial f}{\partial z''} \bigg|_{z''=g(z')} \right] \cdot dg = 0.$$

The matrix $\partial f / \partial z''$ is invertible in a sufficiently small neighborhood of a (we will assume U to be that small), hence dg can, at every point of U', be linearly expressed in terms of dz', i.e. is a $\mathbb{C}$-linear map. Thus, g is a holomorphic map. ■

C o r o l l a r y. *Let D be a domain in $\mathbb{C}^n$, and let $f: D \to \mathbb{C}^n$ be a holomorphic map with* $\det \partial f / \partial z(a) \neq 0$ *at a point $a \in D$. Then there are neighborhoods $U \ni a$ and $V \ni b = f(a)$ such that the map $f: U \to V$ is bijective, and such that the map f^{-1}: $V \to U$ inverse to it is holomorphic also.*

■ Consider in a neighborhood of the point $(a,b) \in \mathbb{C}_z^n \times \mathbb{C}_w^n$ the holomorphic map $F(z,w) = (f_1(z) - w_1, \ldots, f_n(z) - w_n)$. We have $F(a,b) = 0$ and

$$\det \frac{\partial F}{\partial z}(a,b) = \det \frac{\partial f}{\partial z}(a) \neq 0.$$

By the implicit function Theorem (Theorem 1), there are neighborhoods $U \ni a$, $V \ni b$ such that $\{F(z,w)=0\} \cap (U \times V)$ is the graph of a holomorphic map $z = g(w)$. Clearly, $g = f^{-1}$. ■

D e f i n i t i o n 1. A map f: $D \to G$ between domains in $\mathbb{C}^n$ is called *biholomorphic* if it is holomorphic, bijective, and if the inverse map f^{-1}: $G \to D$ is holomorphic also. Domains D, G for which such maps (biholomorphisms) exist are called *biholomorphically equivalent*.

The holomorphy of f^{-1} is, as in the one-dimensional case, a consequence of the other two conditions, and the condition $\det \partial f / \partial z \neq 0$ is not only sufficient, but also necessary in order that f be a local biholomorphism (see below).

D e f i n i t i o n 2. A set $M \subset \mathbb{C}^n$ is called an (embedded) *complex manifold of dimension p* if for each point $a \in M$ there are a neighborhood $U \ni a$ in $\mathbb{C}^n$ and functions $f_1, \ldots, f_{n-p} \in \mathcal{O}(U)$ such that

$$M \cap U = \{z \in U: f_1(z) = \cdots = f_{n-p}(z) = 0\},$$

and $\operatorname{rank}_a f = n - p$, where $f = (f_1, \ldots, f_{n-p})$. If, moreover, $M \subset G$, where G is an open subset of $\mathbb{C}^n$, and M is closed in G (i.e. $\overline{M} \cap G = M$), then M is called a *complex submanifold* of G. It is clear that every complex manifold in $\mathbb{C}^n$ is a complex submanifold of a neighborhood of it. The dimension of M is denoted by $\dim_{\mathbb{C}} M$, or simply by $\dim M$; the number $n - \dim M =: \operatorname{codim} M$ is called the (complex) *codimension* of M. By the implicit function Theorem, every complex manifold M in $\mathbb{C}^n$ is also a smooth (C^∞) manifold, of real (topological) dimension $\dim_{\mathbb{R}} M = 2\dim_{\mathbb{C}} M$. A map f: $M \to N$ between complex manifolds (in $\mathbb{C}^n$ and $\mathbb{C}^m$, respectively) is called holomorphic if it is holomorphic in a neighborhood of M in $\mathbb{C}^n$, and biholomorphic if it is also bijective and if the inverse map f^{-1}: $N \to M$ is holomorphic also.

T h e o r e m 2. *Let* f: $D \to \mathbb{C}^m$ *be a holomorphic map, defined in a neighborhood* D *of a point* $a \in \mathbb{C}^m$, *such that* $\operatorname{rank}_z f \equiv r$ *in* D. *Then there are neighborhoods* $U \ni a$ *in* D, $V \ni f(a)$ *in* $\mathbb{C}^m$ *and biholomorphic maps* ϕ: $U \to \tilde{U} \subset \mathbb{C}^n$, ψ: $V \to \tilde{V} \subset \mathbb{C}^m$ *such that*

1) $f(U)$ *is an* r-*dimensional complex submanifold in* V;

2) *there is an* r-*dimensional complex plane* $L \ni a$ *such that* f *biholomorphically maps* $L \cap U$ *onto* $f(U)$;

3) *the map* $\psi \circ f \circ \phi^{-1}$: $\tilde{U} \to \tilde{V}$ *has the form*

$$(\zeta_1, \ldots, \zeta_n) \to (\zeta_1, \ldots, \zeta_r, 0, \ldots, 0) \in \mathbb{C}^m;$$

in particular, U *is foliated on the* $(n-r)$-*dimensional complex submanifold* $f^{-1}(w) \cap U$ *with respect to the parameter* $w \in f(U)$, *and* U *is biholomorphically equivalent to the manifold*

$$(L \cap U) \times (f^{-1}(f(a)) \cap U).$$

■ For convenience we assume $a = f(a) = 0$ (Figure 17). By requirement, the functions $f_1, \ldots, f_m$ (the components of f) and the coordinates $z_1, \ldots, z_n$ can be renumbered such that the determinant of the matrix $\partial f' / \partial z'(0) = (\partial f_k / \partial z_j(0))_{j,k=1}^{r}$ is not zero. By Corollary 1, there is a neighborhood $U \ni 0$, $U \subset D$, which is biholomorphically mapped by $\zeta = \phi(z) := (f'(z), z'')$ onto a neighborhood of $\tilde{U} \ni 0$ in $\mathbb{C}^n_\zeta$. Without loss of generality we may assume that $\tilde{U}$ has the form $\tilde{U}' \times \tilde{U}''$, where $\tilde{U}' \subset \mathbb{C}^r_{\zeta'}$, $\tilde{U}'' \subset \mathbb{C}^{n-r}_{\zeta''}$. By construction, the map $f \circ \phi^{-1}$ can be represented as $w = g(\zeta) := (\zeta', g_{r+1}(\zeta), \ldots, g_m(\zeta))$. Since

$\text{rank}_z f \equiv r$ in U, and since the rank is invariant under biholomorphic coordinate changes, all $\partial g_k / \partial \zeta_j$, $k,j > r$, vanish identically in $\tilde{U}$, hence the g_k, $k > r$, are independent of ζ'', i.e. $g_k(\zeta) = g_k(\zeta',0) =: h_k(\zeta')$, $k > r$. (In terms of the original map f this means that in U the functions f_k, $k > r$, can be holomorphically expressed in terms of $f_1, \ldots, f_r$.) This clearly implies that $f(U) = f \circ \phi^{-1}(\tilde{U})$ is the complex manifold that is the graph of $w''' = h(w')$ in $V := \tilde{U}' \times \mathbf{C}^{m-r}$, where $w' = (w_1, \ldots, w_r)$, $w''' = (w_{r+1}, \ldots, w_m)$, and $h = (h_{r+1}, \ldots, h_m)$.

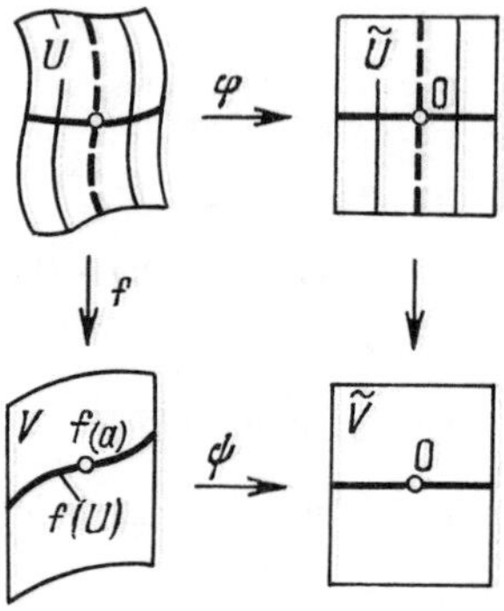

Figure 18.

For L we take the plane $z'' = 0$. In the space of variables $\zeta = \phi(z)$ it corresponds to $\zeta'' = 0$, on which $f \circ \phi^{-1}$ takes all values in $f(U)$, since $f \circ \phi^{-1}$ is independent of ζ''. By construction, $f\colon L \cap U \to f(U)$ is a biholomorphic map.

Finally, we denote by ψ the map $w \mapsto (w', w''' - h(w'))$, which biholomorphically maps V onto $\tilde{V}$ ($= V$ if $f(a) = 0$). Then $\psi \circ f \circ \phi^{-1}(\zeta) = (\zeta', 0''')$, as required. ■

C o r o l l a r y 1. *Let $f\colon D \to G$ be a one-to-one holomorphic map between domains in $\mathbf{C}^n$. Then $f^{-1}\colon G \to D$ is also a holomorphic map (hence, f is a biholomorphism).*

■ As is well known, the conditions imply that f^{-1} is continuous in G; in particular, locally bounded. Put $J = \det \partial f / \partial z$. By Theorem 2, $J \not\equiv 0$ in D, hence $\log |J|$ is a plurisubharmonic function in D. If $E = f(Z_J)$ is the image of the zero set of J, then, by the Corollary to Theorem 1, f^{-1} is holomorphic in $G \setminus E$.

But E is a polar set, since the function $\log |J \circ f^{-1}|$, which is plurisubharmonic in G, equals $-\infty$ everywhere on it (see A1.2). Hence by Corollary A1.4, f^{-1} is holomorphic in all of G. ■

Corollary 2. *A holomorphic map* $f: D \to G$ *between domains in* $\mathbb{C}^n$ *is locally biholomorphic if and only if* $\det \partial f / \partial z$ *is zero free in* D.

■ In one direction this is a consequence of Theorem 1. The converse statement follows from holomorphy of f^{-1} and the equation

$$\det \frac{\partial f^{-1}}{\partial w}(f(z)) = \frac{1}{\det \frac{\partial f}{\partial z}(z)}$$

al all points at which $\det \partial f / \partial z \neq 0$ (it suffices to prove this equation for tangential linear maps, and for such maps it is well known from linear algebra). ■

A2.3. Complex manifolds in $\mathbb{C}^n$. Let a complex manifold M in $\mathbb{C}^n$ be defined in a neighborhood of a point $a \in M$ by a holomorphic function $f = (f_1, \ldots, f_{n-p})$ with $\operatorname{rank}_a f = n - p$. A vector $v \in \mathbb{C}^n$ is tangent to M at a (as a tangent vector to a smooth manifold) if $(df_k)_a(v) = 0$, $k = 1, \ldots, n-p$, i.e. if the derivatives of all f_k at a in the direction of v vanish. The totality of all such vectors forms the tangent space T_aM (the tangent plane in the usual geometric sense is $a + T_aM$). Since all functions $(df_k)_a(z)$ are complex linear, T_aM is a complex subspace in $\mathbb{C}^n$. This property turns out to be characteristic for complex manifolds in $\mathbb{C}^n$.

Levi-Civita's theorem. *A smooth* (C^1) *manifold* M *in* $\mathbb{C}^n$ *is complex if and only if all tangent planes to* M *are complex.*

■ It suffices to prove the statement in a neighborhood of a fixed point, which we will take as the coordinate origin. If T_0M is a complex subspace of $\mathbb{C}^n$, there exists a unitary transformation mapping it onto the coordinate plane of the variable $z' = (z_1, \ldots, z_p)$, where $2p = \dim_{\mathbb{R}} M$. By the classical implicit function Theorem, in some neighborhood $U = U' \times U'' \ni 0$ the manifold M is given by the equation $z'' = g(z')$, where $g: U' \to U''$ is a smooth map. Hence the tangent plane to M at an arbitrary point $a \in M \cap U$ is given by

$$z'' = (dg)_{a'}(z') = \frac{\partial g}{\partial z'}(a') \cdot z' + \frac{\partial g}{\partial \bar{z}'}(a') \cdot \bar{z}'.$$

If T_aM is complex, it must contain with (z',z'') also the vector $i(z',z'')$. Substituting it in the equation defining T_aM and cancelling i we thus obtain $(\partial g/\partial \bar{z}'(a'))\cdot\bar{z}' = 0$ on T_aM. Since T_aM projects onto $\mathbb{C}^p_{z'}$ (i.e. $z' \in \mathbb{C}^p$ is arbitrary), $\partial g/\partial \bar{z}'(a') = 0$. Thus, if all T_aM are complex, $\partial g/\partial \bar{z}' \equiv 0$ in U', i.e. g is holomorphic in U', and hence $M \cap U$ is a complex manifold in $\mathbb{C}^n$. The converse has been proved above. ■

We now consider some examples of complex manifolds in $\mathbb{C}^n$.

Examples. a) A smooth map $f: D \to \mathbb{C}^m$ of a domain $D \subset \mathbb{C}^n$ is holomorphic if and only if its graph $\Gamma_f := \{(z,f(z)) : z \in D\}$ is a complex submanifold of $D \times \mathbb{C}^m$ (see the proof of the Levi-Civita theorem): the condition on a map of being complex differentiable is equivalent to the geometric condition on the tangent planes to its graph of being complex.

b) Let $\alpha_1, \ldots, \alpha_m$ be pairwise distinct complex numbers, and let M be the set in $\mathbb{C}^2$ defined by

$$z_2^k = c(z_1-\alpha_1)\cdots(z_1-\alpha_m), \quad c \neq 0,$$

or, for short, by $f(z) := z_2^k - p(z_1) = 0$. The differential of the defining function is

$$(df)_a = -p'(a_1)\,dz_1 + ka_2^{k-1}dz_2.$$

If $a \in M$ and $a_2 = 0$, then $p(a_1) = 0$, hence $p'(a_1) \neq 0$ (all roots of p are simple). Hence $\operatorname{rank}_z f \equiv 1$ on M, i.e. M is a one-dimensional complex manifold in $\mathbb{C}^2$. A particular instance is the manifold $M: z_1^2 + z_2^2 = 1$. It is the "complexification" of the unit sphere $x_1^2 + x_2^2 = 1$, which is the intersection of M and the real subspace $\mathbb{R}^2 \subset \mathbb{C}^2$. As distinct from the sphere, M is unbounded, since the equation $z_2^2 = 1 - z_1^2$ is solvable with respect to z_2 for arbitrary z_1.

c) The set $A: z_2^2 = z_1^3$ in $\mathbb{C}^2$ (the "semicubic parabola") is defined by the function $f = z_2^2 - z_1^3$. Its differential vanishes at $z = 0$ only, hence $A \setminus \{0\}$ is a one-dimensional complex manifold in $\mathbb{C}^2$. The set A is the image of the complex plane $\mathbb{C}$ under the holomorphic map $\phi: \zeta \to (\zeta^2, \zeta^3)$. Since $\phi: \mathbb{C} \to A$ is bijective, A is a topological manifold. However, if g is a function holomorphic and equal to zero in a neighborhood of 0 on A, then $g\cdot\phi(\zeta) \equiv 0$ in a neighborhood of 0 in $\mathbb{C}_\zeta$. Therefore, by Taylor's formula,

$$\frac{\partial g}{\partial z_1}(0)\cdot 2\zeta + \frac{\partial g}{\partial z_2}(0)\cdot 3\zeta^2 + \cdots \equiv 0$$

and hence $\partial g/\partial z_1(0) = \partial g/\partial z_2(0) = 0$. Thus, there is no neighborhood of

$z = 0$ in which A is a complex manifold. This can also be seen geometrically: if A would be a complex manifold, then the plane $\mathbb{C}_1$: $z_2 = 0$ would be tangent to A at 0 (the distance from $z \in A$ to $\mathbb{C}_1$ is equal to $|z_2| = |z_1|^{3/2} = o(|z|)$). Therefore, in a small neighborhood of 0 the orthogonal projection of A on $\mathbb{C}_1$ would be bijective; however, it is not: the two distinct points $(z_1, \pm\sqrt{z_1^3}) \in A$ are projected into $(z_1,0)$, $z_1 \neq 0$.

A2.4. Real manifolds in $\mathbb{C}^n$. The real subspaces in $\mathbb{C}^n$ can be classified according their relation to the complex structure. Clearly, the complex subspaces are distinguished by the following property:

An $\mathbb{R}$-linear subspace in $\mathbb{C}^n$ is complex if and only if it contains with every vector v also the vector iv.

Thus, if L is an $\mathbb{R}$-linear subspace in $\mathbb{C}^n$ and $iL := \{iv\colon v \in L\}$, then $L^c := L \cap iL$ is the maximal complex subspace of $\mathbb{C}^n$ belonging to L; it is called the *complex component* of L.

If $L^c = \{0\}$, the subspace L is called *totally real* (such are, e.g., the real subspace $\mathbb{R}^n \subset \mathbb{C}^n$, $z_2 = \bar{z}_1$ in $\mathbb{C}^2$, etc.). Clearly, the orthogonal complement of L^c in L with respect to the real scalar product Re $(\cdot,\cdot)$ is a totally real subspace in $\mathbb{C}^n$, hence

Every $\mathbb{R}$-linear subspace in $\mathbb{C}^n$ can be represented as the direct sum of a complex subspace and a totally real subspace.

Among the odd-dimensional $\mathbb{R}$-linear subspaces $L \subset \mathbb{C}^n$ there are subspaces with maximal complex structure, i.e. subspaces for which the codimension of the complex component, $\dim_{\mathbb{R}} L - 2\dim_{\mathbb{C}} L^c$, is equal to 1. Such planes are called *maximal complex*. A typical example is a real hypersurface in $\mathbb{C}^n$.

Besides an intrinsic partial complex structure an $\mathbb{R}$-linear subspace $L \subset \mathbb{C}^n$ has a, so-to speak, extrinsic domain of influence in the ambient space. This is the $\mathbb{C}$-linear subspace $L \oplus iL$ in $\mathbb{C}^n$, obtained from linear combinations with complex coefficients of vectors of L; it is called the complex hull (envelope) of L. It is clear that in order to construct it, it suffices to construct the hull of the totally real component of L: if $L = L^c \oplus L'$, then $L \oplus iL = L^c \oplus (L' + iL')$. In particular, this implies that

Every maximal complex subspace $L \subset \mathbb{C}^n$ is a real hyperplane in its complex hull $L \oplus iL$.

Usually, a (complex or real) plane in $\mathbb{C}^n$ is a subset defined by a system of, in general inhomogeneous, $\mathbb{C}$- or $\mathbb{R}$-linear equations. The terminology introduced above will be used also for planes, by considering the corresponding homogeneous equations. The notions introduced also make sense in any $\mathbb{C}$-linear space, and we will use them without further notice.

A smooth manifold in $\mathbb{C}^n$ of real codimension k is locally defined by a system of equations $\rho_1 = \cdots = \rho_k = 0$, where the ρ_j are real valued functions of corresponding smoothness and with $\mathbb{R}$-linearly independent differentials (the latter is written as $d\rho_1 \wedge \cdots \wedge d\rho_k \neq 0$, for short). Similar to real subspaces, smooth manifolds in $\mathbb{C}^n$ can be classified according to their relation to the complex structure, reflected in the structures of their tangent spaces. We find maximal compatibility with the complex structure in $\mathbb{C}^n$ in the case of complex manifolds: for them all tangent planes are complex (A2.3). Their opposites are the totally real manifolds, i.e. the manifolds in $\mathbb{C}^n$ of class C^1 for which all tangent planes are totally real. This is a class of fairly "fine" manifolds: their real dimension in $\mathbb{C}^n \approx \mathbb{R}^{2n}$ is at most n, and, moreover, by intersecting with a complex manifold in $\mathbb{C}^n$ they give a set of at most half the dimension, as can be seen from the proof of the following Lemma.

L e m m a. *Let $A \subset \mathbb{C}^n$ be a connected complex manifold, and let $M \subset \mathbb{C}^n$ be a totally real manifold. Then $M \cap A$ is nowhere dense in A; if* $\dim A > 1$, *then $A \setminus M$ is connected.*

■ Let $0 \in A \cap M$. Since T_0M is a totally real space, there is a linear transformation of $\mathbb{C}^n$ mapping T_0M into $\mathbb{R}^n \subset \mathbb{C}^n$. Hence we may immediately assume that $T_0M \subset \mathbb{R}^n$. Consequently, in a neighborhood of 0 the manifold M belongs to the set $\tilde{M}$: $y = \phi(x)$, where $z = x + iy$ and ϕ is a real valued vector function of class C^1. On the other hand, T_0A can be bijectively projected onto some coordinate plane $\mathbb{C}_I$, $\# I = p = \dim A$; we may assume $I = (1, \ldots, p)$. Then A is defined, in a neighborhood of 0, by a system $z'' = f(z')$, where $z = (z', z'')$, $z' \in \mathbb{C}_I$, and f is a holomorphic vector function in a neighborhood of 0 in $\mathbb{C}_I$. Thus, in some neighborhood of 0 on $A \cap M$ we have $z'' = f(z')$, $y = \phi(x)$, which imply $x'' = \operatorname{Re} f(x' + i\phi'(x', x''))$. Since $|\phi| = o(|x|)$ as $x \to 0$, the implicit function Theorem in a neighborhood of 0 on $A \cap M$ implies that the equation $x'' = \psi(x')$ holds for a certain $\psi \in C^1$. Hence the projection of $A \cap M$ in $\mathbb{C}_I$ belongs to the totally real manifold $y' = \phi'(x', \psi(x'))$ of real dimension p in $\mathbb{C}^p$. This clearly implies both assertions of the Lemma (see A6.1). ■

The degree of "complexity" of a manifold M at a point $a \in M$ is reflected by the complex component $T_a^c M = (T_a M) \cap i(T_a M)$ of the tangent space. The plane $a + T_a^c M$ is called the *complex tangent plane*, and the elements of $T_a^c M$ are called *complex tangent vectors*, to M at a. The complex dimension of $T_a^c M$ is called the CR-dimension of M at a; it is denoted by $\operatorname{CRdim}_a M$. Manifolds of constant CR-dimension (which is then denoted by CRdim) play a large part in modern complex analysis; they are called Cauchy-Riemann manifolds (for short, CR-manifolds). Complex and totally real manifolds in $\mathbb{C}^n$ are extreme examples of Cauchy-Riemann manifolds.

By analogy with holomorphic functions, on smooth manifolds in $\mathbb{C}^n$ there are defined so-called CR-functions, i.e. functions satisfying the Cauchy-Riemann conditions in complex tangential directions (see, e.g., [174], [181]). If $a \in M$ and if coordinates are chosen such that $T_a^c M = \mathbb{C}_{1 \ldots p}$, the tangential Cauchy-Riemann conditions at a for a function can be written as

$$\frac{\partial f}{\partial \bar{z}_j}(a) = 0, \quad j = 1, \ldots, p.$$

Smooth odd-dimensional manifolds $M \subset \mathbb{C}^n$ for which the totally real component of every tangent space has dimension 1 (i.e. $\dim_{\mathbb{R}} M = 2 \operatorname{CRdim} M + 1$) are called *maximal complex*. Their relation with complex manifolds can be seen in the following

P r o p o s i t i o n 1. *Let M be a smooth manifold in $\mathbb{C}^n$ with boundary S. If M is a complex manifold in $\mathbb{C}^n$, then S is maximal complex.*

■ Let $T_a M$ be the tangent space to (M,S) at a point $a \in S$. Since $T_a M$ is the limit of $T_{a_j} M$ as $g \to a$, $a_j \in M$, and since all $T_{a_j} M$ are complex subspaces in $\mathbb{C}^n$, $T_a M$ is complex also. By the definition of manifold with boundary, the tangent space to S at a is a real hyperplane in $T_a M$, hence maximal complex. ■

The relation between maximal complex manifolds and real hypersurfaces and CR-functions can be seen in the following

P r o p o s i t i o n 2. *1) Let M be a maximal complex manifold in $\mathbb{C}^n$ and let*

$a \in M$. Then there is a neighborhood $U \ni a$ in $\mathbb{C}^n$ such that $M \cap U$ can be diffeomorphically projected into a real hypersurface of the complex hull of T_aM.

2) Let Γ be a hypersurface in $\mathbb{C}^n$ and let $f: \Gamma \to \mathbb{C}^m$ be a smooth map. The graph of f, i.e. the set $\Gamma_f = \{(z, f(z)): z \in \Gamma\}$ in $\mathbb{C}^p \times \mathbb{C}^m$ is a maximal complex manifold if and only if all components of f are CR-functions on Γ.

■ Statement 1) clearly follows from the fact that locally M can be diffeomorphically projected into its tangent plane, and from the fact that this tangent plane is maximal complex.

2) If f satisfies the tangential Cauchy-Riemann conditions, the restriction onto each T_a^cM of its differential is complex linear. This implies that at each point of Γ_f the complex tangent plane has dimension at least $p-1$. Since $\dim_{\mathbb{R}} \Gamma_f = 2p-1$, this implies that Γ_f is maximal complex. Conversely, let Γ_f be maximal complex and let $a \in \Gamma$. Since the projection $T_{(a,f(a))}\Gamma_f \to T_a\Gamma$ is one-to-one, the image of the complex component is a $(p-1)$-dimensional complex plane in $T_a\Gamma$, i.e. $T_a^c\Gamma$. Passing from projection to lifting we thus obtain that the graph of the restriction onto $T_a^c\Gamma$ of $(df)_a$ is a complex plane, which elementarily implies (see A2.3) that $\bar{\partial} f|_{T_a^c\Gamma} = 0$, i.e. f satisfies at a the tangential Cauchy-Riemann conditions. ■

A3. Projective spaces and Grassmannians

A3.1. Abstract complex manifolds. Recall that a topological manifold of dimension N is a Hausdorff space X in which every point has a neighborhood homeomorphic to an open set in $\mathbb{R}^N$. Thus, for every point $x \in X$ there are a neighborhood $U \ni x$ and a homeomorphism $\phi: U \to V \subset \mathbb{R}^N$. A *complex chart* on an even-dimensional manifold X is a pair (U, ϕ), where U is open in X and $\phi: U \to V$ is a homeomorphism onto an open set $V \subset \mathbb{C}^n$. Here, U is called a coordinate neighborhood and the components of ϕ are called *complex coordinates* in U. On intersections of two charts a neighboring relation arises, i.e. maps $\phi_2 \circ \phi_1^{-1}$: $\phi_1(U_1 \cap U_2) \to \phi_2(U_1 \cap U_2)$ between open sets in $\mathbb{C}^n$. Two charts (U_1, ϕ_1) and (U_2, ϕ_2) are called holomorphically compatible if $\phi_2 \circ \phi_1^{-1}$ is a biholomorphic map or if $U_1 \cap U_2$ is empty. A *complex atlas* on a Hausdorff space X is a family $\mathcal{U} = \{(U_j, \phi_j)\}$ of pairwise holomorphically compatible complex charts on X covering X (i.e. $X = \cup U_j$). Two complex atlases $\mathfrak{U}$, $\mathfrak{U}'$ on X are called equivalent, $\mathfrak{U} \sim \mathfrak{U}'$, if all charts (U,ϕ) of $\mathfrak{U}$ are holomorphically compatible with all charts (U',ϕ') of $\mathfrak{U}'$. A *complex structure* Σ on X is, by definition, an equivalence class of complex atlases on X. The pair (X, Σ), the Hausdorff space plus a complex structure on it, is called a *complex manifold* (of dimension n if the charts are homeomorphic to domains in $\mathbb{C}^n$); it is usually denoted by a single symbol, the structure Σ being assumed known (or of no importance at all). In order to specify a complex structure on X it suffices to give one complex atlas $\mathfrak{U} \in \Sigma$ (then $\Sigma = \{\mathfrak{U}: \mathfrak{U}' \sim \mathfrak{U}\}$).

Clearly, every n-dimensional complex manifold is infinitely differentiable; in particular, it is a topological manifold of dimension $2n$, hence the standard objects of analysis, such as smooth functions and maps, tangent and cotangent spaces and bundles, etc., can be naturally defined on it.

A map $f: X \to Y$ between complex manifolds (of dimensions n and m, with complex atlases $\{(U_i,\phi_i)\}$ and $\{(V_j,\psi_j)\}$, respectively) is called *holomorphic* if it is holomorphic with respect to the local complex coordinates, i.e. if all maps $\psi_j \circ f \circ \phi_i^{-1}$ between the corresponding open sets in $\mathbb{C}^n$ and $\mathbb{C}^m$ are holomorphic. In particular, a function $f: X \to \mathbb{C}$ is holomorphic if all functions $f \circ \phi_i^{-1}$ are holomorphic in the corresponding $\phi_i(U_i) \subset \mathbb{C}^n$. A map f is called *biholomorphic* (a biholomorphism) if it is one-to-one and if f^{-1} is also holomorphic. Two complex manifolds X, Y are called *biholomorphically equivalent* if there is a biholomorphic map $f: X \to Y$.

A set M on an n-dimensional complex X is called a p-dimensional complex submanifold in X (see A2.2) if M is closed in X and if for every point $a \in M$ there are a coordinate neighborhood (U,z), $U \ni a$, and functions $f_1, \ldots, f_{n-p} \in \mathcal{O}(U)$ such that

$$M \cap U = \{\zeta \in U: f_1(\zeta) = \cdots = f_{n-p}(\zeta) = 0\}$$

and $\operatorname{rank}_a f = n - p$, where $f = (f_1, \ldots, f_{n-p})$. By the implicit function Theorem, (U,z) can be chosen such that $M \cap U$ is given by a system of holomorphic equations $z'' = \psi_U(z')$, where $z' = (z_1, \ldots, z_p)$, $z'' = (z_{p+1}, \ldots, z_n)$. The holomorphic map $z \mapsto z'$, defined on U, "projects" $M \cap U$ onto the open set $U' = z(U') \subset \mathbb{C}^p$, and hence gives the complex chart $(M \cap U, z')$ on M. The inverse map transforms z' into the point on $M \cap U$ with local coordinates

$(z',\psi_U(z'))$. Therefore, if (V,w) and $(M\cap V,w')$ are charts on X and M, respectively, thus related, then the neighboring relation between $(M\cap U,z')$ and $(M\cap V,w')$ is the holomorphic map

$$z'\mapsto(z',\psi_U(z'))\mapsto(w',\psi_V(w'))\mapsto w'$$

(the second arrow is the biholomorphic change of coordinates (z',z'') to (w',w'') in $U\cap V$). Thus, a complex structure is canonically defined on a complex submanifold $M\subset X$; it is called the complex structure *induced* by X. It is easy to verify that the holomorphic functions on M are precisely the functions satisfying on M the tangential Cauchy-Riemann conditions (i.e. the CR-functions). If f is holomorphic on M, it has a holomorphic continuation into each neighborhood U on X constructed above (constant with respect to the variable z''). Thus, every function holomorphic on M is locally the restriction of functions holomorphic in open subsets of X.

A3.2. Complex projective space $\mathbb{P}_n$. We consider in the domain $\mathbb{C}^{n+1}\setminus\{0\}=:\mathbb{C}_\star^{n+1}$ the following equivalence relation: $z\sim w$ if $z=\lambda w$ for some $\lambda\in\mathbb{C}_\star=\mathbb{C}\setminus\{0\}$. The equivalence class of a point z is the punctured complex line $\{\lambda z:\lambda\in\mathbb{C}_\star\}$; the set of these classes $[z]$ is denoted by $\mathbb{P}_n$. The quotient map Π: $\mathbb{C}_\star^{n+1}\to\mathbb{P}_n$, under which $z\to[z]$, is called canonical projection. The *homogeneous coordinates* of a point $[a]\in\mathbb{P}_n$ are the coordinates of an arbitrary point $z\in\mathbb{C}_\star^{n+1}$ belonging to the class $[a]$; they are defined up to a common factor $\lambda\in\mathbb{C}_\star$.

We give two more interpretations of the set $\mathbb{P}_n$. Assigning to each class $[z]$ the complex line $\mathbb{C}\cdot z$ we see that $\mathbb{P}_n$ is the set of all complex lines in $\mathbb{C}^{n+1}$ passing through the coordinate origin. Furthermore, each class $[z]$ uniquely determines a set

$$\gamma_z=\left\{e^{i\phi}\frac{z}{|z|}:0\leqslant\phi<2\pi\right\}$$

on the unit sphere S^{2n+1} in $\mathbb{C}^{n+1}$, hence $\mathbb{P}_n$ is the set of all circles γ_z on S^{2n+1}.

It is most easy to introduce a topology in $\mathbb{P}_n$ using the last model: a neighborhood of a point $[a]\in\mathbb{P}_n$ consists of all points $[z]\in\mathbb{P}_n$ for which γ_z belongs to a neighborhood of γ_a in S^{2n+1}. Since the sphere is compact and Π: $S^{2n+1}\to\mathbb{P}_n$ is, by definition, continuous, $\mathbb{P}_n$ is also compact.

Now we endow $\mathbb{P}_n$ with a complex structure. Let $z_0, \ldots, z_n$ be coordinates in $\mathbb{C}^{n+1}$ (this enumeration is most convenient). Define the neighborhood

$$U_j = \{[z_0, \ldots, z_n]: z_j \neq 0\}$$

and the maps

$$\phi_j: [z] \rightarrow \left[\frac{z_0}{z_j}, \ldots, \frac{z_{j-1}}{z_j}, \frac{z_{j+1}}{z_j}, \ldots, \frac{z_n}{z_j}\right].$$

Then the $\phi_j: U_j \rightarrow \mathbb{C}^n$ are homeomorphisms, and

$$\phi_j^{-1}: (\zeta_1, \ldots, \zeta_n) \mapsto [\zeta_1, \ldots, \zeta_{j-1}, 1, \zeta_{j+1}, \ldots, \zeta_n].$$

The neighboring relations are clearly holomorphic (e.g., $\phi_k \circ \phi_0^{-1}(\zeta) = \phi_k([1, \zeta_1, \ldots, \zeta_n])$). Thus, the charts (U_j, ϕ_j) form a complex atlas and define on $\mathbb{P}_n$ a complex structure. The complex manifold thus constructed is called n-dimensional complex projective space, and is also denoted by $\mathbb{P}_n$.

Every coordinate neighborhood U_j is biholomorphic to $\mathbb{C}^n$. We identify U_0: $z_0 \neq 0$ and $\mathbb{C}^n$, and call the map

$$\phi_0^{-1}: (\zeta_1, \ldots, \zeta_n) \mapsto [1, \zeta_1, \ldots, \zeta_n]$$

the canonical imbedding of $\mathbb{C}^n$ in $\mathbb{P}_n$, or the transition from affine coordinates (in $\mathbb{C}^n$) to projective (homogeneous) coordinates. The complement of $\mathbb{C}^n$ in projective space is called the "hyperplane at infinity", H_0: $z_0 = 0$; it is a complex submanifold in $\mathbb{P}_n$ of codimension 1. It consists of the points $[0, z_1, \ldots, z_n]$, where $(z_1, \ldots, z_n) \in \mathbb{C}_*^n$, hence is biholomorphic to $\mathbb{P}_{n-1}$. In turn, $\mathbb{P}_{n-1}$ splits into the dense open set $\{z_1 \neq 0\} \approx \mathbb{C}^{n-1}$ and the manifold $z_1 = 0$, biholomorphic to $\mathbb{P}_{n-2}$, etc. As a result we obtain the construct

$$[0, \ldots, 0, 1] = \mathbb{P}_0 \in \mathbb{P}_1 = \mathbb{P}_0 \cup \mathbb{C}^1 \subset \cdots \subset \mathbb{P}_n = \mathbb{P}_{n-1} \cup \mathbb{C}^n$$

(at each stage we "glue" the cell $\mathbb{C}^{k+1}$ of higher dimension to $\mathbb{P}_k$). In particular, $\mathbb{P}_1$ is the usual Riemann sphere $\mathbb{C} \cup \{\infty\}$, $\mathbb{P}_2$ is $\mathbb{C}^2$ "glued at infinity" to the Riemann sphere, etc.

As can be easily seen, the complex line $L = \{a + \lambda v: \lambda \in \mathbb{C}\} \subset \mathbb{C}^n$ has the unique limit point $[0, v_1, \ldots, v_n]$ at infinity. This point is determined by the direction of L only (it does not depend on a), hence $H_0 \approx \mathbb{P}_{n-1}$ can be regarded as the set of all complex directions in $\mathbb{C}^n$.

Linear (in particular, unitary) transformations of $\mathbb{C}^{n+1}$ map complex lines to complex lines, hence induce biholomorphic maps of $\mathbb{P}_n$ onto itself (*automorphisms* of $\mathbb{P}_n$). In homogeneous coordinates such a map has the form $[z]\mapsto[l_0(z),\ldots,l_n(z)]$. In affine coordinates in $\mathbb{C}^n$ $(=U_0)$ this transformation can be viewed as

$$\zeta\mapsto[1,\zeta]\mapsto[l_0(1,\zeta),\ldots,l_n(1,\zeta)]\mapsto\left[\frac{l_1(1,\zeta)}{l_0(1,\zeta)},\ldots,\frac{l_n(1,\zeta)}{l_0(1,\zeta)}\right]$$

(the first arrow denotes transition to homogeneous coordinates, the second - to an automorphism in homogeneous coordinates, and the third denotes return to affine coordinates). Hence the automorphisms of $\mathbb{P}_n$ are called fractional-linear (we can show that every biholomorphic map of $\mathbb{P}_n$ onto itself has this form). For $n = 1$ these are the usual fractional-linear maps of the Riemann sphere.

It is convenient to use in $\mathbb{P}_n$ the following metric. By definition, the angle between two complex lines $\mathbb{C}a$ and $\mathbb{C}b$ in $\mathbb{C}^{n+1}$ is equal to $\arccos|(a/|a|, b/|b|)|$. It is the minimum of the angles between two rays $\mathbb{R}v$ and $\mathbb{R}w$ on these lines (it is assumed if $\operatorname{Im}(v,w) = 0$; moreover, we can always take a in place of v). The *distance* between two points $a = [\tilde{a}]$, $b = [\tilde{b}]$ in $\mathbb{P}_n$ (for the time being we will denote it by $|a,b|$) is defined as $\arccos|(\tilde{a}/|\tilde{a}|, \tilde{b}/|\tilde{b}|)|$, i.e. as the angle between the corresponding complex lines in $\mathbb{C}^{n+1}$. The fact that it is a distance can be seen from Figure 18, which clarifies the triangle inequality: let $a,b,c\in\mathbb{P}_n$, $|a,b|\geqslant|a,c|$, $|a,b|\geqslant|b,c|$, let $c = [\tilde{c}]$, and let $\tilde{a},\tilde{b}$ be the points of $\mathbb{C}^{n+1}_{*}$, from the classes a, b, respectively, nearest to $\tilde{c}$. Then $|a,c| = \angle\tilde{a}\tilde{c}$, $|b,c| = \angle\tilde{b}\tilde{c}$, and if c' is the point on the real plane $\mathbb{R}\tilde{a}+\mathbb{R}\tilde{b}$ nearest to $\tilde{c}$, then $|a,b|\leqslant\angle\tilde{a}\tilde{b} = \angle\tilde{a}c' + \angle c'\tilde{b}\leqslant\angle\tilde{a}\tilde{c} + \angle\tilde{c}\tilde{b} = |a,c| + |b,c|$, as required. The diameter of $\mathbb{P}_n$ in this metric is $\pi/2$.

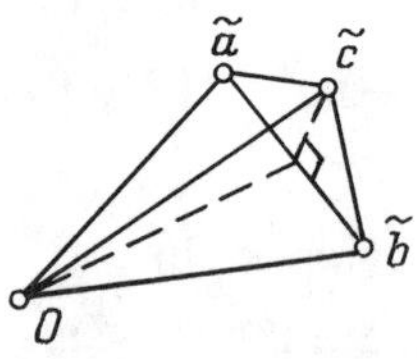

Figure 18.

The notion of orthogonality in $\mathbb{P}_n$ (denoted by $\perp$) is naturally taken from $\mathbb{C}_*^{n+1}$, since the equation $(z,w) = 0$ is independent of the representatives of the classes $[z]$, $[w]$. The definition of distance implies:

Two points $a,b \in \mathbb{P}_n$ are orthogonal if and only if they are diametrically opposite: $a \perp b \Leftrightarrow |a,b| = \pi/2$.

A3.3. Complex planes in $\mathbb{P}_n$. By definition, a complex plane in $\mathbb{P}_n$ of dimension p is a set $\{[z]: l_1(z) = \cdots = l_{n-p}(z) = 0\}$, where the l_j are linearly independent complex linear functions in the homogeneous coordinates. As can easily be seen, each such plane is a compact p-dimensional complex submanifold in $\mathbb{P}_n$, biholomorphically equivalent to $\mathbb{P}_p$. For $p \neq 0$ the diameter of every p-dimensional complex plane in $\mathbb{P}_n$ is equal to the diameter of $\mathbb{P}_n$.

For each fixed point $a \in \mathbb{P}_n$ the diametrically opposite (orthogonal) points form a complex hyperplane, $a^\perp$ (e.g., if $a = [1,0,\ldots,0]$ is the coordinate origin in $\mathbb{C}^n \subset \mathbb{P}_n$, this hyperplane is the hyperplane at infinity $H_0 = \mathbb{P}_n \setminus \mathbb{C}^n$). In general, for any p-dimensional complex plane $L \subset \mathbb{P}_n$ the set $L^\perp = \{z \in \mathbb{P}_n: z \perp a$ for all $a \in L\}$ orthogonal to it is an $(n-p-1)$-dimensional complex plane in $\mathbb{P}_n$, which corresponds to the orthogonal complement of $\Pi^{-1}(L)$.

The automorphisms of $\mathbb{P}_n$ corresponding to unitary transformations of $\mathbb{C}^{n+1}$ (we will call them *unitary automorphisms* of $\mathbb{P}_n$) clearly preserve distance and orthogonality in $\mathbb{P}_n$, i.e. are isometric transformations. Such transformations ("rotations" of $\mathbb{P}_n$) are abundant: any two p-dimensional complex planes in $\mathbb{P}_n$ can be mapped onto each other by unitary automorphisms; in particular, every hyperplane in $\mathbb{P}_n$ can by a rotation of $\mathbb{P}_n$ be "corrected at infinity", i.e. mapped onto H_0.

Fix an arbitrary p-dimensional complex plane L in $\mathbb{P}_n$. Any $(p+1)$-dimensional plane L' intersects $L^\perp$ at a unique point (cf. the corresponding subspaces in $\mathbb{C}^{n+1}$). If $L' \neq L''$ are two such $(p+1)$-dimensional planes, then $L' \cap L'' = L$, i.e. the points $L' \cap L^\perp$ and $L'' \cap L^\perp$ are distinct. Thus, in $\mathbb{P}_n \setminus L$ there is defined a map

$$\pi_L: \mathbb{P}_n \setminus L \to L^\perp,$$

under which each point $a \in \mathbb{P}_n \setminus L$ is put into correspondence with the point of intersection of $L^\perp$ and the $(p+1)$-dimensional plane L' containing L and a. This map is called *projection of* $\mathbb{P}_n$ *from* L *onto* $L^{\perp}$, and is a holomorphic map of rank $n-p-1$ (this readily follows from the homogeneity of $\mathbb{P}_n$ and the rank Theorem). In particular, in the affine part $\mathbb{C}^n$, the projection $\mathbb{P}_n \setminus [1,0,\ldots,0] \to H_0$ is "blowing up" $\mathbb{C}^n_*$ from the coordinate origin to infinity (Figure 19): it puts each complex ray $\mathbb{C}_* v \subset \mathbb{C}^n \subset \mathbb{P}_n$ into correspondence with its limit point at infinity $[0,v]$. The projection from a point at infinity $a = [0,v]$ $\mathbb{C}^n \subset \mathbb{P}_n$ is within the ordinary projection along the vector v on the hyperplane $v^{\perp}$ (in the Figure this is shown in the model of the real projective plane by a disk for which diametrically opposite points of the boundary are identified).

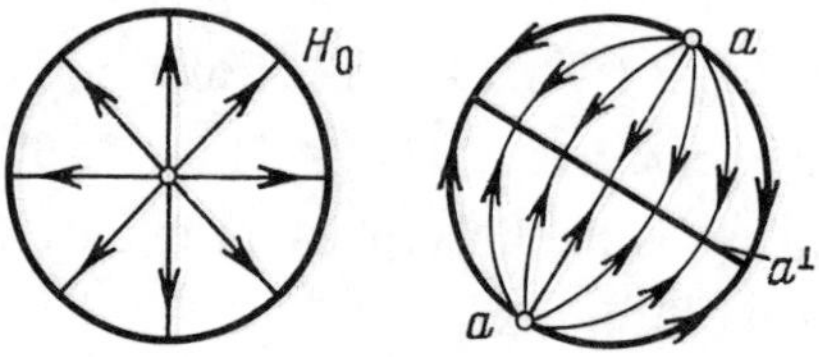

Figure 19.

A3.4. The Grassmannians $G(k,n)$. By definition, the set $G(k,n)$ is the set of all k-dimensional complex linear subspaces in $\mathbb{C}^n$. By choosing a basis in such a subspace we can assign a nonsingular $k \times n$ matrix to it, whose rows are the vectors in the basis. Such matrices can be regarded as points in the space $\mathbb{C}^{kn}$, and the nonsingular matrices in $\mathbb{C}^{kn}$ form a dense open set $\mathrm{St}(k,n) \subset \mathbb{C}^{kn}$, the so-called Stiefel manifold of k-dimensional bases in $\mathbb{C}^n$. Conversely, to every nonsingular matrix $A \in \mathrm{St}(k,n)$ corresponds the k-dimensional linear subspace of $\mathbb{C}^n$ spanned over the columns of A; this defines a map

$$\psi\colon \mathrm{St}(k,n) \to G(k,n).$$

Two matrices A_1, A_2 represent the same element $L \in G(k,n)$ (i.e. $\psi(A_1) = \psi(A_2)$) if and only if $A_2 = BA_1$ for some nonsingular $k \times k$ matrix B (this is the matrix of transition between the respective bases in the subspace $L \subset \mathbb{C}^n$). The map ψ naturally endows $G(k,n)$ with a topology: $U \subset G(k,n)$ is open if and only if $\psi^{-1}(U)$ is open. Moreover, by means of ψ we can introduce a complex structure on $G(k,n)$. This is done as follows.

Consider an ordered multi-index $I = (i_1, \ldots, i_k)$, $1 \leqslant i_j \leqslant n$. Let, for a general $k \times n$ matrix Z, $(Z)_I$ denote the square matrix obtained from the columns of Z with indices $i_1, \ldots, i_k$ (in this order). We define the set $U_I \subset G(k,n)$ by the condition: $L \in U_I$ if $\det(Z)_I \neq 0$ for some (hence for any) matrix Z representing L, i.e. belonging to $\psi^{-1}(L)$. Geometrically, U_I is the set of k-dimensional subspaces $L \subset \mathbb{C}^n$ that are one-to-one projected onto the coordinate plane $\mathbb{C}_I$ of variables $z_{i_1}, \ldots, z_{i_k}$ in $\mathbb{C}^n$. Each point $L \in U_I$ corresponds to the unique matrix Z for which the $k \times k$ unit matrix $E = E_k$ occupies the $(Z)_I$ positions. The columns of such a matrix Z with indices not belonging to I form the $k \times (n-k)$ matrix $(Z)_J$, $J = (j_{k+1}, \ldots, j_n)$, which can be arbitrary. In this way we obtain a parametrization of the points in U_I by $k \times (n-k)$ matrices, i.e. a one-to-one correspondence $\phi_I\colon U_I \to \mathbb{C}^{k(n-k)}$, for each I. E.g., the points of $U_{1 \cdots k}$ can be one-to-one represented by matrices (E, W) with W an arbitrary $k \times (n-k)$ matrix. Consequently, we have constructed complex charts covering $G(k,n)$. The neighboring relations are, as can be easily verified, holomorphic, hence these charts form on $G(k,n)$ a complex atlas, hence a complex structure. The complex manifold thus constructed is called the complex *Grassmann manifold*, or simply the *Grassmannian*, and is denoted by $G(k,n)$ also.

Projective spaces are particular instances of Grassmannians: by construction, $\mathbb{P}_n = G(1,n+1)$. Sets of complex planes in $\mathbb{P}_n$ can also be interpreted as Grassmannians, since to each k-dimensional complex plane in $\mathbb{P}_n$ corresponds a $(k+1)$-dimensional subspace of $\mathbb{C}^{n+1}$, i.e. a point of $G(k+1,n+1)$. In particular, the hyperplanes in $\mathbb{P}_n$ form the Grassmannian $G(n,n+1)$, which is isomorphic to $\mathbb{P}_n$ itself: to each hyperplane

$$a_0 z_0 + \cdots + a_n z_n = 0$$

in $\mathbb{P}_n$ uniquely corresponds the point $[a_0, \ldots, a_n] \in \mathbb{P}_n$ (it is not difficult to show that this correspondence is biholomorphic).

Linear, in particular unitary, transformations of $\mathbb{C}^n$ map k-dimensional subspaces onto k-dimensional subspaces, hence induce biholomorphic maps of $G(k,n)$ onto itself (automorphisms). The group of unitary automorphisms acts transitively on $G(k,n)$, since every k-dimensional subspace in $\mathbb{C}^n$ is unitarily equivalent to $\mathbb{C}_{1 \cdots k}$.

A3.5. Incidence manifolds and the σ-process. The correspondence between the points of $G(k,n)$ and k-dimensional subspaces of $\mathbb{C}^n$ can be clearly visualized using "the graph" of this correspondence. Consider in $\mathbb{C}^n \times G(k,n)$ the set $I(k,n)$ of pairs (z,L) for which $z \in L$,

$$I(k,n) := \{(z,L) : z \in L\}.$$

If $v_1, \ldots, v_k$ is a basis on L, the condition $z \in L$ is equivalent to $z \wedge v_1 \wedge \cdots \wedge v_k = 0$, hence $I(k,n)$ is in local coordinates defined by a system of polynomial identities. In order to describe these we consider the chart $U_{1 \cdots k} \subset G(k,n)$, whose points are represented by $k \times n$ matrices (E, W) with W an arbitrary $k \times (n-k)$ matrix. The equations for $I(k,n)$ mean, in the coordinates z, W, that the vector z is a linear combination of the rows of the matrix (E, W). By considering the first k columns only we see that the coefficient at the i-th row in this linear combination is z_i. Thus, above $U_{1 \cdots k}$ the set $I(k,n)$ is defined by the system

$$z_j = z_1 w_{1j} + \cdots + z_k w_{kj}, \quad j = k+1, \ldots, n,$$

(we number the w_{ij} as in the matrix (E, W)). It is clear that this is a complex manifold in $\mathbb{C}^n \times U_{1 \cdots k} \approx \mathbb{C}^n \times \mathbb{C}^{k(n-k)}$ of codimension $n-k$ (and dimension $k + k(n-k)$). The remaining affine charts $U_I \subset G(k,n)$ are mapped onto $U_{1 \cdots k}$ by a unitary automorphism of $G(k,n)$, while the set $I(k,n)$ is mapped onto itself by compatible unitary automorphisms of $\mathbb{C}^n$ and $G(k,n)$. Thus, $I(k,n)$ is a complex submanifold in $\mathbb{C}^n \times G(k,n)$ of codimension $n-k$. It is called the *incidence manifold* and is extensively used in complex geometry (see below A6.4, p.3.8, p.10.5, p.13.5).

The fiber of $I(k,n)$ above an arbitrary point of U_I is a subspace $L \subset \mathbb{C}^n$ that can be one-to-one projected onto $\mathbb{C}_I$. The only coordinates acting in these subspaces in $\mathbb{C}^n$ are $z_I = (z_{i_1}, \ldots, z_{i_k})$, hence $I(k,n) \cap \{L \in U_I\}$ is biholomorphically equivalent to $\mathbb{C}^k \times U_I$. The matrix of transition from coordinates z_I to coordinates z_J on intersections of subsets in $I(k,n)$ clearly has holomorphic entries, hence $I(k,n)$ together with its projection on $G(k,n)$ is a complex vector bundle of rank k (which is trivial above every U_I). This is the so-called *universal vector bundle of rank k*, which is extensively used in modern differential geometry. In the case $\mathbb{P}_n = G(1,n+1)$, the bundle $I(1,n+1) \to \mathbb{P}_n$ is linear (the fibers are one-dimensional), the transition functions are $g_{jk} = z_j / z_k$ on $U_j \cap U_k$, hence $I(1,n+1)$ is dual to the bundle of a hyperplane section above $\mathbb{P}_n$.

The projection of $I(k,n)$ on the first factor in $\mathbb{C}^n \times G(k,n)$ has a complicated structure, and we will only investigate the case $k = 1$. Above a fixed point $z \in \mathbb{C}^n_\star$ lies the unique point $(z, \mathbb{C}\cdot z)$ on $I(1,n)$, i.e. the projection $I(1,n) \to \mathbb{C}^n$ is one-to-one above $\mathbb{C}^n_\star$. Above the coordinate origin $0 \in \mathbb{C}^n$ lies the whole projective space $\mathbb{P}_{n-1}$ in $I(1,n)$ (more precisely, $\{0\} \times G(1,n) \approx \mathbb{P}_{n-1}$), which is a submanifold of $I(1,n)$ of codimension 1. Under transition from $\mathbb{C}^n$ to $I(1,n)$ the coordinate origin is, as it were, blow up to $\mathbb{P}_{n-1}$, and $I(1,n)$ can be visualized as $\mathbb{C}^n$ in which projective space $\mathbb{P}_{n-1}$ is "glued" at the position of the 0, as a submanifold. Transition from $\mathbb{C}^n$ to $I(1,n)$ is usually called the σ-process. If $(w_1, \ldots, w_n)$ are homogeneous coordinates in $\mathbb{P}_{n-1}$, then in $\mathbb{C}^n \times \mathbb{P}_{n-1}$ the set $I(1,n)$ is defined by $z \wedge w = 0$, i.e. by the system $z_j w_k = z_k w_j$, $j,k = 1, \ldots, n$. Hence the σ-process is also called the quadratic transformation.

3.6. Imbedding of Grassmannians into $\mathbb{P}_N$. Grassmann manifolds allow holomorphic imbeddings into projective spaces as complex submanifolds, and this largely simplifies their study and use. Complex submanifolds in $\mathbb{P}_N$, as well as the abstract complex manifolds biholomorphic to them, are called algebraic, more precisely smooth complex projective algebraic, manifolds (all words are necessary!).* We show that all $G(k,n)$ are such algebraic manifolds.

P r o p o s i t i o n. *There is a canonical holomorphic imbedding of $G(k,n)$ into $\mathbb{P}_N$, with $N = \binom{n}{k} - 1$, under which every unitary automorphism of $G(k,n)$ has a continuation to a unitary automorphism of $\mathbb{P}_N$.*

■ We use the correspondence between k-dimensional linear subspaces of $\mathbb{C}^n$ and decomposable k-vectors from the space $\wedge^k \mathbb{C}^n \approx \mathbb{C}^{N+1}$ (for the terminology of exterior algebra see A4.1). Let L be such a subspace and let $v^1, \ldots, v^k$ be a basis in L. Then the k-vector $v = v^1 \wedge \cdots \wedge v^k$ is distinct from zero, and if $w^1, \ldots, w^k$ is another basis in L, then $w^1 \wedge \cdots \wedge w^k$ is obtained from v by multiplication by a complex number (the determinant of the matrix of transition between these bases). Thus, to a plane $L \subset \mathbb{C}^n$ corresponds the punctured complex line $\mathbb{C}_\star \cdot v$ in $\wedge^k \mathbb{C}^n$; moreover, the lines corresponding to distinct L', $L \subset \mathbb{C}^n$ do not intersect. By transition to the projectivization, Π: $\mathbb{C}^{N+1}_\star \to \mathbb{P}_N$, we obtain an imbedding

*Also: algebraic varieties.

$$\Phi\colon G(k,n) \to \mathbb{P}_N.$$

An analytic description of this imbedding can be obtained from the matrix representation of the elements of $G(k,n)$. For each matrix $Z \in \mathrm{St}\,(k,n)$ (i.e. nonsingular $k \times n$ matrix) the exterior product of its columns, $z^1 \wedge \cdots \wedge z^k$, in $\wedge^k \mathbb{C}^n$ can be expanded with respect to a canonical basis, and this expansion has the form $z^1 \wedge \cdots \wedge z^k = \sum'_{\sharp I = k} Z_I e_I$, where the prime denotes that the summation extends over ordered multi-indices only, and $e_I = e_{i_1} \wedge \cdots \wedge e_{i_k}$ are the elements of the canonical basis. As can be readily seen, the coefficients Z_I are equal to $\det (Z)_I$, where $(Z)_I$ is the $k \times k$ matrix formed from the columns of Z with indices $i_1, \ldots, i_k$ (in this order). By transition to projectivization we obtain a map $Z \mapsto [Z_I] \in \mathbb{P}_N$. The homogeneous coordinates Z_I of the image of Z are called the *Plücker coordinates* of the k-dimensional plane $L \subset \mathbb{C}^n$ spanned over the rows of Z. They are determined by the plane L itself up to a common complex factor. By construction, $\Phi(L) = [Z_I]$. In affine coordinates of neighborhoods $U_J \subset G(k,n)$ the representation of linear subspaces of $\mathbb{C}^n$ by matrices Z with $(Z)_J = E$ is one-to-one, hence the definition of the complex structure on $G(k,n)$ implies that Φ is holomorphic. In $U_{1 \ldots k}$ with matrix coordinates $Z = (E, W)$ the determinant Z_I for $I = (1, \ldots, i-1, i+1, \ldots, k, j)$ equals, up to sign, w_{ij}, hence $\Phi\colon\ U_{1 \ldots k} \to \mathbb{P}_N$ is a map of maximum possible rank $k(n-k)$. The same holds in the other U_J, hence Φ is a holomorphic imbedding of $G(k,n)$ into $\mathbb{P}_N$.

In matrix representation, any unitary transformation of $G(k,n)$ has the form $Z \to Z \cdot A$, where A is a unitary $n \times n$ matrix. To $Z \cdot A$ corresponds the k-vector

$$z^1 A \wedge \cdots \wedge z^k A = \sum_{\sharp I = k}{}' Z_I e_{i_1} A \wedge \cdots \wedge e_{i_k} A = \sum_{\sharp I = \sharp J = k}{}' Z_I a_{IJ} e_J,$$

hence the vector of Plücker coordinates, (Z_I), is multiplied by a square matrix $\mathcal{A}$, independent of Z, as a result of the action of A. In this way we obtain a map $A \to \mathcal{A}$ of the group $\mathbb{U}(n)$ of unitary transformations into the group of linear transformations of $\mathbb{C}^{N+1}$, and this map is continuous. It is clear from the construction that this is a group homomorphism, hence the image of $\mathbb{U}(n)$ consists of nonsingular $\mathcal{A}$. This and the compactness of $\mathbb{U}(n)$ readily imply that the image of $\mathbb{U}(n)$ belongs to $\mathbb{U}(N+1)$ (if ϕ is a nonunitary transformation of $\mathbb{C}^{N+1}$, then all limit transformations of the sequence of iterates ϕ, $\phi \circ \phi$, ..., are singular). Thus, to each $A \in \mathbb{U}(n)$ corresponds a unitary matrix $\mathcal{A} \in \mathbb{U}(N+1)$, hence a unitary automorphism of $\mathbb{P}_N$. Under such automorphisms of $\mathbb{P}_N$ the image of $G(k,n)$ is mapped onto itself (by construction). ■

We will often assume that $G(k,n) \subset \mathbb{P}_N$, identify $G(k,n)$ with its canonical image. The Plücker coordinates will be numbered lexicographically; then $U_{1 \cdots k}$ is mapped into the affine part U_0: $z_0 \neq 0$ in $\mathbb{P}_N$, and $G(k,n) \setminus U_{1 \cdots k}$ lies in the hyperplane at infinity H_0: $z_0 = 0$.

As an e x a m p l e we investigate the equation defining the image of $G(2,4)$ under its canonical imbedding into $\mathbb{P}_5$. The dimension of $G(2,4)$ is 4, hence the image is a hyperplane. In the affine part $U_{12} \subset G(2,4)$ the points are uniquely parametrized by the matrices

$$\begin{bmatrix} 1 & 0 & \zeta_{13} & \zeta_{14} \\ 0 & 1 & \zeta_{23} & \zeta_{24} \end{bmatrix},$$

to which correspond the Plücker coordinates

$$(z_0, \ldots, z_5) = (1, \zeta_{23}, \zeta_{24}, -\zeta_{13}, -\zeta_{14}, \zeta_{13}\zeta_{24} - \zeta_{14}\zeta_{23}).$$

Thus, within the affine part of $\mathbb{P}_5$ the manifold $G(2,4)$ is given by the single equation $\zeta_5 = \zeta_1\zeta_4 - \zeta_2\zeta_3$. This readily implies (see p.7.2) that all of $G(2,4)$ (which is given by the closure in $\mathbb{P}_5$ of its affine part) is given by the equation $z_0z_5 - z_1z_4 + z_2z_3 = 0$. From this it can be seen that $G(2,4)$ is unitarily equivalent to the hyperquadric $z_0^2 + \cdots + z_5^2 = 0$ (i.e. is obtained from it by a unitary automorphism of $\mathbb{P}_5$).

The affine part $U_{1 \cdots k} \subset G(k,n)$ is biholomorphically equivalent to the space $\mathbb{C}^{k(n-k)}$, hence $G(k,n)$ is a compactification of $\mathbb{C}^{k(n-k)}$. However, this compactification is essentially more complicated than $\mathbb{P}_{k(n-k)}$. E.g., the points at infinity of $G(2,4)$ form on $H_0 \subset \mathbb{P}_5$ the "cone" $z_1z_4 = z_2z_3$, and this cone is not a complex manifold in any neighborhood of $[0, \ldots, 0, 1]$.

A4. Complex differential forms

A4.1. Exterior algebra. An important role in the study of linear structures is played by exterior (Grassmann) algebras of linear spaces. We will briefly recall the basic definitions, referring to [58], [92], [189] for details.

Let L be a linear space over a field $\mathbb{K}$ $(=\mathbb{R}$ or $\mathbb{C})$ with a fixed basis $e_1, \ldots, e_N$. By definition, $\wedge^p L$ $(= \wedge^p_{\mathbb{K}} L)$ is the $\mathbb{K}$-linear space with basis

$$\{e_I := (e_{i_1}, \ldots, e_{i_p}): 1 \leqslant i_1 < \cdots < i_p \leqslant N\};$$

as can be seen, its dimension is $\binom{N}{p} = N!/p!(N-p)!$. By definition, $\wedge^1 L = L$. Put $\wedge L := \oplus_{p=1}^N \wedge^p L$. We can introduce in this $\mathbb{K}$-linear space an anticommutative exterior product

$$\wedge : (\wedge^p L) \times (\wedge^q L) \rightarrow \wedge^{p+q} L,$$

under which $e_I \wedge e_J := \epsilon_{IJ} e_{(I \cup J)'}$, where the prime denotes the ordered multi-index formed from the given elements and ϵ_{IJ} is equal to 0 if $I \cap J$ is nonempty and equal to the sign of the permutation $I \cup J \rightarrow (I \cup J)'$ in the opposite case. The operation $\wedge$ is extended by linearity to nonbasis elements. With this multiplication each basis vector $e_I \in \wedge^p L$ can be represented as $e_{i_1} \wedge \cdots \wedge e_{i_p}$. The linear space $\wedge L$ with the exterior product $\wedge$ is called the *exterior algebra* of the linear space L (we can show that it is independent of the choice of a basis in L). The elements of $\wedge^p L$ are called p-vectors, and the common name for the elements of $\wedge L$ is *polyvector*. A scalar product (which is Hermitian if $\mathbb{K} = \mathbb{C}$) can be naturally introduced in $\wedge L$, such that all basis vectors e_I have length 1 and are mutually orthogonal.

There is a strong relation between linear subspaces in L and polyvectors. To each p-dimensional subspace $L' \subset L$ with basis $v^1, \ldots, v^p$ we can associate the p-vector $v^1 \wedge \cdots \wedge v^p \in \wedge^p L$ (such p-vectors are called *decomposable*). If $w^1, \ldots, w^p$ is another basis in L', the anticommutativity of the exterior product implies that $w^1 \wedge \cdots \wedge w^p = (\det A) \cdot v^1 \wedge \cdots \wedge v^p$, where A is the matrix of transition from the basis v to w. In other words, the p-vector corresponding to L' is uniquely determined up to a scalar factor. If $\mathbb{K} = \mathbb{R}$, this gives a one-to-one correspondence between oriented subspaces in L of dimension p and decomposable unit p-vectors in $\wedge_{\mathbb{R}} L$.

For us the main exterior algebras are the algebra $\wedge_{\mathbb{R}} \mathbb{C}^n$ of real polyvectors in $\mathbb{C}^n$ and the complex algebra $\wedge_{\mathbb{C}} (\mathbb{C}^n)^*_{\mathbb{R}}$, where $(\mathbb{C}^n)^*_{\mathbb{R}}$ is the space of complex valued $\mathbb{R}$-linear functions in $\mathbb{C}^n$ with basis $z_1, \bar{z}_1, \ldots, z_n, \bar{z}_n$. If $v^1, \ldots, v^p$ is a complex basis of a $\mathbb{C}$-linear subspace $\Lambda \subset \mathbb{C}^n$, then the vectors $v^1, iv^1, \ldots, v^p, iv^p$ form an $\mathbb{R}$-linear basis in it, and the given order of writing the elements of the basis determines on Λ an orientation which is independent of the choice of the complex basis $v^1, \ldots, v^p$ (the determinant of the matrix of transition between the corresponding $\mathbb{R}$-linear bases is positive, see A2.1). This orientation is called the *canonical orientation* of the complex subspace. The polyvectors

$$v^1 \wedge iv^1 \wedge \cdots \wedge v^p \wedge iv^p$$

and linear combinations with positive coefficients of them are called *positive* polyvectors. Thus there is a one-to-one correspondence between $\mathbb{C}$-linear spaces of dimension p and decomposable positive unit p-vectors. As we see, the algebra $\wedge_{\mathbb{R}} \mathbb{C}^n$ reflects the geometrical $\mathbb{R}$-linear structure in $\mathbb{C}^n$ and its relation with the complex structure; it is convenient in geometrical questions. The algebra $\wedge_{\mathbb{C}} (\mathbb{C}^n)^*_{\mathbb{R}}$ reflects the dual structure of the $\mathbb{R}$-linear functions in $\mathbb{C}^n$, and complex differential forms on complex manifolds are defined using it.

A4.2. Differential forms. We assume that the reader is acquainted with the general theory of differentiable manifolds; in particular, with the simplest properties of differential forms (a good handbook on this theme is R. Narasimhan's book [92], see also [58], [189]). We will mainly deal with forms on manifolds lying in $\mathbb{C}^n$ or in $\mathbb{P}_n$. We briefly recall the basic notions.

Let M be a differentiable manifold, $TM = \cup_{a \in M} T_a M$ its tangent bundle, and $T^* M = \cup T^*_a M$ its cotangent bundle. Next to real tangent and cotangent vectors we will consider also elements of the spaces

$$\mathbb{C} T_a M \supset T_a M \quad \text{and} \quad \mathbb{C} T^*_a M \supset T^*_a M$$

with the same bases, but with complex coefficients. For each s, $0 \leqslant s \leqslant \dim M$, there is defined on M the $\mathbb{C}$-linear bundle

$$\wedge^s \mathbb{C} T^* M = \bigcup_{a \in M} (\wedge^s \mathbb{C} T^*_a M).$$

The elements of $\wedge^s \mathbb{C} T^*_a M$ are the skewsymmetric $\mathbb{C}$-multilinear forms (functions) on the direct product $(\mathbb{C} TM)^s$. As a basis in this space we can take the forms

$$(dx_J)_a := (dx_{j_1})_a \wedge \cdots \wedge (dx_{j_s})_a, \quad J = (j_1, \ldots, j_s), \quad j_1 < \cdots < j_s,$$

where x are local coordinates in a neighborhood of a. The value of the form $(dx_J)_a$ on an ordered tuple of tangent vectors $v^1, \ldots, v^s$ is equal to

$\det\,(dx_{j_k}(v_i))_{i,k=1}^{s}$ (this can be taken as a definition). In view of the skewsymmetry this expression depends only on the exterior product

$$v = v^1 \wedge \cdots \wedge v^s \in \wedge^s \mathbf{C}T_aM;$$

we put $<\phi, v> = \phi(v^1, \ldots, v^s)$, by definition. The sections of the bundle $\wedge^s \mathbf{C}T_a^*M$ are called complex differential forms of degree s, or simply (differential) s-forms. Locally they can be written as $\phi = \sum'_{\sharp J = s} \phi_J dx_J$, where the ϕ_J are functions and the prime denotes summation over ordered multi-indices J. A form vanishes at a point a if all its coefficients ϕ_J vanish at this point. The support of a form ϕ (defined on the whole manifold M) is taken to be the closure of the set of points at which it does not vanish; it is denoted by supp ϕ. A form is called of compact support if supp ϕ is compact.

Complex conjugation has, in local coordinates, the form $\overline{\sum' \phi_J dx_J} = \sum' \overline{\phi}_J\, dx_J$; a form ϕ is real (i.e. takes real values at the elements of TM) if $\overline{\phi} = \phi$. For any form ϕ the forms $\mathrm{Re}\,\phi = (\phi + \overline{\phi})/2$ and $\mathrm{Im}\,\phi = (\phi - \overline{\phi})/2i$ are real, and $\phi = \mathrm{Re}\,\phi + i\,\mathrm{Im}\,\phi$.

If ρ: $M \to N$ is a smooth (C^1) map between smooth manifolds, and if ϕ is a form on N, then its *pull-back* $\rho^*\phi$ is the form on M defined as follows: if $\phi = \sum'_{\sharp J = s} \phi_J(y) dy_J$ in local coordinates y in a neighborhood $V \subset N$, and if ρ_j are the components of ρ in these coordinates, then in $\rho^{-1}(V)$ we have

$$(\rho^*\phi)(x) := \sum_{\sharp J = s}' (\phi_J \circ \rho)(x)\, d\rho_{j_1}(x) \wedge \cdots \wedge d\rho_{j_s}(x).$$

Smoothness of a differential form (on a sufficiently smooth manifold) is defined in terms of its coefficients. The set of all differential forms of degree s and class C^∞ on an infinitely differentiable manifold M is denoted by $\mathcal{E}^s(M)$. The subset of it consisting of the forms of compact support is denoted by $\mathcal{D}^s(M)$. It is convenient to also introduce the spaces

$$\mathcal{E}(M) = \oplus \mathcal{E}^s(M) \text{ and } \mathcal{D}(M) = \oplus \mathcal{D}^s(M).$$

We introduce the following strong topologies in these spaces: $\phi^j \to^{\mathcal{E}} \phi$ if, in any coordinate neighborhood, the corresponding coefficients converge, together with all partial derivatives, uniformly on compact subsets; $\phi^j \to^{\mathcal{D}} \phi$ if $\phi^j \to^{\mathcal{E}} \phi$ and if there is a compact set $K \subset M$ such that supp $\phi^j \subset K$ for all j.

Let now M be a complex manifold and (U,z) a complex coordinate neighborhood. To the latter correspond real coordinates (x,y), $z = x+iy$. Passing from x,y to $z,\bar{z}$ we obtain representations of vector fields and differential forms in complex coordinates. Related to this is the decomposition of $\mathbf{C}TM$ as the direct sum $T_{1,0}M \oplus T_{0,1}M$; sections of $T_{1,0}M$, fields $\sum a_j \partial / \partial z_j$, are called fields of type (1,0), while sections $\sum b_j \partial / \partial \bar{z}_j$ of $T_{0,1}M$ are called fields of type (0,1). We also introduce a supplementary grading in $\wedge \mathbf{C}T^*M$. A differential form ϕ is called a form of *bidegree* (p,q) if it is locally representable as

$$\phi = \sum_{\substack{\#I=p \\ \#J=q}}' \phi_{IJ} dz_I \wedge d\bar{z}_J.$$

Forms for which the bidegree is defined are called *bihomogeneous*. Any s-form ϕ can be represented as $\phi = \sum_{p+q=s} \phi^{p,q}$ (decomposition with respect to bidegree), where $\phi^{p,q}$ has bidegree (p,q) and is called the (p,q)-component of ϕ. The infinitely differentiable forms of bidegree (p,q) form the subspaces

$$\mathcal{E}^{p,q}(M) \subset \mathcal{E}^{p+q}(M) \text{ and } \mathcal{D}^{p,q}(M) = \mathcal{D}(M) \cap \mathcal{E}^{p,q}(M).$$

It is clear that under holomorphic maps between complex manifolds the bidegrees of forms (under transition to pull-backs) is preserved.

The exterior differentiation operator d on a complex manifold can be represented as $d = \partial + \bar{\partial}$. The operator $\bar{\partial}$ increases the bidegree by (0,1), and acts in local coordinates by the rule

$$\bar{\partial}\left[\sum{}' \phi_{IJ} dz_I \wedge d\bar{z}_J\right] = \sum{}' \left[\sum \frac{\partial \phi_{IJ}}{\partial \bar{z}_k} d\bar{z}_k\right] \wedge dz_I \wedge d\bar{z}_J;$$

∂ acts similarly, and $\overline{\partial \phi} = \bar{\partial}\bar{\phi}$. The operator $d^c := i(\bar{\partial} - \partial)$, and also d, is real, i.e. maps real forms to real forms. It can be immediately verified that

$$d^2 = \partial^2 = \bar{\partial}^2 = (d^c)^2 = \partial\bar{\partial} + \bar{\partial}\partial = 0 \text{ and } dd^c = 2i\partial\bar{\partial}.$$

We give two simple statements concerning the relation between bidegrees and the complex structure.

L e m m a 1. *Let $M \subset \mathbf{C}^n$ be a CR-manifold of CR-dimension p (i.e. at all points $\zeta \in M$ the complex tangent planes $T_\zeta^{\mathbf{C}}M = T_\zeta M \cap iT_\zeta M$ have the same complex dimension p), and let* $\dim_{\mathbf{R}} M = 2p+q$. *Let $\phi = \phi^{r,s}$ be a differential form of* *bidegree (r,s) in a neighborhood of M, with* max $(r,s) > p+q$. *Then the restriction to M of ϕ (i.e. to $TM \subset T\mathbb{C}^n$) vanishes. In particular, if M is a complex manifold (i.e. $q=0$) and $r+s=2p$ but $(r,s) \neq (p,p)$, then $\phi^{r,s}|_M = 0$.*

■ We can choose coordinates in a neighborhood of an arbitrary point $\zeta \in M$ such that $T_\zeta^c M$ becomes the plane $\mathbb{C}^p_{z'}$ of variables $z_1, \ldots, z_p$, and $T_\zeta M$ becomes the plane of variables $z_1, \ldots, z_p$ $x_{p+1}, \ldots, x_{p+q}$. Then the restrictions to $T_\zeta M$ of the differentials $dz_{p+q+1}, \ldots, dz_n$ vanish, hence $dz_I|_{T_\zeta M} = d\bar{z}_I|_{T_\zeta M} = 0$ if $\# I > p+q$. Since $\phi = \sum'_{\# I = r, \# J = s} \phi_{IJ}\ dz_I \wedge d\bar{z}_J$, under the condition max $(r,s) > p+q$ the restriction to $T_\zeta M$ of every term of this decomposition of ϕ vanishes. Since ζ is arbitrary, $\phi|_M = 0$. ■

The property proved is a local property, hence it is true on arbitrary complex manifolds and not only in $\mathbb{C}^n$.

L e m m a 2. *Let M be a p-dimensional complex manifold, u a function, and ψ a form of bidegree $(p-1,p-1)$ and of class $C^1(M)$. Then, everywhere on M,*

$$du \wedge d^c\psi = d\psi \wedge d^c u.$$

If ϕ is a form of bidegree $(p-1,p-1)$, then for any functions $u,v \in C^1(m)$ we have

$$du \wedge d^c v \wedge \phi = dv \wedge d^c u \wedge \phi.$$

■ By definition,

$$du \wedge d^c\psi = i(\partial u \wedge \bar{\partial}\psi - \bar{\partial}u \wedge \partial\psi - \partial u \wedge \partial\psi + \bar{\partial}u \wedge \bar{\partial}\psi).$$

The third and fourth terms on the righthand side have bidegree $(p+1,p-1)$ and $(p-1,p+1)$, respectively, and by Lemma 1 such forms on M vanish. Since $\partial\psi$ is a form of odd degree, $-\bar{\partial}u \wedge \partial\psi = \partial\psi \wedge \bar{\partial}u$, hence

$$du \wedge d^c\psi = i(\partial u \wedge \bar{\partial}\psi + \partial\psi \wedge \bar{\partial}u).$$

The form $d\psi \wedge d^c u$ transforms the same way.

Similarly,

$$du \wedge d^c v \wedge \phi = i(\partial u \wedge \bar{\partial}v - \bar{\partial}u \wedge \partial v) \wedge \phi =$$

$$= i(\partial u \wedge \bar{\partial}v + \partial v \wedge \bar{\partial}u) \wedge \phi = dv \wedge d^c u \wedge \phi. \quad ■$$

A4.3. Integration of forms. Stokes' theorem. Integrals of differential forms over an m-dimensional oriented manifold M of class C^1 are defined in a standard manner in terms of local coordinates and partition of unity. If (U,x) is a coordinate neighborhood on M, an m-form ϕ in this neighborhood is represented as $f(x)dx_1\wedge\cdots\wedge dx_m$, where f is a function of the same smoothness class as ϕ. If the support of ϕ belongs to U, we put, by definition,

$$\int_M \phi = \int_{x(U)} f(x)dx_1 \cdots dx_m;$$

on the righthand side stands a multiple integral over the domain $x(U)$ in $\mathbb{R}^m$. That this local definition is correct follows from the formula of change of variables in a multiple integral (see, e.g., [92]). For a form ϕ of arbitrary support, $\int_M\phi$ is defined as $\sum\int_M\lambda_j\phi$, where $\{\lambda_j\}$ is a smooth partition of unity subordinated to some covering of M by coordinate neighborhoods (as can be readily verified, this definition does not depend on the choice of a partition of unity). Note that when the orientation of M is reversed, the integral changes sign.

A differential form ϕ of degree m on an m-dimensional oriented Riemannian manifold M is called a volume form if for any positively oriented and orthogonal basis $v^1,\ldots,v^m$ in T_aM, for an arbitrary point $a\in M$, we have $\phi(v^1,\ldots,v^m)=1$ (as a rule: the volume of the unit cube is 1). If local coordinates $x_1,\ldots,x_m$ are orthonormal with respect to the Riemannian metric S on M (i.e. $S(\partial/\partial x_j, \partial/\partial x_k)=\delta_{jk}$) and positively oriented, then in these coordinate the volume form has the representation $dx_1\wedge\cdots\wedge dx_m$, as in $\mathbb{R}^m$ with the Euclidean metric. If they are not orthonormal, a positive coefficient must be placed in front: $\phi=\lambda(x)dx_1\wedge\cdots\wedge dx_m$, $\lambda>0$. It is easy to understand that there is a, moreover unique, volume form on any oriented Riemannian manifold. (This form is traditionally denoted by dV_M or simply by dV, although there is no differentiation involved.) The volume form on an m-dimensional Riemannian manifold M induces a Borel measure, which, as follows from A6.2, coincides with the Hausdorff measure $\mathcal{H}_m$, i.e. $\mathcal{H}_m(U)=\int_U dV$ for any open set $U\subset M$.

If M is an m-dimensional oriented Riemannian manifold with volume form dV, and if a vector field $v^1,\ldots,v^m$ on M forms, at every point, a positively oriented basis of the tangent space, moreover $|v^1\wedge\cdots\wedge v^m|=1$, then any m-form ϕ on M can be represented as $\phi=\langle\phi,v\rangle dV$, where $v=v^1\wedge\cdots\wedge v^m$ and $\langle\phi,v\rangle=\phi(v^1,\ldots,v^m)$ (this clearly follows from that fact that $\wedge^m TM$ is one-dimensional and from the definition of volume form). The polyvector $(v^1 \wedge \cdots \wedge v^m)(x)$ uniquely determines the tangent plane $T_x M$ and an orientation of it, and is independent of the choice of the given field; it will be denoted by $\tau_M(x)$. Thus, for any integrable m-form ϕ on M we have

$$\int_M \phi = \int_M <\phi, \tau_M> d\mathcal{H}_m. \qquad (\star)$$

The righthand side can be taken as definition of integral of a differential form over a Riemannian manifold. Although the standard definition is simpler, this representation is more convenient for obtaining estimates. Formula $(\star)$ establishes a relation between so-called integrals of the second kind (of forms) and integrals of the first kind (with respect to measures).

The main theorem relating integration and differentiation is Stokes' theorem. We state it for manifolds with piecewise smooth boundary. For our purposes every open subset M of an oriented ambient manifold $\tilde{M}$ whose boundary $\tilde{M} \setminus M$ locally belongs to the union of a finite number of smooth hypersurfaces with pairwise transversal intersections on $\tilde{M}$ is such a manifold. On the smooth parts of the boundary $bM := \tilde{M} \setminus M$ we assume an orientation given, compatible with that of M. An integral over bM is understood to be the integral over the oriented manifold of its smooth points. The local representation implies that it is defined, e.g., for any continuous differential $(m-1)$-form ϕ of compact support on $\tilde{M}$.

T h e o r e m. *Let M be a smooth oriented m-dimensional manifold with piecewise smooth boundary bM, and let ϕ be a form of degree $m-1$ and class $C^1(M \cup bM)$, of compact support on $\tilde{M}$. Then*

$$\int_M d\phi = \int_{bM} \phi.$$

As usual, if $M \cup bM$ is compact we do not need ϕ to be compactly supported. Using a partition of unity on $\tilde{M}$ the statement is naturally reduced to the local statement. In the case of a smooth boundary the proof in $\mathbb{R}^m$ is standard (see, e.g., [92]); the general case clearly reduces to this one by local approximation of piecewise smooth hypersurfaces by smooth ones.

A4.4. Fubini's theorem. Let N, M be oriented C^1-manifolds of dimension n and

m, respectively, $n \geqslant m$, and let $\rho: N \to M$ be a C^1-map of maximum rank m. By the rank Theorem, each fiber $N_y = \rho^{-1}(y)$, $y \in M$, of ρ is a C^1-manifold of dimension $n-m$. The orientations of N and M naturally determine an orientation in the fibers of ρ: we say that a basis $v^{m+1}, \ldots, v^n$ in $T_a N_y$ is positive if it is part of a positive basis $v^1, \ldots, v^n$ in $T_a N$ such that $\rho'(v^1), \ldots, \rho'(v^m)$ is a positive basis in $T_{\rho(a)}M$ (here ρ' is the tangent map). Such an orientation in the fibers of ρ is called canonical.

If ψ is an $(n-m)$-form of compact support on N with, say, continuous coefficients (with respect to local coordinates), then on M is defined the function $\int_{N_y}\psi$, also continuous and of compact support, which can be integrated over M after multiplication by an arbitrary locally integrable m-form ϕ. The result is the repeated integral $\int_M(\int_{N_y}\psi)\phi$. On the other hand, to ϕ uniquely corresponds its pull-back $\rho^*\phi$, which is a form of degree m on N. Multiplying it by ψ we obtain an integrable form of compact support on N, and the integral $\int_N(\rho^*\phi)\wedge\psi$.

In order to prove that these integrals are equal, we note that this is a local statement. By the rank Theorem we can choose positively oriented local coordinates $x_1, \ldots, x_n$ in a neighborhood of a point a on N and $y_1, \ldots, y_m$ in a neighborhood $V \ni \rho(a)$ on M such that in these coordinates ρ is the projection $(x_1, \ldots, x_n) \to (x_1, \ldots, x_m)$ (i.e. $y_j = x_j$, $j = 1, \ldots, m$) or, abbreviated, $(x', x'') \mapsto x'$. We may assume that the domain $x(U) \subset \mathbb{R}^m$ has the form $U' \times U''$, where $U' \subset \mathbb{R}^m_{x'}$ and $U'' \subset \mathbb{R}^{n-m}_{x''}$. On the fibers of ρ all $dx_j = 0$ for $j > m$, hence the restriction of ψ to each $\mathbb{N}_y$ is $\tilde{\psi}(x)dx_{m+1}\wedge \cdots \wedge dx_n$, and

$$\int_{N_y}\psi = \int_{U''}\tilde{\psi}(x)dx_{m+1} \cdots dx_n.$$

Within V we have

$$\phi = \tilde{\phi}(y)dy_1\wedge \cdots \wedge dy_m, \quad \text{while } \rho^*\phi = \tilde{\phi}(x')dx_1\wedge \cdots \wedge dx_m.$$

By the classical Fubini theorem, our repeated integral equals

$$\int_{y(V)}\left[\int_{U''}\tilde{\psi}(x',x'')dx_{m+1} \cdots dx_n\right]\tilde{\phi}(y)dy_1 \cdots dy_m =$$

$$= \int_{U'}\left[\int_{U''}\tilde{\psi}(x',x'')dx_{m+1} \cdots dx_n\right]\tilde{\phi}(x')dx_1 \cdots dx_m =$$

$$= \int_{U' \times U''} \tilde{\psi}(x',x'')\tilde{\phi}(x')dx_1 \cdots dx_n$$

(ψ is supported in U, the coordinates x, x'', y are positively oriented in U, N_y, and V, respectively; by definition they give the canonical orientation in the fibers). But this also equals the integral of $(\rho^\star\phi)\wedge\psi$ over U, since $(\rho^\star\phi)\wedge\psi = \tilde{\phi}(x')\,\tilde{\psi}(x',x'')$ $dx_1 \cdots dx_n$. The local result proved and partition of unity give equality of the multiple integral and the repeated integral in the general case. Thus we have the following analog of Fubini's theorem for differential forms.

Theorem. *Let N and M be oriented manifolds of class C^1, of dimension n and m, respectively, $n \geqslant m$, and let ρ: $N \to M$ be a C^1-map of maximum rank m at all points of N. On the fibers N_y the map ρ is assumed to induce the canonical orientation. Then, for every continuous $(n-m)$-form ψ of compact support on N and every locally integrable m-form ϕ on M we have*

$$\int_N (\rho^\star\phi)\wedge\psi = \int_M \left[\int_{N_y} \psi\right]\phi.$$

Note that if the m-form ϕ on M vanishes nowhere, then any n-form Φ on N is representable as $\Phi = (\rho^\star\phi)\wedge\psi$ (is, so to speak, divisible by ϕ). The form ψ is determined up to a term α for which $(\rho^\star\phi)\wedge\alpha \equiv 0$. If N is a Riemannian manifold, ψ can be determined uniquely by the condition $|\psi| = |\psi|_{N_y}|$ at every point $x \in N_y$, $y \in M$ (i.e. ψ is tangent to all fibers of ρ); this ψ is denoted by Φ / ϕ. Of special interest is the case when M is also a Riemannian manifold and ϕ is the volume form on M. Then $dV_N = (\rho^\star dV_M)\wedge(dV_N / dV_M)$. Since, by definition, dV_N / dV_M is the $(n-m)$-form tangent to the fibers of ρ, at every point $x \in N_y$ it is proportional to dV_{N_y}; moreover, the proportionally coefficient is (by the choice of orientations) equal to $|dV_N / dV_M|$ (norm in the space $\wedge^{n-m}T_x^\star N$). Thus, the "scalar" form of Fubini's theorem for maps between Riemannian manifolds is as follows: for any continuous function f of compact support on N we have

$$\int_N f dV_N = \int_M \left[\int_{N_y} f \left|\frac{dV_N}{dV_M}\right| dV_{N_y}\right] dV_M.$$

A4.5. Positive forms. Forms $\prod_{j=1}^{p} i(l_j \wedge \bar{l}_j)/2$ in $\mathbf{C}^n$ with $l_j \in (\mathbf{C}^n)^*$ linear functions are called principal positive forms, while linear combinations with nonnegative coefficients of them are simply called positive forms. A differential form ϕ on a complex manifold Ω is called *positive* if $(\phi)_a$ is a positive form on $T_a\Omega \approx \mathbf{C}^n$ for all $a \in \Omega$; positive differential forms in $\mathbf{C}^n$ given by $\lambda \prod_1^p i(dl_j \wedge d\bar{l}_j)/2$ with λ a nonnegative function are called *principal*. Note that the exterior product of positive forms is positive, and that the property of being positive is independent of the choice of local holomorphic coordinates.

Examples. (a) The form

$$\omega = \frac{i}{2}\sum dz_j \wedge d\bar{z}_j = \frac{i}{2}\partial\bar{\partial}|z|^2$$

in $\mathbf{C}^n$ is clearly positive.

(b) After multiplication by $|z|^4$ the form $\omega_0 = (i/2)\partial\bar{\partial}\log|z|^2$ in $\mathbf{C}_*^{n+1}$ gives

$$\frac{i}{2}\left[|z|^2\partial\bar{\partial}|z|^2 - \left[\sum_0^n \bar{z}_j dz_j\right] \wedge \left[\sum_0^n z_k d\bar{z}_k\right]\right].$$

At the point $(1,0,\ldots,0)$ this is equal to $(i/2)\sum_1^n dz_j \wedge d\bar{z}_j$, hence ω_0 is positive at this point, hence in all of $\mathbf{C}_*^{n+1}$ since it is invariant under unitary transformations and dilatations.

(c) Let ρ be a real valued function on a complex manifold Ω and let $(d\rho)_a = (dx_1)_a$ in local holomorphic coordinates $(z_1,\ldots,z_n)$. Then $(d^c\rho)_a = (dy_1)_a$, hence $(d\rho \wedge d^c\rho)_a = (i/2)(dz_1 \wedge d\bar{z}_1)_a$ is a positive form. Since the property of being positive is independent of the choice of holomorphic coordinates, $d\rho \wedge d^c\rho = 2i\partial\rho \wedge \bar{\partial}\rho$ is a positive differential form in Ω.

(d) Let u be a real valued function of class C^2 on a complex manifold Ω. In a neighborhood of a point $a \in \Omega$ we have, in local coordinates,

$$i\partial\bar{\partial}u = i\sum \frac{\partial^2 u}{\partial z_j \partial \bar{z}_k} dz_j \wedge d\bar{z}_k.$$

By a unitary change of coordinates the Hermitian matrix $(\partial^2 u/\partial z_j \partial \bar{z}_k)(a)$ is brought to diagonal form, hence we may immediately assume that $(i\partial\bar{\partial}u)_a = \sum \lambda_j (idz_j \wedge d\bar{z}_j)_a$. This form is positive if and only if all λ_j are nonnegative, i.e. if and only if the matrix $(\partial^2 u/\partial z_j \partial \bar{z}_k)(a)$ is (weakly) positive definite. Without reduction to diagonal form this condition can be written as

$$\sum \frac{\partial^2 u}{\partial z_j \partial \bar{z}_k}(a) v_j \bar{v}_k \geqslant 0 \text{ for all } v \in \mathbf{C}^n.$$

Thus, positivity of $i\partial\bar{\partial}u = (dd^c u)/2$ is equivalent to u being a plurisubharmonic function.

The examples given show that the positive forms are bound to play an essential part in complex analysis; this is actually true, especially in problems related to estimating integrals.

The value of an arbitrary positive (p,p)-form ϕ on an arbitrary positive $2p$-vector $v = v^1 \wedge iv^1 \wedge \cdots \wedge v^p \wedge iv^p$ is nonnegative. Indeed, if $\phi = \prod_1^p (i/2)(dl_j \wedge d\bar{l}_j)$ in $\mathbf{C}^n$, then, after a suitable unitary transformation of coordinates, $\phi = \lambda \prod_1^p (i/2)(dz_j \wedge d\bar{z}_j) = \lambda dV'$, where dV' is the Euclidean volume form of $\mathbf{C}^p$ and $\lambda \geqslant 0$. Hence, $<\phi, v> = \lambda <dV', v> = \lambda <dV', v'>$, where v_j' are the projections of v^j into $\mathbf{C}^p$ and $v' = v_1' \wedge iv_1' \wedge \cdots \wedge v_p' \wedge iv_p' = \mu e^1 \wedge ie^1 \wedge \cdots \wedge e^p \wedge ie^p$, $\mu \geqslant 0$. Thus $<\phi, v> = \lambda\mu \geqslant 0$. For $(1,1)$- and $(n-1, n-1)$-forms ϕ the converse holds: if $<\phi, v> \geqslant 0$ on all positive polyvectors v, then ϕ is positive. For (n,n)-forms this is trivial: by definition, positivity of such a form $\phi = \lambda dV$ means that λ is a nonnegative function. In particular, the integral of any positive (n,n)-form over an n-dimensional complex manifold is nonnegative.

There are not so few positive forms as it would appear at first glance.

L e m m a. *In the space of differential forms in $\mathbf{C}^n$ of bidegree (p,p) and with constant coefficients there is a basis of principal positive forms. Every differential form in an arbitrary domain $D \subset \mathbf{C}^n$ has a representation in this basis with complex valued functions as coefficients. Every real differential form in D with continuous coefficients has a representation as the difference of two positive forms with continuous coefficients.*

■ Put $\alpha_{jk} = (i/2)\partial\bar{\partial}|z_j + z_k|^2$, $\beta_{jk} = (i/2)\partial\bar{\partial}\ |z_j + iz_k|$, $1 \leqslant j,k \leqslant n$. It can be seen immediately that

$$dz_j \wedge d\bar{z}_k = \frac{1}{2}\alpha_{jk} + \frac{i}{2}\beta_{jk} - \frac{1+i}{4}(\alpha_{jj} + \alpha_{kk}).$$

Substituting this in

$$dz_J \wedge d\bar{z}_K = \pm \prod dz_{j_\nu} \wedge d\bar{z}_{k_\nu'}, \ \# J = \# K = p,$$

we see that the $\mathbf{C}$-linear hull of the principal positive forms in the finite-dimensional space of all (p,p)-forms in $\mathbf{C}^n$ with constant coefficients coincides with all of this space, hence this space has a basis of principal positive forms. For $p = 1$ the set $\{\alpha_{jk}, \beta_{jk}\}$ can be taken as such a basis, since the number of forms in this set, $2n^2$, coincides with the dimension of the space of $(1,1)$-forms in $\mathbf{C}^n$.

If $\phi = \sum' \phi_{JK} dz_J \wedge d\bar{z}_K$ has bidegree (p,p), and if $\{\alpha_1, \ldots, \alpha_N\}$ is a positive basis in the space of forms with constant coefficients, then by decomposing $dz_J \wedge d\bar{z}_K$ with respect to this basis and after grouping we obtain $\phi = \sum \phi_j \alpha_j$, where the ϕ_j are linear combinations of the original coefficients with constant complex coefficients which are independent of ϕ. If ϕ is real, the ϕ_j are all real (uniqueness of decomposition with respect to a basis). Representing ϕ_j as $\phi_j^+ - \phi_j^-$ with $\phi_j^{\mp} = \max(\pm\phi_j, 0)$, we obtain a representation

$$\phi = \sum \phi_j^{\star} + \alpha_j - \sum \phi_j^- \alpha_j$$

as the difference of two positive forms, which are continuous if ϕ is continuous. ■

By the Lemma, the convex cone of positive forms in the space of (p,p)-forms with constant coefficients has interior points (these may be called strictly positive (p,p)-forms). It is easy to show (see [164]) that $\omega^p = \omega \wedge \cdots \wedge \omega$ is one such form.

A5. Currents

A5.1. Definitions. Positive currents. Currents on a manifold X of class C^∞ are analogs of distributions (generalized functions), when as test space the space $\mathcal{D}(X)$ of all complex valued differential forms of class C^∞ and of compact support on X is taken, endowed with the strong topology (see A4.2). By definition, *currents* on X are continuous linear functionals on $\mathcal{D}(X)$; they form the space $\mathcal{D}'(X)$ dual to $\mathcal{D}(X)$. The value of a current T on a test form ϕ is denoted by $T(\phi)$ or $<T,\phi>$. The space $\mathcal{D}'(X)$ is complete in the weak-⋆ topology: $T_j \to T$ if $T_j(\phi) \to T(\phi)$ for all $\phi \in \mathcal{D}(X)$. We say that a current T has *dimension* m if $T(\phi) = 0$ for all forms ϕ of degrees $s \neq m$; the space of all currents of dimension m on X is denoted by $\mathcal{D}_m'(X)$. Currents of dimension 0 are ordinary distributions on X. A current is called *real* if $T(\phi) \in \mathbb{R}$ for all real ϕ. A current T has *singularity* $\leqslant k \leqslant \infty$ (*smoothness* $\geqslant -k$) if it can be extended to a continuous linear functional on the space of all forms of class $C^k(X)$ and of compact support. For such currents multiplication by a function of class C^k is naturally defined: $<fT,\phi> := <T,f\phi>$. Currents of singularity 0 (i.e. continuous linear functionals on the space of continuous forms of compact support) are called *currents of measure type*. We say that a current vanishes on an open set $U \subset X$ if $T(\phi) = 0$ for all $\phi \in \mathfrak{D}(X)$ with supports in U. The smallest closed subset in X outside which T vanishes is called the support of T; it is denoted by supp T. The space of currents with compact support on X is denoted by $\mathcal{E}'(X)$; it is naturally dual to the space $\mathcal{E}(X)$.

Examples. (a) If M is a smooth m-dimensional manifold, embedded in X, oriented, and of locally finite volume, the current of integration over M is defined by

$$<[M],\phi> = \int_M \phi, \quad \phi \in \mathfrak{D}^m(X),$$

and $<[M],\phi> = 0$ for forms of degrees $s \neq m$. It is clearly a current of measure type, real, and of dimension m. Its support coincides with the closure of M in X.

(b) Let ψ be an s-form with locally integrable coefficients on a smooth N-dimensional manifold X. The current $[\psi]$ of dimension $N-s$ is defined by the rule

$$<[\psi],\phi> = \int_X \psi \wedge \phi, \quad \phi \in \mathfrak{D}^{N-s}(X).$$

Of course, this is also a current of measure type; its support coincides with the essential support of the form ψ.

(c) Let ψ be an s-form in $\mathbb{R}^N$ with generalized (distribution) coefficients: $\psi = \sum' \psi_I dx_I$, $\psi_I \in \mathfrak{D}'$. The current $[\psi]$ is defined by the formula $<[\psi],\phi> = \sum' <\psi_I,\phi_J> \epsilon_{IJ}$, where $\phi = \sum'_{\#J=N-s} \phi_J dx_J$ and the numbers ϵ_{IJ} ($=0$ or ± 1) are determined by the conditions $\epsilon_{IJ}\, dx_1 \wedge \cdots \wedge dx_n = dx_I \wedge dx_J$.

In essence, the last example is all-embracing; the proof of the following statement is a simple exercise.

Lemma 1. *Any current T of dimension m on an open subset $G \subset \mathbb{R}^N$ has the representation $[\psi]$, where ψ is a differential form of degree $N-m$ with generalized* *coefficients.*

In relation to this, the number $N - m$ is called the *degree* of the current T. The general representation of a continuous linear functional on a space of continuous functions readily implies that the coefficients of a current of measure type (written as a generalized differential form) are indeed locally finite Borel measures.

On a complex manifold Ω currents, just like forms, decompose with respect to bidegrees. For $T \in \mathcal{D}'(\Omega)$ we define the current $T_{p,q}$ by the condition $T_{p,q}(\phi) = T(\phi^{p,q})$, where $\phi^{p,q}$ is the (p,q)-component of the form ϕ. Clearly, $T = \sum T_{p,q}$ (decomposition with respect to bidegrees, or Dolbeault decomposition). The currents $T_{p,q}$ are called currents of bidimension (p,q). Their set is denoted by $\mathcal{D}'_{p,q}(\Omega)$. Clearly, $T = 0$ if and only if all $T_{p,q} = 0$. Locally, any current of bidimension (p,q) can be represented by a generalized differential form of bidegree $(n-p,n-q)$, where $n = \dim\Omega$, hence $(n-p,n-q)$ is called the *bidegree* of the current $T_{p,q}$.

For smooth maps $\rho: X \to Y$ we can naturally define the pull-backs of differential forms, $\rho^*\phi$ (see A4.2), and images of currents, $\rho_* T$, by the rule: $<\rho_* T,\phi> = <T,\rho^*\phi>$. If ρ is not proper, in general the form $\rho^*\phi$ for a $\phi \in \mathcal{D}(Y)$ is not of compact support, and the definition of $\rho_* T$ need refinement (e.g., $\rho_* T$ is defined for all currents with compact support). Note that dimensions (but not degrees) of currents are preserved under smooth maps (when passing to the images), while under holomorphic maps between complex manifolds the bidimensions are preserved also.

A current T of bidimension (p,p) on a complex manifold Ω is called (weakly) *positive* if $T(\phi) \geqslant 0$ for all positive (p,p)-forms $\phi \in \mathcal{D}(\Omega)$. As an example we can give a current $[\sum \mu_j \psi_j]$, where $\mu_j \geqslant 0$ are measures and ψ_j are positive continuous forms in Ω.

L e m m a 2. *In local holomorphic coordinates the coefficients of an arbitrary positive current are complex measures.*

■ The statement being local, we will prove it in $\mathbf{C}^n$. Let $\{\alpha_j\}$ be a basis of principal positive forms in the space of (p,p)-forms in $\mathbf{C}^n$ with constant coefficients, and let $\{\beta_j\}$ be the basis dual to it in the space of $(n-p,n-p)$-forms, i.e. $\alpha_j \wedge \beta_k = \delta_{jk} dV$ (the β_j need not be positive). A current T of bidimension (p,p) in a domain $D \subset \mathbf{C}^n$ can be represented as a generalized differential form

$$T = \sum{}' T_{JK} dz_J \wedge d\bar{z}_K, \quad \sharp J = \sharp K = n - p.$$

Decomposing $dz_J \wedge d\bar{z}_K$ with respect to the basis $\{\beta_j\}$ and substituting this in the formula we obtain $T = \sum T_j \beta_j$. If T is positive, then for any nonnegative function $\lambda \in \mathfrak{D}(D)$ we have $T_j(\lambda) = \langle T, \lambda\alpha_j \rangle \geqslant 0$, i.e. the generalized functions T_j take nonnegative values on nonnegative functions of compact support. It is well-known that in this case T_j is a positive measure (more precisely, $T_j(\phi) = \int \phi d\mu_j$ for a certain positive Borel measure μ_j). Returning to the original basis $\{dz_J \wedge d\bar{z}_K\}$ we see that the T_{JK} are $\mathbb{C}$-linear combinations of the measures T_j, i.e. the T_{JK} are complex measures. ■

A5.2. The operators $d, \bar\partial$, and integral representations. Differential operators on currents can be defined in terms of generalized differential forms and generalized derivatives. E.g.,

$$d\left[\sum_I{}' T_I dx_I\right] = \sum_I{}' \sum_j \frac{\partial T_I}{\partial x_j} dx_j \wedge dx_I$$

with respect to local coordinates x. An equivalent global definition is as follows. Let T be a current of degree s on a manifold Ω. Then dT is the current of degree $s+1$ acting by the rule

$$\langle dT, \phi \rangle = (-1)^{s+1} \langle T, d\phi \rangle, \quad \phi \in \mathfrak{D}(\Omega).$$

If Ω is a complex manifold, the operators $\partial, \bar\partial$, and d^c are defined similarly. E.g.,

$$\langle \bar\partial T, \phi \rangle = (-1)^{s+1} \langle T, \bar\partial \phi \rangle, \quad \phi \in \mathfrak{D}(\Omega).$$

Clearly these operators are continuous on the space of currents.

E x a m p l e s. (a) If $T = [M]$, where M is an oriented manifold of codimension s and with boundary bM in Ω, then Stokes' theorem clearly implies that $d[M] = (-1)^{s+1}[bM]$. In particular, if D is a domain in Ω with piecewise smooth boundary Γ, then $d[D] = -[\Gamma]$.

(b) Let D be a domain on a complex manifold Ω whose boundary $\Gamma = bD$ is piecewise smooth, and let f be a holomorphic function in D of class $C^1(\bar{D})$. Define f to be zero outside $\bar{D}$, and consider this function as a current $[f]$ of degree 0. By definition,

$$<\bar{\partial}[f],\phi> = -\int_D f\bar{\partial}\phi^{n,n-1} = -\int_D d(f\phi^{n,n-1}) =$$

$$= -\int_\Gamma f\phi^{n,n-1} = <-f\Gamma^{0,1},\phi>$$

hence, up to a "geometrical" factor $\Gamma^{0,1}$, the function $\bar{\partial}[f] = -f\Gamma^{1,0}$ is the boundary value of f on Γ.

(c) Let Γ be a closed hypersurface in Ω of class C^1, let $f \in C^1(\Gamma)$ satisfy on Γ the tangential Cauchy-Riemann conditions (see A2.4), and let $\tilde{f}$ be some smooth continuation of f to Ω. Then

$$\int_\Gamma \tilde{f}\bar{\partial}\phi = \int_\Gamma \bar{\partial}(\tilde{f}\phi) - \int_\Gamma (\bar{\partial}\tilde{f}) \wedge \phi, \quad \phi \in \mathcal{D}^{n,n-2}(\Omega).$$

Since $\bar{\partial}(\tilde{f}\phi) = d(\tilde{f}\phi)$, by Stokes' theorem the first integral at the righthand side vanishes. Thus, $\int_\Gamma f\bar{\partial}\phi = 0$ for all $\phi \in \mathcal{D}^{n,n-2}(\Omega)$ if and only if $\int_\Gamma (\bar{\partial}\tilde{f}) \wedge \phi = 0$ for all $\phi \in \mathcal{D}^{n,n-2}(\Omega)$. In view of the arbitrariness in ϕ, the latter is equivalent to $\bar{\partial}\tilde{f}|_\Gamma = 0$, which are the tangential Cauchy-Riemann conditions for f on Γ. Thus, these conditions can be written in integral form: $\int_\Gamma f\bar{\partial}\phi^{n,n-2} = 0$. The lefthand side is equal to $<f\Gamma^{0,1},\bar{\partial}\phi> = <\bar{\partial}(f\Gamma^{0,1}),\phi>$, hence the tangential Cauchy-Riemann conditions in generalized form mean that the current $f\Gamma^{0,1}$ is $\bar{\partial}$-closed (a similar statement holds for arbitrary CR-manifolds, cf. [181]).

(d) The Bochner-Martinelli form in $\mathbb{C}^n$ $k_{BM} = c_n\partial|z|^{2-2n} \wedge \omega(z)^{n-1}$, where $\omega = (i/2)\,\partial\bar{\partial}|z|^2$, has locally integrable coefficient. Hence it defines in $\mathbb{C}^n$ a current, K_{BM}, of bidegree $(n,n-1)$. Up to a constant the function $|z|^{2-2n}$ is a fundamental solution of the Laplace equation, i.e. $\Delta|z|^{(2-2n)} = \tilde{c}_n \cdot \delta_0$, where δ_0 denotes the delta-function at 0 in $\mathbb{C}^n$, $n > 1$. Since for any (generalized) function ψ the equation $\partial\bar{\partial}\psi \wedge \omega^{n-1} = (-i/2)(\Delta\psi)\omega^n/n$ holds (this can be easily verified by comparing the bihomogeneous components), $\bar{\partial}K_{BM} = \delta_0 dV$ for a suitable choice of the constant c_n (dV is the Euclidean volume form in $\mathbb{C}^n \approx \mathbb{R}^{2n}$). If D is a bounded domain in $\mathbb{C}^n$ with piecewise smooth boundary $\Gamma \not\ni 0$, and if χ denotes the characteristic function of D, then

$$\bar{\partial}(\chi K_{BM}) = \bar{\partial}\chi \wedge K_{BM} + \chi\delta_0 dV = -\Gamma^{0,1} \wedge K_{BM} + \chi\delta_0 dV.$$

All these currents have singularity $\leqslant -1$ and are supported in $\bar{D}$, hence they can act on any function $f \in C^1(D)$. As a result we obtain a generalized Bochner-Martinelli integral representation:

$$f(0) = \int_\Gamma f k_{BM} - \int_D \bar{\partial}f \wedge k_{BM} \quad (0 \in D).$$

For $n = 1$ the kernel $k_{BM} = dz/(2\pi iz)$, and this representation is Cauchy's generalized integral formula (see, e.g., [175], [185]).

(e) If D is a bounded domain in $\mathbb{C}^n$ with twice differentiable boundary Γ, the above can be repeated while replacing $\partial|z|^{2-2n}$ by $d^c g$, where g is the Green function of D with pole at $0 \in D$. Since $dd^c\psi \wedge \omega^{n-1} = 2i\partial\bar{\partial}\psi \wedge \omega^{n-1} = (\Delta\psi)\omega^n/n$ and $\Delta g = -c_n\delta_0$, $c_n > 0$, we obtain, following the same scheme as above,

$$c_n f(0) = -\int_\Gamma f d^c g \wedge \omega^{n-1} + \int_D df \wedge d^c g \wedge \omega^{n-1}.$$

By Lemma 2, A4.2,

$$df \wedge d^c g \wedge \omega^{n-1} = dg \wedge d^c f \wedge \omega^{n-1}.$$

Since $g|_\Gamma = 0$, and $dg, d^c g$ have coefficients that are integrable in D, Stokes' formula implies

$$-c_n f(0) = \int_\Gamma f d^c g \wedge \omega^{n-1} + \int_D g dd^c f \wedge \omega^{n-1} =$$

$$= \int_\Gamma f d^c g \wedge \omega^{n-1} + \frac{1}{n}\int_D (\Delta f) \cdot g\omega^n,$$

with a positive constant c_n depending on n only. This is the complex way of writing Green's formulas in domains D in $\mathbb{C}^n$.

A5.3. Regularization. Fix a nonnegative function $\lambda \in C^\infty(\mathbb{R}^N)$, vanishing outside the unit ball, depending on $|x|$ only, and such that $\int \lambda dV = 1$. For each generalized differential form $T = \Sigma' T_I dx_I$ in a domain $G \subset \mathbb{R}^N$, and for each $\epsilon > 0$, in the open set

$$G_\epsilon = \{x \in G: \operatorname{dist}(x, \partial G) > \epsilon\}$$

the form

$$(T)_\epsilon = \Sigma' T_I^\xi \left[\epsilon^{-N}\lambda\left[\frac{\xi - x}{\epsilon}\right]\right] dx_I$$

is defined; it is the coefficientwise convolution with the function $\epsilon^{-N}\lambda(x/\epsilon)$. In the case when the T_I are locally integrable in G, the coefficient of $(T)_\epsilon$ at dx_I is equal to

$$\epsilon^{-N}\int T_I(\xi)\lambda\left[\frac{\xi - x}{\epsilon}\right] dV(\xi) = \int T_I(x + \epsilon\xi)\lambda(\xi) dV(\xi).$$

This operation, replacement of T by $(T)_\epsilon$, is called regularization (more specifically, ϵ-regularization); it is extensively used in analysis. We give some well-known properties of regularization (the proofs easily follow from the definition).

1) If $T \in \mathcal{D}'(G)$, then $(T)_\epsilon \in \mathcal{E}(G_\epsilon)$, i.e. $(T)_\epsilon$ is infinitely differentiable in G_ϵ.

2) Regularization preserves degrees of forms; in $\mathbb{C}^n$ $(=\mathbb{R}^{2n})$ it also preserves bidegrees.

3) $\mathrm{supp}(T)_\epsilon$ belongs to the ϵ-neighborhood of $\mathrm{supp}T$.

4) $\lim_{\epsilon\to 0}(T)_\epsilon = T$ in the weak-$\star$ topology (i.e. $<(T)_\epsilon, \phi> \to <T,\phi>$ for every $\phi \in \mathcal{D}(G)$).

5) $<(T)_\epsilon,\phi> = <T,(\phi)_\epsilon>$ if $\mathrm{supp}\,\phi \in G_\epsilon$.

6) $D(T)_\epsilon = (DT)_\epsilon$ for any linear differential operator D with constant coefficients in $\mathbb{R}^N$; in particular, $d(T)_\epsilon = (dT)_\epsilon$ and $\bar\partial(T)_\epsilon = (\bar\partial T)_\epsilon$.

A5.4. The $\bar\partial$-problem and the jump Theorem. In the proof of the Harvey-Lawson theorem (p.19.6) we used a most simple Lemma concerning solvability of the $\bar\partial$-problem in $\mathbb{C}^n$, and a Theorem concerning the representation of a CR-function on a hypersurface as the difference of (jump between) the limit values of two holomorphic functions, analogous to Sokhotskii's formula for Cauchy-type integrals (see [175], [180], [181]).

L e m m a 1. *Let g be a generalized $\bar\partial$-closed formula of bidegree (0,1) and with compact support in $\mathbb{C}^n$. Then there is in $\mathbb{C}^n$ a generalized function f such that $\bar\partial f = g$; moreover, if $n > 1$, then f has compact support also.*

■ We first assume that $g = \Sigma g_j d\bar z_j$ is a form of class C^1, and put

$$f(z) = \frac{1}{2\pi i}\int_{\mathbb{C}_n} \frac{g_n(z',\zeta_n)d\zeta_n\wedge d\bar\zeta_n}{\zeta_n - z_n}.$$

Cauchy's generalized integral formula (see [175] or A5.2) implies that $\partial f / \partial \bar z_n = g_n$. By differentiation under the integral sign and by taking into account that $\partial g_n / \partial \bar z_j = \partial g_j / \partial \bar z_n$ (the condition of g being $\bar\partial$-closed), the same formula gives $\partial f / \partial \bar z_j = g_j$, hence $\bar\partial f = g$.

If $\mathrm{supp}\, g \subset \{|z| \leqslant R\}$, then $\bar\partial f = 0$ for $|z| > R$, i.e. f itself is holomorphic. If $n > 1$ and $|z_1| > R$, the integral defining f vanishes, hence $f(z) = 0$ for

$|z_1| > R$. By the uniqueness Theorem, $f(z) = 0$ for $|z| > R$.

In the general case we replace g by an ϵ-regularization (which preserves being $\bar\partial$-closed), and find an f_ϵ in $|z| < R + \epsilon$ such that $\bar\partial f_\epsilon = (g)_\epsilon$. The limit as $\epsilon \to 0$ of f_ϵ exists (in the space of generalized functions) and is the required solution. We can also immediately define f as the convolution of g_n and $1/z_n$. We will not prove these simple limit transitions. ■

The solvability of the $\bar\partial$-problem in products of domains in the plane is somewhat more complicated. Let $U = U_1 \times \cdots \times U_n \subset \mathbf{C}^n$, and let g be a current of bidegree (0,1) and with compact support in $\mathbf{C}^n$ which is $\bar\partial$-closed in U. If $g = \Sigma g_j d\bar z_j$ is of class C^1, we put

$$h_n(z) = \frac{1}{2\pi i}\int_{\mathbf{C}_n} \frac{g(z',\zeta_n)}{\zeta_n - z_n} d\zeta_n \wedge d\bar\zeta_n.$$

Since $\partial h_n / \partial \bar z_n = g_n$, the form $g - \bar\partial h_n = \Sigma_1^{n-1} g_{j,n} d\bar z_j$ is also $\bar\partial$-closed in U, implying, in particular, that the $g_{j,n}$ are holomorphic with respect to $z_n \in U_n$. The function

$$h_{n-1}(z) = \frac{1}{2\pi i}\int_{\mathbf{C}_{n-1}} \frac{g_{n-1,n}(z'',\zeta_{n-1},z_n)}{\zeta_{n-1} - z_{n-1}} d\zeta_{n-1} \wedge d\bar\zeta_{n-1}$$

is holomorphic with respect to $z_n \in U_n$, and $\partial h_{n-1} / \partial \bar z_{n-1} = g_{n-1,n}$. Hence the form $g - \bar\partial h_n - \bar\partial h_{n-1} = \Sigma_1^{n-2} g_{j,n-1}\, d\bar z_j$ has coefficients that are holomorphic with respect to $(z_{n-1}, z_n) \in U_{n-1} \times U_n$. By continuing this process we obtain a function $h = h_1 + \cdots + h_n$ such that $g = \bar\partial h$ in U. In the general case when g is a current we can, as in Lemma 1, use regularizations and limit transitions as $\epsilon \to 0$ (the existence of the limits is ensured by the construction of the functions and by the fact that g is a current of compact support in $\mathbf{C}^n$).

The jump Theorem readily follows from the solvability of the $\bar\partial$-problem; we formulate it in its simplest local version, which suffices for the applications in this book.

L e m m a 2. *Let Γ be a real hypersurface of the form $y_n = \phi(z', x_n)$, $x_n + iy_n = z_n$, in $\mathbf{C}^n$, $\phi \in C^1(\bar U' \times I)$, where $U' = U_1 \times \cdots \times U_{n-1}$ is a bounded domain in $\mathbf{C}^{n-1}$ and I is an interval on the x_n-axis, and let f be a C^1-function on Γ satisfying the tangential Cauchy-Riemann conditions. Then there are functions*

$h_{\pm} \in C(U_{\pm} \cup \Gamma)$, holomorphic in the domains $U_{\pm}$: $z' \in U'$, $x_n \in \operatorname{int} I$, $\pm(y_n - \phi(z', x_n)) > 0$, such that $f = h_+ - h_-$ on Γ.

■ Orient Γ as part of the boundary of U_-. By Example (c), A5.2, the current $g = f\Gamma^{0,1}$ of bidegree (0,1) is $\bar\partial$-closed in $U = U' \times (\operatorname{int} I \times \mathbb{R})$. By requirement, this current in $\mathbb{C}^n$ has compact support $\bar\Gamma$. By what was proved above, there is a (generalized) function h such that $\bar\partial h = g$ in U. Sokhotskii's formula and Example (b), A5.2, readily imply that the function h_n occurring in the first construction step for h is a Cauchy-type integral with respect to z_n:

$$h_n(z) = \frac{1}{2\pi i} \int_{\gamma_{z'}} \frac{f(z', \zeta_n) d\zeta_n}{\zeta_n - z_n},$$

where $\gamma_{z'}$ is the arc $y_n = \phi(z', x_n)$, $x_n \in I$, for fixed $z' \in U'$. The current $g - \bar\partial h_n$ has continuous coefficients, and h has continuous extensions to Γ from both sides. By Example (b), A5.2, $f = h_+ - h_-$ on Γ. ■

Now we prove the Lemma concerning solvability of the $\bar\partial$-problem in a halfspace, which we used in p.19.5.

L e m m a 3. *Let T be a current of measure type, $\bar\partial$-closed, and of bidegree (0,1), in the halfspace U: $x_1 > 0$ in $\mathbb{C}^n$. Let, moreover, the closure of* $\operatorname{supp} T$ *in $\mathbb{C}^n$ be compact. Then there is a function S, locally integrable in U, such that $\bar\partial S = T$, such that the closure in U of the support of S is compact in $\mathbb{C}^n$, and such that S is uniformly bounded on compact subsets in $U \setminus$* $\operatorname{supp} T$.

■ We first assume that the generalized form $\sum T_j d\bar z_j$ corresponding to T has coefficients of class C^1 (by requirement the T_j are measures in U). Then, similar to Lemma 1, the continuous function

$$S(z) = \frac{1}{2\pi i} \int_{\mathbb{C}_n} \frac{T_n(z', \zeta_n) d\zeta_n \wedge d\bar\zeta_n}{\zeta_n - z_n}$$

is the required solution. Note that in any domain $D \subset\subset U$ the norm of S in $L^1(D)$ does not exceed $C \|T_n\|_{L^1(D)}$, where the constant C depends on the diameter of $\operatorname{supp} T_n$ only.

In the general case we must use regularization. If S_ϵ is the above constructed solution of $\bar\partial S_\epsilon = (T)_\epsilon$ in $U_\epsilon: x_1 > \epsilon$, then the fact that the estimate given above holds uniformly in ϵ (which follows since T is of measure type) implies that the functions S_ϵ have, as $\epsilon \to 0$, a limit S; moreover, $S \in L^1_{\text{loc}}(U)$ and $\bar\partial S = T$. Since S is holomorphic outside supp T, the fact that S is locally integrable implies that it is uniformly bounded on compact subsets from $U \setminus \text{supp}\, T$. ■

Ending this Section we note a feature regarding regularity of solutions of the $\bar\partial$-problem for currents T of bidegree (0,1). If S_1, S_2 are two solutions of the equation $\bar\partial S = T$, then $S_1 - S_2$ is a holomorphic function. Therefore, if some solution S has, in a domain D, a certain smoothness, then any other solution has in D the same smoothness.

A6. Hausdorff measures

A good exposition of the theory of Hausdorff measures can be found in Federer's monograph [153]. There is no such handbook in the Russian language, hence below we give all properties of such measures necessary to us (see also [55], [106]).

A6.1. Definition and simplest properties. Let E be an arbitrary subset of a metric space. Consider a covering of E by at most countably many balls B_j (it is immaterial whether these are open or closed) of radii r_j, respectively, and compute the sum Σr_j^α, where $\alpha \geqslant 0$ is a fixed number. The infimum of such sums, $m_\alpha(E)$, will to some extent have to reflect the property that E is a set of "dimension α"; however, this quantity is too coarse (for, say, the unit ball in $\mathbb{R}^m$ it equals 1 for all $\alpha \leqslant m$). Therefore we first fix $\epsilon > 0$, and define the quantity

$$\mathcal{H}^\epsilon_\alpha(E) := \inf\left\{c_\alpha \sum_j r_j^\alpha : E \subset \cup B_j,\ r_j < \epsilon\right\}.$$

For integer $\alpha = m$ we take the constant $c_\alpha > 0$ equal to the volume of the unit ball in $\mathbb{R}^m$ (in particular, $c_0 = 1$, $c_1 = 2$, $c_2 = \pi$, etc.); for noninteger α we take it to be the corresponding expression with the gamma function (for us the constants do not play a role at all, but this normalization is commonly adopted); the index α will always be nonnegative. As ϵ decreases the quantities $\mathcal{H}^\epsilon_\alpha(E)$ monotone increase, hence have a (finite or infinite) limit:

$$\mathcal{H}_\alpha(E) := \lim_{\epsilon \to 0} \mathcal{H}^\epsilon_\alpha(E).$$

This number is called the *Hausdorff measure* of order (dimension) α (or the $\mathcal{H}_\alpha$-measure, or simply the α-measure) of E. Clearly, $\mathcal{H}_\alpha \geqslant c_\alpha m_\alpha(E)$. Since the determination of m_α is more simple, it is useful to note that

$$\mathcal{H}_\alpha(E) = 0 \text{ if and only if } m_\alpha(E) = 0$$

(this is also obvious). For $\alpha = 0$ all $r_j^0 = 1$, and Σr_j^0 is the number of elements of the covering $\{B_j\}$. Hence $\mathcal{H}_0(E) = \sharp\, E$, the number of elements of E.

As was seen, Hausdorff measures are defined for arbitrary sets in a metric space (with values in $[0, +\infty]$). For us the properties of $\mathcal{H}_\alpha$ from the point of view of measure theory do not have special significance, important is that the scale of Hausdorff measures properly reflects the intuitive idea of dimension. We give some simple properties of the measures $\mathcal{H}_\alpha$.

1. Subadditivity: $\mathcal{H}_\alpha(\cup_1^\infty E_k) \leqslant \Sigma_1^\infty \mathcal{H}_\alpha(E_k)$, and if $E = \cup_1^\infty E_k$ is a locally finite union of pairwise disjoint compact sets, then $\mathcal{H}_\alpha(E) = \Sigma_1^\infty \mathcal{H}_\alpha(E_k)$.

2. Homogeneity: if $E \subset \mathbb{R}^m$ and $tE := \{tx : x \in E\}$ for $t > 0$, then $\mathcal{H}_\alpha(tE) = t^\alpha \mathcal{H}_\alpha(E)$.

3. If $\mathcal{H}_\alpha(E) < \infty$, then $\mathcal{H}_\beta(E) = 0$ for all $\beta > \alpha$. If $\mathcal{H}_\alpha(E) > 0$, then $\mathcal{H}_\gamma(E) = \infty$ for all $\gamma < \alpha$. The number $\inf:\{\alpha: \mathcal{H}_\alpha(E) = 0\}$ is called the *Hausdorff* or *metric dimension* of E.

4. If $f: X \to Y$ is a continuous map between metric spaces that uniformly satisfies a Lipschitz condition (i.e. $\rho_Y(f(x), f(x')) \leqslant C\rho_X(x, x')$ for some constant C and all $x, x' \in X$), then $\mathcal{H}_\alpha(f(E)) \leqslant C^\alpha \mathcal{H}_\alpha(E)$ for all $E \subset X$. In particular, under projections $\mathbb{R}^N \to \mathbb{R}^m \subset \mathbb{R}^N$ Hausdorff measures do not increase. The properties 1-4 readily follow from the definition.

5. The properties $\mathcal{H}_\alpha(E) = 0$ or $\mathcal{H}_\alpha(E) = \infty$ for subsets of a smooth manifold M do not depend on the choice of a metric in M which is compatible with the smooth structure on M. This clearly follows from 4. We show yet another useful consequence of 4.

6. P r o p o s i t i o n. *Let M be a connected m-dimensional smooth (C^1) manifold, and let E be a closed subset of M such that $\mathcal{H}_{m-1}(E) = 0$. Then the set $M \setminus E$ is connected.*

■ We first investigate the case when M is a convex domain D in $\mathbb{R}^m$. Let

$a,b \in D \setminus E$, and let $\rho: D \setminus \{a\} \to S^{m-1}$ be the smooth map $x \mapsto (x-a)/|x-a|$ to the unit sphere in $\mathbb{R}^m$. Since $a \notin E$, $\rho|_E$ satisfies uniformly on E a Lipschitz condition. By property 4, $\mathcal{H}_{m-1}(\rho(E)) = 0$, i.e. almost all rays issuing from a do not intersect with E. Since $b \in D \setminus E$, there is in D a ball $B \ni b$ disjoint from E. By what was proved, there is a ray L, issuing from a, not intersecting with E, but intersecting with B. If $c \in B \cap L$, then the polygonal line $[a,c] \cup [c,b]$, which belongs to D since D is convex, does not intersect with E.

The general case is easily derived from what has been proved and the path connectedness of M. ■

A notion of dimension must be naturally compatible with the operation of taking a direct product. In general, for Hausdorff measures this is a complicated matter, hence we restrict ourselves to the following simple statement, which suffices for our purposes.

7. Proposition. *Let E be a set in $\mathbb{R}^k$ such that $\mathcal{H}_\alpha(E) = 0$. Then $\mathcal{H}_{\alpha+m}(E \times \mathbb{R}^m) = 0$.*

■ In view of the countable additivity it suffices to prove that $\mathcal{H}_{\alpha+m}(E \times K) = 0$ for the unit cube $K \subset \mathbb{R}^m$. Fix an arbitrary $\epsilon > 0$. Let $\{B_j\}$ be a covering of E by balls of radii r_j, respectively, such that $\Sigma r_j^\alpha < \epsilon$. For every j, cover K by cubes K_{ji} with sides of length r_j and parallel to the coordinate axes such that the interiors of these cubes are pairwise disjoint and such that their number, N_j, is at most $(2/r_j)^m$. Then

$$E \times K \subset \bigcup_j \bigcup_{i=1}^{N_j} (B_j \times K_{ji});$$

moreover, $B_j \times K_{ji}$ has diameter $\leqslant c(m) r_j$. Since

$$\sum_j \sum_{i=1}^{N_j} (c(m) r_j)^{\alpha+m} \leqslant c_1(m) \sum_j r_j^\alpha (N_j r_j^m) \leqslant c_2(m) \sum_j r_j^\alpha < c_2(m)\epsilon,$$

and ϵ is arbitrary, $\mathcal{H}_{\alpha+m}(E \times K) = 0$. ■

A6.2. $\mathcal{H}_m$ on an m-dimensional manifold. We first show that in $\mathbb{R}^m$ the measure $\mathcal{H}_m$ coincides with ordinary Lebesgue measure vol_m.

L e m m a. *For every Lebesgue measurable set $E \subset \mathbb{R}^m$ the equation $\mathcal{H}_m(E) = \mathrm{vol}_m(E)$ holds.*

■ It is clear that $\mathcal{H}_m$ is an outer measure, i.e. $\mathcal{H}_m(E)$ is equal to $\inf \mathcal{H}_m(U)$, taken over all open sets $U \supset E$. Hence it suffices to prove the statement for open sets. Every open set $U \subset \mathbb{R}^m$ can be represented as a union of cubes with pairwise disjoint interiors. By slightly shrinking each cube we obtain a locally finite union of pairwise disjoint closed cubes K_j with total volume arbitrarily close to the volume of U. The homogeneity and semiadditivity of $\mathcal{H}_m$ imply that $\mathcal{H}_m(U)$ is also arbitrarily close to $\mathcal{H}_m(\cup K_j) = \Sigma\, \mathcal{H}_m(K_j)$. Thus, it remains to prove that the unit cube K: $0 < x_k < 1$, $k \leqslant m$, in $\mathbb{R}^m$ has Hausdorff measure $\mathcal{H}_m$ equal to 1.

For any covering of K by balls B_j of radii r_j the estimate

$$c_m \sum r_j^m = \sum \mathrm{vol}_m B_j \geqslant \mathrm{vol}_m K = 1$$

holds, hence $\Theta := \mathcal{H}_m(K) \geqslant 1$. The standard coverings of K by balls of identical radii 2^{-j} imply that $\Theta < \infty$. By homogeneity,

$$\mathcal{H}_m(tK) = \Theta t^m = \Theta\, \mathrm{vol}_m(tK).$$

This equation, the semiadditivity of $\mathcal{H}_m$, and the countable additivity of vol_m imply $\mathcal{H}_m(U) \leqslant \Theta\, \mathrm{vol}_m U$ for any open set $U \subset \mathbb{R}^m$ (again using standard coverings by cubes).

Fix an arbitrary $\delta > 0$ and choose $\epsilon = 2^{-j+1}$ such that $\mathcal{H}_m^\epsilon(K) \geqslant \Theta - \delta$. Partition K into equal cubes with sides of length ϵ and parallel to the coordinate axes, and inscribe in each cube a closed ball B_i of radius $\epsilon / 2$, $i = 1, \ldots, 2^{m(j-1)}$. The volume of every B_i is $c_m 2^{-jm}$, and the volume of $\cup_i B_i$ is $c_m 2^{-jm} \cdot 2^{m(j-1)} = c_m 2^{-m}$; in particular, the latter is independent of ϵ. Put $U = K \setminus \cup_i B_i$. Then $\mathrm{vol}_m U = 1 - c_m 2^{-m}$, hence

$$\Theta - \delta \leqslant \mathcal{H}_m^\epsilon(K) \leqslant c_m 2^{-m} + \mathcal{H}_m(U) \leqslant c_m 2^{-m} + \Theta \mathrm{vol}_m U =$$

$$= \Theta + c_m 2^{-m}(1 - \Theta).$$

This implies $\Theta - 1 \leqslant \delta \cdot 2^m / c_m$, hence $\Theta = 1$ (since δ is arbitrary). ■

Let now M be a Riemannian manifold of dimension m (e.g. a smooth manifold embedded in $\mathbb{R}^N$, or in $\mathbb{P}_n$, with the induced metric). If $a \in M$, and if local coordinates x in a neighborhood $U \ni a$ on M are chosen such that they are orthonormal at a, then the coordinate map $U \to \mathbb{R}^m$ is almost isometric; more precisely, for every $\epsilon \in (0,1)$ there is a neighborhood $V \ni a$ in U such that for any distinct points $b, c \in V$ the inequality

$$1-\epsilon < \frac{\operatorname{dist}(b,c)}{|x(b)-x(c)|} < 1+\epsilon$$

holds. By property 4, $|1-\mathcal{H}_m(E)/\mathcal{H}_m(x(E))| < C\epsilon$ for every $E \subset U$. These estimates and the Lemma clearly imply that $|1-\mathcal{H}_m(E)/\operatorname{vol}_m E| < C'\epsilon$. Using the additivity of vol_m and $\mathcal{H}_m$ (property 1) we thus obtain the following

P r o p o s i t i o n. *On an m-dimensional Riemannian manifold M the measure* $\mathcal{H}_m$ *coincides with outer Lebesgue measure.*

As a consequence we obtain that $\mathcal{H}_m|_M$ is a locally finite measure, and that $\mathcal{H}_m(U) > 0$ for every nonempty open subset $U \subset M$.

A6.3. The Lemma concerning fibres.

L e m m a. *Let M,N be Riemannian manifolds of class* C^1, *let* $f\colon N \to M$ *be a smooth map, and let E be a subset in N such that* $\mathcal{H}_\alpha(E) = 0$ *for an* $\alpha \geqslant m = \dim M$. *Then* $\mathcal{H}_{\alpha-m}(E \cap f^{-1}(x)) = 0$ *for almost all* $x \in M$.

■ The statement being local, it suffices to consider the case when $N \subset \mathbb{R}^n$, $M \subset\subset \mathbb{R}^m$ is an open set, and f satisfies uniformly on E a Lipschitz condition.

Since $\mathcal{H}_\alpha(E) = 0$, for every $\epsilon > 0$ there is a covering of E by balls B_j with radii $r_j < \epsilon$ such that $c_\alpha \Sigma r_j^\alpha < \epsilon$. For each point $x \in M$ the inequality $\mathcal{H}^\epsilon_{\alpha-m}(E \cap f^{-1}(x)) \leqslant c_{\alpha-m} \Sigma_x r_j^{\alpha-m}$ holds, where Σ_x indicates that summation is over those j for which $B_j \cap E \cap f^{-1}(x)$ is nonempty. We want to integrate this inequality over M. Since a priori the lefthand side can be a nonmeasurable function, instead of the ordinary integral with respect to Lebesgue measure dV we take the upper integral $\int^*$ (by definition, $\int^*_M \phi dV = \inf\{\int_M g dV\colon g$ integrable and $g \geqslant \phi\})$. So,

$$\int_M^* \mathcal{H}^\epsilon_{\alpha-m}(E \cap f^{-1}(x)) dV \leqslant c_{\alpha-m} \sum_j r_j^{\alpha-m} \int_{f(B_j)} dV \leqslant$$

$$\leqslant c_{\alpha-m} \sum_j r_j^{\alpha-m} c_m (C r_j)^m < C'\epsilon.$$

Since $\mathcal{H}^\epsilon_{\alpha-m} \to \mathcal{H}_{\alpha-m}$, monotone increasing as $\epsilon \to 0$, this inequality and B. Levi's theorem implies that $\int_M^* \mathcal{H}_{\alpha-m}(E\cap f^{-1}(x))dV = 0$, hence the integrand vanishes almost everywhere on M. ■

R e m a r k. It can be similarly proved that if $\mathcal{H}_\alpha(E) < \infty$, then also $\mathcal{H}_{\alpha-m}(E\cap f^{-1}(x)) < \infty$ for almost all $x \in M$.

A6.4. Sections and projections.

L e m m a. *Let E be a set in $\mathbb{C}^n$ such that $\mathcal{H}_{2p+1}(E) = 0$ for some integer $p < n$. Then there is an $(n-p)$-dimensional plane $L \ni 0$ in $\mathbb{C}^n$ such that $\mathcal{H}_1(E\cap L) = 0$; moreover, almost every plane $L \in G(n-p,n)$ has this property.*

■ Consider in $\mathbb{C}^n \times G$, $G = G(n-p,n)$, the incidence manifold $I = I(n-p,p)$ consisting of all pairs (z,L) for which $z \in L$ (see A3.5). Let π_1 be the natural projection to $\mathbb{C}^n$, and let $\tilde{E} := \pi_1^{-1}(E\setminus\{0\}) \cap I$ be the lift of $E\setminus\{0\}$ to I. Every fiber $\pi_1^{-1}(z) \cap I$, $z \neq 0$, is isomorphic to $G(n-p-1,n-1)$; in particular, has complex dimension $(n-p-1)\cdot p$. Locally the projection $\pi_1 \colon I \to \mathbb{C}^n$ is isomorphic to a direct product and projection to a factor, hence $\tilde{E}$ has Hausdorff α-measure zero, with $\alpha = 2p+1+2(n-p-1)\cdot p = 2(n-p)\cdot p+1$. On the other hand, let π_2 be the natural projection to G. Then Lemma A6.3 applies to $\pi_2 \colon I \to G$. According to it, for almost all $L \in G$ the fiber $\pi_2^{-1}(L) \cap \tilde{E}$ has β-measure zero, where $\beta = \alpha - 2p(n-p) = 1$. It remains to note that π_1 projects $\pi_2^{-1}(L) \cap I$ biholomorphically onto L, and that $\pi_2^{-1}(L) \cap \tilde{E}$ is projected by it onto $L \cap (E\setminus\{0\})$. ■

C o r o l l a r y. *Let E be a locally closed set in $\mathbb{C}^n$ such that $\mathcal{H}_{2p+1}(E) = 0$ for some integer $p < n$, and let $0 \in E$. Then there are arbitrarily small neighborhoods*

$$U' \subset \mathbb{C}^p, \quad U'' \subset \mathbb{C}^{n-p}, \quad U' \times U'' = U \ni 0,$$

and a unitary transformation $l \colon \mathbb{C}^n \to \mathbb{C}^n$ such that the set $l(E) \cap \overline{U}$ is closed in $\overline{U}$ and disjoint from the compact set $\overline{U}' \times \partial U''$. In particular, the projection $\pi \colon l(E) \cap U \to U'$ is a proper map. Moreover, for almost all unitary transformations l there is a neighborhood U such that $l(E) \cap \overline{U}$ has the properties listed.

A set E is called locally closed if it is closed in a neighborhood of it. A map between topological spaces is called proper if the pre-image of every compact set is a compact set (see p.3.1). If l is a transformation as in the Corollary, then for any compact set $K' \subset U'$ the set

$$\pi^{-1}(K') \cap l(E) \cap U = (K' \times U'') \cap l(E)$$

is closed and bounded in $\mathbb{C}^n$, i.e. compact. Hence it suffices to prove the first part of the statement.

■ Let L be an $(n-p)$-dimensional plane from the Lemma, let l be a unitary transformation mapping L to $\mathbb{C}^{n-p}_{z''}$, and let V be an arbitrary neighborhood of 0 such that $E \cap V$ is closed in V. Since $\mathcal{H}_1(E \cap L) = 0$, and since lengths do not increase under "radial projection" $z \mapsto |z| \in \mathbb{R}_+$, there is an $r > 0$ such that $\{|z| \leqslant r\} \subset V$ and such that E is disjoint from the sphere $L \cap \{|z| = r\} \subset \mathbb{C}^{n-p}_{z''}$ (see the proof of Theorem A1.4). Take for U'' the ball $|z''| < r$ in $\mathbb{C}^{n-p}_{z''}$. Since $E \cap V$ is closed in V, there is an $r' \in (0,r)$ such that E is also disjoint from the set $\{|z'| \leqslant r'\} \times \partial U''$. Take for U' the ball $|z'| < r'$ in $\mathbb{C}^p_{z'}$. Then $E \cap \overline{U}$ is closed in $\overline{U}$, where $U = U' \times U''$, and E is disjoint from the compact set $\overline{U}' \times \partial U''$, on which $|z''| = r$ and $|z'| \leqslant r$. ■

References

1. S. ABHYANKAR, *Local analytic geometry.* Academic Press, 1964.
2. W.A. ADKINS, *Local algebraicity of some analytic hypersurfaces.* Proc. Amer. Math. Soc., 1980, **79**, 546-548.
3. L.A. AIZENBERG, A.P. YUZHAKOV, *Integral representations and residues in multidimensional complex analysis.* Amer. Math. Soc., 1983 (translated from the Russian).
4. H. ALEXANDER, *Polynomial approximation and hulls in sets of finite linear measure in* $\mathbb{C}^n$. Amer. J. Math., 1971, **93**, 65-74.
5. H. ALEXANDER, *Continuing 1-dimensional analytic sets.* Math. Ann., 1971, **191**, 143-144.
6. H. ALEXANDER, *Volumes of images of varietes in projective space and in grassmanian.* Trans. Amer. Math. Soc., 1974, **189**, 237-249.
7. H. ALEXANDER, *On zero sets for the ball algebra.* Proc. Amer. Math. Soc., 1982, **86**, 71-74.
8. H. ALEXANDER AND R. OSSERMAN, *Area bounds for various classes of surfaces.* Amer. J. Math., 1975, **97**, 753-769.
9. H. ALEXANDER, B.A. TAYLOR AND J.L. ULLMAN, *Areas of projections of analytic sets.* Invent. Math., 1972, **16**, 335-341.
10. E. AMAR, *Sur le volume des zeros des fonctions holomorphes et bornées dans le boule de* $\mathbb{C}^n$. Proc. Amer. Math. Soc., 1982, **85**, 47-52.
11. L. BUNGART, *Integration on real analytic varieties.* J. Math. Mech., 1966, **15**, 1039-1054.
12. J. BECKER, *Continuing analytic sets across* $\mathbb{R}^n$. Math. Ann., 1972, **195**, 103-106.
13. J. BECKER, *Parametrization of analytic varieties.* Trans. Amer. Math. Soc., 1973, **183**, 265-292.
14. J. BECKER, *Holomorphic and differentiable tangent spaces to a complex analytic variety.* J. Diff. Geometry, 1977, **12**, 377-401.
15. V.K. BELOSHAPKA, *On a metric property of analytic sets.* Math. USSR-Izv., 1976, **10**, 1333-1338. (Original: Izv. Akad. Nauk SSSR, 1976, **40**, 1409-1414.)

16. H. BEHNKE AND P. THULLEN, *Theorie der Funktionen mehrerer komplexen Veränderlichen.* Springer, 1970.

17. B. BERNDTSSON, *Zeros of analytic functions of several variables.* Ark. Math., 1978, **16**, 251-262.

18. J. BESNAULT AND P. DOLBEAULT, *Sur les bords d'ensembles analytiques complexes dans* $\mathbb{P}^n(\mathbb{C})$. In: Proc. Symp. Pure Math., **34**, Amer. Math. Soc., 1981, 205-213.

19. D. BESCHERON, *Multiplicité d'intersection et formules integrales.* Lecture Notes in Math., 1975, **482**, 168-179.

20. E. BISHOP, *Mappings of partially analytic spaces.* Amer. J. Math., 1961, **83**, 209-242.

21. E. BISHOP, *Partially analytic spaces.* Amer. J. Math., 1961, **83**, 669-692.

22. E. BISHOP, *Condition for the analyticity of certain sets.* Michigan Math. J., 1964, **11**, 289-304.

23. T. BLOOM, C^1*-functions on a complex analytic variety.* Duke Math. J., 1969, **36**, 283-296.

24. E. BOMBIERI, *Addendum to my paper: "Algebraic values of meromorphic maps".* Invent. Math., 1970, **11**, 163-166.

25. D. BURGHELEA AND A. VERONA, *Local homological properties of analytic sets.* Manuscr. Math., 1972, **7**, 1, 55-66.

26. J.E. BJÖRK, *On extension of holomorphic functions satisfying a polynomial growth condition on algebraic varieties in* $\mathbb{C}^n$. Ann. Inst. Fourier, 1975, **24**, 4, 157-165.

27. N.T. VAROPOULOS, *Sur les zeros des classes de Hardy* $\mathbb{H}^p(\Omega)$. C.R. Acad. Sc. Paris, 1978, **287**, 627-628.

28. A.N. VARCHENKO, *The integrality of the limit of the curvature integral along the boundary of an isolated singularity of a surface in* $\mathbb{C}^3$. Russian Math. Surveys, 1978, **33**, 6, 263-264. (Original: Uspekhi Mat. Nauk, 1978, **33**, 6, 199-200.)

29. C. WATANABE, *A remark on the theorem of Bishop.* Proc. Japan Acad., 1969, **45**, 243-246.

30. J. WERMER, *The hull of a curve in* $\mathbb{C}^n$. Ann. of Math., 1958, **68**, 550-561.

31. W. WIRTINGER, *Eine Determinantenidentität und ihre Anwendung auf analytische Gebiete in Euklidischer und Hermitescher Massbestimmung.* Monatsch. für Math. und Physik, 1936, **44**, 343-365.

32. V.S. VLADIMIROV, *Methods of the theory of functions of several complex*

variables. M.I.T., 1966 (translated from the Russian).

33. R.C. GUNNING, *Lectures on complex analytic varieties I. The local parametrization theorem.* Princeton Univ. Press, 1970.
34. R.C. GUNNING, *Lectures on complex analytic varieties II. Finite analytic mappings.* Princeton Univ. Press, 1974.
35. R. GUNNING AND H. ROSSI, *Analytic functions of several complex variables.* Prentice-Hall, 1965.
36. M. GILMARTIN, *Nondifferentiability of retractions of* $\mathbb{C}^n$ *to subvarieties.* Proc. Amer. Math. Soc., 1965, **16**, 1028-1029.
37. G.M. GOLUZIN, *Geometric theory of functions of a complex variable.* Amer. Math. Soc., 1969 (translated from the Russian).
38. H. GRAUERT AND R. REMMERT, *Singularitäten komplexer Mannigfaltigkeiten und Riemannsche Gebiete.* Math. Zeitschr., 1957, **67**, 103-128.
39. H. GRAUERT AND R. REMMERT, *Komplexe Räume.* Math. Ann., 1958, **136**, 245-318.
40. PH. GRIFFITHS, *Topics in algebraic and analytic geometry.* Princeton Univ. Press, 1974.
41. PH. GRIFFITHS, *Complex differential and integral geometry and curvature integrals, associated to singularities of complex analytic varieties.* Duke Math. J., 1978, **45**, 427-512.
42. PH. GRIFFITHS AND J. KING, *Nevanlinna theory and holomorphic mappings between algebraic varieties.* Acta Math., 1973, **130**, 145-220.
43. PH. GRIFFITHS AND J. HARRIS, *Principles of algebraic geometry.* Vol. 1-2, Wiley, 1978.
44. L. GRUMAN, *The area of analytic varieties in* $\mathbb{C}^n$. Math. Scand., 1977, **41**, 365-397.
45. L. GRUMAN, *La géométrie globale des ensembles analytiques dans* $\mathbb{C}^n$. Lecture Notes in Math., 1980, **822**, 90-99.
46. S.M. HUSEIN-ZADE, *The monodromy groups of isolated singularities of hypersurfaces.* Russian Math. Surveys, 1977, **32**, 2, 23-69. (Original: Uspekhi Mat. Nauk, 1977, **32**, 2, 23-65).
47. J.P. DEMAILLY, *Construction d'hypersurfaces irreducibles avec lieu singulier donné dans* $\mathbb{C}^n$. Ann. Inst. Fourier, 1980, **30**, 3, 219-236.
48. J.P. DEMAILLY, *Formules de Jensen en plusieurs variables et applications arithmétiques.* Bull. Soc. Math. France, 1982, **110**, 75-102.
49. J.P. DEMAILLY, *Sur les nombres de Lelong associés a l'image direct d'un*

courant positif fermé. Ann. Inst. Fourier, 1982, **32**, 37-66.

50. G. DLOUSSKY, *Enveloppes d'holomorphie et prolongements d'hypersurfaces*. Lecture Notes in Math., 1977, **578**, 217-235.
51. G. DLOUSSKY, *Analyticité separée et prolongements analytiques*. Lecture Notes in Math., 1978, **693**, 179-202.
52. E.P. DOLZHENKO, *Elimination of singularities of analytic functions*. Uspekhi Mat. Nauk, 1963, **18**, 4, 135-142 (in Russian).
53. R.N. DRAPER, *Intersection theory in analytic geometry*. Math. Ann., 1969, **180**, 175-204.
54. A.YA DUBOVITSKII, *On differentiable maps of the n-dimensional cube into the k-dimensional cube*. Mat. Sb., 1953, **32**, 2, 443-464 (in Russian).
55. L. CARLESON, *Selected problems on exceptional sets*. v. Nostrand, 1967.
56. H. CARTAN, *Variétés analytiques réeles et variétés analytiques complexes*. Bull. Soc. Math. France, 1957, **85**, 77-99.
57. H. CARTAN, *Sur les fonctions de plusieurs variables complexes: les espaces analytiques*. In: Proc. Intern. Congress Math. 1958, Cambridge Univ. Press, 1960, 33-52.

58a. H. CARTAN, *Calcul différentiélle*. Hermann, 1967.

58b. H. CARTAN, *Formes différentiélle*. Hermann, 1967.

59. K. KATO, *Sur le théorème de P. Thullen et K. Stein*. J. Math. Soc. Japan, 1966, **18**, 211-218.
60. V.E. KATSNEL'SON AND L.I. RONKIN, *On the minimum volume of an analytic set*. Siberian Math. J., 1975, **15**, 3, 370-378. (Original: Sibirsk. Mat. Zh. 1974, **15**, 3, 516-528.)
61. CH. KISELMAN, *Densité des fonctions plurisousharmoniques*. Bull. Soc. Math. France, 1979, **107**, 295-304.
62. J.R. KING, *The currents defined by analytic varieties*. Acta Math., 1971, **127**, 185-220.
63. J.R. KING, *Global residues and intersections on a complex manifold*. Trans. Amer. Math. Soc., 1974, **192**, 163-199.
64. A.N. KOLMOGOROV AND S.V. FOMIN, *Elements of the theory of functions and functional analysis*. Graylock, 1957-1961 (translated from the Russian).
65. J. KOREVAAR, J. WIEGERINCK AND R. ZEINSTRA, *Minimal area of zero sets in tube domains of* $\mathbb{C}^n$. Univ. of Amsterdam, 1982, **81-13**.
66. M. CORNALBA AND B. SHIFFMAN, *A counterexample to the 'Transcendental*

Bezout problem'. Ann. of Math., 1972, **96**, 402-406.

67. A.I. KOSTRIKIN, *Introduction to algebra.* Springer, 1982 (translated from the Russian).

68. S. COEN, *Some consequences of theorem A for complex spaces.* Ann. Mat. Pura ed Appl., 1975, **106**, 119-153.

69. R.O. KUJALA, *Generalized Blaschke conditions on the unit ball in* $\mathbb{C}_p$. In: Value Distribution Theorie, part A. M. Dekker, 1974, 249-261.

70. N. KUHLMANN, *Über holomorphe Abbildungen komplexer Räume.* Arch. Math., 1964, **15**, 2, 81-90.

71. N. KUHLMANN, *Bemerkungen über eigentliche holomorphe Abbildungen.* Math. Ann., 1977, **226**, 171-181.

72. K. LAMOTKE, *Die Homologie isolierter Singularitäten.* Math. Zeitschr., 1975, **143**, 27-44.

73. K. LAMOTKE, *The topology of complex projective varieties after S. Lefschetz.* Topology, 1981, **20**, 1, 15-51.

74. LE DUNG TRANG, *Sur un critère d'equisingularité.* Lecture Notes in Math., 1974, **409**, 124-161.

75. LE DUNG TRANG, *Calculation of Milnor number of isolated singularity of complete intersection.* Functional Anal. Appl., 1974, **8**, 2, 127-131. (Original: Funkts. Anal. Prilozh. 1974, **8**, 2, 45-49).

76. P. LELONG, *Propriétés métriques des variétés analytiques complexes definies par une équation.* Ann. Sci. Ecole Norm. Sup. (3), 1950, **67**, 393-419.

77. P. LELONG, *Integration sur un ensemble analytique complexe.* Bull. Soc. Math. France, 1957, **85**, 239-262.

78. P. LELONG, *Fonctions plurisousharmoniques et formes différentielles positives.* Gordon and Breach, 1968.

79. P. LELONG, *Application des courants positifs fermés à la géométrie analytique.* Bull. Soc. Math. France, 1974, **38**, 9-25.

80. P. LELONG, *Sur la structure des courants positifs fermés.* Lecture Notes in Math., 1977, **578**, 136-156.

81. L. LEMPERT, *Boundary behavior of meromorphic functions of several variables.* Acta Math., 1980, **144**, 1-25.

82. J. LERAY, *Le calcul différentiel et intégrale sur une variété analytique complexe.* Bull. Soc. Math. France, 1959, **87**, 81-180.

83. B. MALGRANGE, *Sur les fonctions différentiables et les ensembles analytiques.* Bull. Soc. Math. France, 1963, **91**, 113-127.

84. B. MALGRANGE, *Analytic spaces.* Enseign. Math. (2), 1968, **14**, 1-28.

85. B. MALGRANGE, *Ideals of differentiable functions.* Oxford Univ. Press, 1966.

86. D. MUMFORD, *Algebraic geometry I. Complex projective varieties.* Springer, 1976.

87. J. MILNOR, *Singular points of complex hypersurfaces.* Annals Math. Studies, **61**, Princeton Univ. Press, 1968.

88. N. MOK, Y.T. SIU AND S.T. YAU, *The Poincaré-Lelong equation on complete Kähler manifolds.* Compos. Math., 1981, **44**, 183-218.

89. R.E. MOLZON, B. SHIFFMAN AND N. SIBONY, *Average growth estimates for hyperplane sections of entire analytic sets.* Math. Ann., 1981, **257**, 43-59.

90. R. NARASIMHAN, *Compact analytic varieties.* Enseign. Math. (2), 1968, **14**, 75-98.

91. R. NARASIMHAN, *Introduction to the theory of analytic spaces.* Lecture Notes in Math., 1966, **25**.

92. R. NARASIMHAN, *Analysis on real and complex manifolds.* North-Holland, 1968.

93. T. NISHINO, *Sur les ensembles pseudoconcaves.* J. Math. Kyoto Univ., 1962, **1**, 2, 225-245.

94. K. NOMIZU AND B. SMYTH, *Differential geometry of complex hypersurfaces.* J. Math. Soc. Japan, 1968, **20**, 498-521.

95. M. OKADA, *Un théorème de Bezout transcendent sur* $\mathbb{C}^n$. J. Funct. Anal., 1982, **45**, 236-244.

96. V.P. PALOMODOV, *Multiplicities of holomorphic mappings.* Functional Anal. Appl., 1967, **1**, 3, 218-226. (Original: Funkts. Anal. Prilozh., 1967, **1**, 3, 54-65).

97. YI-CHUAN PAN, *Analytic sets of finite order.* Math. Zeitschr., 1970, **116**, 271-298.

98. I.I. PRIWALOW, *Randeigenschaften analytischer Funktionen.* Deutsch. Verlag Wissenschaft., 1956 (translated from the Russian).

99. G. DE RHAM, *On the area of complex manifolds.* In: Global Analysis. Princeton Univ. Press, 1966, 141-148.

100. R. REMMERT, *Projektionen analytischer Mengen.* Math. Ann., 1956, **130**, 410-441.

101. R. REMMERT, *Holomorphe und meromorphe Abbildungen komplexer Räume.* Math. Ann., 1957, **133**, 328-370.

102. R. Remmert, *Über die wesentlichen Singularitäten analytischer Mengen.* Math. Ann., 1953, **126**, 263-306.

103. R. Remmert and K. Stein, *Eigentliche holomorphe Abbildungen.* Math. Zeitschr., 1960, **73**, 159-189.

104. J. Riibentaus, *Removable singularities of analytic functions of several complex variables.* Math. Zeitschr., 1978, **158**, 45-54.

105. O. Riemenschneider, *Über den Flächeninhalt analytischer Mengen und die Erzeugung k-pseudokonvexer Gebiete.* Invent. Math., 1967, **2**, 307-331.

106. C.A. Rogers, *Hausdorff measures.* Cambridge Univ. Press, 1970.

107. H. Rossi, *Vector fields on analytic spaces.* Ann. of Math., 1963, **78**, 455-467.

108. H. Rossi, *Continuation of subvarieties of projective varieties.* Amer. J. Math., 1969, **91**, 565-575.

109. W. Rothstein, *Über die Fortsetzung analytischer Flächen.* Math. Ann., 1951, **122**, 424-434.

110. W. Rothstein, *Zur Theorie der Singularitäten analytischer Funktionen und Flächen.* Math. Ann., 1953, **126**, 221-238.

111. W. Rothstein, *Der Satz von Casorati-Weierstrass und ein Satz von Thullen.* Arch. Math., 1954, **5**, 338-343.

112. W. Rothstein, *Zur Theorie der analytischen Mannigfaltigkeiten im Raume von n komplexen Veränderlichen.* Mat. Ann., 1955, **129**, 96-138; 1957, **133**, 271-280; 1957, **133**, 400-409.

113. W. Rothstein, *Bemerkungen zur Theorie komplexer Räume.* Math. Ann., 1959, **137**, 304-315.

114. W. Rothstein, *Zur Theorie der analytischen Mengen.* Math. Ann., 1967, **174**, 8-32.

115. W. Rothstein, *Das Maximumprinzip und die Singularitäten analytischer Mengen.* Invent. Math., 1968, **6**, 2, 163-184.

116. W. Rudin, *A geometric criterion for algebraic varieties.* J. Math. Mech., 1968, **17**, 671-683.

117. W. Rudin, *Function theory in polydiscs.* Benjamin, 1969.

118. W. Rudin, *Function theory in the unit ball of* $\mathbb{C}^n$. Springer, 1980.

119. H. Rutishauser, *Über Folgen und Scharen von analytischen und meromorphen Funktionen mehrerer Variabeln, sowie von analytischen Abbildungen.* Acta. Math., 1950, **83**, 249-325.

120. A. Sadullaev, *Criteria for analytic sets to be algebraic.* Functional Anal.

Appl., 1972, **6**, 1, 78-79. (Original: Funkts. Prilozh. Anal. 1972, **6**, 1, 85-86.)

121. A. SADULLAEV, *An analog of Jensen's formula for functions of several variables.* Izv. Akad. Nauk YzSSR, **2**, 1975, 19-22 (in Russian).

122. A. SADULLAEV, *An estimate for polynomials on analytic sets.* Math. USSR Izv., 1982, **20**, no. 3, 493-502. (Original: Izv. Akad. Nauk SSSR, 1982, fB46, **524-534**).

123. U. CEGRELL, *Removable singularities for plurisubharmonic functions and related problems.* Proc. London Math. Soc., 1978, **36**, 2, 310-336.

124. U. CEGRELL, *Relations between removable singularity sets for plurisubharmonic functions and postive, closed (1,1)-currents.* Arch. Math., 1978, **30**, 422-426.

125. N. SIBONY AND P.M. WONG, *Some results on global analytic sets.* Lecture Notes in Math., 1980, **822**, 221-237.

126. N. SIBONY AND P.M. WONG, *Some remarks on the Casorati-Weierstrass theorem.* Ann. Polon. Math., 1981, **39**, 165-174.

127. R.R. SIMHA, *Certain cones are not set-theoretic complete intersections.* Arch. Math., 1976, **27**, 169-171.

128. R.R. SIMHA, *Über die kritischen Werte gewisser holomorpher Abbildungen.* Manuscr. Math., 1970, **3**, 97-104.

129. H. SKODA, *Sous-ensembles analytiques d'ordre fini ou infini dans* $\mathbb{C}^n$. Bull. Soc. Math. France, 1972, **100**, 358-408.

130. H. SKODA, *Valeurs au bord pour les solutions de l'operateur d'', et characterization des zeros des fonctions de la classe de Nevanlinna.* Bull. Soc. Math. France, 1976, **104**, 225-299.

131. G. SPRINGER, *Introduction to Riemann surfaces.* Chelsea, reprint, 1981.

132. J. STUTZ, *Analytic sets as branched coverings.* Trans. Amer. Math. Soc., 1972, **166**, 241-259.

133. J. STUTZ, *Equisingularity and local analytic geometry.* In: Proc. Symp. Pure Math., **30**, 1. Amer. Math. Soc., 1979, 77-84.

134. G. STOLZENBERG, *Uniform approximation on smooth curves.* Acta Math., 1966, **115**, 185-198.

135. G. STOLZENBERG, *Volumes, limits, and extensions of analytic varieties.* Lecture Notes in Math., 1966, **19**.

136. YUM-TONG SIU, $\mathcal{O}^N$*-approximable and holomorphic functions on complex spaces.* Duke Math. J., 1969, **36**, 451-454.

137. YUM-TONG SIU, *Analyticity of sets associated to Lelong numbers and the extension of closed positive currents.* Invent. Math., 1974, **27**, 1-2, 53-156.

138. M. TADOKORO, *Sur les ensembles pseudoconcaves generaux.* J. Math. Soc. Japan, 1965, **17**, 281-290.

139. P. TWORZEWSKI, *Some conditions for the algebraicity of analytic sets.* Bull. Acad. Polon. Sci., 1981, **29**, 545-548.

140. P.R. THIE, *The Lelong number of point of a complex analytic set.* Math. Ann., 1967, **172**, 269-312.

141. P.R. THIE, *The area of an analytic set in complex projective space.* Proc. Amer. Math. Soc., 1969, **21**, 553-554.

142. P.R. THIE, *The Lelong number of a complete intersection.* Proc. Amer. Math. Soc., 1970, **24**, 319-323.

143. W. THIMM, *Über starke und schwache Holomorphie auf analytischen Mengen.* Math. Zeitschr., 1961, **75**, 426-448.

144. J.G. TIMOURIAN, *The invariance of Milnor's number implies topological triviality.* Amer. J. Math., 1977, **99**, 437-446.

145. D. TOLEDO AND L. TONG YUE LIN, *Duality and intersection theory in complex manifolds I.* Math. Ann., 1978, **237**, 41-77; *II.* Ann. of Math., 1978, **102**, 519-538.

146. P. THULLEN, *Über die wesentlichen Singularitäten analytischer Funktionen und Flächen im Raume von n komplexen Veranderlichen.* Math. Ann., 1935, **111**, 137-157.

147. P. THULLEN, *Bemerkungen über analytische Flächen im Raume von n komplexen Veranderlichen im Zusammenhang mit dem zweiten Cousinschen Problem.* Math. Ann., 1969, **183**, 1-5.

148. H. WHITNEY, *Local properties of analytic varieties.* In: Diff. and Combinator. Topology. Princeton Univ. Press, 1965, 205-244.

149. H. WHITNEY, *Tangents to an analytic variety.* Ann. of Math., 1965, **81**, 496-549.

150. H. WHITNEY, *Complex analytic varieties.* Addison-Wesley, 1972.

151. R.O. WELLS, JR., *Differential analysis on complex manifolds.* Springer, 1980.

152. F. PHAM, *Introduction à l'étude topologique des singularités de Landau.* Gauthier-Villars, 1967.

153. H. FEDERER, *Geometric measure theory.* Springer, 1969.

154. G. FISHER, *Complex analytic geometry.* Lecture Notes in Math., 1976, **538**.

155. O. FORSTER, *Riemannsche Flächen.* Springer, 1977.

156. O. FORSTER AND K.L. RAMSPOTT, *Über die Anzahl Erzeugenden von*

projektiven Steinschen Moduln. Arch. Math., 1968, **19**, 417-422.

157. H. FUJIMOTO, *On the contiunation of analytic sets.* J. Math. Soc. Japan, 1966, **18**, 51-85.

158. H. FUJIMOTO, *Riemann domains with boundary of capacity zero.* Nagoya Math., J., 1971, **44**, 1-15.

159. O. FUJITA, *Sur les familles d'ensembles analytiques.* J. Math. Soc. Japan, 1964, **16**, 379-405.

160. K. FUNAHASHI, *On the extension of analytic sets.* Proc. Japan Acad., 1978, **54**, 1, 24-26.

161. H. HAMM, *Lokale topologische Eigenschaften komplexer Räume.* Math. Ann., 1971, **191**, 235-252.

162. F.R. HARVEY, *Three structure theorems in several complex variables.* Bull. Amer. Math. Soc., 1974, **80**, 633-641.

163. F.R. HARVEY, *Removable singularities for positive currents.* Amer. J. Math., 1974, **96**, 67-78.

164. F.R. HARVEY, *Holomorphic chains and their boundaries.* In: Proc. Pure Math., **30**, 1, Amer. Math. Soc. 1977, 309-382.

165. F.R. HARVEY AND J.R. KING, *On the structure of positive currents.* Invent. Math., 1972, **15**, 47-52.

166. F.R. HARVEY AND A.W. KNAPP, *Postivive (p,p)-forms, Wirtinger's inequality and currents.* In: Value Distribution Theory, part A. M. Dekker, 1974, 43-62.

167. F.R. HARVEY AND H.B. LAWSON, JR., *On boundaries of complex analytic varieties I.* Ann. of Math., 1975, **102**, 223-290.

168. F.R. HARVEY AND H.B. LAWSON, JR., *On boundaries of complex analytic varieties II.* Ann. of Math., 1977, **106**, 213-238.

169. F.R. HARVEY AND J. POLKING, *Extending analytic objects.* Comm. Pure Appl. Math., 1975, **28**, 701-727.

170. F.R. HARVEY AND B. SHIFFMAN, *A characterization of holomorphic chains.* Ann. of Math., 1974, **99**, 553-587.

171. R.M. HARDT, *Slicing and intersection theory of chains associated with real analytic varieties.* Acta Math., 1972, **129**, 75-136.

172. G.M. HENKIN, *The Lewy equation and analysis on pseudoconvex manifolds, I.* Russian Math. Surveys, 1977, **32**, 3, 59-130. (Original: Uspekhi Mat. Nauk, 1977, **32**, 3, 57-118.)

173. G.M. HENKIN, *Lewy equation and analysis on pseudoconvex manifolds, II.* Math. USSR-Sb, 1977, **31**, 63-94. (Original: Mat. Sb., 1976, **102**, 1, 71-108.)

174. G.M. Henkin and E.M. Chirka, *Boundary properties of holomorphic functions of several complex variables.* J. Soviet Math., 1976, **5**, 612-687. (Original: Sovrem. Probl. Mat., 1975, **4**, 13-143.)

175. L. Hörmander, *An introduction to complex analysis in several variables.* North-Holland, 1973.

176. A. Hirschowitz, *Sur la géométrie analytique au-dessus des grassmanniens.* C.R. Acad. Sci. Paris, 1970, **271**, 1167-1170.

177. Chee Pak Soong, *The Blaschke condition for bounded holomorphic functions.* Trans. Amer. Math. Soc., 1970, **148**, 249-263.

178. Chee Pak Soong, *On the generalized Blaschke condition.* Trans. Amer. Math. Soc., 1970, **152**, 227-231.

179. Chee Pak Soong, *A uniqueness set for all $H^p(B_n)$ with $p > 0$.* J. Austral. Math. Soc., 1978, **26**, 65-69.

180. E.M. Chirka, *Analytic representation of CR-functions.* Math. USSR Sb., 1975, **27**, 527-553. (Original: Mat. Sb., 1975, **98**, 4, 591-623.)

181. E.M. Chirka, *Currents and some of their applications.* Appendix to the Russian translation of [164] (in Russian).

182. E.M. Chirka, *Regularization and $\bar{\partial}$-homotopy on a complex manifold.* Soviet Math. Dokl., 1979, **20**, 1, 73-76. (Oroginal: Dokl. Akad. Nauk SSSR, 1979, **244**, 300-303.)

183. E.M. Chirka, *On removable singularities of analytic sets.* Soviet Math. Dokl., 1979, **20**, 5, 965-968. (Original: Dokl. Akad. Nauk SSSR, 1979, **248**, 47-50.)

184. E.M. Chirka, *Regularity of the boundaries of analytic sets.* Math. USSR Sb., **45**, 1983, 3, 291-335. (Original: Mat. Sb., 1982, **117**, 3, 291-336.)

185. B.V. Shabat, *Introduction to complex analysis.* Vol. 1-2, Nauka, Moscow, 1976 (in Russian).

186. B.V. Shabat, *Distribution of values of holomorphic mappings.* Amer. Math. Soc., 1985 (translated from the Russian).

187. Ph. Charpentier, *Sur la formule de Jensen et les zeros des fonctions holomorphes dans le polydisque.* Math. Ann., 1979, **242**, 27-46.

188. I.R. Shafarevich, *Basic algebraic geometry.* Springer, 1977 (translated from the Russian).

189. L. Schwartz, *Analyse.* Vol. 1-2, Hermann, 1970.

190. B. Shiffman, *On the removal of singularities of analytic sets.* Michigan Math. J., 1968, **15**, 111-120.

191. B. SHIFFMAN, *Local complex analytic curves in an analytic variety.* Proc. Amer. Math. Soc., 1970, **24**, 432-437.

192. B. SHIFFMAN, *On the continuation of analytic curves.* Math. Ann., 1970, **184**, 268-274.

193. B. SHIFFMAN, *On the continuation of analytic sets.* Math. Ann., 1970, **185**, 1-12.

194. M. SCHNEIDER, *Vollständige Durchschnitte in Steinschen Mannigfaltigkeiten.* Math. Ann., 1970, **186**, 191-200.

195. K. SPALLEK, *Differenzierbare und holomorphe Funktionen auf analytischen Mengen.* Math. Ann., 1965, **161**, 143-162.

196. K. SPALLEK, *Über Singularitäten analytischer Mengen.* Math. Ann., 1967, **172**, 249-268.

197. W. STOLL, *Einige Bemerkungen zur Fortsetzbarkeit analytischer Mengen.* Math. Zeitschr., 1954, **60**, 287-304.

198. W. STOLL, *Über die Fortsetzbarkeit analytischen Mengen endlichen Oberflächeninhaltes.* Arch. Math., 1958, **9**, 167-175.

199. W. STOLL, *The growth of the area of a transcendental analytic set.* Math. Ann., 1964, **156**, 47-78; 1964, **156**, 144-170.

200. W. STOLL, *Normal families of non-negative divisors.* Math. Zeitschr., 1964, **84**, 154-218.

201. W. STOLL, *The multiplicity of a holomorphic map.* Invent. Math., 1966, **2**, 15-58.

202. H. EL MIR, *Sur le prolongement des courants, positifs, fermés, de masse finie.* C.R. Acad. Sci. Paris, 1982, **294**, 181-184.

203. M. HERVÉ, *Several complex variables: local theory.* Oxford Univ. Press & Tata Inst. Fundam. Research, 1963.

204. R. EPHRAIM, *C^1-preservation of multiplicity.* Duke Math. J., 1976, **43**, 797-803.

205. A.N. YUZHAKOV AND A.K. TSIKH, *The multiplicity of zero of a system of holomorphic functions.* Siberian Math. J., 1979, **19**, 3, 489-492. (Original: Sibirsk. Mat. Zh., 1978, **19**, 3, 693-697.)

206. S.S.T. YAU, *Kohn-Rossi cohomology and its application to the complex Plateau problem I.* Ann. of Math., 1981, **113**, 67-110.

References added in proof

1. E. BEDFORD, *Levi flat hypersurfaces in $\mathbb{C}^2$ with prescribed boundary: stability.* Ann. Sc. Norm. Super. Pisa, 1982, **9**, 4, 529-510.
2. J. BECKER, *On the critical degree of differentiability of a complex planar curve.* Trans. Amer. Math. Soc., 1982, **269**, 1, 339-350.
3. B. BENNETT AND S.S.T. YAU, *Milnor number and classification of isolated singularities of holomorphic maps.* Lecture Notes in Math., 1982, **949**, 1-34.
4. C.A. BERENSTEIN AND B.A. TAYLOR, *On the geometry of interpolating varieties.* Lecture Notes in Math., 1982, **919**, 1-25.
5. B. GAVEAU, *Integrals de courbure et potentiels sur les hypersurfaces analytiques de $\mathbb{C}^n$.* Lecture Notes in Math., 1982, **919**, 108-122.
6. B. GAVEAU AND J.P. DEMAILLY, *Majoration statistique de la courbure d'une variété analytique.* Lecture Notes in Math., 1983, **1028**, 96-124.
7. G. KENNEDY, *Griffiths' integral formula for the Milnor number.* Duke Math. J., 1981, **48**, 1, 159-165; 1982, **49**, 1, 249.
8. CH. KISELMAN, *Stabilité du nombre de Lelong par restriction à une sousvariété.* Lecture Notes in Math., 1982, **919**, 324-336.
9. R. LANGEVIN AND T. SHIFRIN, *Polar varieties and integral geometry.* Amer. J. Math., 1982, **104**, 3, 553-605.
10. P. LELONG, *Ensembles analytiques complexes définis comme ensembles de densité et controles de croissence.* Invent. Math., 1983, **723**, 465-489.
11. R.E. MOLZON, *Potential theory in Nevanlinna theory and algebraic geometry.* Lecture Notes in Math., 1983, **1039**, 361-375.
12. R.E. MOLZON, *Blaschke conditions for holomorphic mappings.* Indiana Univ. Math. J., 1984, **33**, 3, 419-433.
13. R.E. MOLZON, *Potential theory on complex projective space: Application to characterization of pluripolar sets and growth of analytic varieties.* Ill. J. Math., 1984, **28**, 1, 103-119.
14. R.E. MOLZON AND G. PATRIZIO, *Meromorphic maps in the Nevanlinna class.* Proc. Amer. Math. Soc., 1984, **91**, 3, 395-398.
15. S.I. PINCHUK, *On the boundary behavior of analytic sets and algebroidal mappings.* Soviet Math. Dokl., 1983, **27**, 1, 82-85. (Original: Dokl. Akad. Nauk SSSR, 1983, **268**, 2, 296-298.)
16. H. SKODA, *Prolongement des courants, positifs, fermés de masse finie.* Invent. Math., 1982, **66**, 3, 361-376.

17. E. FORTUNA AND S. LOJASIEWICZ, *Sur l'algebricité des ensembles analytiques complexes.* J. Reine Angew. Math., 1981, **329**, 215-220.

18. W. FULTON AND R. LAZARSFELD, *Connectivity and its applications in alebraic geometry.* Lecture Notes in Math., 1981, **862**, 26-92.

19. M. HAKIM AND N. SIBONY, *Ensemble des zeros d'une fonction holomorphes bornee dans la boule unité.* Math. Ann., 1982, **260**, 4, 469-474.

20. H.A. HAMM, *Zur Homotopietyp Steinscher Räume.* J. Reine Angew. Math., 1983, **338**, 121-135.

21. M. SCHNEIDER, *Vollständige, fast-vollständige und mengen-theoretisch-vollständige Durchsnitte in Steinschen Mannigfaltigkeiten.* Math. Ann., 1982, **260**, 2, 151-172.

22. H. EL MIR, *Sur le prolongement des courants positifs fermés à travers des sousvariétés rulles.* C.R. Acad. Sci. Paris, 1982, **295**, 6, 419-422.

INDEX